Separations Division 2022

Held at the 2022 AIChE Annual Meeting

Phoenix, Arizona, USA
13-18 November 2022

ISBN: 978-1-7138-7889-6

Printed from e-media with permission by:

Curran Associates, Inc.
57 Morehouse Lane
Red Hook, NY 12571

Printed with permission by Curran Associates, Inc. (2023)

For permission requests, please contact AIChE
at the address below.

AIChE
120 Wall Street, FL 23
New York, NY 10005-4020

Phone: (800) 242-4363
Fax: (203) 775-5177

www.aiche.org

Additional copies of this publication are available from:

Curran Associates, Inc.
57 Morehouse Lane
Red Hook, NY 12571 USA
Phone: 845-758-0400
Fax: 845-758-2633
Email: curran@proceedings.com
Web: www.proceedings.com

TABLE OF CONTENTS

ADSORBENT MATERIALS

ADSORBENT MATERIALS FOR SUSTAINABLE ENERGY AND CHEMICALS

ADSORBENT MATERIALS: MOFS

ADVANCED INORGANIC MATERIALS FOR MEMBRANE GAS SEPARATION

MOLECULAR AND DATA SCIENCE MODELING OF ADSORPTION II

ADVANCES IN BIOSEPARATIONS

ADVANCES IN FLUID PARTICLE SEPARATIONS

MEMBRANES FOR CO2 CAPTURE II

NUCLEATION AND GROWTH II

ANTIFOULING MEMBRANES FOR WATER PURIFICATION

AREA PLENARY: ADSORPTION AND ION EXCHANGE - IN HONOR OF PROF. ANDREAS SEIDEL-MORGENSTERN (INVITED TALKS)

AREA PLENARY: FUNDAMENTALS AND APPLICATIONS OF ADSORPTION AND ION EXCHANGE

CHARGED POLYMERS FOR MEMBRANE-BASED WATER AND ENERGY APPLICATIONS

CO2 CAPTURE BY ADSORPTION

CONTINUOUS CRYSTALLIZATION PROCESSES

CRYSTALLIZATION OF PHARMACEUTICAL AND BIOLOGICAL MOLECULES

SOLID FORM SELECTION: COCRYSTALS, SALTS, SOLVATES, POLYMORPHS, AND BEYOND

DEVELOPMENTS IN EXTRACTIVE SEPARATIONS: PROCESSES

DEVELOPMENTS IN EXTRACTIVE SEPARATIONS: SOLVENTS

DIFFUSION, TRANSPORT AND DYNAMICS IN ADSORPTION SYSTEMS

DISTILLATION AND ABSORPTION PROCESSES FUNDAMENTALS, DEVELOPMENTS, OPTIMIZATION, AND APPLICATIONS

DIVISION PLENARY: GERHOLD AND KUNESH AWARDS ON SEPARATIONS (INVITED TALKS)

EXPERIMENTAL METHODS IN ADSORPTION

FLUID PARTICLE SEPARATIONS IN ENERGY AND ENVIRONMENTAL SYSTEMS

FRONTIERS IN NEW MATERIALS AND MEMBRANES FOR BIOSEPARATIONS

HIGHLY SELECTIVE SEPARATIONS WITH MEMBRANES

HONORARY SESSION FOR PROF. SUZANA NUNES I

MEMBRANE FORMATION

MEMBRANE MODELING AND SIMULATION

MEMBRANES FOR CO2 CAPTURE I

WATER TREATMENT, DESALINATION, AND REUSE II

MIXED-MATRIX MATERIALS FOR GAS SEPARATION

MODELING AND CONTROL OF CRYSTALLIZATION

MOLECULAR AND DATA SCIENCE MODELING OF ADSORPTION I

MOLECULAR SIMULATIONS FOR DESIGNING ADSORBENTS AND ADSORPTION PROCESSES

NEXT GENERATION BIOMOLECULES AND BIOPROCESSES

NUCLEATION AND GROWTH I

PARTICLE FORMATION AND CRYSTALLIZATION PROCESSES FROM LIQUIDS, SLURRIES, AND EMULSIONS

PLENARY SESSION: CRYSTALLIZATION AND EVAPORATION - AREA 2B (INVITED TALKS)

POSTER SESSION: FUNDAMENTALS AND APPLICATIONS OF ADSORPTION AND ION EXCHANGE

POSTER SESSION: GENERAL TOPICS ON SEPARATIONS

POSTER SESSION: MEMBRANE SEPARATIONS

POSTER SESSION: SEPARATIONS DIVISION

PSA/TSA AND ADSORPTION PROCESSES DESIGN AND SCALE-UP

HONORARY SESSION FOR PROF. RANIL WICKRAMASINGHE II

STRUCTURED ADSORBENTS: BEYOND PELLETS AND BEADS

SURFACE ENGINEERED AND RESPONSIVE MEMBRANES

WATER TREATMENT, DESALINATION, AND REUSE I

MEMBRANE-BASED ORGANIC SOLVENT SEPARATIONS

INNOVATION IN MEMBRANE MANUFACTURING

ADVANCED POLYMER MEMBRANES FOR GAS AND VAPOR SEPARATIONS

FUNDAMENTALS, BIG DATA, MACHINE LEARNING AND HIGH-THROUGHPUT SCREENING FOR BIOSEPARATIONS

TRANSFORMATION FROM BATCH TO CONTINUOUS PROCESSING IN BIOSEPARATIONS

HONORARY SESSION FOR PROF. RANIL WICKRAMASINGHE I

PAT AND PROCESS MONITORING IN CRYSTALLIZATION DEVELOPMENT AND MANUFACTURING

CHROMATOGRAPHIC SEPARATIONS AND SMB

CHARACTERIZATION OF ADSORBENT MATERIALS

SCALE-UP AND INDUSTRIAL APPLICATIONS OF ADSORPTION

ADSORPTION PROCESSES FOR ENERGY STORAGE AND UTILIZATION

ADSORPTION APPLICATIONS FOR SUSTAINABLE ENERGY AND CHEMICALS

<u>HONORARY SESSION FOR PROF. SUZANA NUNES II</u>

HONORARY SESSION FOR PROF. SUZANA NUNES III

339a Synthesis and Characterization of Isomorphically Substituted Alpo-5 Zeolites for Oxygen Adsorption

Natalia Ali, Steven Wilson, Ellen B. Stechel, Ivan Ermanoski, Andrea Ambrosini, Christopher L. Muhich, Shuguang Deng

Zeolites are inexpensive, hydrothermally stable, and well-studied heterogeneous catalysts and sorbents. The effectiveness of zeolites can be improved by isomorphic substitution, the substitution of one framework ion with a non-framework ion during synthesis. Aluminophosphates are a class of zeolites that are more amenable to isomorphic substitution, due to the lack of Si ions that form more rigorous framework structures. One specific aluminophosphate, Aluminiophosphate-5 (AlPO-5) is of particular interest for the study of isomorphic substitution during oxygen adsorption processes, as its unsubstituted framework cannot achieve high quantities of oxygen adsorption. The substitution of framework metals has been theoretically shown to improve the adsorption of oxygen gas in AlPO-5 isomorphically substituted with Si, Ge, Sn, Pd, Pt, Ti, V, Cr, Mn, Zr, Mo, Hf, W, Ce, and Pr. This work developed and evaluated multiple synthesis routes for the production of isomorphically substituted AlPO-5 crystals that preferentially adsorb oxygen at ambient temperature and low pressures.

Five hydrothermal synthesis routes were evaluated in which the composition of the synthesis gel, mixing time, crystallization temperature, or crystallization time were varied to determine the optimal procedure for isomorphic substitution in the AlPO-5 framework. The effect of substituting different metals (including Ce, Mn, Mo, Sn, and Si) into the AlPO-5 framework on oxygen adsorption was studied. The isomorphic substitution was verified with powder X-ray diffraction and X-ray photoelectron spectroscopy. AlPO-5 crystallinity, crystal morphology, and porosity/surface area were characterized via powder X-ray diffraction, scanning electron microscopy, and nitrogen physisorption, respectively. Oxygen uptake and release characteristics of the substituted AlPO-5 samples were evaluated using a thermogravimetric analyzer. Oxygen chemisorption or physisorption was verified by the calculated heat of adsorption, obtained from oxygen adsorption experiments at 273 K, 298 K, 323 K, and 348 K. Oxygen adsorption experiments were conducted at 298 K and 1 bar on a Micromeritics 3 Flex adsorption apparatus after degassing the samples for 24 hours at 673 K. At ambient temperature, substituted AlPO-5 achieved a maximum oxygen capacity of 0.1 mmol O_2/g sample. This work has elucidated the hydrothermal synthesis variables necessary for AlPO-5 framework substitution and demonstrated the feasibility of oxygen sorption on isomorphically substituted AlPO-5 frameworks. Additionally, the understanding gleaned from this research can be leveraged to other microporous material synthesis routes to create a larger class of oxygen-adsorbing sorbent materials.

339b Developing Hierarchical Silica-Salt Composites Via Wet Impregnation for Atmospheric Water Harvesting from Arid Air

Jamie Salinger, Carmen Chen, Molly Essig, Krista Walton

Providing reliable methods for generating potable drinking water for remote locations remains a great, and increasingly growing, humanitarian challenge. Dehumidifiers generate liquid water from humid air, but the high humidity and energy expenditures needed to operate such systems prevent their use in arid regions. Currently, desalination is the leading method of alternative water harvesting in arid regions, but many limitations still exist such as high operating and transportation costs, as well as the requirement to exist near seawater and limiting water access for inland regions. In recent years, adsorption-based water harvesting has become an area of significant interest, in which the abundant water vapor in the atmosphere can be extracted using porous sorbents. Commonly studied sorbents that strongly adsorb water vapor possess microporous structures that limit their water capacity. Hierarchical silicas are water-stable sorbents with an expansive porous structure containing a wide pore size distribution and large macropore volume, that allow for improved water uptake over strictly microporous sorbents and at %RHs above those where pure hygroscopic salts saturate. However, these materials exhibit limited water adsorption at low relative humidity. Here, we present the synthesis, characterization, and water adsorption testing of an optimized hierarchical silica-salt composite in which a hygroscopic salt is incorporated into the sorbent matrix via a wet impregnation method. The composite is capable of adsorbing large quantities of water (> 0.4 g/g) at humidities as low as 5-10% RH. Additional kinetics and cycling tests indicate the silica-salt composite maintains the same performance after multiple cycles.

339d Atomic Layer Deposited Dual-Function Materials for Combined Adsorption and Oxidation of VOCs

Busuyi Adebayo, Ali Rownaghi, Fateme Rezaei

Concentrating ppm-level VOCs in-situ, before their oxidation, offers a practical approach to reduce the catalyst inventory and capital cost associated with VOC emissions abatement. This work reports on the development of novel Ni nanoparticles-deposited mixed-metal oxides ZrO_2-SiO_2 through atomic layer deposition (ALD) method and their application in combined capture and oxidation of benzene, as a model compound of aromatic VOCs. The benzene vapor adsorption isotherms were measured at 25 °C and in the pressure range of 0 to benzene saturation vapor pressure thereof (0.13 bar). In the combined capture-reaction tests, the materials were first exposed to ca. 86100 ppm_v benzene vapor at 25 °C, followed by desorption and catalytic oxidation while raising bed temperature to 250 °C. The textural properties revealed that ALD of Ni or ZrO_2 on SiO_2 decreased surface area and pore volume, while sequential doping with ZrO_2 and then Ni caused the otherwise. The benzene vapor adsorption isotherms followed the type-IV isotherm classification, revealing a combination of monolayer-multilayer and capillary condensation adsorption mechanisms in sequence. At saturation vapor pressure, an average equilibrium adsorption capacity of 15 mmol/g was obtained across the materials. However, the dynamic adsorption capacities were up to 50% less than the corresponding equilibrium uptake for the materials. Benzene desorption temperature was observed around 90 °C, and conversion of 85-95% and TOF of 1.28-16.42 $mmol_{C6H6}/mol_{Ni}/s$ were obtained over the materials, with 3Ni/ZrO_2-SiO_2, prepared with 3 ALD cycles, exhibiting the maximum conversion and TOF indicating synergistic effects of Ni nanoparticles and ZrO_2 support based on the number of ALD cycles. However, the yields of CO_2 and H_2O were about 5% and 40%, respectively. The small value of the CO_2 yield was hypothesized to be due to simultaneous high-temperature adsorption of CO_2 as the catalytic reaction progressed. In conclusion, the high adsorption affinity, low desorption temperature, and high catalytic activity of the materials investigated in this study made them worthwhile for the abatement of BTX. Specifically, we made bold that the 3Ni/ZrO_2-SiO_2 material stands to be one of the best dual-functional materials for the abatement of BTX.

Reference

*Adebayo, et al., ACS Appl. Mater. Interfaces*2020, 12, 35, 39318–39334.

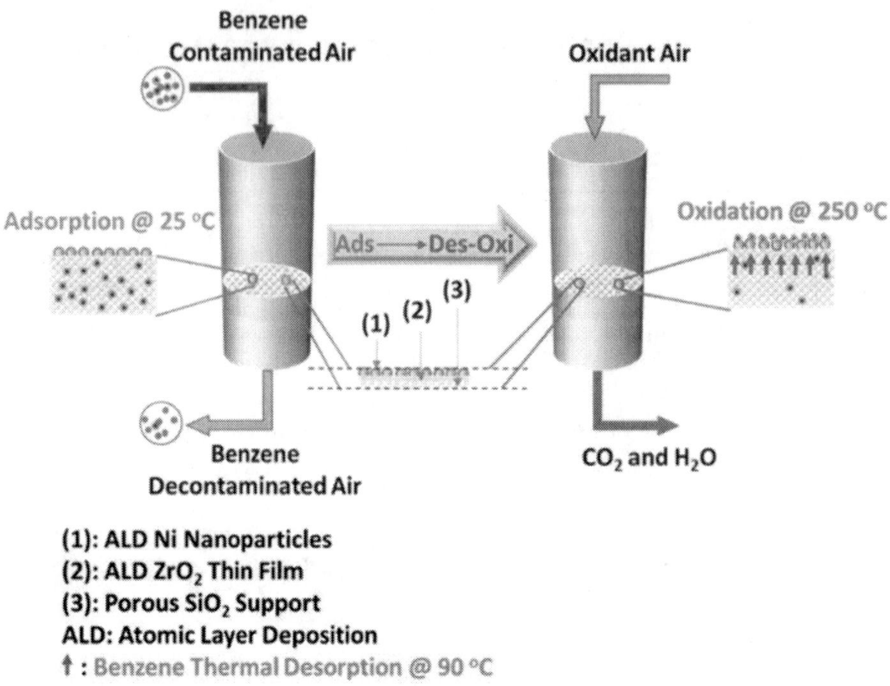
Benzene
Contaminated Air
Oxidant Air
Adsorption @ 25 °C
Ads → Des-Oxi
Oxidation @ 250 °C
(1)
(2)
(3)
Benzene
Decontaminated Air
CO_2 and H_2O
(1): ALD Ni Nanoparticles
(2): ALD ZrO_2 Thin Film
(3): Porous SiO_2 Support
ALD: Atomic Layer Deposition
↑ : Benzene Thermal Desorption @ 90 °C

339e Modification of Zeolites with Organic Phosphonic Acids for Adsorptive Separation of Gases

Xinpei Zhou, John L. Falconer, J. Will Medlin

The adsorptive separation of propylene and propane gases using molecular sieves is a challenging process due to the similarity in molecular sizes for the two molecules. We deposited organic phosphonic acid (PA) monolayers on zeolites to tune the diffusion rates and adsorptive selectivity of various gases including propylene and propane. The PAs with different chemical functional groups, chain lengths, and steric structures were used to modify zeolites. The PA monolayers can hypothetically function as "gate-keepers" by adding a tunable diffusion barrier on the zeolite surface. We used pressure-decay adsorption measurements to demonstrate that PA monolayers on zeolites significantly enhanced the ideal selectivity of C_3H_6/C_3H_8 adsorption by changing the diffusion mechanism based on the properties of the alkyl tail. Moreover, we used pulse method for adsorption measurements of propylene and propane mixture followed by temperature-programmed desorption to determine the mixture selectivity.

For single gas adsorption, on zeolite 5A coated with *n*-octadecylphosphonic acid (ODPA), the kinetic selectivity of C_3H_6/C_3H_8 was initially >8 at 25 °C, whereas for uncoated 5A, it was limited to ~1.2. Kinetic modeling showed that in ODPA-coated 5A, the diffusion of propylene and propane molecules was limited by the PA monolayer at the external surface of zeolite. In contrast, for uncoated 5A, it was controlled by the pore channels of zeolite, so that the enhanced kinetic selectivity from ODPA-coated 5A was related to a different limiting transport mechanism. The kinetic selectivity was not temperature sensitive in the range of 25–150 °C in 5A-ODPA as the diffusion activation energies of propylene and propane were both small. Modification of 5A with other PAs, including those with methyl, *n*-butyl, and *tert*-butyl groups, also increased the kinetic selectivity. However, modification with methylphosphonic acid, which partially penetrated the near-surface region of zeolite, severely lowered diffusion rates of propylene and propane. Coating with *n*-butylphosphonic acid yielded lower kinetic selectivity than ODPA, ostensibly due to its shorter alkyl tail. Coating with *tert*-butylphosphonic acid, a sterically bulky ligand, decreased kinetic selectivity still further.

Moreover, our recent studies on propylene and propane mixture adsorption on zeolite 5A indicated that there was competitive adsorption between propylene and propane and the mixture selectivity of C_3H_6/C_3H_8 was higher than the ideal selectivity, which was ascribed to the stronger interaction forces of zeolites with propylene than propane.

The use of organic films may enable rational design of selective adsorbents based on providing gas-specific resistance at the pore entrance. Coatings with PA monolayers on zeolites has potential to provide another lever to tune the diffusion rates of gases by choosing suitable PAs and zeolites. Such an approach may be broadly applicable to diverse adsorbents, i.e., a kinetically selective coating can enhance performance for an underlying equilibrium-selective material.

339f Effect of SO_2 on CO_2 Capture Performance of Self-Supported Branched Poly(ethyleneimine) Scaffolds

Pavithra Narayanan, Chun-Jae Yoo, Ryan P. Lively, Christopher W. Jones

The separation of CO_2 from post-combustion flue gas is a critical component in global carbon emission control and climate change mitigation. Structured contactors in the form of monoliths for CO_2 capture not only combat the high energy consumption and corrosion rate observed in traditionally used aqueous amines but also offer a low pressure drop in comparison to pellet or powder adsorbents.[1] Amine materials have shown improved sorption capacities in the presence of humidity, compared to other nanoporous materials such as metal organic frameworks (MOFs) and zeolites, which sometimes have reduced performance in the presence of humidity.[2,3] Two critical challenges in developing a cost-effective amine-based adsorbent for CO_2 capture from flue gas are the creation of a support material that contains the CO_2-philic functionality and the combatting presence of acid gas impurities such as SO_2 in the feed stream. In this work, we have synthesized a self-supported monolithic sorbent by crosslinking poly(ethyleneimine) (PEI) at sub-ambient temperatures using an ice templating method.[4] Scanning electron microscopy (SEM) confirms the presence of templating-induced porosity, which is also a function of the temperature of crosslinking and concentration of PEI. Breakthrough experiments with simulated flue gas are performed to show the effect of pore morphology on the breakthrough behavior and sorbent capacity for CO_2 capture in the presence of humidity. The ratio of CO_2 adsorption capacity to SO_2 adsorption capacity for the synthesized PEI scaffolds was 1.5 for pseudo-equilibrium adsorption and increased to 16 when the cycle time was reduced. This in comparison to a ratio of 0.86 obtained for a powdered sorbent of PEI impregnated in an inorganic silica support.[5] Since tertiary amines have a higher affinity for SO_2 than CO_2, and the temperature and degree of crosslinking affect the distribution and relative ratio of primary, secondary, and tertiary amines in the crosslinked PEI scaffolds respectively, these two parameters can be tuned to improve the selectivity for CO_2 by at least 57%. Additionally, with a sorbent lifetime of over 15 cycles in the presence of SO_2, the crosslinked PEI scaffolds are promising materials for CO_2 capture from flue gas with scope for improvement by optimizing the cycle time.

References:

1. Rezaei, P. Webley, Chem. Eng. Sci., 64 (24), 2009, 5182–5191.
2. C.Burtch, I. M.Walton, J. T. Hungerford, C. R. Morelock, Y. Jiao, J. Heinen, Y. Chen, A. A. Yakovenko, W. Xu, D. Dubbeldam, K. S. Walton, Nature Chemistry, 12, 2020, 186-192.
3. Mve. Mfoumou, S. Mignard, T. Belin, Ads. Sci andamp; Technol., 36 (5–6), 2018,1246–1259.
4. J. Yoo, P. Narayanan, C. W. Jones, J. Mater. Chem., A 7 (33), 2019, 19513–19521.
5. Kim, W. Choi, M. Choi, ACS Appl. Mater. Interfaces., 11, 2019, 16586−16593.

339g CO and CO2 Adsorption by a New ZSM-5@MOF-199 Composite Adsorbent

Farshad Feyzbar-khalkhali-Nejad, Ehsan Hassani, Katie D. Leonard, Samuel Sessions, Morteza Taghavi Kouzehkanan, Tae-Sik Oh

Detrimental effects of greenhouse and hazardous gases such as climate change and ozone depletion can be mitigated by CO and CO_2 capture from point sources or ambient air. Adsorption gained extensive attention due to its low energy requirement. Here, we report the development of ZSM-5@MOF-199 composite comprised of a copper-based metal organic framework (MOF) shell and ZSM-5 zeolite core through the solvothermal method. Using techniques such as XRD, surface and pore analysis, and SEM, we confirmed the formation of the composite. We demonstrated that the anion of the copper source affects the physical structure of the formed MOF shell. Single component adsorption isotherms of CO, CO_2, and N_2 were collected at 25 °C to investigate the equilibrium adsorption capacity and selectivity. We systematically controlled the Cu^{+}/Cu^{2+} ratio through auto-reduction under vacuum to enhance the composite's CO adsorption capacity and selectivity. The increase in the Cu^{+}/Cu^{2+} ratio was confirmed using XPS analysis. Moreover, the kinetics of adsorption were investigated under realistic conditions. We investigated the adsorbate-adsorbent interaction using various equilibrium and kinetic models to show the potential of ZSM-5@MOF-199 composite for CO and CO_2 capture from source point processes.

339i Alkyl Amine Incorporation in Functional and Processible Porous Organic Polymers for CO_2 Capture Applications

Ali Sekizkardes, Victor A. Kusuma, Jeffrey Culp, Patrick Muldoon, James S. Hoffman, David Hopkinson

Porous sorbent materials have been considered as one of the potential material technologies for use in CO_2 capture applications. Porous organic polymers (POPs) and particularly their soluble class polymers with intrinsic microporosity (PIMs) have been extensively studied in gas separation applications. PIMs are mostly considered as a membrane material due to their high free volume and easy processability properties. Although PIMs possess high surface area and microporosity desired for a sorbent, they also suffer from low CO_2 adsorption capacity (<9 cc/g, 0.15 bar and 298K) due to relatively large (>1nm) non-polar micropores as well as some mesopores, limiting their use in sorbent applications. Here, we present novel PIM-based chemical sorbent synthesis and characterization. We synthesized and post-synthetically functionalized PIM-1 with carboxylic acid and amide groups.[1-2] The main scope of this study is to create acidic functional groups on the porous media to interact with basic amine molecules to ensure a strong connection between the sorbent and amines. These sorbents not only showed the highest (five fold more than PIM-1) CO_2 capture capacity and CO_2/N_2 selectivity reported in all PIMs to date, they also provide scalable, stable and processible sorbent design, which can be adjusted into a library of different materials. Primary amines append to porous polymers through acid-base and hydrogen bonding interaction, creating a starting point for future amine-based PIMs studies focusing on the sorbent stability and performance. The synthesis, characterization and processibility of the discoverd sorbents will be presented with evaluation of their CO2 capture performance.

References

1. Sekizkardes, A. K.; Hammache, S.; Hoffman, J. S.; Hopkinson D. “Polymers of Intrinsic Microporosity Chemical Sorbents Utilizing Primary Amine Appendance Through Acid–Base and Hydrogen Bonding Interactions” *ACS Appl. Mater. Interfaces* **2019**, *11*, 30987-30991.

2. US20220032268A1, patent granted in Feb **2022**, Hopkinson, D. and Sekizkardes, A.K.

143a Zeolite-13X- and ZIF-8 Filled PVA Mixed Matrix Membranes for Use in CO_2/N_2 and CO_2/CH_4 Separation

Iris Samputu, F Handan Tezel

Continual global warming has become a global concern with policy makers, and particularly the Intergovernmental Panel on Climate Change (IPCC) warning that the earth temperature should not be more than 2°C higher than pre-industrial revolution temperatures. Separation of carbon dioxide (CO_2) from nitrogen (N_2) has therefore become a topic of interest for applications on CO_2 capture from combustion gases. Also, separation of CO_2 from methane (CH_4) has been explored to produce renewable natural gas from landfill gas. This research focuses on separation of CO_2 from CH_4 and N_2 gases using mixed-matrix membranes filled with adsorbents. The mixed matrix membranes were composed of amine-based Poly vinyl alcohol (PVA) polymer membranes that had previously been optimized for CO_2/N_2 and CO_2/CH_4 separation. Two different adsorbent filler materials were used: the metal organic framework zeolitic imidazolate framework-8 (ZIF-8) and zeolite 13X. These fillers were chosen because of the good separation performance they had shown when used with different polymeric materials. The membranes were tested at near atmospheric conditions at a feed pressure of 2.5 atm. The best performing CO_2/N_2 and CO_2/CH_4 separating membranes were then tested at three different feed pressures from 1atm- 3atm to investigate the effect of the feed pressure on CO_2/N_2 and CO_2/CH_4 separation.

143b Kinetic Process Assessment of H_2 Purification over Highly Porous Carbon Sorbents Under Multicomponent Feed Conditions

Qasim Al-Naddaf, Shane Lawson, Ali Rownaghi, Fateme Rezaei

As a universal energy career, the need for pure H_2 is ever increasing due to its ubiquitous role in petrochemical refining, metal reduction, and the up-and-coming fuel cell market. Hydrogen produced from steam methane reforming (SMR) is typically laden with impurities such as CO_2, CO, and CH_4. A full assessment of the efficacy of potential sorbents for H_2 purification requires a thorough evaluation of thermodynamic and kinetic behaviors associated with multicomponent H_2 upgrading by pressure swing adsorption (PSA). As such, in this study we assessed three commercially available activated carbons with high surface area and pore volume for the PSA upgrading of H_2 from simulated SMR off-gas stream consisting of $H_2/CO/CH_4/CO_2$ (75/5//5/15 vol%). In addition to high-pressure adsorption isotherms for pure gases, H_2 purity, recovery, and productivity were estimated from PSA runs, while actual $(CO+CH_4+CO_2)/H_2$ selectivity values were estimated from breakthrough experiments. For the best performing material, the results demonstrated H_2 purity and recovery of 99.7 and 65%, respectively with a productivity of 20.85 mol_{H2}/kg.h and actual $(CO+CH_4+CO_2)/H_2$ selectivity of 73. Moreover, the affinity of the different adsorbates toward the activated carbons presented from the most adsorbed to the least adsorbed gas was in the order of $CO_2 > CH_4 > CO >> H_2$.

143c Tuning Shape and Position of Step-Shaped Isotherms of Flexible MOFs through Ligand Functionalization for Improved CO_2/CH_4 Separation

Lukas Bingel, Krista Walton

Natural gas is a key green energy solution towards alleviating climate change. One of the key challenges to using natural gas is its energy-intensive purification process. Adsorption-based technologies offer an energy-efficient way to extract methane as an energy source from other components in natural gas like carbon dioxide. Metal-organic frameworks (MOFs) are a highly tunable class of adsorbents known for their modular assembly of metal nodes and organic linkers. More than 500,000 possible MOF structures have been predicted in literature so far, each with different chemical modifications affecting their adsorption behavior. Among these materials, third generation porous framework materials with structural flexibility - also known as flexible MOFs - have gained considerable attention. Flexible MOFs have been shown to exhibit a gate-opening behavior in the presence of an external stimuli, such as guest gas molecules. This gate-opening behavior is characterized by the MOF transitioning between a closed, non-microporous structure and an open, porous phase through linker rotation. As a result of this behavior, flexible MOFs have step-shaped adsorption isotherms and can adsorb and release gas molecules over a narrow pressure range, allowing for efficient pressure swing adsorption and regeneration applications. Ligand functionalization has a dual effect on flexible MOF materials: it provides adsorption sites for a particular component and it changes the shape and position of the step-shaped adsorption isotherm. Here, different amounts of nitrogen sites were introduced into the structure of the flexible MOF ZIF-7 to provide high-energy adsorption sites for carbon dioxide, while tuning the shape and position of the stepped isotherms. This modification results in a functionalized surface that is more selective towards carbon dioxide than methane, and impacts the isotherm shapes for the two components differently. Binary adsorption of methane and carbon dioxide was studied using breakthrough experiments to investigate the persistence of the preferred adsorption of CO_2 under real mixture conditions. Experimentally calculated CO_2/CH_4 selectivity was compared to the selectivity predicted using the Osmotic Framework Adsorbed Solution Theory (OFAST) to investigate the applicability of OFAST for tuned ZIF-7 systems.

143e MgO-CuO-CeO$_x$ Sorbent Bead for a Trace Carbon Monoxide Removal for Fuel-Cell Grade Hydrogen Production

Gina Bang, Seong-min Jin, Kyung-Min Kim, Chang-Ha Lee

Hydrogen is noted as a clean energy source for fuel cells and power generation. Various techniques to separate and purify hydrogen mixtures have been developed to produce high-purity hydrogen (99.99+% level) [1]. Many studies have reported performance degradation of fuel cells due to trace amounts of CO in hydrogen, let alone H_2 purity. Adsorptive cyclic processes such as P(V)SA and TSA can produce high purity H_2 from various H_2 mixture off-gases, but the steep decrease of recovery is expected to produce H_2 with less than 3 ppm CO for fuel-cell grade [2]. Currently, selective catalytic oxidation is mainly studied as CO removal for fuel-cell grade hydrogen production, but this reaction is accompanied by other gases, which has the disadvantage of requiring an additional separation process. Therefore, the removal of trace CO in H_2 is still a great challenge in producing H_2 with ppb level of CO.

Herein, we present the MgO-CuO-CeO$_x$ sorbent beads with very strong adsorption affinity at an extremely low pressure of CO. The adsorbents were synthesized by a combustion-assisted method. The spherical composite beads with hundreds of micrometers in diameter were highly porous, and the metals were homogeneously dispersed in the bead. The morphological characterization of spherical MgO-CuO-CeO$_x$ sorbent bead was conducted via SEM, HR-TEM, XRD, and a gas sorption analyzer. In addition, the ratio of the surface copper to bulk copper in the beads was measured through a dissociative H_2 TPR method, and the adsorption capacity and mechanism were also analyzed by adding XPS analysis.

Adsorption isotherms of CO on the as-prepared composites with various metal ratios were measured by a volumetric method. The high and strong adsorption capacity was observed, especially up to 100 ppm CO. The regeneration conditions and cyclic stability were evaluated by cyclic thermogravimetric tests. Based on the analysis results, the direction of developing metal composites for various physisorption-chemisorption mechanisms accompanied by CO adsorption as well as CO removal is suggested.

[1] G Bang, DK Moon, JH Kang, YJ Han, KM Kim, CH Lee, Chem. Eng. J., 411 (2021)

[2] YW You, DG Lee, KY Yoon, DK Moon, SM Kim, CH Lee, Int. J. Hydrogen Energy, 37 (2012)

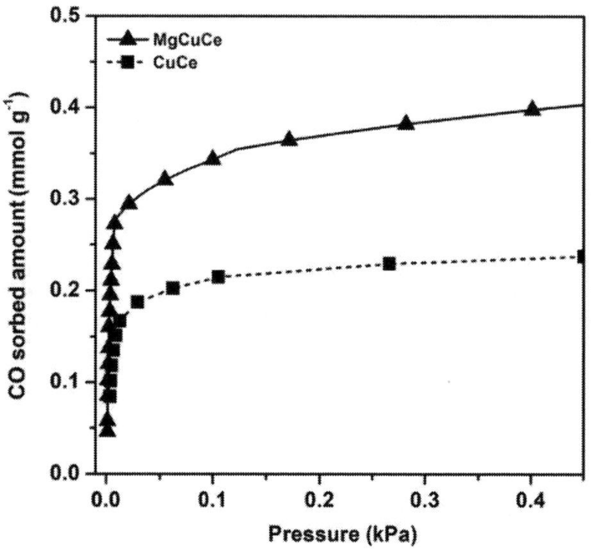

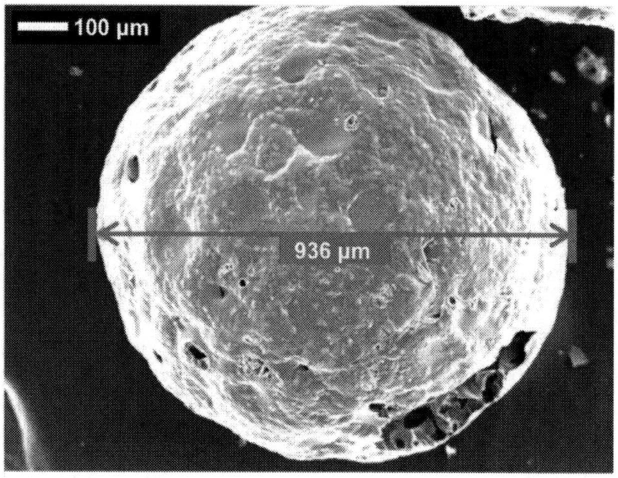

143f Novel Adsorbent Design for Energy-Efficient and Economical Lithium Recovery from Unconventional Resources

Hoon Choi, Marisa Moss, Elizabeth Kolbe, Stefan Haugen, Zbyslaw Owczarczyk, Bryan S. Pivovar, Eric Karp

Lithium (Li) is one of the critical materials, which is widely used for energy storage materials. Because of the high global demand and anticipated growth for Li, it is a key strategic goal to secure the domestic supply chain. This leads to exploiting unconventional Li resources including geothermal brines, oil andamp; gas formation water, mine waters, etc. that are located within the U.S. However, the current lithium mining and separation processes using conventional hydrometallurgical and direct lithium extraction approaches are energy- or chemical-intensive to exploit these resources. Thus, it is critical to develop an efficient separation process to exploit these resources for Li. To that end, adsorption can be effective for the purification of Li from multicomponent separations. In this presentation, we will discuss the designing of novel adsorbents with zwitterionic functional groups for the efficient chromatographic process to separate Li from these unconventional resources. Experimental and simulation data of the zwitterionic adsorbents and the dynamic mechanisms of Li separations will be discussed. Finally, the use of these designed adsorbents for Li recovery in a scaled Simulated Moving Bed (SMB) process will be presented.

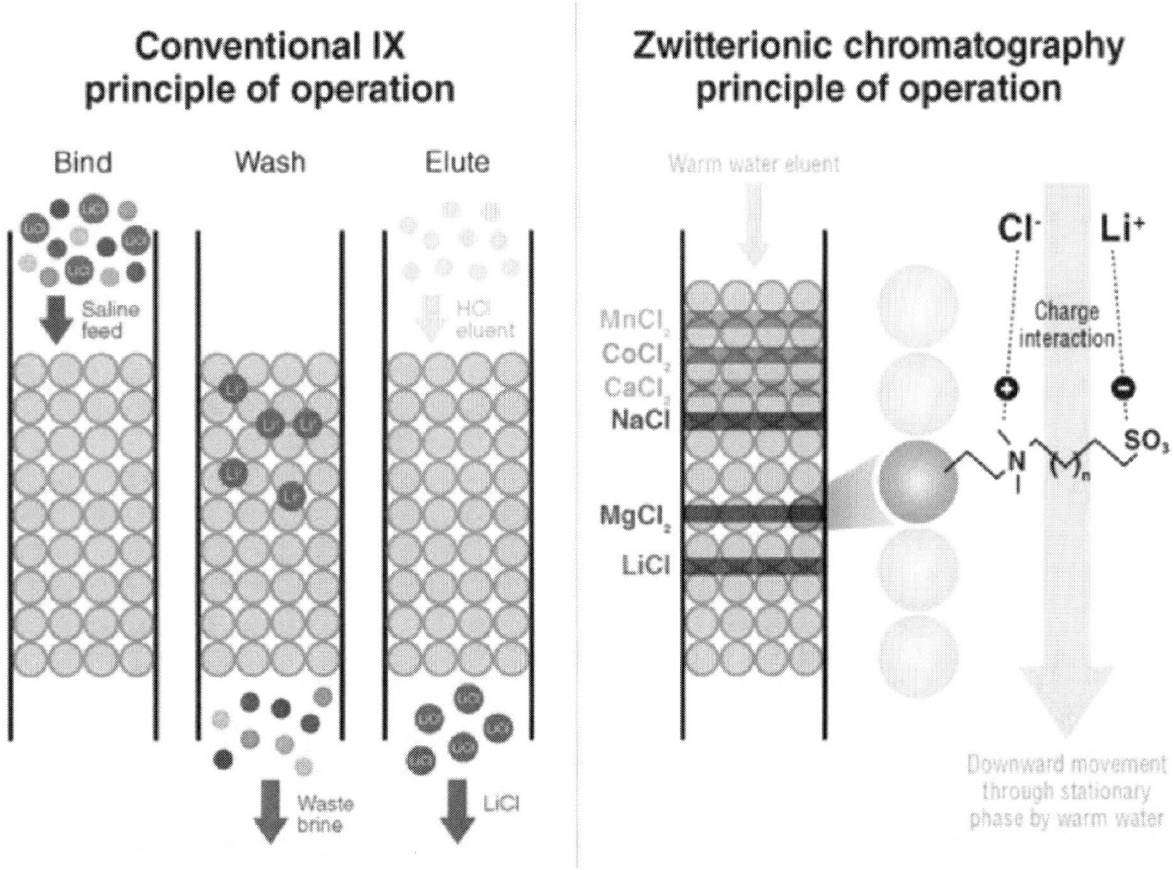

143g Development of Double-Functionalized Halloysite Nanotubes with Improved Adsorption Capacity for Carbon Capture

Oluwole Ajumobi, Azeem Farinmade, Julia A. Valla, Vijay T. John

Amine-functionalized porous supports as adsorbents for CO_2 capture has been widely studied however, improving the adsorption capacity of these amine functionalized adsorbents is the current focus. Recently, double amine-functionalization of porous adsorbent to increase the amine density using different amine sources has been studied. The pores of the support are covalently functionalized with amines from alkoxy silanes and polymeric amines to form a class 1/class 2 hybrid adsorbent. However, the polymeric amine and silanes both compete for space the same space. In this study, we show a simple concept of dual amine functionalization of halloysite nanotubes (HNT) to form a hybrid adsorbent with no competition for anchoring site between the two amine sources. Halloysite nanotubes are natural aluminosilicate hollow tubular clay materials with inner diameter (lumen) of 15 – 30 nm and length of 0.5 – 3 μm. First, the silanol external surface of HNT is functionalized with 3-Aminopropyl triethoxysilane to covalently bind the amine group via silylation. Afterwards, 30wt% polyethyleneimine is loaded into the available lumen of the already functionalized HNT (A_HNT) by vacuum suction. The final morphology (A_HNT30PEI) is such that the HNT lumen is infiltrated with PEI while the external surface is tethered with covalently bonded amine group. Characterization result show that the structure of the HNTs remained unaltered after silane modification. Adsorption of CO_2 was performed using thermogravimetric analysis (TGA) at 45 °C and atmospheric pressure. The results show that the adsorption capacity of A_HNT30PEI is equivalent to that of A_HNT and HNT30PEI (30 wt% PEI in HNT) combined. This shows that we can leverage on the dual surface modification of HNT to create a class 1/class 2 hybrid adsorbent with enhanced CO_2 capture capability.

143h Amine Functionalization of Hyper-Cross-Linked Polymer (HCPs) for CO_2 Separation from Biogas

Olusola Johnson, Babu Joseph, John Kuhn

Climate change mitigation, lowering greenhouse gas emissions, and expanding renewable energy sources to match the growing population are among the most pressing current technological challenges. Biogas is a valuable renewable energy source generated from the anaerobic digestion of biodegradable organic matter, made up of carbon dioxide (CO_2) and methane (CH_4), two potent greenhouse gases. Although biogas could be directly utilized in combustion engines, CO_2 separation from CH_4 is necessary to increase biogas' calorific value so that the biomethane can be injected into the natural gas grid or used as a vehicle fuel [1].

Hyper-cross-linked polymeric (HCP) resins are organic porous materials used in CO_2 adsorption, primarily because of their properties such as low density, high specific surface area, and high mechanical stability [2]. HCPs have tunable porous structures and rigid networks that prevent the porous wall from collapsing owing to their high degree of crosslinking, making them excellent solid adsorbents. Recently, amine-modified HCP resins have been proposed and evaluated for CO_2 separation from flue gases and air [3]. However, its use in biogas upgrading and the economic feasibility evaluation of such adsorbents' practical application has not been explored in detail [4]. This work studied the use of PEI-impregnated resin sorbent for CO_2 separation from biogas. The sorbent exhibited the highest adsorption capacity of 2.7 $mmol_{CO2}/g_{ads}$ at 30% PEI loading, increasing to 2.9 $mmol_{CO2}/g_{ads}$ in the presence of moisture, and remained stable for several adsorption-desorption cycles. In-situ DRIFTS studies showed that CO_2 adsorption on PEI-impregnated sorbent is consistent with the zwitterion reaction mechanism, and the sorbent could be regenerated completely at 100°C. The upgrading cost of biogas is primarily dominated by the operating cost of regeneration and the adsorbent cost. Economic feasibility analysis projected that PEI-impregnated resin sorbent requires less capital and operating cost than conventional biogas upgrading technologies. Therefore, PEI-functionalized HCPs resins are promising for CO_2 separation from biogas.

References:

[1] X. Zhao, A. Naqi, D.M. Walker, T. Roberge, M. Kastelic,, B. Joseph, J.N. Kuhn, Sustainable Energy andamp; Fuels (2019) 53949. 10.1039/C8SE00344K.

[2] H.R. Penchah, H.G. Gilani, A. Ghaemi, Journal of Chemical andamp; Engineering Data (2020) 490513. 10.1021/ACS.JCED.0C00541.

[3] J. Wang, M. Wang, W. Li, W. Qiao, D. Long, L. Ling, AIChE Journal (2014) 97280. 10.1002/AIC.14679.

[4] O. Johnson, B. Joseph, J.N. Kuhn, Journal of Industrial and Engineering Chemistry (2021) 25563. 10.1016/J.JIEC.2021.07.038.

465a Metal-Organic Frameworks As Suitable Candidates for Atmospheric Water Harvesting

Saumil Chheda, Nikita Hanikel, Xiaokun Pei, Hao Lyu, WooSeok Jeong, Joachim Sauer, Omar Yaghi, Joern Siepmann, Laura Gagliardi

Atmospheric water harvesting (AWH) has recently emerged as a potential solution for the rising global water crisis.[1,2] Adsorption-based water harvesters capture water from the atmosphere in an adsorption-assisted step while a second thermal desorption step regenerates the sorbent for the next cycle. The performance of the AWH device critically depends on the water affinity and water adsorption characteristics of the adsorbent amongst several other factors. Metal-organic frameworks, porous and crystalline materials built from inorganic metal oxide nodes connected by organic linkers, are suited as sorbents for AWH owing to their tunable water adsorption properties ranging between extremely hydrophilic sorbents, such as aluminosilicates and silica gels, and hydrophobic materials, such as all-silica zeolites and polymer gels. MOF-303, a rod-based MOF in which infinite Al(OH) rods are connected by 1-*H*-pyrazole-3,5-dicarboxylate ($PZDC^{2-}$) linkers, is posited as a state-of-the-art sorbent for AWH with a reported water uptake of 0.7 $Lkg^{-1}day^{-1}$ in the arid climate of the Mojave desert (upto 10% *R.H.* (relative humidity), 10).[1]

In our recent work,[2] we elucidated the molecule-by-molecule water uptake in MOF-303, concomitant with its water adsorption isotherm, using a combination of periodic density functional theory (DFT) calculations and single-crystal X-ray diffraction. MOF-303 exhibits an initial small S-shaped step in its isotherm followed by a larger steep step in the isotherm at *R.H.* = 13% where the initial step can reduce the water output from the harvester by 20%. The higher hydrophilicity of MOF-303 is attributed to the presence of alternating hydrophilic cavities in the MOF, where each hydrophilic cavity is constituted by opposite pairs of linker pyrazole functionalities and their three neighboring μ_2-OH groups in the inorganic rods. The hydrophilic cavities were found to strongly bind upto three water molecules to the framework through multiple H-bonds with the neighboring H-bond donors and acceptors. Water molecules adsorbed into the primary sites were found to further seed other water molecules to form isolated clusters in the MOF cavities, which subsequently formed interconnected water networks in the MOF pores at the saturated water loading. The role of linkers in strongly binding the initial water molecules to the framework was used to modulate the "initial" water affinity of the MOF by replacing the $PZDC^{2-}$ linkers with 2,4-furandicarboxylate (FDC^{2-}) linkers, forming isoreticular MOF-333, to suppress the initial S-shaped step in its isotherm. The elimination of the S-shaped step in the isotherms of MOF-333 with a moderate shift in its step to *R.H.* = 22% is expected to increase the water output from a MOF-333-based AWH device.

The design of the next-generation sorbents for AWH further necessitates an accurate prediction of the water adsorption properties of the MOF. While DFT-based water binding energies allow to gauge the water affinity of a given MOF for water adsorption, the water uptake in the MOF under different water partial pressures can be predicted using force-field-based Monte Carlo (MC) simulations. However, MC simulations for water adsorption in MOFs are challenging due to the choice of force field parameters which can accurately describe the MOF-water interactions, the different water models available for describing water-water interactions, and the structural flexibility of MOFs in presence of guest molecules. Here, MC simulations in the *NpT*-Gibbs ensemble (GEMC) are used to predict the water adsorption isotherms in MOF-303 and MOF-333.

References:

(1) Hanikel, N.; Prévot, M. S.; Fathieh, F.; Kapustin, E. A.; Lyu, H.; Wang, H.; Diercks, N. J.; Glover, T. G.; Yaghi, O. M. Rapid Cycling and Exceptional Yield in a Metal-Organic Framework Water Harvester. *ACS Cent. Sci.* **2019**, *5*, 1699–1706.

(2) Hanikel, N.; Pei, X.; Chheda, S.; Lyu, H.; Jeong, W.; Sauer, J.; Gagliardi, L.; Yaghi, O. M. Evolution of Water Structures in Metal-Organic Frameworks for Improved Atmospheric Water Harvesting. *Science* **2021**, *374*, 454–459.

Note: This material is based upon worked supported by the Defense Advanced Research Projects Agency under Contract No. HR0011-21-C-0020. Any opinions, findings and conclusions or recommendations expressed in this material are those of the author(s) and do not necessarily reflect the views of the Defense Advanced Research Projects Agency.

465b Deducing Information about Defects in MOFs Using Their Water Stability

Shubham Jamdade, Salah Eddine Boulfelfel, Hanjun Fang, David Sholl

Metal–organic framework (MOF) materials are nanoporous materials with potential uses in a diverse set of applications. To be commercially viable, MOFs should ideally possess water stability. Unfortunately, many MOFs are unstable in water. The vast majority of computational studies of MOFs consider these materials to be rigid and defect free, and in many cases the resulting structures are expected to be hydrophobic. Systematic experimental studies, however, have shown many MOFs are hydrophilic. We will show that insights from experiments of this kind can be used to infer when local defects are present in MOFs and use this observation in combination with detailed molecular simulations to assess the impact of local defects and flexibility in a variety of materials for which defects had not been considered previously.

465c Formulation and Processing of Magnetic-MOF Composite Structured Monoliths for Light Olefin/Paraffin Separation

Khaled Baamran, Ali Rownaghi, Fateme Rezaei

Separation of olefin/paraffin pairs has been identified as the application where low-energy separation replacements could provide significant reduction in overall energy consumption of the process. Among all the alternative separation techniques examined so far, adsorption appears to be the most promising and energy-efficient approach for light hydrocarbons separation. In that respect, development of smart sorbents that can be regenerated when triggered by external stimuli such as magnetic field can overcome the poor energy utilization of the current adsorbents investigated for this separation. In this work, we report the development of novel magnetic Fe_2O_3@MOF sorbents with core-shell structure and assessment of their adsorption performance in C_2H_6/C_2H_4 separation. The electromagnetic properties of the materials were tuned by varying the Fe_2O_3 nanoparticles composition and particle size, and their effects on adsorption capacity and selectivity of the composite sorbents were systematically investigated. Adsorbent regeneration was carried out under an external magnetic field with varied strength to identify its impact on desorption kinetics and product purity. The sorbent materials were formulated into monolithic contactors via additive manufacturing method. Additionally, multicomponent dynamic adsorption experiments were carried out to gain a better understanding of the materials performance under more realistic conditions. This work builds on previous studies that demonstrate the suitability of novel stimuli-responsive sorbents for next generation of separation processes.

465d Carbon Dioxide Capacity Retention on Elastic Layered Metal Organic Frameworks Subjected to Hydrothermal Cycling

Christian Lastoskie

Flexible MOFs, also known as soft porous crystals (SPCs), are a subset of MOFs that possess both a highly ordered network and structural transformability. In contrast with rigid MOFs, which retain their structure and their porosity irrespective of environmental factors, SPCs undergo structural transformations that are dependent on external stimuli such as temperature, mechanical pressure, or guest adsorption, on account of their bi-stable or multi-stable attributes. This facet of SPCs has led to the observation of previously unanticipated gas adsorption phenomena.

A subset of SPCs are the so-called elastic layered metal-organic frameworks (ELMs). ELMs are composed of metal vertex ions, connecting ligands, and charge-balancing counter-ions that are arranged into two-dimensional sheets that in turn assemble into three-dimensional stacked structures. These materials show a latent porosity for the adsorption of gas molecules above a specific pressure, termed the "gate pressure", that results in an expansion of the interlayer spacing between the two-dimensional sheets and a corresponding jump in the adsorption isotherm that cannot be classified in accordance with the conventional IUPAC isotherm designations. The exotic adsorption characteristics of ELMs are not observed in more familiar commercial adsorbents such as activated carbons or zeolites. Nor are they found in the adsorption isotherms of MOFs with rigid pore structures. These unusual features confer upon ELMs potential advantages for CO_2 capture, inasmuch as they combine a high selectivity for separation of CO_2 from gas mixtures with a low energy requirement for adsorbent regeneration and CO_2 recovery.

Adsorption of carbon dioxide on elastic layered metal-organic frameworks (ELMs) was investigated during and after exposure to water. Two ELM variants, ELM-11 and ELM-12, were contacted with water vapor and the impact of cyclical exposure on the CO_2 capacity of the adsorbents was observed. ELM-11 was found to lose CO_2 capacity with each successive exposure to water, whereas ELM-12 retained CO_2 capacity through four exposure cycles. Density functional theory calculations were performed to interpret these observations. Changing the counter-ion from the simple tetrafluoroborate (BF_4^-) to the larger and more complex trifluoromethanesulfonate ($CF_3SO_3^-$) anion expands the number of potential binding sites for adsorbate molecules. While CO_2 directly competes with other adsorbates for binding sites in ELM-11, CO_2 does not directly compete with other adsorbates in ELM-12 due to its preference for direct interaction with both fluorine and oxygen atoms in $CF_3SO_3^-$.

465e Trivalent metals on MOFs for SO2 adsorption

Chengzhai Wang, Krista Walton

SO_2 presents a threat to human health, industrial processes, and the environment. Metal-organic frameworks (MOFs) have been shown to be promising candidates for SO_2 adsorption due to their large surface area and structural tunability. However, developing MOFs with high SO_2 capacity especially at low partial pressure as well as excellent selectivity remains a challenge. Here, we report three pairs of Al-MOFs with ligands containing different N, O, and S functional sites. The introduced N, O, and S functional sites will enhance the SO_2 capacities for these MOFs at various degrees due to the different levels of binding strength between them with adsorbed SO_2 molecules. Powder X-ray diffraction results confirmed that these modified MOFs possessed the same topology and show strong chemical stability to SO_2. SO_2 adsorption from a stream of 0.1% SO_2 in balance N_2 was measured using a home-built breakthrough system. The results showed that all functionalized Al-MOFs presented higher SO_2 loadings than parent CAU-10. Interestingly, MOFs with heterocyclic functional sites exhibited high uptake of SO_2 than their corresponding ones with aryl functional sites. For example, CAU-10-3,5-PDC adsorbed 0.41 mmol/g of SO_2, higher than that of CAU-10-NH_2 (0.25 mmol/g). This talk will present SO_2 adsorption results for these pairs of Al-MOFs, and the mechanism governing the adsorption process will be explored through Attenuated total reflection infrared spectroscopy (ATR-IR) experiments. The results suggest that the introduced functional sites can act as favorable adsorption sites for SO_2. Additional density functional theory (DFT) simulation reveals which type of functional sites exhibit the best SO_2 adsorption behavior. Overall, this study provides insights into establishing basic rules for the design of Al-MOFs with N, O, or S-containing functional groups for SO_2 adsorption.

465f Investigation of the Properties of MOFs for Adsorption of PFOS: Comparative Study of MIL 101 (Cr), MIL 101 (Cr)-AC, MIL 101 (Cr)-NH_2 , and MIL-53(Al) BPDC with Activated Carbon

Jasneet Pala, Tin Le, Medha Kasula, Milad Esfahani

With the introduction of Per- and Polyfluoroalkyl Substances (PFAS) in the early 90s for various applications such as firefighting foams, non-stick cookware, and stain-resistant fabrics, there has been growing concern about its existence in water due to its harmful health effects on human and animals over time. The chemical and thermal stability of PFAS makes the breakdown difficult and PFAS environmental persistence makes the removal challenging. Among all techniques, the adsorption methodology has proved to be an efficient and economical way to the reduction of PFAS in water. Different adsorbent including classical activated carbon and new generation of adsorbent such as metal organic frameworks (MOFs) have been used for the removal of short-chain ($C<6$) and long-chain ($C>7$) PFAS. In this work, we performed perfluorooctane sulphonic acid (PFOS) adsorption at different pH (4-10) using different adsorbent including superfine Activated Carbon (SFAC) and various MOFs including MIL 101 (Cr), MIL 101 (Cr)-AC, MIL 101 (Cr)-NH_2, MIL-53(Al)-biphenyl-4,4'-dicarboxylic acid (Al-BPDC) and performed comparative analysis to investigate the effective parameters governing the PFOS adsorption with respect to MOF physicochemical properties and adsorption process operation conditions. The adsorbents characteristics such as morphology and size (using Scanning Electron Microscopy), crystallinity (using X-ray powder diffraction (XRD)), surface area and pore volume (using Brunauer–Emmett–Teller (BET)), surface charge, and functional groups (using Fourier-transform infrared spectroscopy (FTIR)) were analysis and were used for justification of adsorption performance and adsorption mechanism of each adsorbent. Based on performance, SFAC with around 93% PFOS removal (pH 4-6) showed supreme performance over other MOFs followed by Al-BPDC (86%) > MIL 101-AC (80%) > MIL 101 (65%) > MIL 101-NH_2 (20%). This behavior was explained based on the the presence of extensively large pore size in SFAC that would adsorb PFOS and hydrophobic interactions between SFAC and PFOS that made it an excellent source for PFOS adsorption. On the other hand, the interaction between MOFs and PFOS was mostly dominated by electrostatic interaction.

465g Understanding Small Molecule Binding Potential Landscapes Near Open-Metal Sites in M-MOF-74 (M=Mg, Mn, Ni, Zn)

Ishan Pandey, Chau-Chyun Chen, Joshua Howe

Metal–organic frameworks (MOFs), a class of materials of interest for their highly customizable and tunable properties, often exhibit high adsorption selectivity and capacity, motivating substantial research interest in energy-related applications such as adsorptive separations. Among MOFs, those with undercoordinated metals, or "open metal sites" (OMS), have drawn particular interest for the ability to leverage exposed metal cations for highly selective and tunable adsorptive separations. Because MOFs with OMS are often of interest for leveraging the unique properties at these exposed metals, classical force field-based modeling approaches have struggled to rigorously account for the electronic details relevant for fully describing adsorption. Additionally, adsorption may affect structural changes in the adsorbent that are often neglected but may play a role in adsorption behavior. In this work, we use density functional theory (DFT) to study adsorption of small molecules H_2, N_2, CO_2, and NO in M-MOF-74s, a class of OMS MOFs with a range of d-occupations from among 3d transition metals. We studied binding potentials of these small molecules as a function of distance from the open metal sites in Mg ($3d^0$)-, Mn ($3d^5$)-, Ni ($3d^8$)-, and Zn ($3d^{10}$)-MOF-74s, investigating different structural constraints ranging from complete structural optimization of adsorbate and adsorbent to partial geometrical optimization to single point (static) calculations. We characterize path-dependent energetics and structural changes to facilitate model development for adsorption of small molecules in OMS MOFs. This work presents progress in both rational design of materials for gas separations and development of next-generation force-fields to capture the structural and energetic changes relevant to adsorptive performance in OMS MOFs.

465h Effect of Topology and Structural Tunability on Selective Capture of CO_2 Using Zeolitic Imidazolate Frameworks

Rimita Bose, Parasuraman Selvam, Niket Kaisare

A typical post-combustion flue gas predominantly consists of N_2, CO_2, H_2O and O_2,[1] with the concentration of CO_2 being 15% and N_2 as the major component. Presence of moisture and acid gas impurity, it has become a daunting task to selectively separate CO_2 from the mixture. Commercial CO_2 capture technology, which involves amine-based solvents or alkaline solutions, suffers from high energy penalty. Hence, it is essential to develop cost-effective and energy-efficient CO_2 capture technologies. In this regard, the discovery of metal-organic frameworks (MOFs) has opened up new possibilities as the adsorption-based CO_2 capture from the multi-component gas mixture is increasingly considered a promising technology.

MOFs possess high surface area, large pore volume, excellent structural flexibility and tremendous tuneable functionality. However, they often suffer largely from a lack of chemical stability and high synthesis cost. Among the MOF family of materials, the zeolitic imidazolate frameworks (ZIFs), possessing similar topologies to zeolites, are considered to possess high thermal and good chemical stability and can be synthesized with ease in comparison to many other MOFs. For our study, we have selected a few ZIF-class of materials, e.g., ZIF-7 and ZIF-11, with zinc as a metal centre and benzimidazole as a linker. ZIF-7 and ZIF-11 use the same precursors but different synthesis methods, resulting in materials with varied topology, viz., SOD and RHO, respectively. Consequently, we use these as exemplary materials to investigate the effect of topology on CO_2 adsorption in detail. Furthermore, we consider an exceptionally stable material, i.e., ZIF-71 with RHO topology and Cl-substituted imidazole linker (4,5-dichloroimidazole), for comparison. All the three materials employed in this study are stable in humidity and that ZIF-71 shows outstanding chemical stability even with underexposure of humid CO_2 and/or SO_2.[2]

Preliminary CO_2 adsorption studies on ZIF-7 and ZIF-71 indicate that the former can show significant potential for CO_2 capture with an uptake capacity at 300 K and 1.2 bar is 9 wt% (~ 2 mmol/g) while for the latter it is 3 wt% (~ 0.7 mmol/g). Under similar experimental conditions, ZIF-7 not only has a 3 times higher uptake than the corresponding ZIF-71 but also possesses an interesting gate-opening phenomena triggered by CO_2 molecule leading to an S-shape adsorption isotherm under atmospheric pressure. This structural flexibility could further improve the selectivity of CO_2/N_2 which is essential for separation processes. On the other hand, since ZIF-71 is not affected by the presence of impurities, this material could prove to be a better candidate for CO_2 capture from gas streams containing impurities. In this context, our future direction would be to conduct dynamic breakthrough experiments under a 15% CO_2/N_2 mixture using a packed bed of ZIF-7, ZIF-11 and ZIF-71. The outcome of this experiment would enable us to correlate the effect of functional group, topology and structural flexibility in the relation to the gas mixture separation. Furthermore, we can also hypothesize that ZIF-7 would probably have a longer CO_2 retention time owing to its CO_2 selectivity, while the presence of benzene ring and -Cl group as functionality in ZIF-11 and ZIF-71 will tend to interact differently with CO_2 molecule causing distinctive breakthrough time.

References

1. Sumida, K., Rogow, D.L., Mason, J.A., McDonald, T.M., Bloch, E.D., Herm, Z.R., Bae, T.-H. and Long, J.R., Carbon Dioxide Capture in Metal-Organic Frameworks. *Chem. Rev.,* **112**, 724-781 (2012).
2. Bhattacharyya, S., Han, R., Kim, W.-G., Chiang, Y., Jayachandrababu, K.C., Hungerford, J.T., Dutzer, M.R., Ma, C., Walton, K.S., Sholl, D.S. and Nair, S., Acid Gas Stability of Zeolitic Imidazolate Frameworks: Generalized Kinetic and Thermodynamic Characteristics. *Chem. Mater.,* **30**, 4089-4101 (2018).

518a Carbon Molecular Sieve Hollow Fiber Membranes Derived from Dip-Coated Precursor Hollow Fibers Comprising Nanoparticles

Yuhe Cao, Kuang Zhang, Chen Zhang, William J. Koros

We report a novel approach to economically fabricate dual-layer composite nanoparticle-containing carbon molecular sieve (CMS) hollow fiber membranes having excellent gas separation performance. The economically favored process comprises dip coating engineered support layer fibers to form a dense skin layer in a CMS precursor. The coated fibers are then pyrolyzed to form a high-performance CMS hollow fiber membrane. The nano-particle containing composite carbon molecular sieve hollow fiber membranes showed very attractive selectivities and productivities. This work shows how to produce high performance CMS hollow fiber membrane by coating, which exceeds that achieved with the standard sol-gel support stabilization technique.

518b Tailoring Molecular Structures of Hybrid Carbon Molecular Sieve Membranes for Natural Gas Processing

Hyun Jung Yu, Jong Suk Lee

Carbon molecular sieve (CMS) membranes are promising for excellent gas separation performance owing to their peculiar bimodal pore size distribution consisting of ultramicropores (pore diameter (d) < 0.7 nm) and micropores (0.7 nm < d < 2 nm). However, separating N_2/CH_4 mixture is still challenging because the relationship between CMS structure and N_2/CH_4 separation efficiency is poorly understood. In this presentation, I will report highly N_2-selective hybrid carbon molecular sieve (CMS) membranes fabricated by pyrolyzing rationally designed polyimide (PI) and ladder-structured polysilsesquioxane (LPSQ) blends [1]. The N_2/CH_4 single gas and N_2/CH_4 mixed gas separation properties of CMS PI/LPSQ membranes are demonstrated by varying the PI/LPSQ ratios as well as pyrolysis temperatures. Furthermore, the effect of the SiO_x phases in hybrid CMS membranes on the N_2/CH_4 separation properties is elucidated using DFT calculations.

[1] H.J. Yu, J.H. Shin, A.S. Lee, S. S. Hwang, J.-H. Kim, S. Back, J. S. Lee. Tailoring selective pores of carbon molecular sieve membranes towards enhanced N_2/CH_4 separation efficiency. J. Membr. Sci. 620 (2021) 118814.

518c Ceramic-Carbonate Hollow Fiber Membranes with Improved Mechanical Strength for High Temperature CO_2 Separation

Tianjia Chen, Jerry Lin

Oxygen-carbonate ionic conducting ceramic-carbonate dual-phase membranes are permeable to CO_2 only, not other gases, at high temperatures (>600°C). They have shown promises for applications in separation and reaction processes for hydrogen production and CO_2 capture. Ceramic-carbonate dual-phase hollow fiber membranes offer superb performance of CO_2 separation at high temperatures but suffer from poor mechanical strength. This paper reports a new structure of ceramic-carbonate dual-phase hollow fiber membrane with high CO_2 permeance and improved mechanical strength for CO_2 separation. The hollow fiber supports of about 1.4 mm OD and 1.0 mm ID are made of $Ce_{0.8}Sm_{0.2}O_{2-\delta}$ (SDC) and $La_2NiO_{4+\delta}$ (LNO) materials. The final hollow fiber membrane contains a SDC-carbonate outer layer and a SDC/LNO inner layer, forming a special double-ring structure in the middle of the membrane. The CO_2 flux of the membrane is 2.3 $ml \cdot min^{-1} \cdot cm^{-2}$ at 750 °C. In addition, the ultimate load and bending strength of the new SDC-SDC/LNO-carbonate hollow fiber membrane are respectively 3.6 and 1.9 times higher than that of pure single-layer SDC-carbonate hollow fiber membrane. This improvement of the mechanical strength can be attributed to special function of the double-ring structure.

518d Moisture-Resistant Graphene-Based Nanolaminate Membranes for Hydrogen Purification Enabled By Charge Neutralizing Nanofillers

Behnam Ghalei

Introduction

More than 90% of 70 million tons of worldwide demand for hydrogen (H_2) is currently generated *via* steam reforming and the water-gas-shift (WGS) reaction of fossil fuel sources[1]. This produces mixture of H_2 and carbon dioxide (CO_2) saturated with steam from which H_2 is subsequently purified by selectively adsorbing the CO_2 using aminated solvents. Unlike amine absorption processes, membrane separation requires no changes in phase behavior and is, in principle, the most energetically cost-effective route for primary purification of H_2 and CO_2 [2]. However, polymeric membranes do not have adequate separation capability of CO_2 from H_2 due to the high level of CO_2 solubility in organic materials [3]. Costly metallic and ceramic membranes, operating on the principles of selective adsorption or molecular sieving, have extraordinary selectivity for H_2 against CO_2; ~1000 and ~300, respectively, and sufficient rates of permeance.

Ultra-thin graphene oxide (GO) has been proposed as a step-change membrane material to separate H_2 and CO_2. High selectivities (up to 1000) and triple-digit gas permeation unit (GPU) values (1 GPU = 3.35×10^{-10} mol m^{-2} s^{-1} Pa^{-1}) were reported [4]. However, when the GO membranes are exposed to a humid environment, the hydrated GO sheets become negatively charged and will come apart due to the electrostatic repulsion that promotes the GO membrane's delamination. Such catastrophic swelling is an un-resolved obstacle to the practical implementation of this exciting technology[5]. In this work, we specifically target the stabilization of GO membranes using nanodiamonds (NDs) towards adverse humid conditions while attempting to maintain its overall high performance towards H_2/CO_2 separation[6].

Results and discussion

GO membranes were prepared by vacuum filtration method of a suspension of single-layer GO sheets onto both ceramic and polymeric supports. 3 nm-sized NDs were controllably added to the GO solutions before vacuum filtration. The prepared membranes were designated as GOαND, where α (α= 5, 10, 20, 30, and 35) represents the weight concentration of ND particles. The atomic force microscope (AFM) images of the pure GO membrane revealed a smooth surface without visible defects. However, the surface roughness of the composite membrane increased with the addition of ND particles (Fig. 1a, b).

The cross-section view of the GO membrane displays a highly packed morphology with a uniform thickness of about 38±6 nm. The addition of ND particles (30 wt%) to the GO matrix increased the membrane thickness to 75±8 nm (Fig. 1c, d).

On ceramic supports, the produced native GO membranes were found to have performances comparable to or slightly better than those reported in the literature with initial H_2 permeance of around 1150 GPU and ideal gas selectivity of ~284 over CO_2. However, when exposed to a water-saturated equimolar mixed gas feed, the GO membrane performances dramatically deteriorated over a 100 hr test at room temperature. Permeances and selectivity dropped by 55% and 70%, respectively. By contrast, membranes containing 30 wt% of NDs, GO30ND, were found to have permeances thrice the level of native GO membranes (to 3700 GPU) and with a relatively small reduction in ideal gas selectivity (to $a_{H2/CO2}$ ~ 210) when tested with dry gas. However, most significantly, there was only a ~5% and ~10% drop in permeance, and selectivity of GO30ND membranes, when tested extensively with a wet mixed gas feed (Fig. 1e, f).

The ND particles were positively charged (+45 mV), while the GO sheets carried a net negative charge (-48 mV) at pH=7. Thus, the proper assembling of NDs into the GO structure *via* strong electrostatic interactions can be presumed (Fig. 1g). Even though the NDs altered the stacking of the GO laminates, the composite membrane remained intact due to the electrostatic interaction and H_2 bonding between the GO sheets and ND particles. Hydrogen bonding in the GOαND membranes is verified by shifting prominent bonding peaks in FTIR spectra. C1s and N1s X-ray photoelectron spectroscopy (XPS) spectra confirmed the presence of ND in the GO mixtures. The X-ray analysis of GO membrane suggests highly ordered stacking of GO laminates (sharp peak at $2\theta=6.15°$) with the d-spacing of 0.93 nm (Fig. 1h). The intensity of the peak decreased and broadened by the introduction of NDs, which confirms the disruption of GO stacked ordering. Furthermore, the peak shifted slightly with the addition of NDs, and the equivalent d-spacing reached 0.89 nm in GO30ND membranes. The insertion of positively charged ND particles between the GO laminates diminished the negative charge effect, weakened the interlayer electrostatic repulsion, and reduced the d-spacing of GO sheets. Young's modulus and hardness of the GOαND membranes both improved up to ~25% compared to the pure GO membrane. The improved mechanical properties can be explained by the favorable interaction between GO and NDs, which is desired for long-term operation of the membranes.

The incremental addition of ND particles into the GO structure improved its gas permeation in a controlled and significant manner. At the same time, the separation factors were almost at the same level as GO membranes at low filler concentrations, i.e., up to 30 wt% (Fig. 1i). An H_2 permeance ~3741 GPU ($\alpha_{H2/CO2}$=211.6) was observed in the GO30ND membrane; an exceptional ~300% enhancement in the permeance compared to the pure GO membrane. The decrease in the number of GO layers per laminates and the reduction in the crystallite width of the GO laminates are responsible for the permeance enhancement. N_2 adsorption test indicates a 500% enhancement in the pore volume from 0.036 cm^3/g in pure GO to 0.17 cm^3/g by adding 30 wt% ND particles, which facilitates the gas diffusivity in the composite membrane. Generally, compared across a wide spectrum of inorganic materials, the GO30ND membrane show both exceptional H_2 permeance (>3700 GPU) and H_2/CO_2 selectivity (>200).

The evaluation of gas separation properties of the membranes was investigated under mixed gas feeds. Using equimolar gas feeds similar to that from a WGS reaction, H_2 permeance and H_2/CO_2 selectivity of the GO30ND membrane decreased 6% and 13%, respectively. The reduction in the permeance and selectivity of the membranes under the mixed gas condition is, in general, due to the partial hindrance of H_2 molecules transport by highly adsorbed CO_2 molecules.

The stabilizing property of NDs against humidity ascertained from the immersion of GO30ND membranes in water. Here, thicker GO membranes (~200 nm) prepared over a larger area by vacuum filtration on polyethersulfone supports are seen to disintegrate upon immersion, whereas GO30ND membranes are stable over the same period. However, a more significant test of the ND stabilization effect is demonstrated by exposing the membranes cyclically to wet and dry feeds of equimolar H_2/CO_2 mixture (85% relative humidity) (Fig. 1j). Native GO membranes were not able to survive a single full cycle exposure, becoming fully permeable to both gases. Interestingly the membrane gets worse after being exposed to a second dry gas feed, suggesting that significant and irreversible structural reorganization takes place during both wetting and drying of the hygroscopic material. On the other hand, GOaND membranes are found to have a degree of reversibility in membrane properties when exposed to wet and dry gases, though there is a net reduction in both permeance and selectivity over a series of cycles. Although this reduction is minimized with higher ND content, even GO30ND membranes do not have perfect stability against the vigorous cyclic tests. This quasi-reversible variation of membrane performance under cycling or constant humid conditions suggests that the NDs stabilize the GO nanolaminates within the membrane.

Conclusion

In this study, we have demonstrated the benefit of the intentional addition of positively charged carbonaceous nanoparticles, here exemplified through NDs, to negatively charged GO membranes. We demonstrated that NDs with sp^2/sp^3 core-shell structure and positive surface charge could reduce the electrostatic repulsive forces between hydrated GO sheets, where their robust and GO compatible structures are intercalated between the GO laminates, strengthening the membrane structure. As a result, the delamination of GO nanofilms in the presence of humidity is suppressed. The NDs stabilize GO membranes against humidity while tremendously enhancing its intrinsic separation performance. The approach presented in this work will generate high stability 2D material composites, not only for membrane separation but also for other application areas where the field of 2D materials is gaining traction.

References

1. Voldsund, M., Jordal, K. andamp; Anantharaman, R. H2 production with CO2 capture. *Int. J. Hydrogen Energy* **41**, 4969–4992 (2016).
2. Merkel, T. C., Zhou, M. andamp; Baker, R. W. Carbon dioxide capture with membranes at an IGCC power plant. *J. Memb. Sci.* **389**, 441–450 (2012).
3. Zhang, M. *et al.* Electropolymerization of Molecular-Sieving Polythiophene Membranes for Hydrogen Separation. *Angew. Chemie* **131**, 8860–8864 (2019).
4. Kim, H. W. *et al.* Selective Gas Transport Through Few-Layered Graphene and Graphene Oxide Membranes. *Science* **342**, 91–95 (2013).
5. Abraham, J. *et al.* Tunable sieving of ions using graphene oxide membranes. *Nat. Nanotechnol.* **12**, 546–550 (2017).
6. Huang, G. *et al.* Overcoming humidity-induced swelling of graphene oxide-based hydrogen membranes using charge-compensating nanodiamonds. *Nat. Energy* **6**, 1176–1187 (2021).

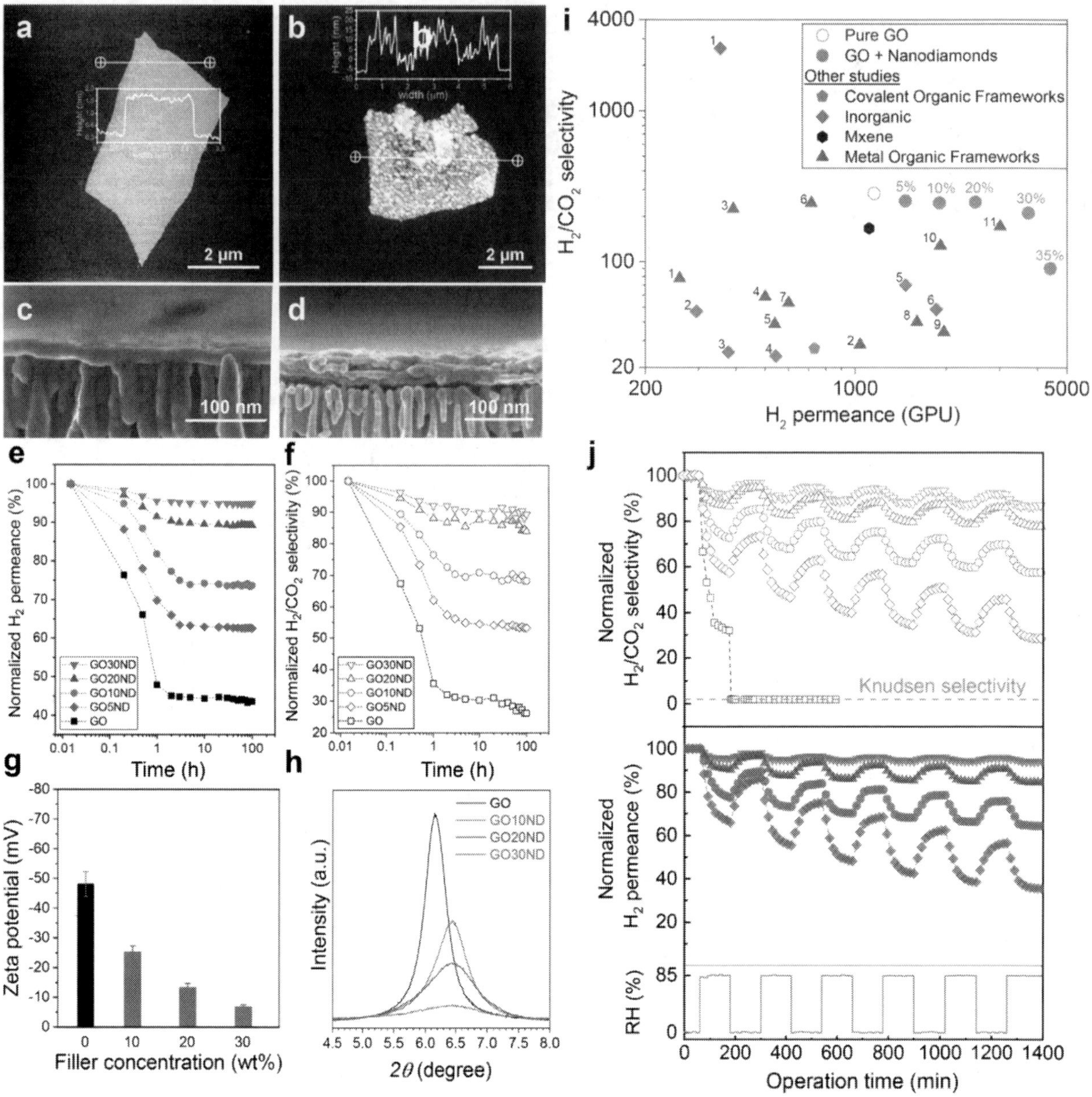
a
2 µm
b
Height (nm)
width (µm)
2 µm
c
100 nm
d
100 nm
i
Pure GO
GO + Nanodiamonds
Other studies
Covalent Organic Frameworks
Inorganic
Mxene
Metal Organic Frameworks
5%
10%
20%
30%
35%
H_2/CO_2 selectivity
H_2 permeance (GPU)
e
Normalized H_2 permeance (%)
GO30ND
GO20ND
GO10ND
GO5ND
GO
Time (h)
f
Normalized H_2/CO_2 selectivity (%)
GO30ND
GO20ND
GO10ND
GO5ND
GO
Time (h)
j
Normalized H_2/CO_2 selectivity (%)
Knudsen selectivity
Normalized H_2 permeance (%)
RH (%)
Operation time (min)
g
Zeta potential (mV)
Filler concentration (wt%)
h
GO
GO10ND
GO20ND
GO30ND
Intensity (a.u.)
2θ (degree)

518e Novel Graphene Oxide-Based Membrane Structure for a Highly Effective Breathable Barrier for Toxic Vapors and Chemical Warfare Agents

Yufeng Song, Cheng Peng, Zafar Iqbal, Kamalesh Sirkar, Gregory W. Peterson

Garments protective against chemical warfare agents (CWAs) or accidentally-released toxic chemicals must protect from toxic gases/vapors by blocking their transport for substantial time and allow moisture transport for breathability. A number of approaches have been studied. A thick mixed matrix membrane having a high loading of metal organic framework HKUST-1 in an elastomeric triblock copolymer completely blocked 2-chloroethyl ethyl sulfide (CEES) (a simulant for sulfur mustard) for quite some time (i.e., high value of permeation lag time) but the moisture vapor transmission rate (MVTR) was low~192 g/m^2-day (minimum needed, 1500-2000 g/m^2-day). Using a highly breathable carbon nanotube (CNT) based membrane blocking biological agents, an alternative approach grew on the top surface of the CNTs a layer of polymer chains which reversibly collapsed in contact with a nerve agent simulant diethyl chlorophosphate (DCP) temporarily shutting CNT pore mouths and dramatically reducing the permeation rate of DCP. However, DCP permeation was restricted to ≈1% of the chemical challenge. Complete blockage for an extended period during use is absent. Our initial studies using a cross-linked graphene oxide (GO) laminate achieved 98.3% rejection of CEES on a steady basis after a short permeation lag time[1]. Our newly developed GO based laminate structure however shows remarkable results namely, complete blockage of ammonia, Mustard (HD), Soman (GD), and a Sarin simulant, dimethyl methyl phosphonate, for 2750 min, 1075 min, 176 min, and 7 days respectively while moisture transmission rate was substantial. We will describe this structure and its performance.

[1]Peng, C., Iqbal, Z., Sirkar, K. K., Peterson, G. W., Graphene oxide-based membrane as

protective barrier against toxic vapors and gases, ACS Appl. Mater. Interfaces. 2020, 12, 11094−11103.

518f Unexpected Gas Separation Selectivity in SSZ-13 Zeolite Membranes

Ninad D. Anjikar, Orhan Talu, Qiang Fu, Sankar Nair, Shaowei Yang

SSZ-13 (CHA structure) is among the most attractive small-pore zeolite membranes for small-molecule gas separations. In a recent permeation study of a few interested gas molecules with SSZ-13 zeolite membranes, unexpected high selectivity was found for a gas pair. Solution-diffusion model (*Permeability* = *Solubility* × *Diffusivity*) is generally applicable for analyzing gas permeation behaviors in zeolite membranes. So the adsorption isotherms and dynamic adsorption behaviors of the molecules were explored with SSZ-13 crystals. The solution diffusion model was used to help interpret where the selectivity comes from. However, selectivity based on adsorption data was about two orders of magnitude smaller than observed membrane selectivity. A hypothesis on the selectivity will be proposed and more experiments will be done to verify the consistency of the observation. This is an interesting phenomenon, and a better understanding could lead to new separation applications with polycrystalline zeolite membranes.

610a Selective Removal of Selenate from Selenate-Sulphate Water Solution By Adsorption on Dispersed Single Atoms on [012] Al_2O_3 Surface: An Ab-Initio Study

Srishti Gupta, Adam Chismar, Christopher Muhich

Consumption of selenium higher than 400μg/day can lead to selenosis which has neurotoxic effects such as lethargy, dizziness, motor weakness and burning and prickling in the extremities. The most common source of high doses of selenium is drinking water, where inorganic selenium is present in selenate (Se(VI)) and selenite (Se(IV)) forms. While both selenium anions are similarly toxic to humans, selenate (Se(VI)) is harder to remove because it has a more stable structure, and is similar to its more prevalent co-contaminant cousin – sulfate. Aluminium oxide is a promising adsorbent for removing selenium oxo-anions from water. However, adsorption of selenate on the Alumina surface occurs in the outer sphere and is controlled by the surface water Hydrogen bonding network, which makes the adsorption non-selective. Here, we investigate the adsorption of selenate and sulphate on nine transition metal single-atom sites dispersed in [012] Al_2O_3 surface using density functional theory to determine if the inclusion of a single atom can induce selenate adsorption selectively over sulphate. Additionally, we elucidate the relative effects of geometric and water network effects from chemical bonding effects. We use a hybrid solvent method which includes implicit and explicit solvation and van der Waals forces, to compare adsorption energies in inner vs. outer sphere configuration. We find that V and Ge Single Atom in Al_2O_3 adsorb selenate exothermically in inner sphere configuration at neutral pH, while this process is endothermic for pure Al_2O_3 surface. We correlate the adsorption energies with physical quantities including bond strength, surface water re-configuration, pH, solvation bond distances, etc. to understand the find the other significant surface phenomenon affecting the adsorption besides the water network.

610b Modeling the Interactions of Heavy Organic Aerosols with Water-Harvesting Metal-Organic Frameworks

Raghuram Thyagarajan, David Sholl

Metal-organic frameworks are porous solids that are highly tunable for several applications including gas storage, chemical separations, catalysis etc. Recently, multiple MOFs have been synthesized with high hydrolytic stability for atmospheric moisture capture. These MOFs possess exceptional water uptake capacity at low vapor pressures as well as high adsorption-desorption cycling stability. However, not much is known about the water stability of these materials in the presence of air-borne contaminants and their impacts on the water harvesting performance. In this work, we model the interactions of heavy organic aerosols found in the atmosphere with these MOFs using density functional theory calculations. We examine the molecular level interactions of water and aerosols with the MOFs inside the pores of both defect-free and defective framework structures.

610c Predicting Adsorption of 1,4-Dioxane in All-Silica Zeolites Using Molecular Simulations

Samiha Sharlin, Tyler R. Josephson

1,4-dioxane is used to manufacture both industrial and consumer products. Its stability, solubility, and widespread use have affected drinking water sources. Consequently, the U.S. Environmental Protection Agency has classified it as an emerging organic contaminant and a probable human carcinogen by all routes of exposure[1]. However, the incompetence of conventional remediation processes demonstrates the need for an effective treatment strategy.

Separation with nanoporous materials requires high capacity and selectivity to adsorb 1,4-dioxane preferentially and exclude water and its co-contaminants[2]. We propose to use hydrophobic all-silica zeolites with optimal pore sizes to selectively adsorb 1,4-dioxane from water. The TraPPE united-atom model for 1,4-dioxane[3] was validated by reproducing its vapor-liquid equilibrium properties using Gibbs ensemble Monte Carlo simulations, with Self-Adapting Fixed-Endpoint Configurational-Bias Monte Carlo (SAFE-CBMC) to sample the flexible ring configurations of 1,4-dioxane. This model and TraPPE-zeo[4] were used to predict 1,4-dioxane adsorption in different zeolite frameworks, investigate the adsorption mechanism, and analyze favorable host-guest interactions into zeolite channels under environmental conditions.

This material is based upon work supported by the National Science Foundation under Grant No. 2138938

[1] Adamson DT; Piña EA; Cartwright AE; Rauch SR; Hunter Anderson R; Mohr T; Connor JA. Sci Total Environ, 596-597:236-245 (2017)

[2] Chen R; Liu C; Johnson NW; Zhang L; Mahendra S; Liu Y; Dong Y; Chen M. Chemical Engineering Journal, 193-202:1385-8947 (2019)

[3] Keasler SJ; Charan SM; Wick CD; Econonmou IG; Siepmann JI, J. Phys. Chem. B 116, 11234-11246 (2012)

[4] Bai P; Tsapatsis M; Siepmann JI, J. Phys. Chem. C 117, 24375-24387 (2013)

610d Accelerating Solvent Selection for Type II Porous Liquids

Chao-Wen Chang, Isaiah Borne, Robin Lawler, Ryan P. Lively, David Sholl

Type II porous liquids are comprised of intrinsically porous molecules, usually porous organic cages, dissolved in a liquid solvent, potentially combining the advantages of higher adsorbate loading and selectivity of solid porous adsorbents with the handling convenience of liquids. Despite growing interest in type II porous liquids, little is known about the solubility trends of porous organic cages or other similar host molecules. This situation means that despite the potential benefits of using type II porous liquids it is currently difficult to find solvent/cage pairs that lead to viable and safe porous liquids for practical use. We demonstrate an efficient screening approach that uses COSMO-RS, a thermodynamic modeling tool which treats solvent implicitly, for high-throughput solubility predictions. In addition, we make predictions of solvent pKa values from a machine-learning model as well as several other features and apply this approach to select solvents from a library of more than 11,000 compounds based on the constraints of forming type II porous liquid. This method is shown to give qualitative agreement with experimental observations for two [4+6] imine based porous organic cages, CC13 (non-polar) and TG-TFB-CHEDA (polar), identifying solvents with higher solubility for these molecules than had previously been known. Ultimately, the algorithm streamlines the down selection of suitable solvents for porous organic cages to enable more rapid discovery of Type II porous liquids.

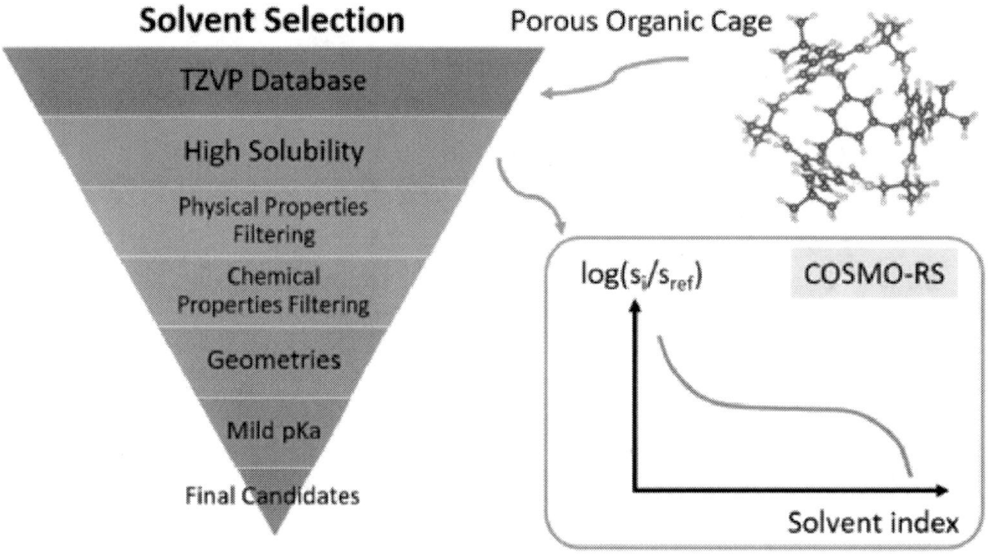

610f Quantitative Insights into the Influence of MOF Defects on Adsorption Properties

Zhenzi Yu, Cai Xuqing, David Sholl

Computationally predicting the adsorption properties of Metal-organic Frameworks (MOF) in a high throughput way has become a useful means of identifying promising materials for chemical separations. In almost all cases, high throughput simulations assume that MOFs are defect-free for computational convenience. However, structure defects are unavoidable during synthesis processes or might be introduced during adsorption processes by acid gases or water. Thorough defect studies have been conducted for some well-known MOFs, for example, UiO-66 and ZIF-8. These studies illustrated that defects can have a strong influence on MOF adsorption properties, physical properties, and mechanical properties.

We developed a python package that can generate missing linker defect structures and dangling linker defect structures at user-specified concentrations in arbitrary MOFs. Capping agents are added based on the local chemical environment. At the DFT level of accuracy, a subset of CoRE MOF database was fully optimized to identify the maximum defect concentration without triggering unit cell change. Structures with lower defect concentrations were optimized using DFT only for metal clusters and capping agents. At the force field level of accuracy, all the defect structures are optimized using UFF4MOF, and based on FF relaxed structure mechanical properties are computed.

We systematically studied the adsorption properties of MOFs at different missing linker defect concentrations. Surface areas and accessible volumes increase as the defect concentration increases, although at high defect concentrations this trend can be reversed due to shrinkage of the material. We used ethene and ethane as examples to test the defect influence on adsorption properties. Loadings and selectivity in MOFs with different defect concentrations are computed using GCMC. Most MOFs have lower uptakes with the increase of defect concentrations at 1 bar, but different adsorbates change in different ways, which complicates the trends for adsorption selectivity.

610g A Molecular-Scale Study of Monazite Beneficiation through Computational and Spectroscopic Techniques

Luke Gibson, Diana Stamberga, Vera Bocharova, Benjamin Doughty, Robert Sacci, Lawrence Anovitz, Vyacheslav S. Bryantsev

Sourcing rare-earth elements (REEs) is an active area of research due to their wide array of applications, ranging from energy to defense sectors.[1] An important step in the processing of REEs is mineral beneficiation, which is a process that separates valuable REE-containing minerals from undesired gangue materials, partly through a method known as froth flotation. During the flotation process, collector agents are introduced to a slurry of ground minerals which selectively bind to the REE-containing mineral and allows for collection. A promising REE-containing mineral for this application is monazite, which can contain various REEs, including Ce, La, and Nd. However, there is limited information on both the crystal morphology and the basic surface chemistries of monazite. Therefore, we have performed density functional theory (DFT) calculations to predict La-monazite crystal morphology via Wulff construction by computing surface energies for several low-index facets in the absence and presence of water. We also evaluate the performance of collector ligands for the predominant facets of monazite using computational and spectroscopic techniques. This work grants fundamental insight into the beneficiation process of monazite and helps guide the design of future collector agents for froth flotation applications.

References

[1] R. Eggert, C. Wadia, C. Anderson, D. Bauer, F. Fields, L. Meinert and P. Taylor, "Rare Earths: Market Disruption, Innovation, and Global Supply Chains", Ann. Rev. Environ. Resour., **2016**, 41, 199-222.

Acknowledgement

This research was supported by the Critical Materials Institute, an Energy Innovation Hub funded by the US Department of Energy, Office of Energy Efficiency and Renewable Energy, Advanced Manufacturing Office.

610h Implementation of Genetic Algorithms to Optimize Metal-Organic Frameworks for CO_2 Capture

Thang Pham, Randall Snurr

Metal-organic frameworks (MOFs) are a class of porous crystalline materials that can be used as adsorbents to capture CO_2 from flue gas. MOFs are highly tunable; by varying their building blocks, MOFs with different chemical and physical characteristics can be designed and synthesized to reach desired properties. Computational approaches have become an important complement to experiments by directing the synthesis toward promising building blocks. In this work, we implemented genetic algorithms (GA) to search for top-performing MOFs for CO_2/N_2 separation. We aimed to provide proof of concept that using GA, we could reduce the number of molecular simulations involved by one to two orders of magnitudes, and still determined most top-performing MOFs. To implement GA, we represented MOFs as a set of discrete variables, which consists of the MOFs' topology, nodular and connecting building blocks. All MOF structures were generated using the topological based crystal constructor (ToBaCCo 3.0) software, and their geometries were optimized using the UFF4MOF force field. Grand canonical Monte Carlo (GCMC) simulations were performed to calculate the CO_2 working capacity between the adsorption (313 K, 1 bar, CO_2:N_2 = 15:85) and the desorption conditions (313 K, 0.1 bar, CO_2:N_2=90:10), and the CO_2/N_2 selectivity at adsorption. The CO_2 working capacity and CO_2/N_2 selectivity were used as the objective functions for our implementation of GA. We used the Latin hypercube sampling to create an initial population for GA. Ranking selection methods (elitism and tournament selection) and genetic operators (crossover and mutation) were used to evolve the population. Important GA parameters, such as the number of MOFs per generation, the number of generations and the mutation probability, were determined. GA performance was tested using both a single objective function and multiple objective functions. Multiple-trajectory GA was used to avoid bad initial sampling and local convergence of GA. Cross-topology GA was used to search for top-performing structures across multiple MOF topologies. Using our implementation of GA, we could determine most of the top-performing MOFs for CO_2/N_2 separation and reduced the number of molecular simulations by more than one order of magnitude.

610i Prediction of Metal-Organic Interactions and Molecular Assembly in High Accuracy and Speed

Cheng Zhu, Samuel Edmund Hoff, Hendrik Heinz

Understanding and predicting metal-organic interactions at the atomic scale is critical for the development of materials for energy conversion, sensors, and therapeutics. Major bottlenecks to-date include scarce experimental data, and lack of facile methods to reliably and fast predict important binding and assembly preferences at the atomic scale. Here we show predictions of adsorption energies of aromatic molecules on platinum surfaces in unrivalled accuracy by molecular dynamics simulations with the Interface Force Field (IFF) in comparison to extensive experimental measurements in highest quality. The accuracy is up to 15 times higher than with density functional theory at 1 million times faster computational speed and allow 2 to 4 times better binding predictions relative to CHARMM-IFF. Quantitative insights into molecular recognition, assembly, the role of various defects, and condensation in the gas phase are described. We demonstrate that the methods are applicable to trillions of unknown metal-organic and metal-electrolyte interfaces for accurate and efficient screening.

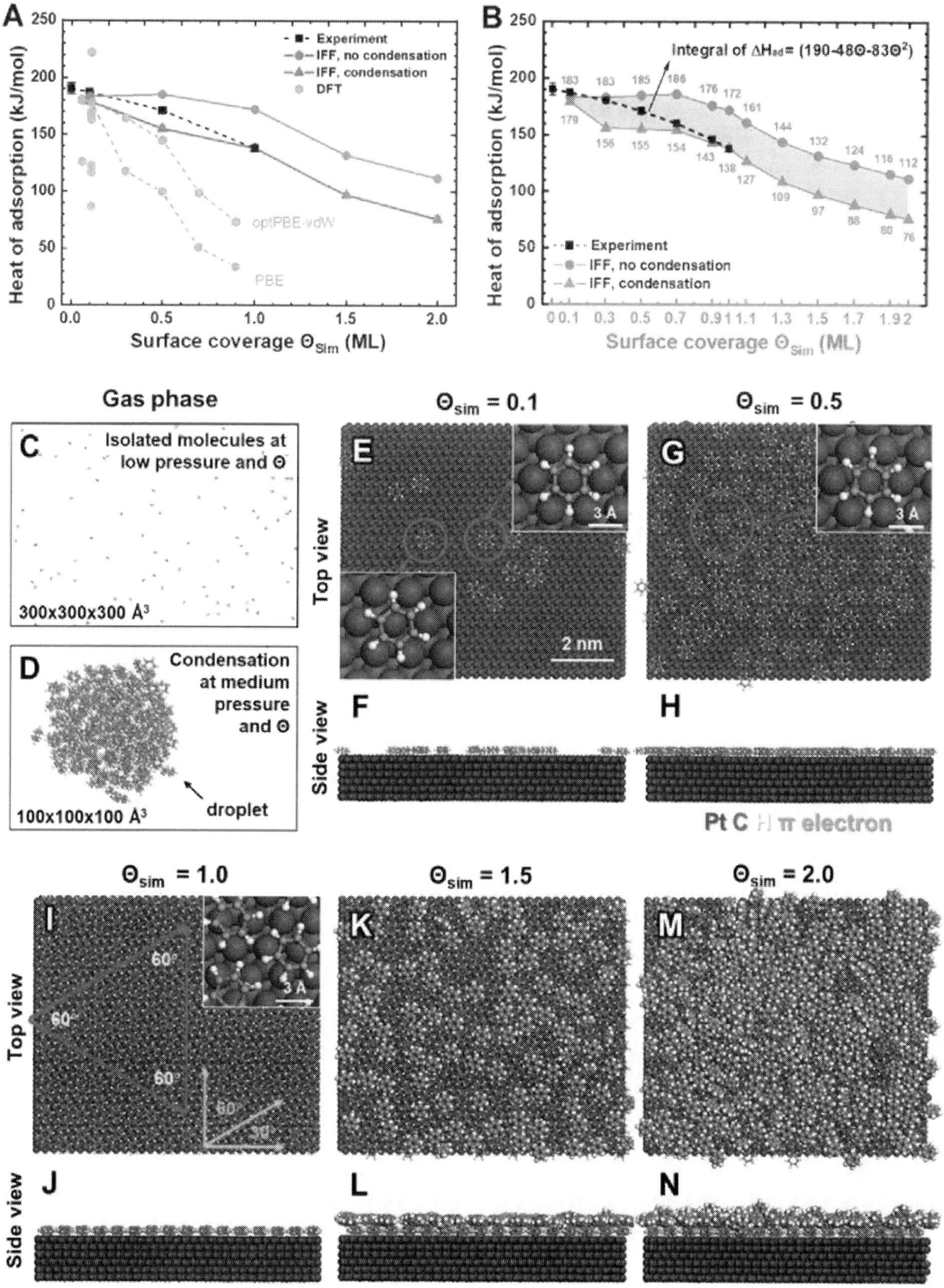
A
B
Heat of adsorption (kJ/mol)
Surface coverage Θ_Sim (ML)
Experiment
IFF, no condensation
IFF, condensation
DFT
optPBE-vdW
PBE
Integral of ΔH_ad = (190-48Θ-83Θ²)
Gas phase
Θ_sim = 0.1
Θ_sim = 0.5
C
Isolated molecules at low pressure and Θ
300x300x300 Å³
D
Condensation at medium pressure and Θ
droplet
100x100x100 Å³
E
F
G
H
Top view
Side view
3 Å
2 nm
Pt C H π electron
Θ_sim = 1.0
Θ_sim = 1.5
Θ_sim = 2.0
I
J
K
L
M
N
60°
30°

340a Adaptive Laboratory Evolution and Rational Engineering Enables Osmolysis-Based Recovery of Biomacromolecules from *Cupriavidus Necator*

Jeremy Adams, Douglas S. Clark

Intracellular biomacromolecules, such as industrial enzymes and biopolymers, represent an important class of bio-derived products in bacterial hosts. A key step in the downstream separation of these biomolecules is lysis of the bacterial cell wall to effect release of cytoplasmic contents. Cell lysis is typically achieved either through mechanical disruption or reagent-based methods, which introduce issues of energy demand, material demand, high costs, and scaling problems. Osmolysis, a cell lysis method that relies on hypoosmotic downshock upon resuspension of cells in distilled water, has been applied for bioseparations such as the purification of polyhydroxbutyrate (PHB) from extreme halophiles and protein products from mammalian cells. However, most industrial bacterial strains are non-halotolerant and relatively resistant to hypoosmotic cell lysis. To overcome this limitation, we sought to increase the susceptibility of non-halotolerant hosts to osmolysis using the lithoautotroph *Cupriavidus necator* as a prototypical strain. *C. necator* was evolved to increase its tolerance to NaCl through adaptive laboratory evolution. The evolved halotolerant strain experienced an osmolytic efficiency of 47% in distilled water following growth in 3% w/v NaCl. The cells were then made more susceptible to osmolysis by knocking out the large-conductance mechanosensitive channel (*MscL*) gene in *C. necator*. Greater than 99% osmolytic efficiency was observed upon osmotic downshock in this strain, indicating that this method serves as a simple and highly effective way to lyse cells for the purification of intracellular biomacromolecules.

340b Isolation Strategies Based on Membrane Processes for Extracellular Vesicles from *Citrus Limon*

Sara Giancaterino, Mariaelena Schiavone, Cristiana Boi

Extracellular vesicles (EVs) are lipidic nanoparticles secreted by all kind of cells. EVs can transport a pool of different biomolecules between close and distal cells, in a site-specific way. They are gaining increasing attention for their potential use in nanomedicine, nutraceuticals and cosmetics, in particular as drug delivery vectors, diagnostic tools and in tissue regeneration. The discovery that also plants do release EVs opened the doors to the use of vegetal vesicles as carriers of bioactive compounds and genes.

The greatest challenge in the field is associated to EVs separation from complex biological fluids. Ultracentrifugation is considered as the golden standard but it is mostly a lab-scale technique with a very low scalability potential. Conversely, alternative techniques as liquid chromatography and filtration processes are rising, allowing to obtain high yield products and feasible application to industrial scale.

This work is aimed at the development of a scalable and inexpensive integrated membrane process for the isolation of EVs from plants and fruits. *Citrus Limon* is selected as source material, due to its beneficial properties on human health, besides being a cheap and highly abundant source. Indeed, *Citrus Limon* juice extracts, represent an innovative and still poorly studied source of EVs that have very promising therapeutic use as they inhibit cancer cell proliferation on different cell lines.

In this process both dead-end and crossflow MF/UF units are compared, both as concentration and purification steps. Followingly, the isolated vesicles are analyzed and characterized with SEC liquid chromatography and both physical and biochemical methods. Finally, active and passive cargo loading techniques are applied to the isolated vesicles to test their encapsulation capability with bioactive phytocomponents, as curcumin.

340c Confocal and Multiphoton Imaging of IgG Gel Layer Formation during Tangential Flow Filtration

Fernanda Cunha, Francesco Rossi, Eduardo Ximenes, Norvin Bruns, Ken K. Qian, McKensie Mason, Brian D. Bowes, Zhao Yu, Dennis T. Yang, Vincent Corvari, Arezoo Ardekani, Gintaras Reklaitis, Michael Ladisch

The concentration of monoclonal antibodies (mAbs) and other biotherapeutic molecules is the last step in the production of injectable biologics. Cell culture results in a protein product that is purified by a sequence of steps consisting of centrifugation to remove cellular material, protein A capture, and then anion, cation, and hydrophobic interaction chromatography, respectively. Final steps are diafiltration and product concentration by tangential flow filtration (TFF) with an ultrafiltration membrane. TFF concentrates the protein from 5 - 10 mg/mL to 200 - 250 mg/mL. Over this concentration range, viscosity increases with increasing antibody concentration from about 1 to 10^2 mPa*s, although shear thinning may occur for the proteins at higher concentrations (Apgar et al., 2020). At these conditions, protein gels are hypothesized to form at the membrane's surface resulting in decreased flux and protein recovery. However, this process is not well understood and has motivated numerous studies on protein accumulation at membrane surfaces.

We report direct measurement of gel formation during concentration of bovine IgG in an operational TFF cell using multiphoton (MP) confocal microscopy to image the TFF membrane channel and to characterize the manner in which protein accumulates within a distance of 1 to 500 μm from the membrane's surface. The TFF flow cassette is fabricated to enable direct observation of gel formation by MP microscopy and is based on a flow cell design derived from previous work, using single photon confocal microscopy (Zuponcic et al., 2021). Bovine IgG (gamma globulin from blood), consisting principly of IgG1 (150 kD) and IgG2 (163 KD), was used as a surrogate for other types of more-expensive antibodies since relatively large amounts of protein (up to 5 g per run) are required for these experiments and since properties of the bovine IgG are well characterized (Butler, 1969, 1983).

While both MP and single photon confocal microscopy enable observation of protein behavior at operational conditions, MP microscopy utilizes laser light in the infrared range to excite target molecules, causing them to fluoresce at a visible wavelength. Since two photons of light result in a single, high-energy photon, this form of MP is referred to as two photon microscopy (Yoshitake et al, 2016). Unlike conventional confocal microscopy, two photon confocal microscopy minimizes background fluorescence, which makes it more suitable to imaging protein gel layers in contact with high protein concentration solutions. Both single and two photon methods enable imaging of protein within a distance of 1 μm from the membrane's surface (Zuponcic et al., 2021).

The work reported in this paper is based on a Millipore Ultracel 30 kD (molecular weight cutoff) Pellicon 3 Module, constructed with a regenerated cellulose membrane, coupled with a D type feed screen that is placed on top of or in-between the flat sheet membrane packets. This screen is designed to induce mixing of the viscous feed (i.e., retentate) passing over the membrane through a 1 mm high feed channel. The feed screen is reported to decrease the concentration of stagnant protein at the membrane surface and to enhance mass transfer (Millipore 2003, 2018, Zou et al., 2016)).

We demonstrate that the TFF membrane cassette may be configured and fabricated to enable observation of protein gel layer development on membrane surfaces and along the axial length of an operational ultrafiltration module. This optical flow cassette (OFC) is fabricated from a commercial

TFF membrane module through careful machining, employed to expose the feed screen and the membrane underneath. A thin, optically transparent plexiglass cover and rubber gasket are positioned and sealed on top of the feed screen and cassette housing to form an optical channel with a 4 mm optical path length, which is compatible with the 10x objective lens of the confocal microscopes. The overall assembly is constructed in an aluminum housing that fits securely on the microscope stage and may be operated at a maximum of 60 psi (this is also the pressure rating of the commercial TFF module). Detection is based on adding a small aliquot of protein, labeled with Alexaflor 488 with a nominal excitation wavelength of 720 or 830 nm (ThermoFisher Scientific, accessed March 10, 2022). The ratio of labeled to unlabeled protein was $25*10^{-4}$ to 1. Scanning generates Z-stack images of the inside of the cassette, with the images being reconstructed into different views using vendor-provided software.

Two photon confocal imaging shows that gel accumulates within spaces of the feed-screen to a depth of 40 to 60 mm. Unlike single photon confocal microscopy, the background is minimized by directing an infrared laser to the desired point, with emission of two photons occurring from the fluorescent label (Ustione, 2011). By scanning the surface of the membrane at various distances away from the membrane, images are formed that show the presence of protein gel at the membrane surface and within the feed screen. Fluorescence occurs at a specific point because two photons are required to excite the label. This avoids fluorescent background from other components of the flow cell becoming visible and generating a background signal that interferes with observation of the target. Background signals otherwise generated include the plexi cover glass (0.3 mm thick), the void fraction filled with buffer and protein (1 mm thick), and a 0.5 mm thick feed screen.

In this paper we report, for the first time, use of modeling combined with MP microscopy for visualization of the bovine IgG gel layer on a TFF membrane surface at process conditions within an operational cassette where bulk protein concentrations of 250 mg/mL may occur. Together with separate measurements of flux, retentate concentration, transmembrane pressure, and flow rates, these images are helping to inform development of a first-principles, computational model for predicting conditions that maximize protein concentration and recovery. The combination of visualization, process data, and modeling, carried out in a manner not previously possible, sets the stage for understanding mechanisms and impacts of gel layer formation on TFF performance.

References

Apgar, J. R., A. S-P. Tam, R. Sorm, S. Moesta, A. C. King, H. Yang, K. Kelleher, D. Murphy, A. M. D'Antona, G. Yan, X. Zhong, L. Rodriguez, W. Ma, D. E. Ferguson, G. J. Carven, E. M. Bennett, L. Lin, "Modeling and mitigation of high-concentration antibody viscosity through structure-based computer-aided protein design," PLOS ONE, doi: 10.1371/journal.pone.0232713; 1-26 (2020).

John E. Butler, Bovine Immunoglobulins: A Review, J. Dairy Science, 52(12), 1895-1909 (1969).

J.E. Butler, Bovine Immunoglobulins: An augmented Review, Veterinary Immunology and Immunopathology, 4: 43-152(1983).

Millipore, TB032 Rev C, Protein Concentration and Diafiltration by Tangential Flow Filtration, 24 pp. (2003).

Millipore, EMD Millipore, Lit No MS_PF3654N00, Pellicon 3 Casettes with Ultracel Membrane Performance Guide, 28 pp. (2018).

Jessica Zuponcic, Francesco Rossi, Eduardo Ximenes, Norvin Bruns, Steven Geng, Yinying Tao,

Vincent Corvari, Gintaras V. Reklaitis and Michael Ladisch, Confocal Imaging of Protein Gel Layer Formation during Tangential Flow Filtration to Inform Process Conditions Reducing Protein Losses and Increasing Protein Concentration, Annual AIChE Meeting (Boston, MA), in Advances in New Modalities (Large Molecule, Oligos, Peptides, ADCs, Gene therapy, etc.) Product and Process Development (Nov. 9, 2011).

ThermoFisher Scientific, Fluorescent Probes for Two-Photon Excitation Microscopy – Note 1.5, Technical Reference Library, The Molecular Probes Handbook. (Accessed March 10, 2022)

Ustione A., and D. W. Piston, A simple introduction to multiphoton microscopy, Journal of Microscopy, 243(3), 221-226 doi: 10.1111/j.1365-2818.2011.03532.x (2011)

Yoshitake, T., M. G. Giacomelli, L. C. Cahill, D. B. Schmolze, H. Vardeh, B. E. Faulkner-Jones, J. L. Connolly and J. G. Fujimoto, Direct comparison between confocal and multiphoton microscopy for rapid histopathological evaluation of unfixed human breast tissue, Journal of Biomedical Optics, 21(12), 126021-1-12 (2016).

Irene Yu Zou, S Reynolds, A. Barlett, D. Kinzlmaier, M. Kozlov, High Viscosity TFF Cassette Deisgn and Application in Ultrafiltration Process, Poster by MilliporeSigma, Lit No. PS2076EN00 (2016).

340d Predicting Biomolecule Partitioning in Aqueous Two-Phase Systems through Thermodynamic Modeling

Maximilian Wessner, Christoph Brandenbusch

Aqueous Two-Phase Extraction (ATPE) using an Aqueous Two-Phase System (ATPS) has long been shown to be a promising alternative in the purification of high-value biomolecules (e.g., biopharmaceuticals) from fermentation broth. Although offering a high biocompatibility, a gentle separation profile due to low interfacial tension and high efficiencies, ATPE has not yet made it into an industrial application. Reasons are several challenges, typically arising from state-of-the-art ATPS selection: (1) The high experimental effort required to identify a suitable ATPS by trial-and-error screenings. (2) The lack of knowledge about the influence of the ATPS solutes (e.g. phase formers, displacement agent and excipient) on the stability and activity of the high-value biomolecule. (3) The lack of suitable models that can predict the biomolecule partitioning in the ATPS, thus allowing for rapid preselection of potential ATPS and operating conditions.

To address the first two challenges, we recently presented a physically sound approach that enables the selection of optimized ATPS while maintaining conformational and colloidal stability of the biomolecule, enabling high yields, and running the process within an optimal process window.

Within this work, we now extend our approach towards predicting biomolecule partitioning within ATPS by thermodynamic modeling using the electrolyte perturbed-chain statistical associating fluid theory (ePC-SAFT) equation of state to further decrease time- and cost-intense experiments. A new, heterosegmental modeling approach is introduced, allowing for consideration of the biomolecule in ePC-SAFT and prediction of its phase behavior. For the fitting of biomolecule pure-component and binary interaction parameters, static light scattering data of aqueous biomolecule solutions is used. The predictive capability of the new approach was evaluated by predicting the partitioning behavior of Immunoglobulin G (IgG) in two different ATPS (1. monosodium glutamate-polyethylene glycol 2000 ATPS and 2. trisodium citrate-polyethylene glycol 2000 ATPS) containing different concentrations of sodium chloride as displacement agent. The predictions for the partitioning of IgG in both ATPS show a very good agreement with the experimental data, thus emphasizing the high potential of the new approach. Calculations can be performed in a few hours instead of several days for the respective experimental investigations. Prospectively, this will drastically reduce the experimental effort and pave the way for an accelerated ATPE development for downstream processing of high-value biomolecules.

340e Enzyme Purification and Sustained Enzyme Activity By Fusion with Phase-Separating Intrinsically Disordered Protein

Xinyi Li, Augene Park, Grant S. Murphy, Karla Camacho Soto, Benjamin S. Schuster

In recent decades, biocatalysis using enzymes to synthesize organic molecules has become more prominent for academic studies and pharmaceutical manufacturing. Enzymes are preferred catalysts due to their high selectivity and specificity for chemical reactions, which make the overall synthesis easier to control. However, enzymes are typically used in their crude state since it is time-consuming and cost-prohibitive to purify them using conventional chromatographic methods. Impurities in the crude enzyme solutions might result in unwanted byproducts. Hence, a nonchromatographic purification method would greatly benefit biocatalytic processes. In this work, a synthetic membraneless compartment has been created with enzymes fused to an intrinsically disordered protein region—an Arg/Gly-rich domain (RGG) of LAF-1 protein from *Caenorhabditis elegans* P granules. The RGG domain is capable of undergoing liquid-liquid phase separation and forming liquid condensates triggered by changes in temperature or salt concentration. Leveraging this phase separation property of RGG, we purified enzymes-RGG fusions through centrifuging the liquid condensates and resolubilizing protein pellet in a new buffer. Three enzymes from different classes have been successfully fused to RGG domains—kinases, ATP-dependent ligases, and phosphorylases. We have observed that enzymes purified with our centrifugation method have an increased purity compared to cell lysate. Furthermore, we performed enzymatic reactions utilizing purified fusion proteins to investigate and evaluate the enzyme activities. Results from enzyme assays have indicated that enzyme fusions purified by the centrifugation method exhibit enzyme activity which is at least comparable to that of crude enzymes. To conclude, we believe that this protein purification method by centrifugation is promising in overcoming the difficulties in the development of bioseparation and biocatalysis approaches in pharmaceutical synthesis.

340g ChromaWeb™ platform, a linearly-scalable, high-resolution membrane chromatography cassette for purification of viral vectors.

Gastón de los Reyes, Erik Blomquist, Danielle Marquis

Gene therapy treatments are expected to grow rapidly with at least 4 treatments already approved and many more in the pipeline. Therefore, there is a need for large-scale viral particle isolation without the need for ultracentrifugation or chemical methods. Chromatographic purification would be highly desirable, but the large size of viral vectors — 25~350 nm — requires adsorptive media having pores in the micron scale. This is the domain of adsorptive membranes and certain monoliths, but not of conventional chromatographic media whose pores are only in the 20~40 nm range.

Membranes have very fast kinetics by virtue of having very small diffusional distances because convective flow occurs through the adsorptive pores. Thus, the dynamic binding capacity is almost constant with a weak dependence on residence time, but with the inevitable tradeoff of lower binding capacity and high hydraulic resistance (HR) compared to conventional beads. The high HR demands very shallow beds of 5~15 mm, leading commercially available membrane chromatography devices to have relatively poor chromatographic resolution. Furthermore, none are linearly scalable. Therefore, there is a need for membrane chromatography devices that overcome these limitations.

The ChromaWeb modular chromatography platform is a stackable, planar, linearlyscalable cassette having a novel flow distribution network that delivers a very narrow residence time distribution (RTD) in spite of having a shallow bed and modules in parallel. As a result, ChromaWeb cassettes have HETPs on the order of 30 µm, at least 10 times smaller than the best commercially available devices, and peak asymmetry very close to 1 (vs. > 3). The modular and linearly-scalable structure of the ChromaWeb cassette enables results based on 10 mL lab scale cassettes to be directly translatable to clinical-scale (100~500 mL) and process-scale cassettes (1~10 L).

In this presentation we describe the structure of the ChromaWeb cassette, including its novel distribution network, and how it is capable of delivering linear-scalability and high resolution. Experimental results will be presented with a model protein on a HyperProductive ® bind-and-elute process at a 10·mL scale demonstrating very high productivity, high eluate concentration, and low buffer consumption.

The ChromaWeb platform is in its early development stage at SPF Technologies, and in search of commercial partners to bring it to market.

340h How to Drive a Bio Macromolecule DNA Faster through Nano Pores

Aniruddha Deb, Dr. Partho S.G. Pattader

The aim of separating DNA fragments in human genome research has stimulated efforts to understand molecular processes underlying electrophoretic separation rate resulting in rapid development of electrophoretic techniques in recent years. We used Gaussian white noise to enhance the mobility of **1kbp DNA band** to ~ **86 %** and **6kbp DNA** band to ~**113 %** in comparison to conventional known agarose gel electrophoresis. The experimental setup was similar to conventional agarose gel electrophoresis phenomena under the influence of bias voltage along with the addition of Gaussian noise in the form of voltage pulses. Displacement of the gel bands are tracked from the well bottom to the middle point of a particular band using ImageJ tracking software. The drift velocity of the DNA induced by activated "**noise-lubricity**", follows the Arrhenius-Eyring-like escape rate at the high power of the external noise. Whereas, at the low power, the cooperative dynamics of the DNA molecules impart super Arrhenius-like behavior. Gaussian white noise overcomes the pinning barrier of the molecules, highlighting the symmetry breaking for free random motion of macromolecules. The developed model is interpreted in terms of mobility, interaction and dynamics of the DNA molecules in gel electrophoresis setup. Simulation based on Langevin framework predicted a similar enhancement of mobility across the gel interface matrix. The simulation results also pointed on the increasing mobility fluctuations with increasing power of noise which truly resonates with our findings from the experimental results (as shown in graphical plot attached). This work suggests strategies for noise-activated faster DNA fingerprinting and set up the basis for advanced research on resonance-induced super mobility for the isolation of a specific protein fragment from a crowd.

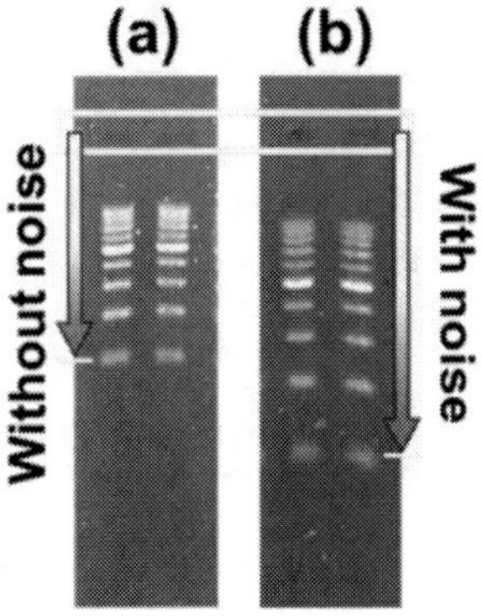

(a) Represents conventional gel doc image and (b) Represents the gel doc image showing faster mobility of DNA bands when Gaussian white noise is used.

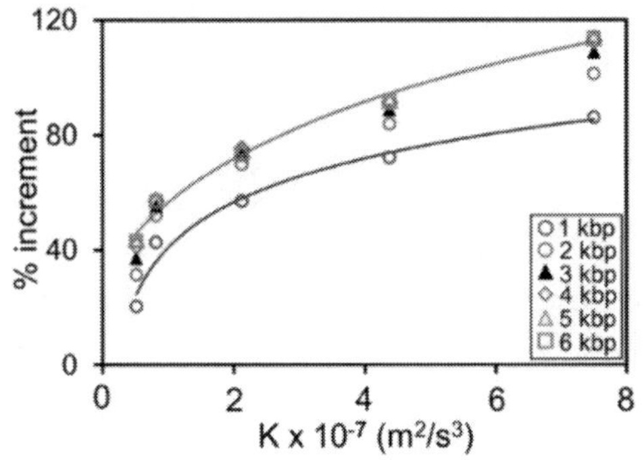

Faster screening with Gaussian noise depicting 1-6 kbp DNA bands

569a Hydrocyclone Separators: 1891-2022 and Beyond

Andre Benard

The objective of this presentation is to develop a quantitative and qualitative understanding of multiphase fluids in hydrocyclone separators. The theory will be used to evaluate the efficacy of using hydrocyclones for solid/fluid separation (clarification and thickening); gas/hydrocyclones separation (degassing); liquid/gas separation (mist control); liquid/liquid separation; deoiling water; and, dehydration of oil. The presentation will illustrate the role that hydrocyclone separators have played in the development of 20th Century technologies related to mineral and chemical processing as well as potential commercial applications for the 21st Century. The potential for hydrocyclones for downhole oil/water separation; and, mud separators in drilling operations will be highlighted.

Brennen, C. E., 2005, Fundamentals of Multiphase Flow, Cambridge University Press.

Bretney, E., 1891, Water Purifier, U.S. Patent 453,105.

Cocco, R., *et al.*, 2014, "Introduction to Fluidization: Back to Basics", CEP, AIChE.

Huang, Y., *et al.*, 2020, "Theoretical Study on Centrifugal Coupling Characteristics of Self-Rotation and Revolution of Particles in Hydrocyclones", Separation and Purification Technology 244:116552.

Jing, J. *et al.*, 2021, "Numerical Simulation Study of Offshore Heavy Oil Desanding by Hydrocyclones, Separation and Purification Technology, 258, 118051.

Liu, Y., *et al.*, Three-Phase Hydrocyclone Separator: A Review, Chemical Engineering Research and Design 100(2015) 554-560.

Ni L., *et al.*, 2019, Optimizing Geometric Parameters in Hydrocyclones for Enhanced Separation: A Review and Perspective, https://doi.org/10.1080/154 22119.2017.1421558.

Manninen, M., *et al.*, 1996, On the Mixture Model for Multiphase Flow, Espoo, Technical Research Centre of Finland, VTT Publication 288.

Soo, S.L., 1989, Particulates and Continuum: Multiphase Fluid Dynamics, Hemisphere.

569b Magnetically-Enabled Recovery of Beads in Microchannels: Quadrupole Magnet Flow Sorters As Enhanced System Design

Cristina González-Fernández, Jenifer Gomez Pastora, Eugenio Bringas, Jeffrey Chalmers, Inmaculada Ortiz

Functionalized magnetic beads (MBs) have proven valuable to recover compounds (including toxins, metal ions, etc.) from complex media (e.g., biofluids) in microfluidic systems. Specifically, the retrieval of target compounds using functionalized MBs relies on their adsorption on the surface of the MBs, which results in the formation of target compound-MB complexes, and the subsequent recovery of these complexes by applying magnetic fields. Despite the promising prospects of this separation strategy, the small size of the particles makes challenging the recovery of the target compound-MB complexes in conventional Y-Y shaped microseparators that use a single permanent magnet to generate the magnetic field.[1,2] In this work, we address the computer-aided design and characterization of a quadrupole magnetic sorter (QMS), which consists of an annular microchannel surrounded by four permanent magnets arranged in quadrupolar orientation, in order to enhance the retrieval of MBs by increasing the driving force for such recovery (i.e., the magnetic force). For that purpose, a computational model that accounts for the magnetic and fluidic forces acting on the MBs was employed by linking an in-house Fortran code with the computational fluid dynamics (CFD) software Flow-3D. Our results revealed that compared to the conventional system, the complete recovery of either (a) MBs flowing at ≈ 360 times higher flow rates or (b) 2.2 times smaller MBs (particle diameter equal to 2.22 μm) can be successfully accomplished in the QMS. Collectively, the enhanced performance of the QMS has been demonstrated, thus, supporting the exploitation of these systems for the efficient magnetic recovery of target compounds from complex media.

References

[1] J. Gómez-Pastora, et al., *Lab Chip.* 18 (2018) 1593-1606.

[2] C. González-Fernández, et al., *Ind. Eng. Chem.* Res. 60 (2021) 16780-16790.

569c Revealing Particle Transport Mechanisms in Microfiltration Processes Using Single-Particle Tracking

Haichao Wu

Filtration membranes are used extensively in processes including water treatment, pharmaceutical sterilization, food/beverage processing, and heterogeneous catalysis. Therefore, it is important to fundamentally understand particle transport mechanisms during filtration processes. While conventional filtration experiments (e.g., flux or concentration measurements as a function of volumetric throughput) provide useful information about mass transport at the ensemble level, they provide limited insight into the microscopic mechanisms that give rise to complex phenomena, including particle remobilization upon flow interruption and membrane fouling. To address this issue, we have developed refractive index matching imaging systems, combined with single-particle tracking methods, allowing the direct visualization of single-particle motion within a variety of porous materials.

Using single-particle tracking methods, we directly visualized particle transport in microfiltration membranes under flow conditions that included stoppage intervals and therefore provided direct evidence for particle remobilization during flow stoppage. Our results suggest that during flow interruption, a sub-population of trapped/retained particles detach from their retention sites and may escape to less trapped areas through Brownian motion within the pore space. This mechanistic information about particle transport remobilization will permit the design of filtration flow profiles that meet customized requirements for particle retention during complex filtration cycles.

In addition, we also visualized the whole membrane fouling processes using single-particle tracking. We found that internal membrane fouling consisted of three stages, fouling site formation, fouling site growth, and fouling site coalescence. Larger particles had a greater chance to be retained initially, nucleating the fouling sites. These sites tended to capture other particles passing through the membrane, causing the fouling sites to grow in size. Eventually, isolated fouling sites grew large enough to coalescence with neighboring sites, blocking a large region of membrane pores, leading to significant fouling. This mechanistic information about internal membrane fouling will assist in the design and optimization of filtration processes to reduce membrane fouling and expand the fundamental understanding of complex mass transport of polydisperse particle distributions.

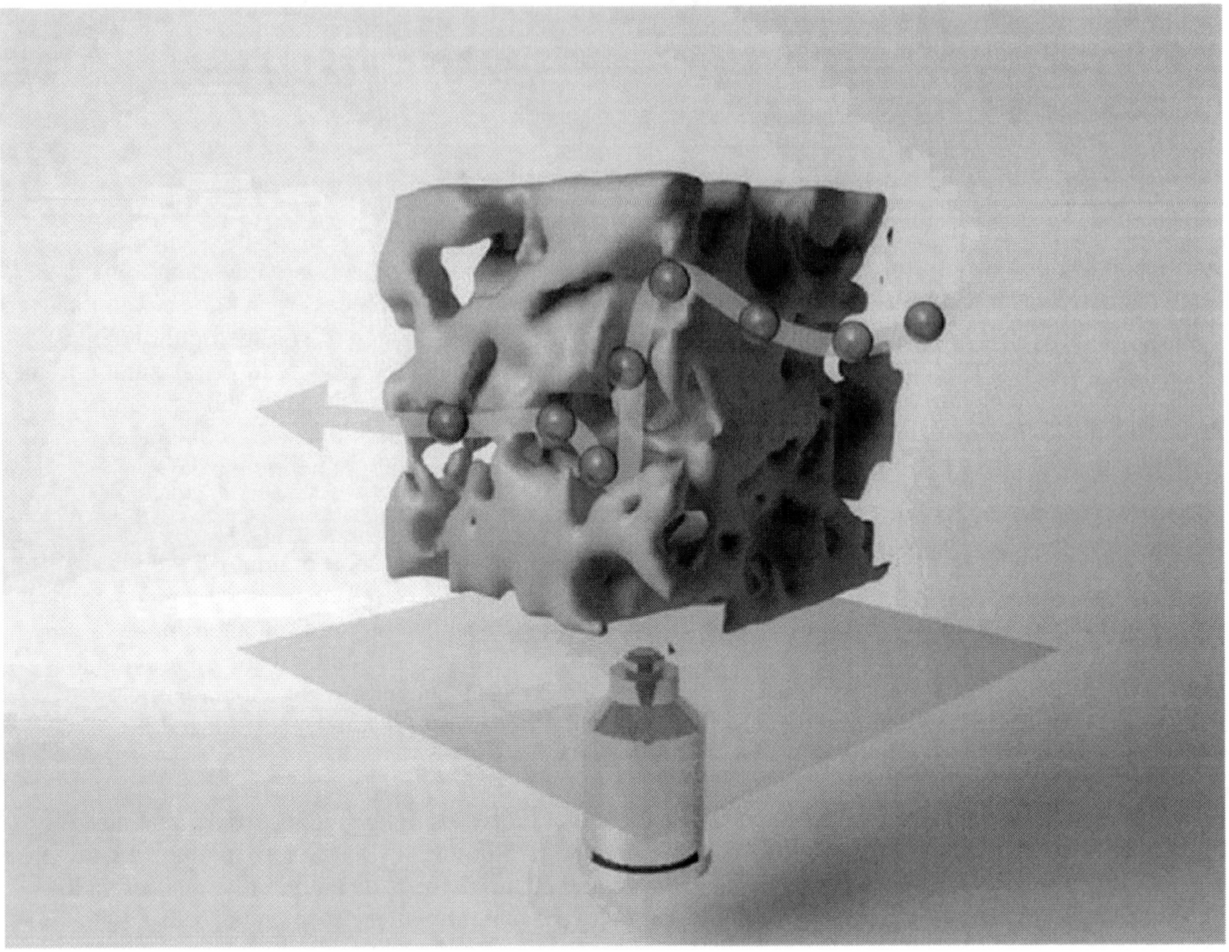

569d Magnetic Self-Assembly of Superparamagnetic Iron Oxide Nanoparticles Studied By in-Situ Small Angle X-Ray Scattering

Xian Wu, Jenifer Gomez Pastora, Jacob Strayer, Jeffrey Chalmers

With the rapid development and significant improvement of nano-material synthesis technologies, a series of superparamagnetic iron oxide nanoparticles (SPIONs) have been produced with special physical properties, such as high surface to volume ratio, biocompatibility and superparamagnetism. Nevertheless, due to the nanometric size of SPIONs, the magnetically driven isolation method to separate those particles from the media is difficult, considering the fact that the Brownian motion and viscous drag are the two key factors that hinder the magnetic motion of the particles.

In our previous work, we reported the ability to recover 5, 15 and 30 nm particles by applying external magnetic fields and gradients through magnetic agglomeration and sedimentation. Our experimental data suggest that the particles agglomerated due to dipole-dipole interactions. The separation process was finally accomplished by the sedimentation of aggregated particles. However, the nano-structural conformation of agglomerated particles when applying an external magnetic field and the mechanisms behind the separation processes are not fully understood.

In order to gain insight into the separation mechanisms, and more specifically, to unravel the SPIONs magnetic aggregation and sedimentation process, we performed the in-situ Small Angle X-ray Scattering (SAXS) analysis, which is an extremely powerful characterization tool that can provide precise structural information of the particle agglomeration. The *in-situ* analysis tracked the dynamic evolution of the particle diameter, volume fraction as well as particle number density. A special magnetic separation sorter was used, allowing the X-ray to pass through the tubing containing the particle suspension while being magnetized by the external field. Multiple variables and parameters were studied, including operation time, the magnetic field gradient and field intensities.

571a Nano-Confined Ionic Liquid Membrane for Highly Efficient CO_2 Capture from Flue Gas

Fan WANG, Huazheng LI, Huanghe Li, Weiwei Xu, Qiaobei Dong, Shiguang Li, Miao Yu

Membrane technology for CO_2 capture has attracted significant attention because of its great potential for low-energy extraction of CO_2 from flue gas. Among various membranes, facilitated transport membranes, incorporated with active species that can reversibly react with CO_2, have shown their strong capability of highly efficiently separating low-concentration CO_2 from other inert components in flue gas, especially with the presence of water vapor, an inevitable component in flue gas. Amino acid ionic liquids (AAIL) may serve as an ideal active species in facilitated transport membranes, owing to their close-to-zero vapor pressure, excellent thermal stability, and amino acid anions that can reversibly react with CO_2 (especially with the presence of water). In this work, we designed and fabricated a nano-confined AAIL membrane utilizing a novel single-walled carbon nanotube and graphene oxide quantum dot (SWCNT/GOQD) framework. Nano-confined space (3-5 nm), combined with rich functional groups on trapped GOQDs in SWCNT mesh, ensures excellent mechanical stability of infiltrated AAIL. The nano-confined AAIL membrane exhibited excellent CO_2 separation performance, with CO_2 permeance as high as 2,400 GPU and CO_2/N_2 selectivity of >2,000 for simulated NGCC flue gas, achieving >95% CO_2 purity by one-stage separation under membrane stage cut condition (65% CO_2 recovery). Excellent membrane stability over 60-100 h was also demonstrated. We expect this new "nano-confined" platform and utilization of AAILs may expand the family of the facilitated transport membranes and greatly facilitate the development of membrane technology for CO_2 capture from flue gas.

571b Elucidating the Role of Water in Amine-Facilitated CO_2 Transport Via Operando Transmission FTIR Spectroscopy

Sarah Pate, Hui Xu, Casey O'Brien

Amine-functionalized polymeric membranes selectively facilitate transport of CO_2 over other gases to separate CO_2 from mixed streams with high permeance and selectivity. Water vapor in the feed gas enhances CO_2 permeation across amine-functionalized facilitated transport membranes (FTMs), but how exactly water influences the CO_2 transport mechanism is not clear.

In this talk, I will describe a new operando transmission FTIR spectroscopy tool that we developed to probe CO_2 transport mechanisms of amine-based FTMs under realistic operating conditions, including in the presence of water vapor. This work describes the design of this tool and demonstrates its functionality through investigating the mechanism of CO_2 transport across polyvinylamine (PVAm), a primary amine FTM commonly used for CO_2 separations. We show that CO_2 permeation across PVAm is enhanced in the presence of water due to the conversion of CO_2 to a reactive intermediate, which is most likely carbamate species based on the infrared band positions. Our results suggest that water catalyzes the formation of carbamate by a three-step mechanism: (1) water protonates PVAm resulting in a hydroxide species, (2) CO_2 reacts with the protonated PVAm to form a zwitterion, and (3) the zwitterion is deprotonated by the hydroxide species produced during the PVAm protonation. This new operando transmission FTIR spectroscopy methodology can be applied to a range of membrane systems (e.g., secondary and tertiary amines) to probe transport mechanisms.

571c Polymer-Graphene Oxide Self-Assembled Membranes for H_2 Purification with Enhanced Selectivity for Carbon Capture Applications

Giacomo Foli, Vincenzo Palermo, Matteo Minelli

2D materials are very attractive for the fabrication and the development of membranes for gas separation due to their peculiar gas transport properties, combined to extremely high aspect ratios. Among the others, graphene and its derivatives are very promising due to their very low permeability to atoms and molecules coupled and the possibility to finely tune its size and assembly. Such strategy can be pursued to create size-selective membranes, e.g. for hydrogen purification, in which the transport of larger molecules such as CO_2 is significantly hindered by the large diffusive paths around the piled 2D materials.

This work is focused on the use of layer-by-layer (LbL) self-assembly technique to fabricate thin coatings on top of polymeric substrates, using graphene oxide (GO) combined to polyelectrolyte species, deposited in an alternated fashion, thus leading to well-ordered structures organized at the nanoscale. The assembly mechanism, indeed, exploits electrostatic interactions of weakly negatively charged GO and polycations, controlling both the growth of the coating and the resulting arrangement of the alternated structures.

Membrane fabrication includes the preparation of a Matrimid polyimide substrate (about 30-40 μm), considered as benchmark material for H_2 separation, then the alternated dip coating deposition of highly diluted aqueous polyelectrolyte solutions and GO dispersions. Intermediate water rinsing steps are performed after each deposition, in order remove the excess of GO or PE attached on the surface and to avoid inhomogeneity on sample, ultimately guarantying the required structural order. The whole process is repeated 10 times (10 bilayers – BLs) to ensure a sufficiently homogeneous structure, in which large molecules are force to follow long selective paths.

The characterization of the obtained membranes by SEM microscopy revealed the effective and homogeneous deposition of the coating, while XRD analysis identified the alternated layer arrangement and provided a quantitative determination of the average thickness of each layer.

Membrane performances are evaluated by direct gas permeation measurements, which show a pronounced sieving ability of the coating, as the gas permeability decreases dramatically as the penetrant kinetic size increases. Therefore, such membranes are able to discriminate the penetrant molecules on size basis, and the very large selectivity makes them very interesting for hydrogen purification (e.g. for H_2/CO_2 or H_2/CH_4 separation).

The gas perm-selectivity performances of such membranes may be conveniently tuned by partially reducing the GO sheets, by direct thermal treatment of the coating (T up to 200°C) causing thus the removal of some oxidized species on the graphenic layers, and leading modified arrangement of the coating (inspected by XRD) and ultimately to different permeabilities. Relevantly, the permeability of small molecules such as H_2 is enhanced almost 30 times, accompanied by one order of magnitude increase in the H_2/CO_2 selectivity, which reaches values as high as 200, overcoming the Robeson's upper bound (Figure 1)

Therefore, a simple and scalable method is proposed for the fabrication of novel H_2 purification membranes with outstanding sieving ability, using a 2D material and a self-assembly technique, able to overcome technological limits on this class of separation processes.

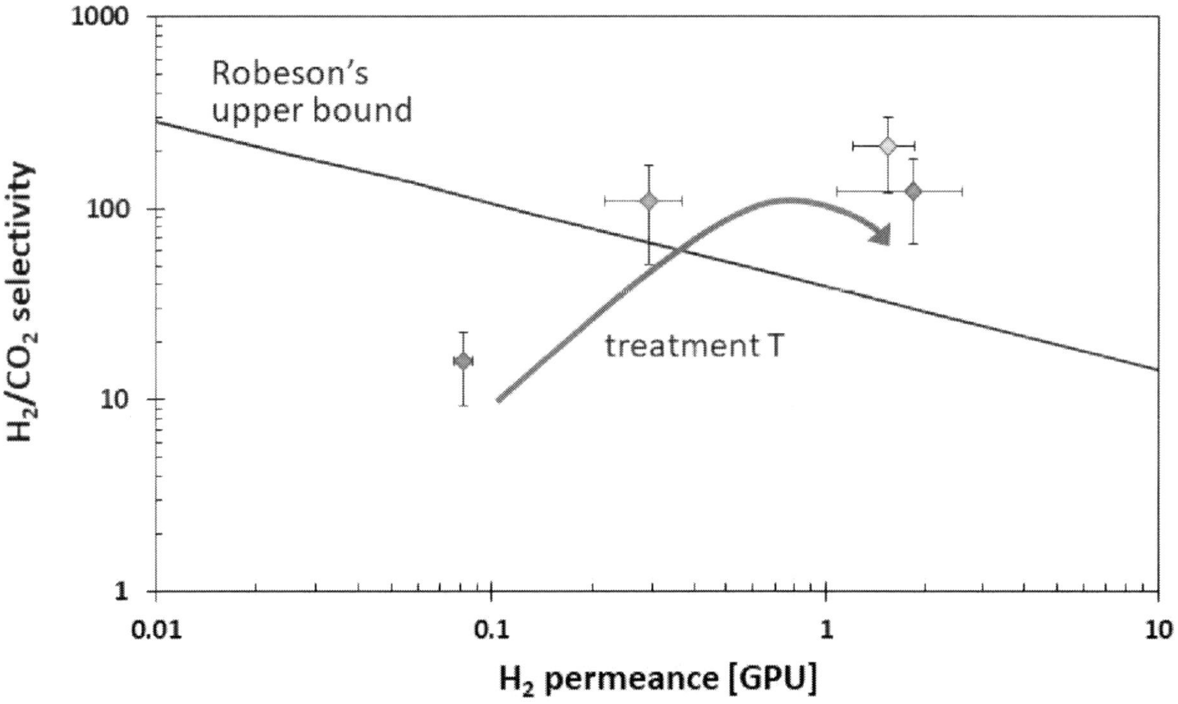

Robeson's
upper bound
treatment T
1000
100
10
1
0.01
0.1
1
10
H2/CO2 selectivity
H2 permeance [GPU]

571d Tuning Ether Motifs in Polymers Membranes for Carbon Capture

Yasemin Basdogan, Zhen-Gang Wang

CO_2 capture for utilization remains important since fossil fuels continue to be the number one energy source around the globe. Membrane technology is highly explored for CO_2 capture over the last decade due to its high energy efficiency and low carbon footprint. Polymer membranes suffer from design challenges due to the negative correlation between two principal features of separation performance (permeability and selectivity). We hypothesize increasing the ether oxygen content would increase the CO_2 permeance, CO_2/N_2 and CO_2/O_2 selectivity. We have investigated five different polymer materials that range in O:C ratio, including polyethylene (PE, O:C=0), polytetramethylene oxide (PTMO, O:C=0.25), polyethylene oxide (PEO, O:C=0.5), poly(1,3-dioxolane) acrylate (PDXLA, O:C=0.67), and polyoxymethylene (POM, O:C=1). First, we used the group contribution (GC) method based on perturbed-chain statistical associating fluid theory equation of state to calculate the solubility. Our results show strong agreement in the absolute CO_2, N_2, and O_2 solubility and the solubility selectivity with the experimental data for PE, PEO, and PTMO. In addition, we predict higher solubility selectivity for the two new candidate polymers, PDXLA and POM. The detailed investigation with molecular dynamics simulations showed POM has higher CO_2/N_2 and CO_2/O_2 diffusivity selectivity compared to the commonly used PEO. Additionally, we studied the effect of adding N_2 phobic functional groups (ex: azo, triazine, and triazole) to the ends of POM polymer membrane. We concluded adding functional groups to the chain ends of the polymer significantly slows down the gas diffusion thus makes these polymer membranes not optimal candidates for neither CO_2/N_2 nor CO_2/O_2 separation. We have also studied the effect of temperature on the gas diffusion in POM polymer membrane. We have showed that the gas diffusion increases with increasing temperature; however, the diffusion selectivity decreases for both CO_2/N_2 and CO_2/O_2 separation. Our future efforts will focus on studying gas diffusion in polymer mixtures to see enhancement of gas diffusion due to synergetic effects from the mixture.

571f A Green Solvent Incorporated Membrane Contactor Based Process for Energy-Efficient CO_2 Capture and Separation

Syed Islam, Md Arifuzzaman, Jingsong Huang, Jacek Jakowski, Rajeev Kumar, Gernot Rother, Vera Bocharova, Ramesh Bhave, Priyesh Wagh, David Sholl, Anisur Rahman, Tomonori Saito, Nirupam Aich, Alexei Sokolov

The climate change due to CO_2 emissions has become a profound environmental concern. This will result in catastrophic consequences including temperature rise, sea level rise, flood, drought, and serious damage in ecosystem. In this work, we have developed a novel scalable energy-efficient and cost-effective hollow fiber membrane contactor-based process for highly efficient CO_2 capture and separation achieved by fundamental understanding of a green solvent-based CO_2 capture and separation. An integrated process is developed using a deep eutectic solvent (DES) in a hollow fiber membrane contactor (HFMC) to provide close interfacial interactions and contact between DES and CO_2. DES, typically a mixture of a hydrogen bond acceptor (HBA) and a hydrogen bond donor (HBD) is a low-cost environmentally friendly physical CO_2 sorbent which requires low energy input for CO_2 separation. It was synthesized using choline chloride and urea as the precursors, and was comprehensively characterized using gas sorption isotherm, viscosity, Differential Scanning Calorimetry (DSC), Karl Fischer titration, Nuclear Magnetic Resonance, and Raman spectroscopy measurements. Commercial low-cost polymer hollow fiber membranes (e. g., microporous polypropylene) are identified based on pore structure, surface area, chemical, and thermal compatibility, and mechanical strength. The DES incorporated HFMC system is used to capture and separate CO_2 from N_2. The results showed selective separation of CO_2 while completely rejecting N_2 of flue gas. Although this green solvent has excellent physicochemical properties for high-performance CO_2 separation, the fundamental understanding of the mechanism of separation of CO_2 using DES is limited. Moreover, how the constituent moieties of DES bind CO_2 and their intermolecular and intramolecular interaction is not well known. This work also utilizes the computational tools such as Density Functional Theory (DFT) calculation and molecular dynamics simulations to elucidate the myriad interactions of DES and CO_2 to understand the binding mechanism for identification of the ideal DES with best CO_2 absorption capacity and selectivity over other gases. This research can provide both fundamental insights about physical solvent-based separation process as well as a pathway for its industrial deployment. A major advantage of HFMC-based process is scalability with a potential for large scale industrial deployment for CO_2 capture and separation from various sources including flue gas emitted from power plant.

571g Process Design and Integration of Membrane-Cryogenic Distillation Based Hybrid Capture Systems for Hydrogen Production Process

Yongjae Song, Mun-Gi Jang, Jin-Kuk KIM

Membrane-based capture processes offer various advantages including modularity configuration, compact size, less noxious emissions, and small carbon footprint, compared to other capture technologies, such as solvent or sorbent processes. Stand-alone membrane-based capture is inherently limited to achieve a very high purity product because of its separation mechanism. However, the shortcoming of non-sharp separation from the membrane can be supplemented with the hybridization of membrane-based capture with other separation technologies, for example, cryogenic distillation. To fully materialize the synergetic benefit of membrane-cryogenic distillation hybrid process, design interactions between two separation mechanisms should be systematically understood. Hence, in this study, process modelling and simulation of membrane-cryogenic distillation hybrid capture systems is carried out when the capture is applied to the post-combustion flue gas stream generated from a SMR(Steam Methane Reforming)-based hydrogen plant. Process modelling and simulation was conducted in a MATLAB® environment which is linked with UniSim® simulator. Genetic Algorithm was applied to determine optimal process configuration and its operating conditions in a holistic manner. Economic trade-off between capital cost and operating expenditure is investigated to understand techno-economic benefits gained from the hybridization concept.

Acknowledgement: This work was supported by Korea Environment Industry andamp; Technology Institute (KEITI) through Environmental Technology Development Program, funded by Korea Ministry of Environment (MOE) (No. 2022000000760)

This work was supported by the National Research Foundation of Korea (NRF) grant funded by the Korea government (MSIT) (No. 2019R1A2C2002263)

468a Two-Step Nucleation Mechanism Drives Crystal Structure Formation By Selective Desolvation

Anish Dighe, Prem Podupu, Paria Coliaie, Meenesh Singh

Nucleation rate and crystal structure can be predicted using two disjointed yet popular theoretical approaches, two-step nucleation theory (TSN) and crystal structure prediction (CSP). The TSN theory is widely used to explain nucleation in crystalline materials in a wide range of solvents. Despite this, the model has not been expanded to predict crystal structures or polymorphisms. CSP techniques, on the other hand, only empirically consider solvent effects. Thus, TSN theory and CSP techniques continue to evolve as separate methods of predicting two critical characteristics of nucleation – rate, and structure. In this study, we resolve this gap by demonstrating for the first time how crystals are formed through the use of TSN theory. In TSN, a sequential desolvation mechanism is proposed, where partial desolvation leads to dense clusters, and selective desolvation of functional groups leads to crystal structure formation. Molecular simulations are used to investigate the effect of specific interactions on the degree of solvation around different functional groups of glutamic acid molecules. Energy landscape and activation barriers, from these molecular simulations indicate sequential and selective desolvation as supersaturation increases. Through a combination of computational and experimental approaches, we demonstrate that crystallization and polymorph selection is governed by a previously unknown phenomenon associated with supersaturation-induced asymmetric desolvation.

468b *In-Silico* Molecular Simulation Methods for Crystal Nucleation in Solution

E. Daniel Cárdenas-Vásquez, James Savino, Erik Santiso

Background: The study of crystal nucleation poses major challenges for both experimental and simulation studies. The stochastic nature of nucleation, combined with its uncanny ability to amplify the effect of minute molecular-level details, make it very difficult to achieve consistent experimental results even for simple systems. Molecular simulation studies are also challenging, as nucleation is a rare event with a free energy barrier that includes a substantial entropic contribution. Additionally, some of the common assumptions in molecular simulations, such as periodic boundary conditions (which can result in serious finite size artifacts) and constant number of molecules, can greatly bias the results. All theories of nucleation focus on parameters of the solid phase (like number of molecules in the solid), but there is evidence from computational studies that parameters from the solution phase could be equally important.

Results: We defined the liquid order parameters for four different solvents (acetonitrile, ethanol, methanol and water) containing urea at five different concentrations. The behavior of each solvent in the nucleation process has a direct correlation with its nucleation mechanism (i.e. one- or two-step mechanism) found in previous studies, showing that we could predict the behavior and/or pathway using the solvent physical and thermodynamic properties. Additionally, these results can be extended to other solutes in the drug discovery industry.

Conclusions: We show some of the first studies that uses solvent order parameters to predict the behavior of the nucleation of crystals in solution. Consequently, we could generalize this procedure to other liquid structures, which will enable us to generate order parameters for any solvent of interest.

468c Selective Construction of $CaCO_3$ Superstructures *Via* Homoporous Interfacial Crystallizer

Xiaobin Jiang, Mengyuan Wu, Gaohong He

Selective construction of $CaCO_3$ Superstructures *via* Homoporous Interfacial Crystallizer

Xiaobin Jiang, Mengyuan Wu, and Gaohong He

Abstract

Biomimetic mineralization and crytallization for highly ordered superstructures is one of the key concerns in crystal engineering. Herein, we propose a ultra-homoporous interfacial crystallizer (UHIC) as a novel interfacial microdevice, which has realized high selective synthesis of $CaCO_3$ superstructures during Gas-Liquid reactive crystallization process. This UHIC with ultra-uniform mass transfer porous channel and interfacial superhydrophilicity render the droplet ultra-homo nucleation site and coordinatively regulate the interfacial nucleation energy. In addition, based on the classic Hagen Poiseuille transfer model, the mixing and diffusion efficiency of the developed multi-scale crystallizer can be significantly intensified when the mass transfer scale decreased to the ultra-uniform 200 nm level (shown in Fig. 1). The proposed interfacial crystallizer was functioning as the aggregation of thousands of parallel microchannels (approximate $6'10^6$ channel/mm^2).

The microchannel in UHIC can control the CO_2 gas flow rate and provide sufficient interfacial diffusion of the $CaCO_3$ pre-nucleation cluster (PNC). By coordinating the PNC diffusion, self-organization and nucleus growth, the synthesised $CaCO_3$ crystals with spherical and polyhedral structures both increased to higher than 96.77 %, and the particle size distribution was very uniform (C.V. =11.7%). UHIC can engineer the biomimetic mineralization procedure by coordinatively regulating the interfacial nucleation at the sub-micron porous channels, which may shed light on the development of material science and crytal engineering.

Fig. 1. Interfacial mixing time ($t_{mix, G}$ for gas phase and $t_{mix, L}$ for liquid phase) of CO_2 versus Reynolds number in different microscale devices (UHIC, classic T-Mixer, crystal kettle).

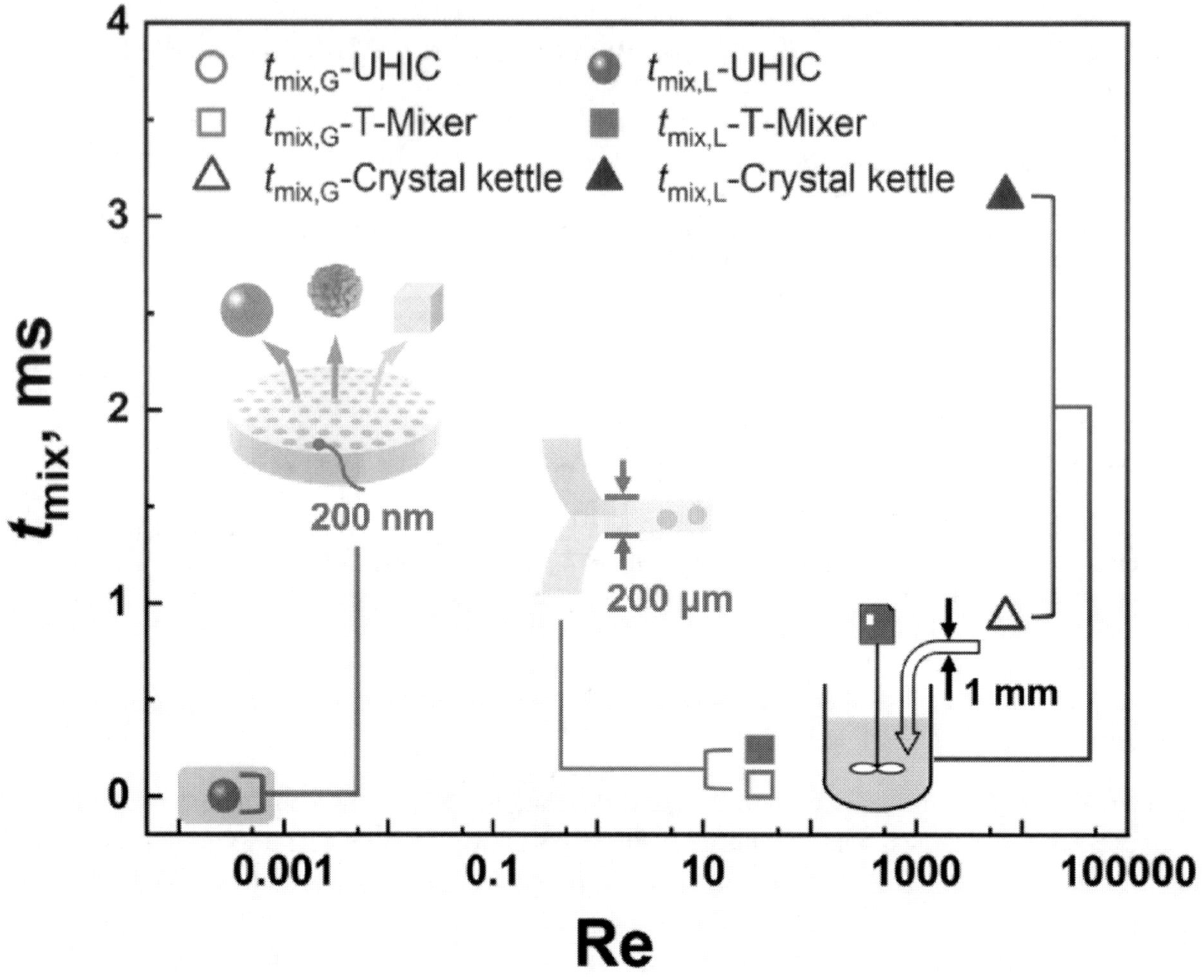

$t_{mix,G}$-UHIC
$t_{mix,L}$-UHIC
$t_{mix,G}$-T-Mixer
$t_{mix,L}$-T-Mixer
$t_{mix,G}$-Crystal kettle
$t_{mix,L}$-Crystal kettle
200 nm
200 μm
1 mm
t_{mix}, ms
Re
0
1
2
3
4
0.001
0.1
10
1000
100000

468e Bimodal Impacts of Modifiers on Nucleation and Growth: Case of Calcium Oxalate Monohydrate

Bryan Alamani, Rica Jan Ondoy

Modifiers (ions, molecules, and macromolecules) are effective at modulating nucleation and growth of crystalline materials. These species are capable of influencing the kinetics and thermodynamics of the crystallization process. Recent studies have show evidence that a number of these modifiers can have non-monotonic behavior (i.e. can act in two or even more ways on the growing crystals). The current work focuses on using calcium oxalate monohydrate (COM) crystals and the impact of divalent ions on its crystallization. Combined bulk and interfacial studies have indicated that these divalent ions have bimodal effect on COM mineralization. Insights on the study may provide further support on the ubiquity of bimodal actions of modifiers on crystallization.

373j Implications of Shear and Cation Choice on Dynamics of Nonliving Natural Organic Matter

Kathlyn Mealio, Holly A. Stretz, Martha J. M. Wells, Katherine Slamen, John Clark

Nonliving natural organic matter (NNOM) is a common material found in environmental water that presents a fouling problem in water treatment membranes. Humic acid (HA) falls under the category of NNOM and displays dynamic particle size characteristics in response to factors such as shear rate, pH, ionic strength, and temperature. These variables have been hypothesized to impact the aggregation and disaggregation of NNOM. The behavior of NNOM found in pond water as a function of shear rate was investigated under constant 0.4M $CaCl_2$ concentrations. A rheometer was used to induce shearing of the NNOM in the range of 10 to 400 sec^{-1}. The Particle Size Distribution (PSD) displayed a complex relationship between shear rate and relaxation behavior. Non-Newtonian behavior of the sample(s) was also explored. Preliminary data shows the disappearance of certain size ranges at specific shear rates and colloidal behavior.

702b Study of MF Membranes for Microalgae Dewatering and Surface Modification for Fouling Mitigation

Erda Deng, Xiaoyi Chen, Darius Rub, Haiqing Lin

Microalgal cultivation is an efficient approach for CO_2 capture and utilization. Microfiltration (MF) membranes have been investigated for concentrating dilute algae (<0.5 g/L) because of their high efficiency. However, the membranes suffer from fouling by organic matters in the algae solutions, and there lacks an understanding of membrane fouling with real algae solutions. In this study, we evaluate the dewatering performance of commercial polyethersulfone (PES) MF membranes in dead-end and cross-flow systems. The membrane surface is also modified with hydrophilic zwitterions using bio-inspired dopamine to improve antifouling properties. The effects of membrane molecular weight cut-off (MWCO), surface hydrophilicity, algae concentration, and hydrodynamics properties on the dewatering performance were thoroughly investigated. The pristine membranes exhibit water permeance of 225 and 80 LMH/bar in the dead-end test against 1 and 10 g/L algae, while the zwitterion modification almost doubles water permeance to 386 and 166 LMH/bar, respectively. Moreover, the zwitterion-modified membrane exhibits 20% higher permeance than the pristine membrane in the long-term cross-flow test, confirming the enhanced antifouling properties. The fouling behaviors in both membranes can be satisfactorily described using the Hermia fouling model. The membranes show permeance lower in the cross-flow tests with a Reynolds (Re) number of ~1,000 than in the dead-end tests with a Re number of 65,000. This result is also consistent with the thickness of the cake layer formed on the membrane surface, i.e., 20 microns after the cross-flow test and 2 microns after the dead-end test. For sustainable operation, the membranes are also cleaned using deionized water or NaOH solutions, and both cleaning methods can recover more than 75% water permeance. Understanding the fouling mechanisms of the membranes with real algae solutions can be critical to designing membranes for this important application.

702c Functionalized Magnetic Metal Organic Framework Thin-Film Nanocomposite Membranes with Real-Time Induced Vibrations for Enhanced Antifouling

Jasneet Pala, Ryan Tracy, Nima Mahmoodi, Milad Esfahani

Membrane fouling is still one of the main limiting factors of membrane separation, even though membrane technologies have been established across the world as one of the most efficient water and wastewater treatment. Fouling reduces membrane separation efficiency by reducing membrane permeability and rejection, shortening membrane lifespan, and increasing process costs due to higher washing requirements. In this work, we report, the electromagnetic-antifouling mechanism based on the surface functionalization of membranes using the metal organic framework (MOF). The generated electromagnetically-induced vibration on the interference of membranes and foulant will hinder and prevent different fouling mechanisms such as cake layer formation, concentration polarization, and pore blocking. The resonance frequency creates the largest vibrations in the membrane; therefore, the foulants absorb the highest energy of vibration, which move them from the membrane surface. We synthesized Fe_3O_4 magnetic nanoparticles (mnp) and then synthesized Fe/Fe_3O_4 MOF using hydrothermal synthesis method. Scanning Electron Microscopy (SEM), X-ray powder diffraction (XRD), size, and zeta potential Brunauer–Emmett–Teller (BET) (for surface area, pore size, pore volume) were used for characterization of fabricated MOFs. The synthesized Fe/Fe_3O_4 MOFs are added to the (i) polyamide (PA) selective layer of thin-film nanocomposite membranes (TFC) and (ii) the support layer of TFC membrane to make mixed matrix membranes. The electromagnetically-induced vibration effect on repellent and movement of foulants (humic acid) are compared for these two conditions. Each foulant has a certain resonance frequency in which it shows the highest vibration. Therefore, it is possible to target specific foulants to be removed from the membrane by producing vibrations with frequencies corresponding to the resonance of foulants. These vibrations do not need high amplitude; even a minimal amplitude (in the range of a millimeter or less) can excite them. We also report the correct frequency of vibration for different foulants.

702d Development of a Modular Shipboard Wastewater Reuse System

Nathaniel Michael, Andrew Wagner

Wastewater disposal in the shipping industry is an ever increasing problem as global regulations become stricter, limiting what can be dumped overboard and where it can discharged. Mainstream Engineering has developed a modular, innovative wastewater processing system that can treat both greywater and blackwater to produce a non-potable, reusable water stream and reduce the amount of water wastes that must be held by over 85%. One of the main drivers is the International Convention for the Prevention of Pollution from Ships (MARPOL) regulations which dictate the level of biological oxygen demand, total suspended solids, pH, coliforms, nitrogen, phosphorus, and chlorine are allowed to be discharged. Current marine sanitation devices (MSDs) and advanced water treatment (AWT) systems are large, bulky, expensive, or are unable to meet the necessary requirements of treating the 40 – 120 gallons/day/person generated on commercial vessels. Due to ever increasing disposal fees, there is a clear need for ships to reduce waste water generation not just for environmental reasons but also economic ones.

Shipboard wastewater is difficult to purify as it includes dishwashing, showering, bathing, laundry, all with their associated cleaning products and contamination. This is a particularly challenging feed due to the high concentration of surfactants (soaps), nitrogen, and solids that must be removed. Many traditional processes are slow, require significant space, or are sensitive to high swings in pH associated with strong acids (toilet cleaner) or strong bases (detergent). We have developed and investigated the use of a novel processing scheme to treat this water without any pretreatment or post-chemical usage focused around purely physical separations. We will present on the wastewater characterization, a variety of membrane/material effectiveness (e.g., PVDF, PAN) and fouling rates, process waste and energy efficiency, and overall system performance and economics with a focus on different control schemes and process models to mitigate long-term degradation.

702e Combined QCM-D/MP-SPR for Early-Stage Membrane Fouling Detection with Unprecedented Senstivity

Thomas Schafer, Iliane Rafaniello

Membrane fouling can seriously affect membrane filtration processes and in particular membrane desalination. As fouling cannot be completely avoided, focus lies on optimizing membrane cleaning intervals in order to increase membrane lifetime while reducing process downtimes and cleaning-costs. Interviews with some of the major Spanish desalination operators revealed that until today operators lack a method sufficiently sensitive for detecting membrane fouling at an early-stage which would avoid significant drops in water flux and permit optimizing process operating conditions such as membrane cleaning protocols. Several methods have been proposed for membrane fouling detection but until today none has proven to be sensitive enough to handle membrane fouling in an anticipative way.

This work presents an approach of overcoming the limitations of conventional methods to detect membrane fouling by combining two advanced surface-sensitive techniques: quartz crystal microbalance with dissipation monitoring (QCM-D) and multi-parameter surface plasmon resonance (MP-SPR). The resulting device yielded an accurate, fast, and non-invasive fouling monitoring of unprecedented sensitivity.

Results on an integrated MP-SPR/QCM-D real-time monitoring device integrated into a RO system will be presented. We could successfully demonstrate that QCM-D/MP-SPR enables characterizing the build-up of even the very first fouling layers at the nanoscale and therefore possessed a sensitivity high enough such as to anticipate the need of membrane cleaning before the filtrated water flux decreases. QCM-D/MP-SPR furthermore allowed to test different membrane cleaning strategies in a time-efficient and material-extensive manner, maximizing in this way cleaning efficiency and prolonging membrane lifetime.

The implementation of the QCM-D/MP-SPR-based monitoring device on a industrial scale will be evaluated, and its potential to increase the process efficiency and mitigate membrane fouling will be critically discussed.

702f – Towards understanding cake formation and filtration characteristics for enhanced media filtration in effluent waste management

Mutiu K Amosa [a,b,*], Gbenga O Komolafe [a], Olatokunbo S Karimu [a], Mohammed S Jami [c], Fatai A Aderibigbe [d], Adewale G Adeniyi [d], Tamrin Tajari [e], Bisola T Bello [d]

[a] *HSEC SBU, Nigerian Upstream Petroleum Regulatory Commission (NUPRC), No. 7, Sylvester Ugoh Crescent, Off Obafemi Awolowo way, Jabi District, Abuja – FCT, Nigeria.*
[b] *National Improved Oil Recovery Centre (NIORC), National Oil and Gas Excellence Centre (NOGEC), NUPRC Annex, 7, Kofo Abayomi Street, Victoria Island, Lagos, Nigeria.*
[c] *Bioenvironmental Engineering Research Centre (BERC), Department of Chemical Engineering & Sustainability, Kulliyyah of Engineering, International Islamic University Malaysia, Kuala Lumpur, Malaysia*
[d] *Department of Chemical Engineering, University of Ilorin, Ilorin, Nigeria.*
[e] *PETRONAS Chemicals Fertiliser Sabah Sdn Bhd, Jalan SOGIP 4, Sipitang Oil & Gas Industrial Park, 89850, Sipitang, Sabah, Malaysia.*

[*]*Corresponding author:* amosa.m.k@nuprc.gov.ng; dhakisalafi@live.com; +2348033981797

Abstract

Understanding the engineering principles behind filter cake's behaviors and its importance in many applications such as wellbore stability, oilfield waterflooding, drilling, produced water re-injection and/or disposal, and all solid-liquid filtration processes is crucial for successful implementation and maintenance of such operations. Many of the problems resulting from these operations are usually due to the inadequate understanding of filter cake mechanisms. Here, the effect of filter media porosity on specific cake resistance (α) from the suspended solids' filtration data of a high-strength effluent is evaluated. The effect of porosity on α is discussed based on the results generated from successive filtration tests. The operating conditions featured transmembrane pressures (TMP) of 40, 80 and 120 kPa and microfiltration membranes with porosities of 0.1 and 0.2 μm. α, compressibility factor (γ), and fouling index (ϑ) are appraised, the data are corrected with Darcy's law of cake filtration, and close correlations of the model was revealed based on the R^2 values. Moreover, the γ values obtained from the slopes of power plots (function of α and pressure gradient) are evaluated. The cakes are observed to be moderately compressible based on the γ values of 0.32 and 0.52 respectively obtained for 0.1 and 0.2 μm MF membranes. Besides, ϑ decreases as the porosity increases with the 0.1 and 0.2 μm MF membranes exhibiting ϑ values of 2.88E+09 and 1.44E+09 cm^{-2}, respectively. Physically at each instant of time, the porosity decreases throughout the cake from the cake surface to the membrane septum where it has the lowest value. It could therefore be concluded that the cake layer serves as a reliable and permeable secondary layer that sustains the flux, hence, sustainable productivity for longer filtration times. Furthermore, the approach of measuring the cake, and other filtration characteristics, provides a wealth of information, which greatly adds to our understanding of the mechanism of flux decline for consideration in every complex effluent treatment system involving filtration unit operations right from the process/plant design stages, for enhanced plant and process sustainability.

Keywords: Specific cake resistance; compressibility factor; fouling index; suspended solids; microfiltration membranes; Low-pressure membranes (LPM)

1. Introduction

In engineering practice and applications, the transport of suspended solid particles contained in aqueous solutions through a porous media where the solid particles are deposited on the pervious medium to form filter cakes, is of paramount importance. Separations of dilute suspensions containing fine particles is being increasingly important in such widely diversified fields as the treatment of industrial effluents, and wastewater reclamation and reuse. Membrane filtration is a key process in such programs but one of the critical issues is membrane fouling which inevitably results in a progressive deterioration in filterability during the filtration. During a filtration process, cakes are compacted while fresh particles are superimposed on the cake surface, hence, gradually increasing the cake thickness. To control particulate fouling at the design stage, as well as monitor it during a plant operation, methods utilised in evaluating the particulate content of feedwater in predicting membrane fouling are essential. Soluble and colloidal materials are assumed to be responsible for membrane pore blockage, whilst suspended solids account mainly for the cake layer resistance [1]. In principle, the governing filtration models are usually employed for the facilitation of the design of membrane processes than any experiment can, yet, experimentations are usually required for validation reasons [2].

Though there exists an extensive number of studies on membrane separations for wastewater streams of diverse chemistries, the accumulated cake behaviour is still yet to be well understood. Particulate fouling in low-pressure membranes such as the microfiltration (MF) and ultrafiltration (UF) membranes is often described by blocking and cake filtration mechanisms while the fouling of high-pressure membranes such as the reverse osmosis (RO) and nanofiltration (NF) is often assumed to occur mainly by cake filtration [3]. However, due to the intrinsic complexity and uncertainty nature usually observed from cake behaviour, extensive study that provides a holistic understanding of the cake with respect to each feedwater and membrane is of great necessity, as a small change in solid fraction can result in permeability variation of several orders of magnitude.

In membrane processes, the flow influences the parking of solids particles which ultimately results in an increased resistance. For compressible cakes, a change in the specific cake resistance with increased pressure are always observed because of the cake particles deformation [4], while the deposition of solid or specific cake resistance during filtration is independent of change in pressure for incompressible cakes [5, 6].

Very few cakes are incompressible as many cakes comprises of highly compressible clays and microbial cells, and as the head loss over the thickness of the cake increases, porosity of the cake reduces because of the particulates being compressed. This consequently results in the porosity distribution to be non-uniform across the cake layer [3, 6, 7].

The structural and physical properties of the cake such as thickness, compressibility, and density have major effects on specific cake resistance thereby influencing filtration performance. Membrane fouling index assumes the retention of particles is constant and specific cake resistance has a time independent permeability and uniform cake porosity throughout the entire depth of the cake when the cake is incompressible [3], however, added complexity usually arises for compressible materials [8]. Cake compressibility could be measured by the compressibility factor γ, which varies from null to one. As γ approaches null, the cake becomes highly incompressible and vice versa. For a compressible cake, porosity reduces as the pressure drop over the cake increases, and the ultimate dense structure will result in an increase in pressure drop [4]. Moreover, as the compressibility factor reduces, the specific cake resistance also drops lower.

When evaluating the sustainability of a membrane process, cake compressibility should be considered. There are sizable number of publications on the treatment of various wastewater streams, but there is dearth of literatures on the cake compressibility analysis of such effluents to better understand the extent at which such processes can be sustained. The current investigation is focused on the mechanistic analysis of the filter cake build-up by suspended solid particles present in a high-strength multicomponent feedwater after an upstream adsorption process that has been detailed in our previous reports [9-14]. The scope of the current investigation focuses on the evaluation of the effects of varying membrane porosities and transmembrane pressures on the specific cake resistance, compressibility and fouling. This will assist the process/environmental engineer to better understand the concept for effective design of membrane processes with enhanced productivity.

2. Materials and Methods

Two commercially available MF membrane cassettes supplied by Sartorius Stedim biotech Sdn. Bhd., Malaysia were employed for the investigation. The membrane properties are shown in Table 1.

Table 1. Physical characteristics of membranes

Characteristic items	Properties
Membrane Material	Polyethersulfone (PESU)
Effective Filtration Area (m^2)	0.1
Membrane Type	Cassette
Pores of the MF membranes (μm)	0.1 (3051545801W-SG),
	0.2 (3051860701W-SG)
Membrane Brand Name	Sartocon Slice cassettes

2.2 Clean Water Filtration (CWF)

Clean water filtration (CWF) was conducted on the two MF membranes of 0.1 and 0.2 μm with distilled water and the permeate volume V was recorded with filtration time t at different transmembrane pressures (TMP) of 40, 80, and 120 kPa. The flux-time data at each TMP were plotted (not shown) and the initial flux capacities of the membranes were measured. These fluxes were utilised as the initial permeate flux at $t = 0$ in the subsequent membrane filtration of the feedwater, as well as the reference points in the model calculations. All filtration experiments continued to steady-state (i.e. until there was no further change in flux) and this was achieved after 1 h filtration time.

2.3 Treatment of Feedwater Using the Hybrid System

Feedwater was sampled from the oil milling industry located at Carey Island, Malaysia. The sample was characterized and preserved under 4 °C. The adsorption reactor was operated with 10 L of wastewater, 60 g/L PAC dose, 200 ± 2 rpm mixing speed, and 68 min. of mixing time. The resultant feedwater was pre-filtered with 5 μm filter to remove the PAC particles from the membrane feed. The pre-filtered feedwater was pumped into the membrane compartment, which houses the MF membrane operated at varying transmembrane pressures (TMP) of 40, 80, and 120 kPa. The permeate volume V at different filtration time t data were recorded at each TMP, and the flux-time plots were utilized in evaluating the fouling propensity of the feedwater. Figure 1 depicts the schematic set-up employed for the hybrid process. All tests were repeated twice to check any instability in the results achieved.

2.4 Analysis of cake build-up

According to Darcy's law, the description of flow of liquid through a porous bed of solids, which allow some fluids to flow through the apparently blocked pores [15], is related to the driving pressure and could be mathematically represented as follows [16, 17]:

$$\text{Flux } (J) = \frac{Volume \text{ (L)}}{Area \text{ (m}^2) \times Time \text{ (h)}} = \frac{1}{A}\frac{dV}{dt} = \frac{\Delta P}{\mu_o R} \quad [1]$$

$$J = \frac{\Delta P}{\mu_0 R} \quad [2]$$

$$R = R_m + R_c \quad [3]$$

where V is the filtrate volume, t is the time, A is the exposed cross-sectional area of the membrane (m^2), ΔP is the pressure drop (kPa) across the membrane plus cake, μ_o is the viscosity (cP) of feed water, R is the porous bed resistance, and R_m and R_c represent the resistances (m^{-1}) due to membrane and cake solids, respectively. R_c could be written in terms of specific cake resistance, α as follow:

$$R_c = \alpha\rho_c\left(\frac{V}{A}\right) \qquad [4]$$

where ρ_c indicates the mass of dry cake solids per volume of filtrate. Combination of above equations yields:

$$\frac{1}{A}\frac{dV}{dt} = \frac{\Delta P}{\mu_0\left[\alpha\rho_c\left(V/A\right)+R_m\right]} \qquad [5]$$

At zero filtration time, it will be zero filtrate volume, then integration of Eq. 5 will subsequently yield [16, 18]:

$$\frac{t}{V/A} = \frac{\mu_0\alpha\rho_c}{2\Delta P}\left(\frac{V}{A}\right)+\frac{\mu_0 R_m}{\Delta P} \qquad [6]$$

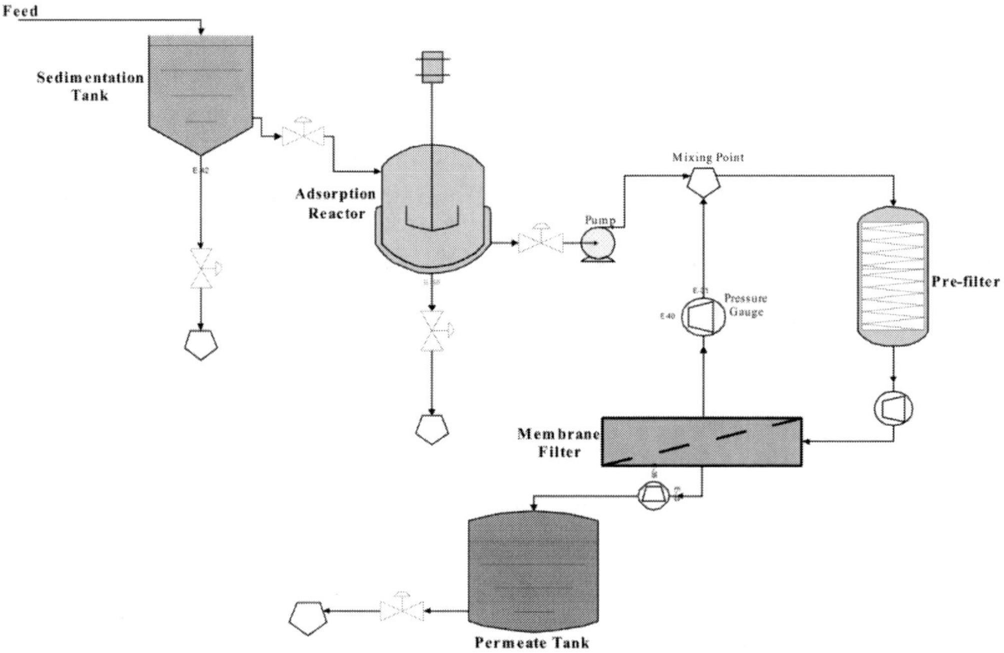

Fig. 1 The bench-scale set-up for wastewater treatment (Adapted from [19])

Eq. 6 is the convenient linearized form of the integrated equation, and plotting $t/(V/A)$ against V/A should result in a straight line [18]. From this plot, the specific cake resistance α, was calculated from the slope while the membrane resistance was estimated from the intercept. Furthermore, α is directly affected by ΔP, the pressure gradient across the cake. The relationship between α, and the power function of ΔP takes the following form [7, 15, 16, 18, 20, 21]:

$$\alpha = \alpha_0\left(\Delta P\right)^{\gamma} \qquad [7]$$

where α_o is an empirical constant and γ is the cake compressibility factor. α is also related to fouling index I and solids concentration, C_b in the feedwater thus [3, 7, 22, 23]:

$$I = \alpha C_b \qquad [8]$$

3. Results and Discussion
3.1 Fluxes at different TMPs

Table 2 shows the effluent characteristics pre- and post-upstream adsorption. The adsorbed feedwater was employed for the filtration process using the MF membranes at TMP range of 40, 80 and 120 kPa. The TMP range, which are within the probable range for a MF membrane were selected based on reports from previous studies [3, 20, 24-30]. The permeate volume *V* at different filtration time *t* data were recorded at each TMP, the fluxes were calculated using equation 1, and the flux-time plots were used to evaluate the fouling propensities of the feedwater. Figures 2 and 3 depict the effect of varying TMP on permeation flux for the two MF membranes.

Table 2. Effluent characterization before and after upstream adsorption

Constituents	Effluent strength (pre-upstream adsorption)	Effluent strength (post-upstream adsorption)
COD (mg/L)	1387	198
Alkalinity (mg/L $CaCO_3$)	1860	880
Hardness (mg/L $CaCO_3$)	1680	440
Manganese as Mn (mg/L)	2.14	0.1
Sulphide as H_2S (mg/L)	0.6	0.05
Total Dissolved Solids (mg/L)	970	820
Turbidity (NTU)	840	53
Silica (mg/L)	58	20
Iron as Fe (mg/L)	ND	ND
Suspended Solids (mg/L)	284	51
pH, units	8.56	9.2

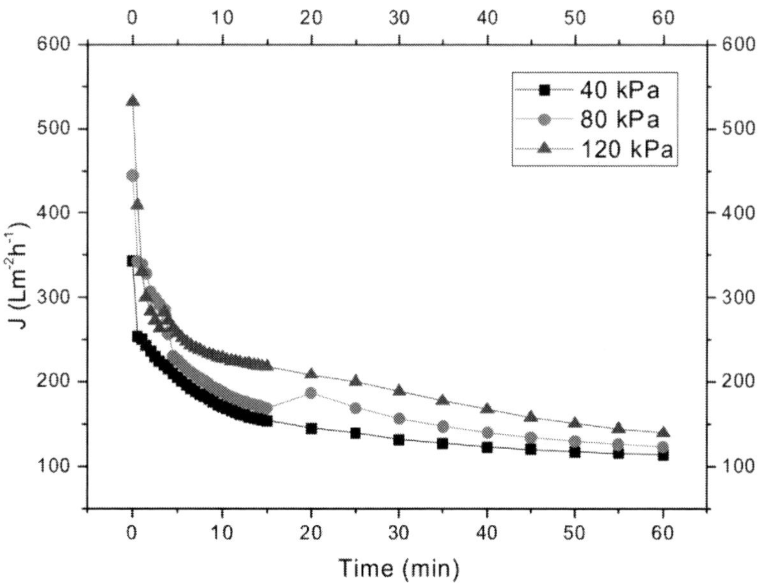

Fig. 2. TMP influence on the flux of 0.2 μm MF membrane

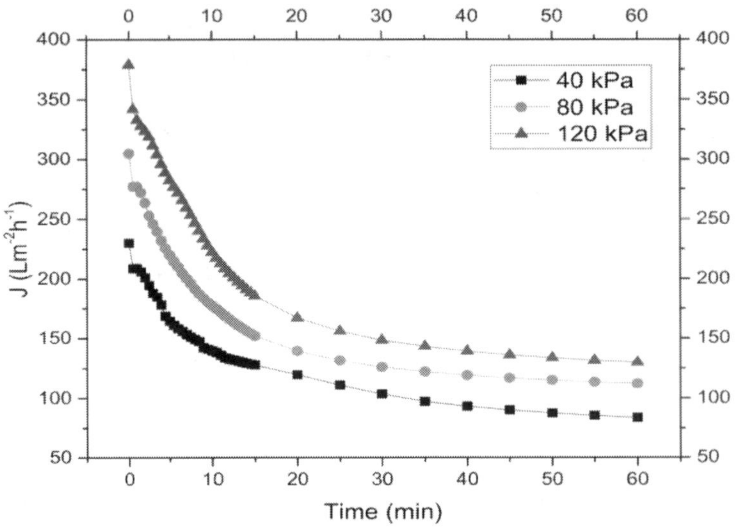

Fig. 3. TMP influence on the flux of 0.1 μm MF membrane

It was observed from the two plots that higher TMPs resulted in higher fluxes for the whole filtration time of 60 min, and the permeation flux (*J*) dramatically decreased with time at the onset of the filtration, but the flux gradually attained relative stability after 20 min of filtration time. This indicates that adsorbed feedwater still contains some constituents that accumulated on the membrane pores thereby resulting in fouling but not necessarily blocking the pores completely. Relatively similar flux-time fouling trends were observed in previous flux decline investigations [31, 32]. It also appeared that the stability of 0.2 μm MF membrane was more established than that of the 0.1 μm MF membrane. This may be due to the fact that 0.1 μm MF membrane experienced more accumulation of foulants on its surface with time compared to the 0.2 μm MF membrane which may have allowed the passage of some of these foulants thereby resulting in lesser accumulation. This conclusion is in agreement with those reached in previous works [3, 33].

3.2 Cake build-up analysis

The transport phenomenon of suspended solid particles in a liquid phase through a porous medium is imperative in the perspective of engineering practice and industrial application. Deposition of such particles on a pervious membrane forms the filter cakes of complex nature. Since it has been established that microfiltration of feedwater is majorly characterized by cake filtration mechanism in our previous study [34], cake build-up analysis of the feed becomes a necessity to better understand the cake's nature. Thus, subsequent to the batch filtration experiments, the flux-time data and solids content of the feedwater was employed in estimating the specific cake resistance (α) with the application of Equation 6. Values of α were determined using the slopes of the plots resulting from equation 6. Figure 4 depicts plots of batch filtration tests for the determination of α for 0.1 μm MF membrane at different TMPs, while Figure 5 represent the plots for 0.2 μm MF membrane.

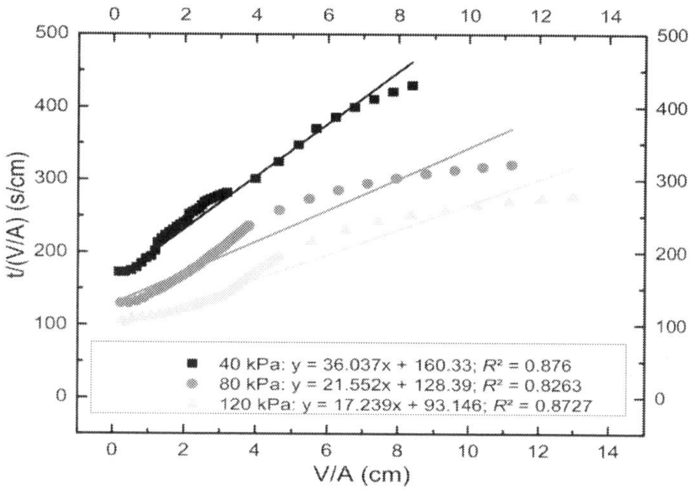

Fig. 4. Batch filtration plots using 0.1μm MF membrane for estimating α

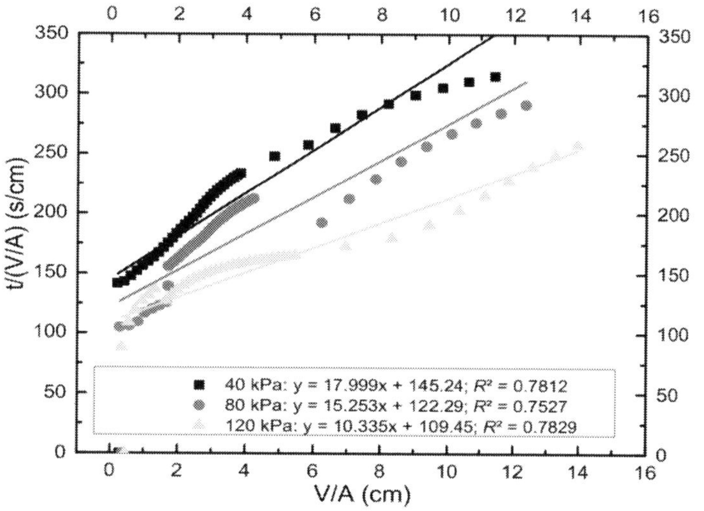

Fig. 5. Batch filtration plots using 0.2μm MF membrane for estimating α

Generally, as membrane filtration continues, the membrane resistance serves as the limiting factor to volumetric flow rate. Nonetheless, more cake layers are formed as more volume of water reaches the membrane surface with time. The solids in the feedwater were retained on the surface of the MF membrane where they build up, forming an increasing thicker cake. And as more feedwater was filtered the solids retained on the membrane could be said to provide most of the filtering action because in cake filtration, the cake is the real filtering element. There was an observation that the slopes increased with the volume filtered, leading to an increase in α when Eq. 6 was applied. This correlates with observations in previous studies [18]. However, α can only increase if the cake could be compressed, and compressible cakes are typically experienced in the filtration of cells and other organic materials. The calculated α from the batch filtration tests for the two membranes unveiled that the specific cake resistances varied with the applied pressures. It could also be said that the α is reversible with the change in pressure drop and this is attributable to the presence of an elastic cake layer due to the compression of the solid particles. This conclusion is based on the fact drawn from the declining pressure drop during the filtration that led to the decrease in the α and vice versa. It could also be said that no breakdown of the original particle or packing of the cake occurred during compression. This observation is in agreement with previous studies [6].

3.3 Cake compressibility analysis

According to Equation 7, the plot for the determination of filter cake compressibility is a power plot, and it is a function of α and pressure gradient. Figures 6 and 7 illustrate the power plots and the two plots exhibited the slopes that correspond to compressibility factors (γ) of 0.32 and 0.52 for the 0.1 and 0.2 µm MF membranes, respectively. These values indicate that the accumulated cakes that resulted from the filtration process is a moderately compressible type. Similar compressibility factor ranges were reported in previous works [15, 20, 35, 36].

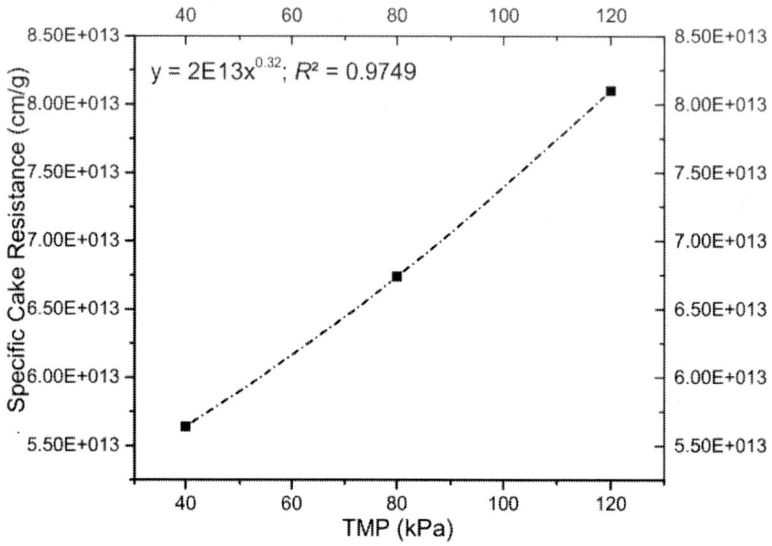

Fig. 6. Power law plots of α as a function of applied pressure for the determination of γ based on 0.1µm MF

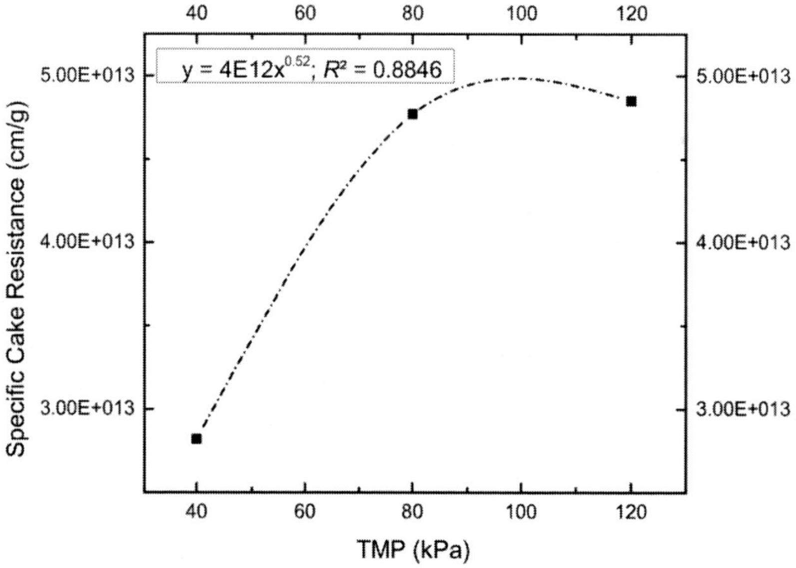

Fig. 7. Power law plots of α as a function of applied pressure for the determination of γ based on 0.2 µm MF

Ideally, as filtration time increases, and more solids particles of the feedwater are filtered as well as transported to the membrane surface by the permeate flow, the cake deposition continues, and the thickness of the cake increases consequent to the solid particles' accumulation. This ultimately results in an analogous increase in the hydraulic pressure resistance across the cake layer leading to a decrease in permeate flux. If the cake is incompressible (i.e., it does not change its volume as pressure builds up) the pressure resistance increases proportionally to the cake thickness. However, since most cakes are compressible, the pressure across the cake typically increases even faster than the cake build-up. In this investigation, the z values are neither close to null nor unity but varies between 0.3 and 0.5 which made it fairly or moderately compressible. This indicates that increase in the pressure across the cake is moderate enough to yield a moderate specific cake resistance, and this allows the filtration to proceed for a longer period without untimely pore blocking. Very compressible cakes (with γ values of close to unity) usually result in early clogging of the membranes and reduce the filtration rate (flux). It could therefore be concluded that higher fluxes attained by the MF membranes is attributable to the nature of cake compression exhibited as similarly hypothesized in earlier investigations [7].

This suggests that MF membranes are suitable for measurement of cake compressibility as it was in consonance with the established principle. It is the rule of thumb that if the cake is composed of compressible particles, then an increase in the applied pressure causes the formation of a denser cake with higher resistance. This usually leads to a marked increase in α as the pressure drop across the cake increases, thus drastically pushing the compressibility factor towards unity, which translates to high cake compressibility. Moreover, it could be asserted that the upstream adsorption process aided in preventing the biofoulants from clogging the membranes and/or passing through them, and this ultimately led to the yielded moderate specific resistances and cake compressibility.

Besides, chemical cleaning of the membranes, as recommended by the membrane manufacturer yielded excellent results. The CWF flux after the first chemical cleaning yielded at least 75% restoration of the membrane flux. Reiterated chemical cleaning of the membrane proved sufficient enough as it was able to bring about more than 96% restoration of CWF flux. It was noticed that the fouling was not severe as the chemical cleaning was able to bring back the membrane into its initial active strength close to 100%.

3.4 Fouling index (ϑ)

As earlier established, the fouling index (ϑ) is related to the specific cake resistance and solids concentration in the feedwater. With this assumption, Schippers and Verdouw [22] came up with the mathematical relation as presented in Eq. 8. The ϑ of the feedwater was estimated for each of the LPM based on the best conditions describing the filtration mechanism. The fouling indices are hereby presented in Table 3 below.

Table 3. Fouling index estimation

LPM	TMP (kPa)	R^2 for Cake Filtration Model	Solids Concentration, C_b (mg/L)	Specific Cake Resistance, α (cm/g)	Fouling Index, ϑ (cm^{-2})
0.2 µm MF	40	0.9167	51	2.82E+13	1.44E+09
0.1 µm MF	40	0.9835	51	5.64E+13	2.88E+09

It is shown in the table that ϑ increased (per square centimetre of the membrane effective area) as the membrane pore reduced. This brought about the 0.1 µm MF membrane having the higher ϑ and α of 2.88E09 cm^{-2} and 5.46E+13 cmg^{-1}, respectively. Expectedly, membranes with higher porosity should have higher ϑ and α values since they have been subjected to similar operating conditions and feedwater chemistry. For these LPMs, the fouling indices can be proposed to be used in the prediction of fouling index of a feedwater having similar chemistry as the one employed for this study.

4. Conclusion

Specific cake resistance, compressibility factor and fouling index exhibited with the MF membranes tested in this study for a specific feedwater were determined. Effects of varying porosities and applied pressures on the variables were evaluated. It was observed that the MF membranes were stable for such measurements as γ values of 0.32 and 0.52 respectively generated by the 0.1 and 0.2 μm MF clearly indicated that the cakes are moderately compressible. Though, high compression always occur in feedwater that are rich in clays and organics [7, 18], hence, the fair compression exhibited by the feedwater is attributable to the fact that higher concentration of the responsible constituents have been removed by the upstream adsorption process. The MF membranes are recommended for characterizing cakes formed from aggregated materials of feedwaters of similar chemistries using a steady-state filtration technique which was attained within 60 min. of filtration time. The results could be employed as requisite data for further investigation, and for the designs of membrane filtration processes, especially for information on cake behaviours and required cleaning techniques that involves similar high-strength wastewater and low-pressure membranes as applied in this study.

References

[1] S. Judd, The MBR book: Principles and applications of membrane bioreactors for water and wastewater treatment, 2nd Edn., Elsevier, UK (2011).
[2] K.-V. Peinemann, S.P. Nunes, L. Giorno, Membranes for Food Applications, WILEY-VCH Verlag GmbH & Co. KGaA, Weinheim, Germany (2010).
[3] S.F. Boerlage, M.D. Kennedy, M.R. Dickson, D.E. El-Hodali, J.C. Schippers, J. Membr. Sci., 197 (2002) 1-21.
[4] T.V. Bugge, M.K. Jørgensen, M.L. Christensen, K. Keiding, Water Res., 46 (2012) 4330-4338.
[5] D.J. am Ende, Chemical Engineering in the Pharmaceutical Industry: R&D to Manufacturing, John Wiley & Sons, Hoboken (2010).
[6] S.G.S. Rodriguez, Particulate and Organic Matter Fouling of Seawater Reverse Osmosis Systems: Characterization, Modelling and Applications. UNESCO-IHE PhD Thesis, CRC Press, Leiden (2011).
[7] S.F. Boerlage, M.D. Kennedy, M.P. Aniye, E. Abogrean, Z.S. Tarawneh, J.C. Schippers, J. Membr. Sci., 211 (2003) 271-289.
[8] P. Kovalsky, M. Gedrat, G. Bushell, T.D. Waite, AIChE J., 53 (2007) 1483-1495.
[9] M.K. Amosa, Environmental Nanotechnology, Monitoring & Management, 4 (2015) 93-105.
[10] M.M. Alam, M.Z. Alam, M.S. Jami, M.K. Amosa, M. Iwata, Waste Biomass Valor, 7 (2016) 583–591.
[11] M.K. Amosa, M.S. Jami, M.F.R. Alkhatib, Waste Biomass Valor, 7 (2016) 109-124.
[12] M.K. Amosa, M.S. Jami, M.F.R. Alkhatib, D.N. Jimat, S.A. Muyibi, Water Qual Expo Health, 7 (2015) 603-616.
[13] M.K. Amosa, M.S. Jami, M.F.R. Alkhatib, T. Tajari, D.N. Jimat, R.U. Owolabi, Cogent Engineering, 3 (2016) 1162384.
[14] S.A. Muyibi, T. Tajari, M.S. Jami, M.K. Amosa, Adv. Environ. Biol., 8 (2014) 590-595.
[15] W. Yuan, A. Kocic, A.L. Zydney, J. Membr. Sci., 198 (2002) 51-62.
[16] G.B. Van den Berg, C.A. Smolders, Desalination, 77 (1990) 101-133.
[17] Y. Bessiere, N. Abidine, P. Bacchin, J. Membr. Sci., 264 (2005) 37-47.
[18] R.G. Harrison, P. Todd, S.R. Rudge, D.P. Petrides, Bioseparations Science and Engineering, Oxford University Press, Inc., New York (2003).
[19] M.K. Amosa, M.S. Jami, M.F.R. Alkhatib, T. Majozi, Environmental Science and Pollution Research, 23 (2016) 22554–22567.
[20] T. Ohn, M. Jami, E. Iritani, Y. Mukai, N. Katagiri, Sep. Sci. Technol., 38 (2003) 951-966.
[21] G. Foley, Membrane Filtration: A Problem Solving Approach with MATLAB®, Cambridge University Press, UK (2013).
[22] J.C. Schippers, J. Verdouw, Desalination, 32 (1980) 137-148.
[23] C. Güell, M. Ferrando, F. López, Monitoring and visualizing membrane-based processes, John Wiley & Sons, Weinheim (2009).
[24] R. Baker, A. Fane, C. Fell, B. Yoo, Desalination, 53 (1985) 81-93.
[25] C.W. Lee, S.D. Bae, S.W. Han, L.S. Kang, Desalination, 202 (2007) 239-246.
[26] Y. Li, J. Wang, W. Zhang, X. Zhang, C. Chen, Chin. Sci. Bull., 56 (2011) 584-590.
[27] K.M. Agbekodo, B. Legube, P. Cote, Journal-American Water Works Association, 88 (1996) 67-74.

[28] P. Fu, H. Ruiz, K. Thompson, C. Spangenberg, Journal-American Water Works Association, 86 (1994) 55-72.
[29] C.-W. Jung, L.-S. Kang, Korean J. Chem. Eng., 20 (2003) 855-861.
[30] F. Springer, S. Laborie, C. Guigui, Sep. Purif. Technol., 108 (2013) 6-14.
[31] M. Cai, S. Wang, H. Liang, Food Sci Biotechnol, 22 (2013) 407-412.
[32] S.S. Rashid, M.Z. Alam, M.B.F.A. Fazli, International Journal of Chemical, Environmental & Biological Sciences (IJCEBS), 1 (2013) 501-506.
[33] T. Mohammadi, M. Kazemimoghadam, M. Saadabadi, Desalination, 157 (2003) 369-375.
[34] M.K. Amosa, M.S. Jami, M.F. Alkhatib, T. Majozi, Sep. Sci. Technol., In press (2016).
[35] M.S. Jami, Nagoya University Japan, Unpublished (2005).
[36] S.-K. Kang, K.-H. Choo, J. Membr. Sci., 223 (2003) 89-103.

702g IMAGE Processing As a Tool in Fouling Characterization for Reverse Osmosis Desalination

Tu Pham Le Phuong, Jisha Ali, Mohamed Soufiane Jouini, Emad Alhseinat

Water scarcity has been the most serious problem to resolve as the need for freshwater keeps increasing over time. The main existing solution is to convert unconventional water resources such as seawater, brackish water i.e. underground water, and wastewater into freshwater by employing desalination technologies. Basically, the technologies are divided into thermal-based processes and membrane filtration processes in which reverse osmosis (RO) membrane filtration is the most favored technique currently. Its advantages include high efficiency, low energy consumption, and low cost. However, due to the complexity of feed water, various matters can deposit on the membrane surface during the filtration process under highly applied pressure, which is called fouling. This generates a cake or gel layer on the membrane which leads to an increase in membrane resistance, reduction of filtration efficiency, and membrane shelf-life. Therefore, monitoring the growth of the fouling layers on the membrane during filtration is compulsory to understand insights into the fouling process. Optical coherence tomography (OCT) is a technique that allows *in situ* and rapid monitoring of the stages of fouling during filtration. However, the applications of OCT in RO membrane filtration are still restricted and primarily focused on biofouling and qualitative analysis. Therefore, this study aims to extend the application of OCT in the observation of scaling i.e. $CaSO_4$ in real hydrodynamic conditions. The acquisition data was processed by a special digital image processing code to extract quantitative analysis data (volume and surface coverage proportion) of the scaling layer and correlate it to the RO performance parameter i.e. permeate flowrate. The digital image process primarily consists of image enhancement by 3D mean filter with a 5-voxels window, image segmentation using combination of thresholding and edge detection methods, and quantification of fouling parameters. The results showed that OCT can detect the concentration polarization and nucleation process at the first hour of the process and the growth of crystals from their nuclei after 1.5 h. After 8 h, when the fouling volume and surface coverage proportion reach 21.473×10^6 (μm^3) and 41.51 (%), respectively, significant reductions of permeate flux, conductivity of concentrate, and Ca^{2+} concentration are recorded 54.52±1.94 ($mL.h^{-1}$), 10.34±0.02 ($mS.cm^{-1}$), and 26.28±0.08 (mM) compared to the initial values of 77.14±0.25 ($mL.h^{-1}$), 10.93±0.42 ($mS.cm^{-1}$), and 27.23±0.12 (mM), respectively. After 24 h operation, when 98.27 (%) of membrane surface is covered by fouling and 188.786×10^6 (μm^3) is obtained as fouling volume, the operation parameters reach 42.34±3.05 ($mL.h^{-1}$), 10.11±0.05 ($mS.cm^{-1}$), and 21.78±0.28 (mM), respectively. The drops of permeate quality is also noticed via the increase of its conductivity, which is 7.85±0.43 ($mS.cm^{-1}$) after 24 h in a comparison to 2.23±0.02 ($mS.cm^{-1}$) as the initial value. In addition, the morphology of scales captured by OCT is compared with a scanning electron microscope (SEM). Therefore, this paper successfully introduces a methodology of using OCT in a combination with digital image processing as a promising and useful technique for quantitative the growth of fouling in membrane filtration processes.

Keywords: Optical coherence tomography (OCT), fouling, membrane-based separation (RO).

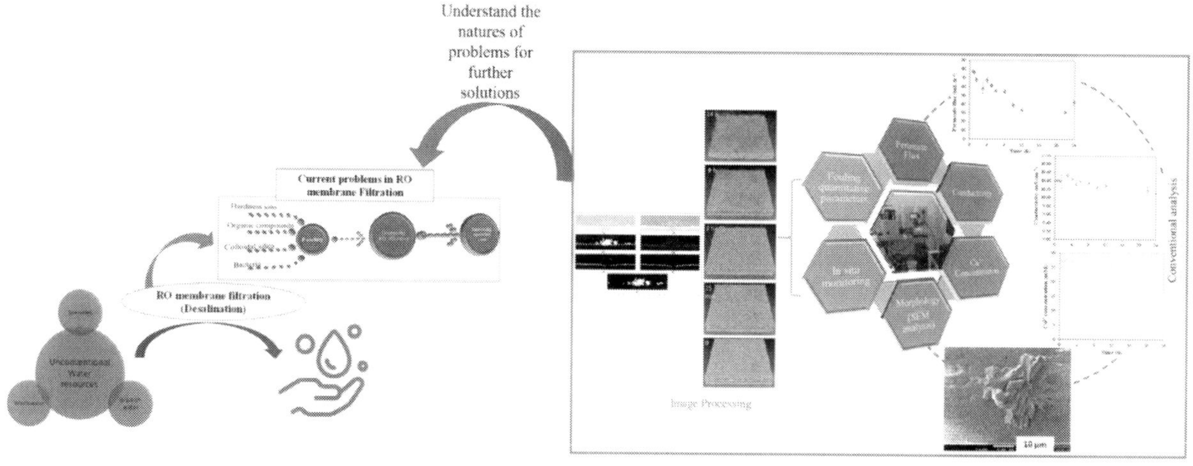
Understand the natures of problems for further solutions
Current problems in RO membrane Filtration
RO membrane filtration (Desalination)
Image Processing
Conventional analysis
10 μm

211a Determination of Pure and Binary Adsorption Isotherms Using Concentration Pulse Chromatography

F Handan Tezel

Knowledge of pure and mixture adsorption isotherms are essential for the design of adsorption separation processes. Volumetric and gravimetric systems have been commonly used for the determination of pure component adsorption isotherms with success. For the determination of mixture adsorption behavior, volumetric and gravimetric systems become quite expensive. One alternative to consider would be the use of concentration chromatography technique for determining binary mixture adsorption isotherms. This technique can also be used for determining the pure component isotherms, as well. The presentation will concentrate on the use of concentration pulse chromatography for determining pure and binary adsorption isotherms, and compares the results obtained to the results from other methods.

211b Advances in the characterization of porous materials for applications in chromatography

Carola Schlumberger, Carlos Cuadrado Collados, Jakob Söllner, Matthias Thommes

During the last two decades, major progress has been made concerning the textural characterisation of porous materials in the gas/dry phase by methods such as gas adsorption and mercury porosimetry. The main drawback of these methods stems from their inability to test wet materials (i.e. in presence of the liquid/mobile phase), thus providing no direct information concerning the state of the stationary phase in typical chromatographic operations. Hence, there is a need to assess the surface area and accessible pore structure in the liquid phase. This is particularly important for functionalised porous materials, where the interaction of surface functional groups with the chosen liquid may affect the effective surface area and pore characteristics for separation processes. Within this context we will report results obtained by the application of novel methodologies based on NMR relaxation measurements, which allow one to characterize solvated porous materials, in combination with complimentary methods based on advanced gas adsorption and liquid intrusion for a comprehensive textural and surface characterization of nanoporous materials.

211c Towards Preparative Chromatographic Classification of Polydisperse Products

Malte Kaspereit, Malvina Supper

The quality of polydisperse products like polymers or nanoparticles depends strongly on their size distribution, that is, their molar weight distribution (MWD) or particle size distribution (PSD). The narrower a product's size distribution, the higher its value, since the more defined are its properties like light absorption and emission, color brilliance, catalytic activity, behavior in biological systems, etc.

At lab-scale, quite defined MWDs and PSDs are attainable by synthesis. However, scale-up to industrial conditions generally leads to broader distributions, necessitating a subsequent classification. The challenge is that, for corresponding products, suitable industrial separation technologies are rare. Most promising for such task is preparative chromatography due to its unprecedented resolution power and proven scalability. Currently, a large collaborative research project funded by the German Science Foundation (CRC 1411 - "Design of Particulate Products" [1]) is devoted to this problem.

Here we present experimental and model-based developments of chromatographic classification for complementary model systems. The first is the reversed-phase separation of molar-weight distributed polyethylene glycols (PEGs). Based on detailed analytical-scale investigations, a computationally efficient process model is established that includes rigorous thermodynamic treatment of retention depending on the most relevant parameters, namely, degree of polymerization, temperature, and mobile phase composition. The model allows highly accurate prediction of chromatograms over a wide range of conditions [2]. After validation, it was applied successfully for designing semi-preparative chromatographic processes to isolate highly pure single PEG homologs of different molecular sizes. Second, the applicability to the separation of nanoparticles and the specific challenges related to this are discussed using selected examples.

Based on the insights, advanced chromatographic operating concepts like column combinations, recycle schemes or multi-column processes are presented and evaluated. The results are of general relevance for developing powerful separations capable of obtaining well-defined products from polydisperse mixtures.

[1] For details https://www.crc1411.research.fau.eu

[2] Supper, Heller, Söllner, Sainio, Kaspereit, *Processes*, 2022 (subm.)

211d Simulated Moving Bed Reactor for Enhancing the Productivity of Equilibrium Limited Reactions

Balamurali Sreedhar, Megan E. Donaldson

Equilibrium-limited reactions are ubiquitous in the chemical industry. Those with highly unfavorable equilibrium constants face very energy intensive operations for conversion and recovery of the product, often needing extensive recycle. Chances are also that such severely equilibrium-limited reactions using conventional reactors are written off at an early stage and are never scaled up to production. In this work, we demonstrate the potential of simulated moving bed reactor (SMBR) in increasing the productivity of strongly equilibrium-limited reactions and possibly rendering them economically feasible. In an SMBR, in-situ chromatographic separation of the products aids in shifting the equilibrium, thereby improving the conversion.

Liquid phase transesterification of propylene glycol methyl ether (DOWANOL™ PM) to propylene glycol methyl ether acetate (DOWANOL™ PMA) using a homogeneous catalyst [1] is considered as a case study to compare the performance of SMBR with that of a conventional batch reactor. Here, DOWANOL™ PM is used both as a reactant and desorbent for the chromatographic separation inside the SMBR. In the SMBR process, an anion exchange resin, AMBERLITE™ IRA904 in chloride form is used as a non-catalytic adsorbent, while sodium alkoxide of PM is supplied as a homogeneous catalyst in the desorbent. Using model based optimization, we demonstrate the advantage of employing a non-catalytic adsorbent in the SMBR for the homogeneously catalyzed reaction. With the ability to independently tune the separation of the product in the SMBR, it is demonstrated that the productivity of the process increases with increasing separation factor. Presence of an optimal product separation factor is observed (around $\alpha = 2.4$) with respect to the amount of DOWANOL™ PM used for reaction and separation. At a fixed conversion of 95%, the SMBR design at this separation factor outperformed batch reactor in increasing the productivity by 85% and reducing DOWANOL™ PM usage to 50% of what is required by the batch reactor.

[1] J. Oh, B. Sreedhar, M. E. Donaldson, A. K. Schultz, T. C. Frank, A. S. Bommarius, Y. Kawajiri, 'Transesterification of Propylene Glycol Methyl Ether in Chromatographic Reactors Using Anion Exchange Resin as a Catalyst,' *J. Chromatography A* **2016**, *1466*, 84-95; DOI:10.1016/j.chroma.2016.08.072

211e Engineering tools for screening and discovery of solid sorbents for CO2 capture

Arvind Rajendran

The past two years have seen major advances in adsorption technology for CO2 capture. Particularly, metal-organic frameworks (MOFs), traditionally considered exotic sorbents, have now been deployed for large-scale post-combustion CO2 capture from industrial flue gas. This has created a renewed interest in screening adsorbents for CO2 capture applications. Over the years, our work in this area has motivated us to develop both computational and experimental methods that allow for the meaningful screening of materials. In this talk, we will summarize recent developments in process modelling and optimization accelerated by machine learning that can be used to screen adsorbents rapidly. Further, we discuss the experimental evaluation of competitive isotherms for complex systems, e.g., CO2+H2O that have important consequences on the practical deployment of the sorbent. We will pay particular attention to breakthrough/ frontal-analysis experiments, an area where Prof. Andreas Seidel-Morgenstern has made major contributions.

211f Optimising adsorption-desorption cycles with flexible Metal Organic Frameworks

Rahul Maity, Mohsen Gholami, Joeri Denayer

Quite a number of Metal Organic Frameworks have a flexible framework. In presence of external stimuli, their structure changes, resulting in a change of pore volume and pore size and thus adsorption properties. This change can be drastic, e.g. the transition from a narrow pore form to a large pore form at a critical pressure or rather gradual i.e. the swelling of a structure with increasing degree of pore filling. Such flexible materials often show unusual isotherms, with sometimes very steep steps and an hysteresis loop in the adsorption and desorption isotherms. In this paper, we analyse how the desorption of molecules from such a flexible MOF with a stepped isotherm showing hysteresis can be made more efficient.

211g Selective Adsorption of Surfactants on Coronavirus Virions

Alexander Neimark, Kolattukudy Santo

COVID-19 pandemic has triggered broad research activities aiming at preventing and curing coronavirus disease. COVID-19 is transmitted by inhaling airborne SARS-CoV-2 virions, which represent biologically active nanoparticles of ~ 120 nm enveloped by a lipid bilayer decorated by a "crown" of Spike protein protrusions. In the respiratory tract, coronavirus virions interact with surfactant films composed of phospholipids and cholesterol that coat lung airways. Active clinical search is underway for physiological and exogenous surfactants that may adsorb on Spike proteins or dilute the lipid envelope affecting the virion ability to bind and penetrate the cells. While the knowledge on the biochemical structure, pathology, and antibody/drug interactions of SARS-CoV2 and its variants is quickly growing, the physico-chemical aspects of the virion interactions with the respiratory system environment and specifically with adsorbing surfactants have been sparsely addressed and are poorly understood. Here, we explore by using coarse-grained molecular dynamics simulations the mechanisms of surfactant adsorption on Spike proteins. With examples of zwitterionic dipalmitoyl phosphatidyl choline, cholesterol, and anionic sodium dodecyl sulfate, we show that surfactants form micellar aggregates that selectively adhere to the specific regions of S1 domain of the Spike protein that are responsible for binding with ACE2 receptors and virus transmission into the cells. We find distinctively higher cholesterol adsorption in comparison with other surfactants, which is consistent with recent experimental studies. Cationic residues, arginine and lysine, which are present in higher quantities in mutated Delta and Omicron variants, preferentially adsorb anionic surfactants that may block direct S1-ACE 2 interactions and hinder binding. Our findings of strong selective adhesion of surfactant aggregates to receptor-binding fragments of Spike proteins may have important implications for informing the clinical search for exogenous therapeutic surfactants for curing and preventing COVID-19 caused by SARS-CoV-2 and its variants.

211h CORE-SHELL AND ENCAPSULATED ADSORBENTS FOR FLOW THROUGH PURIFICATION OF LARGE BIOMOLECULES AND NANOPARTICLES

Giorgio Carta

We examine the structural and functional properties of commercial agarose based core-shell resins and develop models to describe the kinetics of binding for proteins with a broad range of molecular mass in single and multiple component systems. We also describe the development and properties of novel adsorbents obtained by encapsulating ceramic hydroxyapatite particles in inert agarose beads for the adsorption of proteins and nucleic acids impurities.

276a On the Use of the Dual Process Langmuir Model for Predicting Mixed Gas Isotherms of Methane, Ethane and Ethylene Binary and Ternary Mixtures on Activated Carbon

Armin Ebner, Jaclyn Couch, James A. Ritter

The ability of the mixed gas dual process Langmuir (DPL) model to predict both binary and ternary mixtures of methane, ethane, and ethylene on activated from the work of Reich, Ziegler and Rogers[1] is evaluated against predictions carried out by ideal adsorbed solution theory (IAST) either globally or at each Langmuir process individually. Perfect positive correlations are assumed between all three species. Saturation loadings for each of process in all species are assumed to be different and hence not thermodynamically consistent, which is commonly don with the mixed gas DPL model. Predictions are evaluated against experimental data around 130, 350, 680 and 1280 kPa at 28.3 and -60.45 °C. The results show significant agreement between the mixed gas DPL model predictions and the experimental data, even though the experimental are well known to exhibit non-ideal behavior. The results also show that the mixed gas DPL is able to predict just as accurately as the far more complex models, such as the IAST.

276b Comparison of Heterogeneous Langmuirian Models for Representation of Unary, Binary, and Ternary Adsorbate Systems at Adsorption Equilibria

Michael D. Sees, Yinka Otulana, Usman Hamid, Chau-Chyun Chen

Mixed gas adsorption is still a very active area of research, where newly proposed models need to be compared actively to standards in the field. The generalized Langmuir (gL) isotherm, proposed by Hamid *et al.* to address adsorption of different size molecules on heterogeneous adsorbents without requiring evaluation of the Gibbs isotherm, [1] has thus-far been tested mainly against the adsorbed solution theory (AST). The gL isotherm was shown to be thermodynamically consistent, [2] and gave either similar or better representation than both IAST and RAST-aNRTL. [1 - 3] Here, the gL isotherm will be contrasted with two additional models which remain popular in practice: the Sips equation-based loading ratio correlation (LRC) [4 - 6] and the duel-process Langmuir (DPL) isotherm popularized by Ritter and his coworkers at USC. [7 - 11]. The models will be compared based on their representation of the equilibrium loading and isosteric heat of adsorption (where available) of individual components, binary, and ternary mixtures of various gaseous adsorbates on different adsorbents over a range of experimental pressures and temperatures.

1. Hamid, P. Vyawahare, H. Tun, and C.-C. Chen. (2021). Generalization of Thermodynamic Langmuir Isotherm for Mixed-Gas Adsorption Equilibria. *In Review.*
2. Hamid, P. Vyawahare, M.D. Sees, and C.-C. Chen. (2021, November). "Generalized Langmuir Isotherm Addresses Spreading Pressure Dependency of Modified Ideal Adsorbed Solution Theory." Paper presented at the 2021 Annual Meeting of the American Institute of Chemical Engineers, Boston, MA.
3. Hamid, P. Vyawahare, and C.-C. Chen. (2021, November). "Isosteric Heat of Adsorption of Pure Component and Mixed-Gas Adsorption from Generalized Langmuir Isotherm." Paper presented virtually at the 2021 Annual Meeting of the American Institute of Chemical Engineers, Boston, MA.
4. Sips. (1948). On the Structure of a Catalyst Surface. *J. Chem. Phys.* 16, 5, 490 – 495.
5. Sips. (1950). On the Structure of a Catalyst Surface II. *J. Chem. Phys.* 18, 8, 1024 – 1026.
6. M. Yon and P.H. Turnok. (1971). Multicomponent Adsorption Equilibria on Molecular Sieves. *AIChE Symp. Ser.*, 67, 117, 75 – 83.
7. A. Ritter, S.J. Bhadra, and A.D. Ebner. (2011). On the Use of the Duel-Process Langmuir Model for Correlating Unary Equilibria and Predicting Mixed-Gas Adsorption Equilibria. *Langmuir*, 27, 4700 – 4712.
8. J. Bhadra, A.D. Ebner, and J.A. Ritter. (2012). On the Use of the Duel-Process Langmuir Model for Predicting Unary and Binary Isosteric Heats of Adsorption. *Langmuir*, 28, 6935 – 6941.
9. A. Ritter, K.C. Bumiller, K.J. Tynan, and A.D. Ebner. (2019). On the Use of the Duel-Process Langmuir Model for Binary Gas Mixture Components that Exhibit Single-Process or Linear Isotherms. *Adsorption*, 25, 1511 – 1523.
10. J. Tynan, S. Tosso, A.D. Ebner, and J.A. Ritter. (2021). On the Use of Single, Dual, and Three-Process Langmuir Models for Binary Gas Mixtures that Exhibit Unique Combinations of These Processes. *Adsorption*, 27, 637 – 658.

276c On the Use of a New Kinetic Model in a Cyclic Adsorption Process Simulator for Kinetically Limited Gas Separations

Sulaimon Adegunju, Pravin B. C. A. Amalraj, Armin Ebner, James A. Ritter

Current kinetics models, typically LDF models, have been applied on the basis that the presence of a second, slower diffusing, adsorbate affects the equilibrium of the first, faster diffusing, adsorbate in a gas mixture via its gas phase partial pressure. However, this premise is contrary to experimental observations involving either butane or ethane with CO_2 on carbon molecular sieves. The objective of the work is two-fold. First, a new non-equilibrium model that augments the equilibrium driving force in the LDF kinetic model, such that the equilibrium loading of one component depends on the actual loading of the other component in a gas mixture will be presented. Further, in this new model, the equilibrium loading of one component depends on the actual loading of the other component, with both leading to the extended Langmuir model predictions at equilibrium. Second, it will be shown how this new kinetic model performs within the framework of the UofSC dynamic adsorption process simulator (DAPS). Two key kinetically limited binary gas separations with a carbon molecular sieve (CMS) are being examined both experimental and with DAPS using different PSA cycle schedules. These are CO_2-CH_4 and CO_2 N_2 gas mixtures. The latest results will be presented.

276d Understanding the Break-through Curve Measurement When Adsorption Is Fast

Robert DeJaco, Paul N. Patrone, Anthony J. Kearsley

The break-through curve measurement characterizes the relationship between bulk solute—solid properties and solute movement along a packed bed and is a fundamental aspect of large-scale adsorption separations. The bulk properties explain the mechanism of adsorption, while solute movement details the mechanism of separation. A more quantitative description of the relationship between bulk properties and solute movement can accelerate process intensification and high-throughput materials screening.

In this work, we investigate the relationship between an adsorption isotherm and the break-through curve measurement when ε is small, or when mass transfer into the solid is much faster than convection. To accommodate the large number and diversity of adsorbent materials that can be synthesized, we focus on an adsorption isotherm possessing a generic functional form (i.e., smooth, and strictly increasing). We consider a single-solute adsorbing isothermally with negligible axial dispersion. Regular and singular perturbation theory in the limit $\varepsilon \rightarrow 0$, or when mass transfer is much faster than convection, provide a more quantitative and generic interpretation of several popular theoretical approaches, as well as how they are connected. For example, while dispersive waves in equilibrium theory are realized as the leading-order approximation as $\varepsilon \rightarrow 0$, shock waves and linear waves are only valid *at the point* $\varepsilon=0$. More realistic descriptions of the latter waves in the measurement, where ε is positive, can be obtained using boundary layer theory. The influence of the rate expression on the constant-pattern boundary layer, as well as the time scales over which the behavior is dominant, are revealed. In contrast to previous approaches, validity can be assessed quantitatively by the value of ε for a given system of interest. To predict the time at which the center of the constant-pattern boundary layer breaks through the column, its evolution from short times must be resolved. The detailed investigation of the limit $\varepsilon \rightarrow 0$ also identifies new regimes of adsorption isotherms for which solute movement is not yet fully understood.

This work provides a more quantitative understanding of the break-through curve measurement when adsorption is fast. The boundary layer perspective sets the stage for extension to more complex systems and other separation processes, as it does not require extensive expertise with weak solutions.

276e Aspects of Gas Storage: Effect of Confinement on Supercritical Adsorption Behaviour

Simon Eder, Matthias Thommes

During the last decades, major progress was made concerning the understanding of subcritical, low pressure adsorption of fluids like nitrogen and argon at their boiling temperature in nanoporous materials. It was here possible to understand how structural properties affect the shape of the adsorption isotherms, leading to new methods that are now commonly used for characterization. However, within the context of gas storage applications, supercritical high pressure gas adsorption is important. High-pressure adsorption experiments introduce several complexities, both in terms of collecting isotherm data and interpreting the results. A key feature is here that the experimentally determined surface excess adsorption isotherm may exhibits a characteristic maximum at certain pressure. For a given temperature and adsorptive/adsorbent system, the surface excess maximum (and the corresponding (absolute) adsorbed amount) is related to the storage capacity of the adsorbent. However, there is still a lack of understanding how key textural properties such as surface area, pore size, pore volume and pore network characteristics affect the shape of supercritical high pressure adsorption isotherms (and here in particular the position of the surface excess maximum).

In order to address this open questions we have a performed a systematic experimental study assessing the effect of pore size/structure on the supercritical adsorption isotherms of pure fluids such as C_2H_2, CO_2 and H_2 over a wide range of temperatures and pressures (from well above the bulk critical point to the near-critical region) on a series of model materials exhibit well defined pore sizes, i.e. ordered micro- and mesoporous materials such as zeolites, mesoporous molecular sieves (e.g., KIT-6, MCM-41 silica) and a hierarchically structured mesoporous NaY-zeolite. One key result of our experiments is that we find a clear correlation between the pressure of the surface excess maximum (at a given temperature) and pore size. This was further investigated by performing complimentary molecular simulation studies. Our results suggest important structure-property relationships and allowed us to derive a tool for predicting gas storage properties of nanoporous materials at given thermodynamic conditions based on their textural properties. Within this context we have recently also expanded our studies to MOFs and ZIFs.

371q Degradation of Reactive Black 5 Dye Using Nanostructured Carbon Adsorbents

Robert Haxhari, Kevin Zheng, Tyler Zmarzlak, Aldo Bushati, Taejin Kim, Yizhi Meng, Steve Nitodas

Hazardous effluents from global textile industries contain several hundred thousand tons of synthetic dyes per year [1]. A standard azo dye known as reactive black 5 (RB5) provides color for a substantial range of materials in the textile industry; however, water contamination from RB5 can lead to harmful impacts on local ecosystems and human health [1]. The structure of RB5 is inherently stable due to its aromatic and heterocyclic characteristics, rendering it non-biodegradable and mutagenic [2]. As a result, it is necessary for textile industries to employ waste treatment techniques in an effort to remove harmful dyes before they can pollute the environment.

A variety of techniques, such as flocculation and biological degradation methods, have been utilized in the past to facilitate the separation of azo dyes from wastewater. However, these strategies are characterized by major drawbacks, including costly reagents and lengthy removal processes [3]. The aim of the current research is to develop a practical dye removal method for the treatment of RB5-contaminated water. Nanostructured carbon, including activated carbon (AC), graphene nanoplatelets (GNP) and graphene oxide (GO), was investigated as adsorbents for RB5. These materials have been known to exhibit high surface area, stability in aqueous environments and potential for recyclability [2,4,5].

During the course of the experiments, a constant quantity of each studied adsorbent (AC, GNP or GO) was employed and the RB5 degradation capability was tested for 2 hours in RB5 aqueous solutions with concentrations ranging from 10^{-4} to 10^{-5} moles/liter (M) under ultraviolet radiation (365 nm). A TECAN microplate reader was utilized as adsorption spectrophotometer. Representative spectra are shown in Figure 1 for AC at an RB5 concentration of 10^{-4} M. The maximum adsorption peak for RB5 was found to occur at approximately 600 nm and is due to the existence of the nitrogen-nitrogen double 'azo' bond dissociated by the degradation agent [3]. Over the course of two hours, it was observed that the intensity of this peak decreased in the case of all studied adsorbents. Utilizing the value of the decreasing peak over time, it was possible to calculate the remaining RB5 concentration at the end of each experiment, using the calibration curve generated from the pure RB5 concentrations that were studied.

In addition to UV radiation, adsorbents were tested under visible light condition and dark conditions. Finally, equilibrium data of RB5 adsorption on AC, GNP and GO were obtained by carrying out degradation experiments over a 24-hour period, which was deemed to be adequate for the attainment of equilibrium [6]. The results were fitted to the Langmuir and Freundlich isotherm models to evaluate their efficacies. Activated carbon was found to have the highest RB5 degradation capability among the nanostructured carbon adsorbents that were investigated. The results are analyzed and discussed in order to demonstrate the most practical RB5 degradation catalyst that could be employed for RB5 removal in the textile industry.

References:

[1] Bilal, Muhammad, et al. "Toxicological Assessment and UV/TIO2-Based Induced Degradation Profile of Reactive Black 5 Dye." *Environmental Management*, vol. 61, no. 1, 2017, pp. 171–180., https://doi.org/10.1007/s00267-017-0948-7.

[2] Fraga, Tiago José, et al. "Comparative Approach towards the Adsorption of Reactive Black 5 and Methylene Blue by N-Layer Graphene Oxide and Its Amino-Functionalized Derivative." *Adsorption*, vol. 26, no. 2, 2019, pp. 283–301., https://doi.org/10.1007/s10450-019-00156-9.

[3] Ben Mbarek, W., et al. "Rapid Degradation of Azo-Dye Using MN–Al Powders Produced by Ball-Milling." *RSC Advances*, vol. 7, no. 21, 2017, pp. 12620–12628., https://doi.org/10.1039/c6ra28578c.

[4] Karadag, Dogan, et al. "Adsorption Equilibrium and Kinetics of Reactive Black 5 and Reactive Red 239 in Aqueous Solution onto Surfactant-Modified Zeolite." *Journal of Chemical andamp; Engineering Data*, vol. 52, no. 5, 2007, pp. 1615–1620., https://doi.org/10.1021/je7000057.

[5] Kotal, Moumita, et al. "Graphene-Templated Cobalt Nanoparticle Embedded Nitrogen-Doped Carbon Nanotubes for Efficient Visible-Light Photocatalysis." *Crystal Growth andamp; Design*, vol. 20, no. 7, 2020, pp. 4627–4639., https://doi.org/10.1021/acs.cgd.0c00430.

[6] Mengelizadeh, Nezamaddin and Pourzamani, Hamidreza, "Adsorption of Reactive Black 5 Dye from Aqueous Solutions by Carbon Nanotubes and its Electrochemical Regeneration Process", *Health Scope*, vol. 9, no 4, 2020, e102443, doi: 10.5812/jhealthscope.102443.

607a Novel Polyether-Based Amphoteric Ion Exchange Membrane for Electrochemical Systems

Gouree Kumbhar, Geetanjali Shukla, Robert C. Ferrier Jr., Caroline Szczepanski, David Hickey, Chase Bruggeman

Vanadium Redox Flow Batteries (VRFBs) are considered promising candidates for energy storage due to their high energy efficiency. Nafion ion exchange membranes have been used in VRFB applications owing to their good ionic conductivity and excellent chemical and mechanical stability, but Nafion's high cost, excessive swelling and low ion selectivity limits its use for commercialization. Amphoteric ion-exchange membranes have potential for preventing vanadium ion penetration thus increasing ion selectivity. In our current work, we have synthesized an economical crosslinked network to make amphoteric ion exchange membranes. Membranes were synthesized by grafting of epichlorohydrin and propargyl glycidyl ether-based charged copolymer (S-PPGE-ECH) to the PVDF-co-HFP membrane matrix. We varied the copolymer composition in the matrix to study its effect on ionic conductivity. All membranes were characterized using NMR and FTIR spectroscopy to confirm the presence of sulfonate and ammonium charged groups. Gradual increase of copolymer content decreases the contact angle from 85.6° (1% of S-PPGE-ECH) to 67.6° (10% of S-PPGE-ECH) indicating improvement in the hydrophilicity of membrane surfaces with addition of the copolymer. This increase in hydrophilicity resulted in a commensurate increase in swelling ratio from 17.5% to 27.0%, comparable to Nafion. We further saw a commensurate increase in ionic conductivity with an increase in copolymer content up to 0.64 mS/cm. Cation exchange capacity also increased steadily with increasing copolymer content while anion exchange capacity increased only slightly. The increase in exchange capacity was attributed to the presence of $-NR_3$ and-SO_3H groups in the membranes. In this work, we demonstrated control over ionic conductivities by tuning copolymer content without compromising thermal and mechanical stability of the membrane.

607b Synthesis and Performance Characterization of Zwitterionic Poly(arylene ether sulfone) Copolymers for Fouling-Resistant Desalination Membranes

Matthew D. Green, Husain Mithaiwala

Access to clean water resources has accelerated research in the direction of membrane-based separation processes that are often perplexed by acute and pervasive membrane fouling. The properties of the polymeric membranes deployed in these operations play a critical role in understanding the fouling and anti-fouling behavior. Bulk modification with zwitterions to polymeric backbone has been identified as an effective strategy to resist membrane fouling due to their impressive hydrophilicity and antifouling properties. This work demonstrates a reaction pathway to prepare an amphiphilic copolymer poly(arylene ether sulfone-co-sulfobetaine arylene ether sulfone) (PAES-co-SBAES) with controllable charge densities of sulfobetaine zwitterion and linear high molecular weights. The PAES-co-SBAES copolymers were then fabricated into dense membranes for their performance characterization with zeta potential, water contact angle and pervaporation measurements. The overarching goal of this work is to design and optimize the PAES-co-SBAES copolymer model and investigate its structure-property-performance relationship for a real multicomponent fouling and scaling environment.

607c Co-Transport of Lithium, Sodium, and Potassium Ions in Pegylated Sulfonated Polysulfones

Jung Min Kim, Sean Bannon, Patrick M. McCormack, Geoffrey Geise

Lithium (Li^+) is among the most valuable ions for energy storage devices. Li^+-rich geothermal brines are recognized as eco-friendly sources of lithium. However, the composition of these brines makes Li^+ recovery difficult. Electromembrane processes could contribute significantly as a component of a broader separation scheme to isolate Li^+ from complex solutions containing other ions. A major challenge facing this approach is to separate Li^+ from other monovalent ions, and Na^+ and K^+ are two prime examples. Sulfonated polysulfone (sPSf) is a promising material for cation exchange membranes with favorable thermal and mechanical properties. Poly(ethylene glycol) (PEG), a linear polymer containing ether oxygen, produces a strong dipole moment toward cations that hydrate strongly, suggesting favorable interactions with Li^+. Engineering Li^+ selective PEGylated sPSf by combining both of these materials, to leverage their favorable properties, could address challenges facing electromembrane-based monovalent ion separations. Ion transport properties were measured using both single salt measurements (i.e., LiCl, NaCl, or KCl) and measurements where mixtures of the cations were exposed to the membrane. The results inform both the Li^+ selectivity of the modified sPSf material with respect to Na^+ and K^+ and phenomena that occur when the membrane is challenged with mixtures of ions.

607d A Novel Solar Energy Driven Photoelectrodialysis Approach for Brackish Water Desalination

Syed Islam, Jackie Zheng, Nadeesha Lakmali Kothalawala, Prakhar Sharma, Ilia N. Ivanov, Ramesh Bhave, Doo Young Kim, Tomonori Saito, Priyesh Wagh

Although water and energy are the two most essential elements for life, the availability of clean water and energy are among the most important crises of our time. In recent years, a significant effort has been made to find energy efficient alternative water technologies. Among them, electrodialysis stands out with significant promise because of its tunability to varying degrees of salt removal, high water recovery and better fouling control for low salinity water. In this research, we have developed a novel energy efficient photoelectrodialysis process for desalination of brackish water using solar energy. A visible light active, nanostructured high surface area titania film, a photocatalyst that is inexpensive and highly stable in aqueous media is utilized to harness the sunlight, which functions as a photoanode. Ordered mesoporous titania films were synthesized using a scalable surfactant templated sol-gel process also known as evaporation induced self-assembly. To increase the light absorption efficiency and to increase the desalination capacity, a thicker film (1000 nm thickness) was fabricated using a layer-by-layer deposition technique.

To overcome its innate inability to absorb visible light due to its wide band gap, the visible light absorption of mesoporous TiO_2 films was enhanced by nitrogen doping using a plasma-based approach. The operating conditions of the plasma reactor including pressure, power, gas flow rate, and reaction time, are optimized to reduce the band gap of the mesoporous titania films from 3.5 eV to 2.88 eV while maintaining its pore structure. The nanostructure formation, pore accessibility, and visible light absorption, are some of the key features of titania films investigated by various advanced tools such as scanning electron microscopy, X-ray diffraction, UV-visible spectroscopy, X-ray photoelectron spectroscopy, ultra-fast transient absorption spectroscopy, and impedance spectroscopy.

These visible light active titania films are used in the photoelectrodialysis cell that is comprised of three compartments for photoanode, saline water and cathode. The saline water and photoanode are separated by an anion exchange membrane, and the saline water and the cathode are separated by a cation exchange membrane. Desalination of brackish water are performed using the photoelectrodialysis cell under the illumination of a solar simulator and LED lights. The effect of salt concentration in the brackish water from 3,000 to 15,000 ppm on the salt rejection and water recovery are studied. Our results suggest solar energy driven process can be energy efficient for brackish water desalination using solar energy.

607e Obtaining Structure-Function Insight from High-Throughput Membrane Characterization

Jonathan Ouimet, Laurianne Lair, Xinhong Liu, Elvis Eugene, Alexander Dowling, William Phillip

Creating systems and techniques capable of reducing the time, energy, and resources needed to characterize the transport properties of polymer membranes can help increase the rate of material and process development. For example, improved characterization techniques need to address the knowledge gaps related to the interfacial processes that govern solute-solute selectivity and capture the performance of membranes in complex multi-component feed streams. To address this need, a diafiltration apparatus is developed to rapidly characterize membrane performance over a broad range of feed solution compositions. By systematically dosing a high concentration diafiltrate into the stirred cell, a predetermined change in the retentate concentration can be achieved and materials can be characterized up to 12 times faster when compared to a traditional campaign of filtration experiments. The volume and reproducibility of the generated data enables data analytics to inform experimental design. For instance, using the Fischer information matrix, it was determined that parameter precision would increase by monitoring the real time retentate concentration. As such, an inline conductivity probe was incorporated in the experimental apparatus. The technique is validated by characterizing two commercial membranes (i.e., neutral NF90 membranes, charged NF270 membrane). Two primary conclusions result. (1) Incorporating the appropriate governing phenomena identifies a single set of self-consistent transport parameters. (2) Diafiltration experiments can establish relationships between membrane parameters (e.g., solute permeability coefficient) and system properties (e.g., rejection, bulk retentate concentration). We conclude by illustrating how diafiltration experiments can generate structure-function insight by comparing data obtained from functionalized copolymer membranes. Specifically, experiments capture changes to macroscopic system properties (i.e., rejection) that result from systematic modifications to material nanostructure and chemistry (e.g., changes in diamine tether lengths). As such diafiltration experiments can both accelerate material design and, coupled with data analytics, provide insight on governing transport mechanisms.

607f Separation of Urea from Water Solution Using Sulfonated Poly (styrene-isobutylene-styrene) Membranes with Counter-Ion Substitution

Juan Rivera-Diaz, David Suleiman

Water is of vital importance for space exploration. Currently, the International Space Station (ISS) uses a vacuum distillation system for the purification of water from urine; however, this method requires high energy and includes numerous mechanical components, so lower energy alternatives need to be considered. The objective of this work is to test the transport properties of urea throughout polymeric membranes based on sulfonated poly(styrene-isobutylene-styrene) (S-SIBS). The polymer was sulfonated using a sulfonation agent made from sulfuric acid; this functionalization gives the polymer the ability to form hydrophilic ionic nanochannels. The membranes were obtained by membrane casting technique to let the chemical domains thermodynamically self-assemble. The nanochannels of the membranes were chemically modified by the substitution of 12 different counter-ions; by submerging the membranes for 24 h in a 1 M solution of the counter-ion salt. The membranes were characterized by water uptake and swelling tests, FTIR, and TGA. The percentage of sulfonation was quantified by elemental analysis obtained at 66.48%. The presence of the sulfonic groups was verified by FTIR with the appearance of four additional bands (1151, 1124, 1034, 1007 cm^{-1}), representative of the stretching vibrations associated with the sulfonate group. Finally, the urea permeability of the membranes was estimated using the time-lag method in a free diffusion cell. The change in concentration during diffusion was monitored using FTIR; by the implementation of a calibration curve relating the intensity of a small peak nearby the range of 1150 to 1170 cm^{-1} associated with bonds, with urea concentration from 0.1 to 1 M. So far, we have found that there is a relationship between the oxidation state of the counter-ions and the water adsorption capacity of the membranes, affecting the transport properties. The incorporation of counter-ion substitution decreases urea transport. The urea permeabilities range from $1.42x10^{-4}$ to $6.93x10^{-6}$ for unsubstituted and Fe (III) substituted membrane respectively. Some of these membranes are outstanding candidates for the permselective separation of urea from water.

607g Stripping of Metal Ions for Ionic Liquid Recycling in a Self-Optimizing, Continuous Process

Bin Pan, Lanja R. Karadaghi, Richard L. Brutchey, Noah Malmstadt

Ionic liquids (ILs), as green alternatives to some traditional volatile organic compound solvents, are widely employed in a variety of organic reactions and inorganic nanoparticle syntheses; however, the high cost of ILs often hinders their extensive applications. It is essentially favorable to recycle the expensive IL solvent from post-reaction solution for the purpose of minimizing the loss of materials. Also, there are yet very limited studies on IL extraction and separation, especially in the flow. We have designed and implemented a continuous, microfluidic process where extraction to wash the loaded IL and successive phase separation happen.

As mentioned, ILs as solvents can be used in colloidal metal nanoparticle syntheses, which makes metal ions one of the main impurities in the solution. We examined Fe(III) as an impurity in BMIM-Tf2N, a common IL in nanoparticle syntheses, using water as the washing agent to strip Fe ions from the IL phase. In the process, two infusing streams, Fe-loaded IL and water, formed slug flow in a PTFE tubing from a T-junction to allow mass transfer between the two phases. The biphasic slug flow then entered an SLA-printed membrane separator to conduct phase separation, where the aqueous phase (wetting phase) was able to permeate through the hydrophilic nylon membrane while the IL phase (non-wetting phase) was retained. Realtime Fe concentration was monitored by inline spectrophotometry which detected the transmittance at 305 nm, the wavelength of Fe(III) absorbance. The transmittance represented the extraction performance, while separation performance was evaluated by a pair of IR emitters and detectors that could distinguish different slugs in the flow.

To screen the experimental space, we utilized Design of Experiment (DoE) statistics approach. We also defined a parameter "score" to mirror the overall performance of the process as the output of a response function. The resulting fitting response function revealed where high scores resided in the experimental space. We also used a realtime, self-optimizing simplex algorithm to locate the optimal flow rate conditions for high scores with good extraction and perfect separation.

Another kind of IL, Cyphos 104, was applied to the as-established extraction-separation process as an application. Neodymium, one of the rare-earth metals, was dissolved in water first, then was transferred to the Cyphos IL by batch extraction and phase separation. Nd(III)-loaded IL was then washed with 1M nitric acid in flow using a similar configuration as the BMIM-Tf2N system. Nd was stripped off from the IL phase under the optimal flow rate combinations discovered by simplex.

703a Development of Sodium-Borate Adsorbent for CO_2 capture at High Temperature from Flue Gas Stream

Turki Alghamdi, Qasim Al-Naddaf, Fateme Rezaei, Ahmed Al-mamoori

Abstract

Increasing CO_2 emissions to the atmosphere is critical due to the global warming issue. Developing CO_2 adsorbent at high temperature is of paramount importance as its widespread application. In this investigation, Sodium based Borate adsorbent has been developed for CO_2 capture process. Four different Sodium precursors have been utilized, mixed separately with Boric acid, and tested for CO_2 Adsorption simulated the flue gas stream condition. The materials have been tested under temperature range 550-700 ˚C and showed promising results especially between 650- 700 ˚C. The optimum CO_2 uptake for the proposed materials have shown as much as ~5.5 mmol/g. For further investigation, cyclic Adsorption-Desorption process has been tested at 650 and 7000 ˚C for Adsorption and 700 ˚C for Desorption. The materials showed high stability after eight consecutive Adsorption-Desorption cycle. Also, BET surface area, XRD, and SEM techniques have been performed for the proposed materials to study their physio-chemical properties. Such interesting materials that presented high CO_2 capture uptake, fast kinetic, and high stability materials that could be promising candidates for post combustion process.

703b New Adsorbent for Post-Combustion CO_2 Capture: Modification of 13X Zeolite By Molecular Layer Deposition

James A. Ritter, Armin Ebner, Huan Jiang, Charles E. Holland, Ryan Sanders, Richard Ciora, Dong Qiaobei, Brian Sengupta, Miao Yu

The objective of this multidisciplinary project is to develop a transformational molecular layer deposition (MLD) tailor-made sorbent integrated with a tailored pressure swing adsorption (PSA) cycle schedule that can be installed in new or retrofitted into existing pulverized coal (PC) power plants for CO_2 capture with a cost of electricity lower than a supercritical PC with CO_2 capture. The proposed MLD technology builds on pioneering work on precisely altering the properties of an adsorbent by pore misalignment outside the sorbent cavity or cage or by simply coating the interval cavity or cage, or both. The optimized sorbents to date are showing CO_2/N_2 selectivity as high as 130, much higher than state-of-the-art commercial sorbents like 13X zeolite with similar CO_2 capacity under similar adsorption conditions. This perhaps enables the economical achievement of 95% CO_2 purity with 90% CO_2 recovery using a PSA cycle schedule tailored to the MLD adsorbent. The project team is comprised of researchers from UB, RPI, UofSC, Gas Technology Institute (GTI), Trimeric Corporation (Trimeric), and the National Carbon Capture Center (NCCC). The program uses each team member's unique expertise (UB/RPI: sorbent development; UofSC: innovational PSA process integration with sorbents; GTI: field testing processing and engineering; Trimeric: TEA) which is critical to developing the transformational and revolutionary MLD tailor-made sorbent/PSA technology for bench-scale testing, and to achieve DOE cost and performance targets. This presentation will present some of the latest results from this project, including MLD sorbent development, characterization and testing, and/or PSA cycle development and testing, all compared to commercial 13X zeolite. One of the key features of this MLD sorbent is that it is apparently more tolerant to water vapor then commercial 13X zeolite, making it perhaps less energy intensive for capturing CO_2 from a PC power plant.

703c Direct Air Capture (DAC) of CO_2 Using a Tailored Chelating Exchanger at Ambient Temperature

Arup SenGupta

Climate change is considered to be the worst existential threat confronting humanity today and, as always is the case, the suffering of marginalized populations is disproportionately high. An appropriate CO_2 capture technology that engages the public at the grassroots level can be a transformative driver and an equalizer. We, at Lehigh University, have developed a process to capture CO_2 directly from the atmosphere at any location and store as pure CO_2. The proposed technology: (i) is amenable to deployment in local communities including schools, churches, NGOs, and wastewater plants; and (ii) operates at ambient temperature without requiring steam or waste heat. Besides solar driven electricity for electrolysis, no other form of energy is required.

The fundamental scientific tenet of the proposed Direct Air Capture (DAC) of CO_2 stems from the unique weak acid-weak base (WAWB) properties of CO_2 and its interaction with a new class of tailored chelating ion exchangers. When a transition metal cation, namely Cu(II), is covalently attached to a chelating polymer with nitrogen donor atoms (polyamine) through Lewis Acid-Base (LAB) interaction, its two positive charges are not neutralized. The resulting material, referred to as PolyLAB-Cu^{2+}, is essentially an anion exchanger with a very high affinity for carbonate or CO_3^{2-} through both electrostatic and concurrent Lewis acid-base (LAB) interaction. Sorption and desorption processes occur as follows:

CO_2 Sorption and Desorption: Use of OH^- as the counter-ion for PolyLAB-Cu^{2+} leads to irreversible uptake of CO_2 from the atmosphere in the following sequence: i) dissolution of CO_2 in the moisture/humidity at the sorbent interface; ii) transport of non-ionized H_2CO_3 inside the ion exchanger; and iii) rapid neutralization followed by selective binding of CO_3^{2} onto Cu(II). Note that it is a three phase (air-water-exchanger) process where the presence of humidity or moisture in between air and the exchanger facilitates dissolution of CO_2 followed by selective uptake.

Upon exhaustion, passage of 3% Na_2SO_4 at near-neutral pH causes desorption of CO_3^{2-} leading to an exit stream with high alkalinity which again is passed through a weak-acid cation (WAC) exchanger to recover pure CO_2. PolyLAB-Cu^{2+} is subsequently converted into hydroxyl or OH^- form through passage NaOH. Also, the WAC is regenerated with dilute 2-3% H_2SO_4.

Both NaOH and H_2SO_4 are generated through electrolysis of Na_2SO_4 which again is reproduced in the process, thus avoiding external addition of chemicals. Thus, it is essentially a cyclic two-step self-sustaining process where CO_2 is first captured in the first step from the ultra-dilute ambient air and then recovered in the second step as pure CO_2. The entire sorption-desorption cycle operates at ambient temperature and besides electricity for the electrolysis cell, no other chemicals or energy is needed.

Experimental Results: We evaluated various polymers with nitrogen groups that are commercially available and then identified one chelating polymer that is durable, chemically stable and possesses high affinity toward Cu(II). During the last two years, we have carried out an extensive number of laboratory experiments and three key significant findings which make the process quite distinctive are presented below:

5. Under ambient conditions, PolyLAB-Cu^{2+} offered a sorption capacity of 5.1 M CO_2/g dry resin and that is 2.5-3.0 times greater than the capacity of other materials currently in use. Also, these flow-through high capacities are obtained for an empty bed contact time (EBCT) of only 1.0 second.
6. One major downside of direct air capture is the ultra-low concentration of CO_2 in the ambient air (approximately 0.04% or 400 ppm by volume). Understandably, for increased uptake, point-source flue gas streams with higher CO_2 concentrations have been desirable targets. Although counter-intuitive, PolyLAB-Cu^{2+} offers identical CO_2 capture capacity at ambient conditions, with 10% CO_2 and 50% CO_2, the rest being nitrogen. The scientific explanation is being avoided here for the sake of brevity but the observation is very significant from an application viewpoint.

- For flow-through systems, five consecutive runs were carried out with regenerations in between. Capacities did not change or deteriorate. Dilute NaOH and H_2SO_4 produced from the electrolysis cell were used for regeneration and Na_2SO_4 is reproduced at the end of the process and reused.

Path Forward: The proposed process operates at ambient temperature and does not need any thermal energy or waste heat. Consequently, the process is more amenable to implementation around the world including developing countries and marginalized communities with appropriate carbon credit. The work is under progress to put the electrolysis cell in tandem with solar electric panels to make the process a true Negative Emission Technology (NET). Also, the scale-up of the process is relatively straightforward and a skid-mounted prototype is currently under constructio

703d Entrapment of Small Amines in Mesoporous Silica Via Polymeric Coating for the Direct Air Capture of CO_2

Kaleb Friedman, Shailesh Dangwal, Miao Yu

Since the industrial revolution, the concentration of CO_2 in the atmosphere has been continuously rising, resulting in considerable increase in earth temperature. Due to continuous increase in CO_2 emissions, the target of restricting "global warming to 1.5°C" by the Intergovernmental Panel on Climate Change (IPCC) can only be met by negative CO_2 emissions techniques, such as the direct air capture (DAC) of CO_2. Therefore, significant research has been focused on capturing CO_2 from air. Most of the researchers have focused on polymeric amine-based sorbents to ensure thermal stability during temperature swing adsorption. However, it compromises the efficiency of the sorbent material and thus CO_2 uptake capacity. A significant amount of research has been focused on countering this challenge of tradeoff between the thermal stability and CO_2 capacity of sorbent materials. In this work, polymeric coatings were used to seal small amines in a mesoporous silica in order to enhance the thermal stability without compromising CO_2 uptake capacity. Different polymeric coatings, including polyamide and Pebax, were used, which allows for CO_2 diffusion without significant resistance through the coating material during adsorption-desorption steps but prohibits leakage of amines, ensuring long-term thermal stability of the sorbent material. For 10 adsorption-desorption cycles, CO_2 capacity loss was reduced from 6.9% for the base sorbent without a coating to negligible loss for the polyamide coated sorbent.

703e Amine-Impregnated Al_2O_3 Materials for the Direct Air Capture of CO_2 Under Sub-Ambient Conditions

Pranjali Priyadarshini, Guanhe Rim, Cornelia Rosu, Mingyu Song, Fanhe Kong, Ryan P. Lively, Christopher W. Jones

Rising CO_2 emissions due to anthropogenic activities are largely responsible for increasing global temperatures leading to climate change. Capture of CO_2 from air (called direct air capture (DAC)) and its subsequent geological sequestration can be a potential solution to combat the devastating effects of climate change. While significant efforts have been made to develop materials to adsorb CO_2 from air, most of the studies have conducted experiments at ambient temperatures (25 °C) or higher under dry conditions. A significant portion of the natural environment where direct air capture plants can be deployed experience temperatures below 25 °C. In this study, we show the CO_2 adsorption behavior of amine impregnated alumina materials, which are leading materials for possible DAC deployment at warm temperatures, at sub-ambient temperatures.

CO_2 adsorption capacities at 25 °C and 400 ppm were highest for TEPA impregnated alumina samples (2.1 mmol/g), and moderate for PEI-alumina (0.9 mmol/g). The CO_2 capacities for both PEI-, and TEPA-alumina samples decreased at -20 °C (0.45 mmol/g, and 1.1 mmol/g, respectively). The loss of CO_2 uptake at low temperatures was primarily due to the freezing of the chains and their subsequent lower mobility that limits the diffusion of CO_2. TEPA impregnated alumina samples showed a promising working capacity of 0.8 mmol/g after 8 cycles of adsorption-desorption (adsorption at -20 °C and desorption conducted at 60 °C). For both PEI and TEPA samples, the CO_2 was strongly adsorbed at both 25 °C and -20 °C. Introducing moisture (70% RH at -20 °C) improved the CO_2 capacity of PEI impregnated alumina to ~0.8 mmol/g. Reducing the amine content from 40% to 20% decreased the CO_2 capacities at 25 °C as expected (0.6 mmol/g for PEI and 1.6 mmol/g for TEPA). While the CO_2 capacities at -20 °C decreased for both samples (0.43 mmol/g for PEI and 1.1 mmol/g for TEPA), these values are equal to those observed at -20 °C for the 40% amine samples, indicating that there is better amine utilization at slightly lower loadings due to better diffusion of CO_2 along the pores, which may arise due to greater volume available for CO_2 diffusion at low temperatures. These results indicate that alumina impregnated with amines are a potentially promising class of materials for direct air capture at sub-ambient conditions, opening up a variety of opportunities to optimize these materials for the scalable deployment of DAC plants at different environmental conditions.

703f Improving the Hydrothermal Stability of Amine-Grafted MCM-41 Silica Via Incorporation of Aluminum into the Structure of the Support

Masoud Jahandar Lashaki, Hessam Ziaei, Abdelhamid Sayari

In recent years, steam regeneration of CO_2 adsorbents has gained momentum, and is currently used in the near-commercial CO_2 capture technology developed by Svante. Steam acts as both the heat source and the purge gas and is particularly suitable for CO_2 capture in industrial plants where low-heat steam is typically available as waste heat. Steam regeneration is advantageous since it is relatively easy to recover pure CO_2 via steam condensation, following the desorption step. While steam regeneration is conceptually interesting, the exposure of amine-modified silica materials to steam lowers their performance and stability owing to amine leaching, structural degradation of the support, and/or amine deactivation due to certain metal impurities in industrial steam. These shortcomings indicate the need for developing amine-modified silica materials with higher hydrothermal stability. Ordered mesoporous silica materials such as MCM-41 and pore-expanded MCM-41 (PE-MCM-41) possess unique combination of high surface area, tunable pore size with narrow distribution, and high pore volume. However, these materials typically suffer from low hydrothermal stability, limiting their use in cyclic CO_2 adsorption-desorption processes involving steam regeneration. Previous reports indicated that aluminum incorporation into silica framework improves its hydrothermal stability. Despite their high potential, however, reports on the use of amine-modified Al-MCM-41 in cyclic CO_2 adsorption-desorption processes involving steam regeneration remain extremely limited, warranting further investigation. To that end, the impact of aluminum incorporation into the structure of MCM-41 silica on its adsorptive properties and hydrothermal stability during CO_2 capture was studied. Large-pore Si-MCM-41 and Al-MCM-41 (Si/Al = 15) supports with similar pore sizes were synthesized using a novel one-pot synthesis technique developed by our group. Amine functionalization of the supports was conducted using dry and wet grafting of triamine. Amine-tethered Al-MCM-41 adsorbents exhibited higher CO_2 uptakes and faster CO_2 adsorption kinetics relative to their Si-MCM-41 counterparts. Following one three-h steam stripping cycle, significant deterioration of adsorptive properties was observed for Si-MCM-41, particularly at adsorption temperatures of 25 and 50°C. Al-MCM-41, however, showed unprecedented hydrothermal stability with no significant drop in CO_2 uptake and adsorption kinetics. Following steaming, N_2 adsorption measurements at 77 K showed significantly higher reductions in the BET surface area of grafted Si-MCM-41 samples compared to their Al-MCM-41 counterparts. It is proposed that aluminum incorporation into the structure of the support may reduce the restructuring of the grafted materials upon exposure to steam, enhancing their hydrothermal stability.

703g Study of Carbon Mineralization at the Surface of Mg- and Ca-Based Oxides

Colin Lehman-Chong, Ella M. Siefken, Aleksandra Vojvodic

Decarbonization is necessary for us to mitigate the worst effects of climate change. Capture and sequestration of CO_2 will be one part of this multifaceted strategy. One promising route for CO_2 capture uses mine waste rocks containing alkaline Mg and Ca cations to passively capture CO_2 directly from the air and convert it to carbonates. This process is inherently cost effective as the capture does not require blowers or expensive adsorbent materials [1]; however, most widely available waste minerals are kinetically limited in terms of carbonation efficiency. To gain a fundamental understanding of the surface phenomena necessary for improving the kinetics of this carbonation process we use density functional theory (DFT) and ab initio steered molecular dynamics (AISMD). We progress from calculations of bulk mineral oxide thermodynamic properties to calculations of thermodynamics and dynamics in the presence of explicit water on low Miller index surfaces of MgO, CaO, brucite ($Mg(OH)_2$), and more complex silicate minerals. We show that differences in CO_2 adsorption energy under different water surface coverages can be used to explain experimental carbonation activity trends across different materials. Furthermore, AISMD simulations combined with umbrella sampling techniques provide us with results to explain and predict the rates of dissolution processes occurring at these surface-water interfaces. These surface-mediated dissolution processes are a necessary step for carbonation to proceed beyond the surface and into the mineral bulk; therefore, these results are critical for unveiling the chemical fingerprints of reactive mineral surfaces. We demonstrate agreement between calculation and literature experimental material trends and illustrate the difficulties inherent to using bulk descriptors for the prediction of material activity. Furthermore, strong dissolution facet dependence is shown for more complex silicate minerals, suggesting that the mechanisms of carbonation on these silicate minerals are hindered by low reactivity facets being preferentially exposed to air. Methods for improving the reactivity of these more widespread (but less reactive) minerals are proposed based on these mechanistic findings.

References:

1. Hasan, S. N. M. S. andamp; Kusin, F. M. Potential of Mining Waste from Metallic Mineral Industry for Carbon Sequestration. *IOP Conf. Ser. Mater. Sci. Eng.* **458**, (2018).

703h Moisture-Controlled CO_2 Capture and Electrochemical CO_2 Capture

Yuta Kaneko, Matthew D. Green, Klaus S. Lackner

- **Introduction: Two novel approaches for CO_2 capture from air**

Moisture-controlled CO_2 sorption is a novel phenomenon observed in strong-base anion exchange materials (AEMs) with quaternary ammonium cations, so-called moisture-swing materials[1,2]. These materials can sorb and desorb very diluted CO_2 even from the ambient air (~ 400 ppm) by just changing the humidity. It can reduce the cost of Direct Air Capture (DAC) of CO_2 that is necessary to limit global warming to 1.5°C[3]. Moisture-swing for DAC applications have been first introduced by Lackner's group[1,2].

Recently, electrochemical CO_2 capture has also been investigated for CO_2 capture both from flue gas and from air[4,5]. Moisture-controlled and electrochemical CO_2 capture have much in common but differ significantly from temperature-swing or pressure-swing CO_2 capture. For example, their mass transport is driven by potential gradients either in chemical activity or electric fields. Both can perform a loading and unloading swing at constant temperature without introduction of a high-grade heat.

In this research, we apply electrochemical techniques to moisture-controlled CO_2 sorbents to investigate their kinetics. We compare the results of the analyses for these two novel sorbent systems to discuss their relevance in a unified way.

- **Research: Formulate electrochemical analysis for moisture-controlled CO_2 sorption**

In electrochemistry, cyclic voltammetry (CV) is a common technique to investigate the reduction and oxidation processes or electron transfer-initiated chemical reactions[6]. The output current from the analyte is plotted against the linearly increased and decreased voltage input. CV has also been used to characterize electro-swing sorbents for CO_2 capture[4]. The Randles–Sevcik equation can be used to differentiate diffusion-dominated and reaction kinetics-dominated hysteresis by analyzing the functional dependence of the peak current I_p on sweep speed v. If adsorption is the rate determining step, I_p is expected to increase linearly with v. In the other case, I_p scales square root of v, and we can extract the diffusion coefficient of the analyte. We extend this methodology to moisture-controlled CO_2 sorbents and develop theoretical and experimental approaches to such systems.

  - Theory

For moisture-controlled CO_2 sorbents, we refer to the plot corresponding to the CV for electro-swing sorption as CWA (cyclic water activity measurement). The input (x-axis of a CWA plot) should be humidity instead of voltage and CO_2 flux should be used as output (y-axis of a CWA plot) instead of electrical current.

The attached figure shows a theoretically predicted shape of the CWA plot. The main difference from the CV is that diffusive transport of water inside a sorbent precedes that of Dissolved Organic Carbon, which generates the time lag that is not seen in the CV plot. Also, we take all the other differences into consideration to formulate the theoretical CWA plots and establish the protocol to extract kinetic parameters such as ion diffusivities from the plot. Furthermore, we aim to use this plot to differentiate chemical reaction kinetics at surfaces and diffusive transport kinetics inside a sorbent.

- Experiment

We will obtain a CWA plot using a preconditioned commercial anion exchange material. An open-flow experiment will be deployed, where the input humidity can be controlled using mass flow controllers. The CO_2 concentration in the inlet gas and outlet gas will be measured by Infrared Gas Analyzers located before and after the sample chamber.

- **Expected outcome**

We will apply an electrochemical technique to moisture-controlled CO_2 sorbents. It will provide a protocol to analyze the kinetics of moisture-controlled CO_2 sorbents in a quantitative way. Also, it enables us to compare the kinetic behaviors of the two novel CO_2 capture methodologies using the conceptually same plots.

References:

1. *Environ. Sci. Technol.* 2011, 45, 6670–6675
2. *Phys. Chem. Chem. Phys.*, 2013, 15, 504
3. V. Masson-Delmotte, et al., Cambridge University Press, 2021
4. *Energy Environ. Sci.*, 2019, 12, 3530
5. *ACS Catal.* 2020, 10, 13058–13074
6. *J. Chem. Educ.* 2018, 95, 197–206

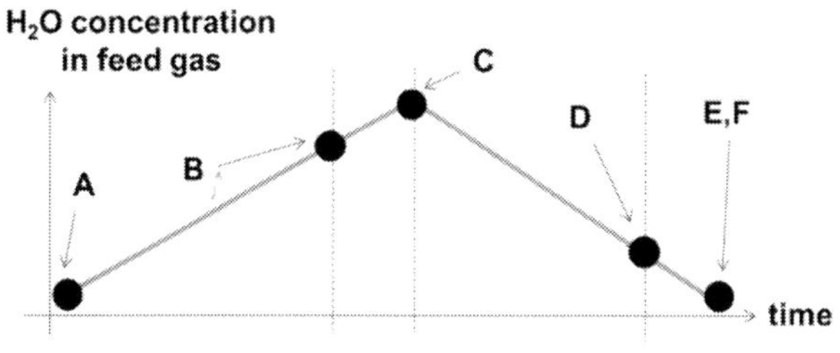

Linear sweep of input humidity level

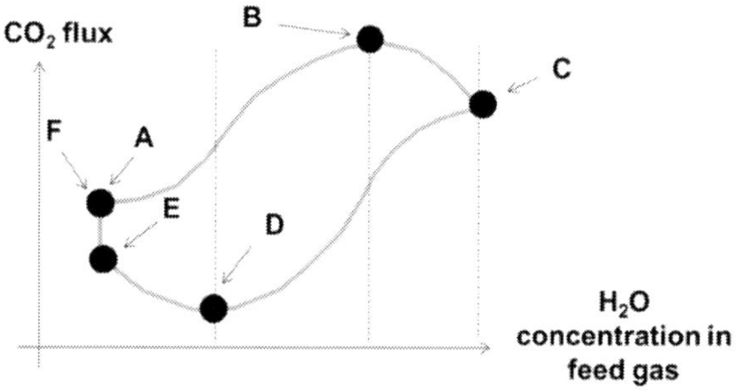

Theoretical CWA plot

703i Process-Based Screening of MOFs for Direct Air Capture (DAC)

Bhubesh Murugappan Balasubramaniam, Ohmin Kwon, Jun Luo, Marco Gibaldi, Tom Woo, Phuc-Tien Thierry, Samuel Lethier, Philip Llewellyn, Cecile Pereira, Veronique Pugnet, Arvind Rajendran

DAC involves separation of CO2 from N2, O2 and H2O in high purities for sequestration. For DAC, a process needs to be developed such that the energy consumed is minimum and the productivity is at its maximum with a constraint on the purity of CO2 to be > 95%. In this work, we consider a temperature – vacuum swing adsorption (TVSA) process for DAC. We consider an amine-based sorbent (APDES – NFC) as a benchmark material. The experimental loadings were fitted to a Toth isotherm model. Detailed energy – productivity optimizations were performed after accounting for H2O sorption. On the other hand, the computational ready experimental MOF (CoRE MOF) database was screened for potential candidates for DAC. CO2 and N2 isotherms were obtained for the CoRE MOFs using GCMC simulations and the isotherms were fit to SSL isotherm models. At 400 ppm, materials which exhibited CO2 pure component capacities greater than 0.3 mmol/g were only chosen for energy – productivity optimizations since capacities lesser than 0.3 mmol/g would not be suitable for DAC under CO2/N2 competitive conditions. For the initial cut, energy – productivity optimizations on CoRE MOFs were performed under dry conditions. The machine-assisted adsorption learner and emulator (MAPLE) framework, an artificial neural network (ANN) platform for rapid simulation of adsorption processes for deployed. The process optimization results and the comparison between different solid sorbents would be presented at the conference.

521a Development of Slug Flow Crystallization for Reactive Crystallization

Mingyao Mou, Anna Berberoglu, Consuelo Del Pilar Vega-Zambrano, Huayu Li, Bing-Shiou Yang, Mo Jiang

In pharmaceutical and chemical industries, many intermediate and essential ingredients need to go through the crystalline phase.[1] One key quality aspect for crystals is morphology, which affects not only product quality, but also process efficiency.[1–4] However, the crystal morphology from reaction synthesis is challenging to control, even with advanced continuous crystallizer designs that already improve quality consistency, controllability, and process flexibility. Also, the process and mechanism for morphology generation are not always clear.

This presentation will elaborate on the development slug flow process and morphology tuning process based on the reactive crystallization of L-Glutamic acid, a widely used amino acid. Specifically, the effect of reaction parameters on crystal morphology is evaluated, including supersaturation, temperature and flow rate. The slug flow configuration is selected, as one of the advanced tubular continuous crystallizers for a uniform reaction environment[2–4]. The crystal growth mechanism is compared between slug flow and traditional batch reactors.

521b Limitations and Optimization Potential of Continuous Chiral Resolution By Simultaneous Preferential Crystallization in Fluidized Bed Crystallizers

Jonathan Gänsch, Nadiia Huskova, Heike Lorenz, Andreas Seidel-Morgenstern

Enantiomers are optical isomers, which behave differently in living organisms but possess identical physical and chemical properties. Consequently, their separation is very important for the pharmaceutical and other life-science industries but at the same time very sophisticated. Preferential Crystallization is an efficient and cost-effective separation technique for chiral resolution, since it allows the direct crystallization of only one desired enantiomer without any chiral auxiliary from a racemic (50:50) solution. However, the inevitable nucleation of the unwanted enantiomer complicates the design of continuous Preferential Crystallization. A continuous and robust process can be achieved by applying Simultaneous Preferential Crystallization in conical-shaped fluidized bed crystallizers [1, 2]. In contrast to the well-studied continuous MSMPR (mixed suspension mixed product removal) concept, the application of fluidized bed crystallizers enables a narrow and adjustable product crystal size distribution, the selective removal of crystal fines and a short and narrow residence time distribution of the mother liquor. The two last mentioned effects support that this process can be operated robust against the nucleation of the unwanted enantiomer. Hence, higher supersaturations can be applied to exceed the current productivity benchmark for continuous crystallization processes for chiral resolution of around 20 g/L/h [3].

A remarkable high productivity of up to 100 g/L/h as well as the adjustable and narrow product crystal size distribution were recently demonstrated within a comprehensive experimental parameter study for the chiral resolution of the amino acid DL-asparagine monohydrate in water [4, 5]. Current efforts aim to study the limitations as well as the optimization potential of this auspicious process. Regarding the first goal, the process is tested for more challenging separation problems. A recent achievement was the successful chiral resolution of the API (RS)-guaifenesin in isopropanol [6], a compound characterized by particular slow crystal growth of needle-like crystals. To achieve the second goal, a reduced process model was developed and validated [5, 7]. This process model is currently applied for the model-based optimization of the operation parameters supersaturation, crystallization temperature, feed flowrate and product removal strategy, as well as for the optimization of the geometrical parameters of the conically shaped fluidized bed crystallizers.

In this contribution, the process as well as the utilized pilot plant will be explained and the good agreement between the experimental results and the process model will be demonstrated. Afterwards, the results of the model-based optimization of the above-mentioned operating parameters and geometrical parameters as well as their experimental validation will be shown and discussed for the substance system DL-asparagine monohydrate in water. The process performance of the separation experiments is evaluated by the productivity, the purity and the product crystal size distribution.

References:

[1] H. Lorenz, E. Temmel and A. Seidel-Morgenstern: **Continuous enantioselective crystallization of chiral compounds.** In: N. Yazdanpanah and Z. K. Nagy (Eds.): The Handbook of Continuous Crystallization, *The Royal Society of Chemistry*, 2020, pp. 422–468.

[2] D. Binev, A. Seidel-Morgenstern and H. Lorenz (2016): **Continuous Separation of Isomers in Fluidized Bed Crystallizers**, *Cryst. Growth Des.* 16, 1409-1419.

[3] T. Köllges and T. Vetter (2018): **Design and Performance Assessment of Continuous Crystallization Processes Resolving Racemic Conglomerates**, *Cryst. Growth Des.* 18, 1686–1696.

[4] E. Temmel, J. Gänsch, H. Lorenz and A. Seidel-Morgenstern (2020): **Systematic Investigations on Continuous Fluidized Bed Crystallization for Chiral Separation**, *Crystals* 10, 394-409.

[5] J. Gänsch, N. Huskova, K. Kerst, E. Temmel, H. Lorenz, M. Mangold, G. Janiga and A. Seidel-Morgenstern (2021): **Continuous Enantioselective Crystallization of Chiral Compounds in Coupled Fluidized Beds**, *Chem. Eng. J.* 422, 129627.

[6] F. Cascella, J. Gänsch, A. Seidel-Morgenstern and H. Lorenz (2022): **Continuous enantioseparation of the API Guaifenesin in a fluidized bed crystallizer**, manuscript in preparation.

[7] M. Mangold, L. Feng, D. Khlopov, S. Palis, P. Benner, D. Binev and A. Seidel-Morgenstern (2015): **Nonlinear model reduction of a continuous fluidized bed crystallizer**, *J. Comput. Appl. Math.* 289, 253-266.

Continuous API Salt Formation *via* Sono-MSMPR Crystallization

Patrick T. Corona, Benjamin M. Cohen

Chemical Process Development

Bristol Myers Squibb Company, NJ, USA

Processes enabling the continuous crystallization of active pharmaceutical ingredients (API) represent attractive means to produce large quantities of crystalline product with reduced costs and improved control of particle size distributions (PSD). Mixed Suspension Mixed Product Removal (MSMPR) represents a common continuous crystallization approach, but process yield and startup times are intrinsically limited by the solubility and desaturation rate at operating conditions. Many APIs are formulated in a designated salt form to enable optimal physical properties such as bioavailability and stability. Salt formation *via* reactive crystallization is commonly used to produce API. Reactive crystallizations occur under conditions with low API solubility and fast desaturation, making them ideally suited for MSMPR crystallizations. However, such processes typically suffer from agglomeration or oiling. In this work, we present studies probing API salt formation in an MSMPR crystallization. For this work, we probe a model salt formation reactive crystallization via methanesulfonic acid addition. In both batch and MSMPR crystallizations, agglomeration is observed during the API crystallization, which results in large particles. To prevent the formation of agglomerates, *in situ* ultrasonication is applied during MSMPR operation, which enables the continuous production of de-agglomerated API crystals. Process parameters (temperature and turnover rate) are adjusted to determine the impact on API particle size. These experiments demonstrate the mutually beneficial nature of sono-MSMPR crystallizations and reactive salt formations.

521d Slug Flow Reactor Simulation for Controllable Residence Time Distribution Using Computational Fluid Dynamics.

Chaeeun Lee, Mingyao Mou, Mo Jiang, Jonggeol Na

In the pharmaceutical industry, the process has traditionally operated in batch type.[1] However, the batch process is not easy to control, which can be challenging to achieve product consistency. Also, batch crystallization has been existed the problems of inefficiency during the process and batch-to-batch variability.[2] Compared to the batch crystallizer, continuous crystallization processes have been expected to improve product quality consistency and manufacturing process efficiency as well as lower capital and production costs.[3-5] Accordingly, many continuous crystallizer designs have been suggested with the aim of controlling the product specification and generating uniform crystals.[6] Among them, slug flow crystallizer does not require additional equipment (e.g., mixer) and reduces common issues (e.g., fouling, clogging) in tubular reactors. For scaling up, the slug flow crystallizer only needs to increase the number of slugs by using longer tubes with the same equipment.[6] Many studies of residence time distribution (RTD) have been conducted on liquid-liquid and liquid-gas slug flow according to the slug size and flow rate.[7,8] However, there are few studies on the effect of RTD due to diffusion through the liquid film, which is also generated depending on the surface tension of the liquid used.

Here, we implemented a gas-liquid slug flow with thin liquid film using computational fluid dynamics (CFD) simulation with multi-phase consideration. Through simulation, the liquid film phenomenon that is generated in a solvent with low surface tension and affects RTD was also considered. Our CFD model was applied the Volume of Fluid (VOF) method that is applicable to form an interface between multiphase flows. The proposed CFD model was verified under the same conditions as the actual experiment. For the reactor, a silicone tubing (Masferflex Transfer Tubing, C-Flex) with an inner diameter of 2.38 mm was used. To capture the moving slugs for validation, a digital single-lens reflex (DSLR) camera (Nikon D5600) was placed 10 cm after the T-junction. CFD model predicted the shape of slug and liquid film accurately. We first discussed the effects of slug flow and liquid film on RTD. Obviously, the slug flow had a narrower RTD than the non-slug flow (one phase laminar flow), and when the thin liquid film is present in the slug flow, it can be confirmed that the RTD spreads slightly due to the diffusion phenomenon through the thin film between each slug. In addition, according to the difference in surface tension, RTD sensitivity analysis was carried out by setting the flow rate as variables for isopropanol and water. The higher the flow rate, the smaller the variance of RTD, but the slug formation was unstable and broken. This phenomenon was observed in the isopropanol solvent in which the film was formed rather than in the water in which the film was not formed, along with the uncertainty of the RTD. Accordingly, we finally performed a comparison of RTD variability according to solvents with various surface tensions. As a result, it was found that the lower the surface tension, the greater the diffusion by the film, and the larger the RTD variance. Through this study, we analyzed the slug flow crystallizer as RTD using CFD simulation. In order to be possible, the controllable RTD, which is considered a benefit of the slug flow, each slug should act as an independent batch rector. The thin film should be minimized by using a solvent with the large surface tension as possible and that an appropriate flow rate should be used for stable slug formation.

[1] Paul, E. L.; Tung, H.-H.; Midler, M. Organic Crystallization Processes. Powder Technol. 2005, 150, 133–143.

Zhang, D., Xu, S., Du, S., Wang, J., andamp; Gong, J. (2017). Progress of pharmaceutical continuous crystallization. *Engineering*, *3*(3), 354-364.

[3] Alvarez, A. J.; Myerson, A. S. Continuous Plug Flow Crystallization of Pharmaceutical Compounds. Cryst. Growth Des. 2010, 10 (5), 2219−2228.

[4] Lawton, S.; Steele, G.; Shering, P.; Zhao, L.; Laird, I.; Ni, X. Continuous Crystallization of Pharmaceuticals using a Continuous Oscillatory Baffled Crystallizer. Org. Process Res. Dev. 2009, 13 (6), 1357−1363.

[5] Ferguson, S.; Morris, G.; Hao, H.; Barrett, M.; Glennon, B. InSitu Monitoring and Characterization of Plug Flow Crystallizers. Chem. Eng. Sci. 2012, 77, 105−111.

[6] Jiang, Mo, and Richard D. Braatz. 'Designs of continuous-flow pharmaceutical crystallizers: developments and practice.' *CrystEngComm* 21.23 (2019): 3534-3551.

[7] Qian, D., andamp; Lawal, A. (2006). Numerical study on gas and liquid slugs for Taylor flow in a T-junction microchannel. *Chemical Engineering Science*, *61*(23), 7609-7625.

[8] Günther, A., Khan, S. A., Thiemann, M., Trachsel, F., andamp; Jensen, K. F. (2004). Transport and reaction in microscale segmented gas–liquid flow. *Lab on a Chip*, *4*(4), 278-286.

521e Continuous Crystallization Process Development of Modafinil: A Step Towards Achieving Integrated Continuous Manufacturing

Shailesh Agrawal, Jean-Christophe Monbaliu, Cornelis Vlaar, Jorge Duconge, Vilmalí López-Mejías, Torsten Stelzer

Although integrated continuous manufacturing (CM) of drug products is already a reality with multiple commercial products available[1], the integrated CM of drug substances (molecule synthesize and purification by crystallization) is generally lacking behind. Recently, we have reported a highly efficient flow synthesis of modafinil[2], a drug used to treat excessive daytime sleepiness associated with narcolepsy, obstructive sleep apnea, or shift work disorder.[3] Modafinil is of particular interest to military and space agencies[4]. This study deals with the purification of modafinil by continuous crystallization (CC) with the ultimate goal to achieve an integrated CM of modafinil. In phase I of the CC development commercially available modafinil was used for (anti-)solvent screening (Crystal16™) including polymorph screening and solid-state characterization (PXRD, DSC, TGA, optical microscopy). All solvents (neat and binary mixtures), except isopropanol, yielded the commercial form I polymorph. Batch trials (50 mL scale) with the most promising solvent methanol (solvent) and water (antisolvent) revealed that relative supersaturation <1.25 (for only cooling) and anti-solvent content < 30% v/v were critical parameters to achieve individual deagglomerated crystals. The crystallization kinetics of commercial and in-house synthesized modafinil were determined by small-scale desupersaturation curve and impurity distribution coefficient experiments followed by a process systems engineering approach that included mathematical modeling. In phase II of the CC development, the process models were used to screen the effect of the critical process parameters (e.g., temperature, solvent/antisolvent ratio, residence time, supersaturation, number of stages) on the critical quality attributes (e.g., purity, yield, crystal size). Finally, the design space of the mixed-suspension mixed-product removal crystallizer (single and multi-stages) at the optimized conditions was determined. The approach allowed to minimize experimental trial-and-error and was supported by subsequent validation experiments to confirm that crude Modafinil obtained as described in literature can be purified by CC. Ultimately, this study represents a vital step towards the realization of integrated CM of modafinil.

1. Vanhoorne, V.; Vervaet, C., Recent progress in continuous manufacturing of oral solid dosage forms. International Journal of Pharmaceutics 2020, 579.
2. Silva-Brenes, D. V.; Emmanuel, N.; López Mejías, V.; Duconge, J.; Vlaar, C.; Stelzer, T.; Monbaliu, J.-C. M., Out-smarting smart drug modafinil through flow chemistry. Green Chemistry 2022, 24 (5), 2094-2103.
3. Billiard, M.; Broughton, R., Modafinil: its discovery, the early European and North American experience in the treatment of narcolepsy and idiopathic hypersomnia, and its subsequent use in other medical conditions. Sleep Medicine 2018, 49, 69-72.
4. Thirsk, R.; Kuipers, A.; Mukai, C.; Williams, D., The space-flight environment: the International Space Station and beyond. Canadian Medical Association Journal 2009, 180 (12), 1216-1220.

341a Streamlining Crystallization Process for a Low Solubility and Highly Solvating API System to Achieve Robust Scale-up Performance

Shujauddin Changi, Andrew McTiernan, Stefanie Roeper, Mei-Hsiu Lai, Christian Zeigler, Hanyu Zhu, Adam Looker, David Willcox

A robust performing crystallization is of utmost importance in the final step of drug substance synthesis, to meet the critical quality attributes (CQAs) for an active pharmaceutical ingredient (API). A case-study is presented involving an API step of a rapidly advancing portfolio molecule, wherein the API had very poor solubility in most solvents that prevented designing a realistically scalable conventional crystallization process. Moreover, the API molecule readily solvated with most solvents and had a complex polymorph landscape. To supply the clinical material needs, early campaigns implemented a first-generation process involving a distillative crystallization to isolate the API, followed by form conversion by drying. As the clinical program advanced, requiring scale-up to manufacture the API at several hundred kilograms, the drying operation performed robustly and always achieved the desired form (neat Form A). However, distillative crystallization processes can be challenging to scale up and this process was evaluated to have a considerable risk of failing the API specifications via generation of new impurities due to thermal stresses and degradation of on-wall solids. There was a critical need to explore an alternate controlled crystallization option. A second-generation process was quickly developed, allowing streamlining of the reaction followed by an antisolvent crystallization and eliminating the need to do distillations. The isolated solid from this process was a new polymorph, THF solvate A, which was classified as a flammable solid under UN Class 4 division 1. Based on the results of polymorph screening and a phase diagram, a slurry conversion unit operation was added to convert the THF solvate A to the desired neat Form A. Thus, a combination of understanding of form landscape in process space, safety assessment, and scale-up data guided the development of a commercial ready process, balancing a fast-moving clinical supply deliverable.

341c Interfacial Structure and Dynamics of Aqueous-Organic Solvent Illuminates the β-Hematin Crystallization Mechanism

Lakshmanji Verma, Peter Vekilov, Jeremy Palmer

Hematin crystallization, which is an essential component of the physiology of malaria parasites and the most successful target for antimalarial drugs, proceeds in mixed organic–aqueous solvents both in vivo and in vitro. Here we employ molecular dynamics simulations to examine the structuring and dynamics of a water–normal octanol mixture (a solvent that mimics the environment hosting hematin crystallization in vivo) in the vicinity of the typical faces in the habit of a hematin crystal. The simulations reveal that the properties of the solvent in the layer adjacent to the crystal are strongly impacted by the distinct chemical and topological features presented by each crystal face. The solvent organizes into at least three distinct layers. We also show that structuring of the solvent near the different faces of β-hematin strongly impacts the interfacial dynamics. The relaxation time of n-octanol molecules is longest in the contact layers and correlates with the degree of structural ordering at the respective face. We show that the macroscopically homogeneous water–octanol solution holds clusters of water and n-octanol connected by hydrogen bonds that entrap the majority of the water but are mostly smaller than 30 water molecules. Near the crystal surface the clusters anchor on hematin carboxyl groups. These results provide a direct example that solvent structuring is not restricted to aqueous and other hydrogen-bonded solutions. Our findings illuminate two fundamental features of the mechanisms of hematin crystallization: the elongated shapes of natural and synthetic hematin crystals and the stabilization of charged groups of hematin and antimalarials by encasing in water clusters. In addition, these findings suggest that hematin crystallization may be controlled by additives that disrupt or reinforce solvent structuring.

341d An Approach to Achieve Clarifying Filtration for the Reactive Crystallization of Poorly Soluble Organic Salts

Lotfi Derdour, Weiling Cai

Abstract

In API manufacturing, removal of foreign particulates prior to the particle forming step is a pre-requisite under GMP conditions and is typically achieved by clarifying filtration of solutions of all the reagents. Drug dissolution in organic solvents requires ubiquitous solvation spheres between the solvent and the drug molecules. Recent trend of increasing lipophilicity of drug candidates can result in poor interaction drug-solvent which may explain the decreasing drug solubility observed in the last few decades[1]. As a result of poor drug solubility, clarifying filtration is becoming increasingly challenging to achieve. In this presentation, we present an approach to achieve clarifying filtration for 2 drug candidates developed as salts and displaying poor solubility in most solvents. The approach is based on identifying co-solvents as solubility enhancers and takes advantage of the salt formation to achieve a full dissolution in a transient state with sufficient solution stability. The approach allowed the design of a GMP-compliant procedures that enable API deliveries to meet project needs.

1. Bergström et al., Adv. Drug Delivery Rev. 2016, 101, 6–21

341e Enabling Crystallization for a Challenging Active Pharmaceutical Ingredient through Use of a Wet-Milling Crystallization Process at Elevated Temperature

Christopher Marton, Zhiwei Yin, Tamar Rosenbaum, Deniz Erdemir, Junhe Ma

Direct crystallization of the desired polymorph of an active pharmaceutical ingredient (API) is an important step in a pharmaceutical manufacturing process. In this work, development to enable direct crystallization for an API that had proven particularly challenging due to the exceptionally slow crystal nucleation and growth kinetics is presented. The first-generation process was composed of two steps, a chiral upgrade crystallization of a different polymorph of the API and then a slurry turnover step to deliver target polymorph. Despite significant development efforts, a direct crystallization of the desired polymorph was hampered by slow crystal growth at moderate supersaturation and oiling risk at high supersaturation.

Additional development work led to the discovery of multiple conformations of the molecule in solution and the effect of these conformations on crystallization kinetics. Proton NMR detected presence of two conformations in solution while variable temperature NMR indicated those two sets of signals merged with increased temperature. This insight enabled direct crystallization of the desired polymorph, leveraging increased temperature to address the molecular conformation, and using in-situ wet milling to provide sufficient surface area for growth. Process development was accelerated using information from inline Process Analytical Technologies. The redeveloped direct crystallization has been demonstrated up to 100 g scale and will be implemented at ~100 kg-scale in the near future.

611a Role of Solid Solutions in Polymorphic Enantiotropy

Fredrik Nordstrom, Brian Linehan, Francesco Ricci

No API or chemical entity is completely pure. Impurities are always present in the API crystallization and tend to be entrapped in the isolated solids to varying degrees. A very common impurity entrapment mechanism in crystallization occurs through the formation of solid solutions wherein the impurity molecules substitutionally replace the host molecule in the crystal lattice of the product. The formation of these solid solutions can, and often will, result in significant changes in the thermodynamic properties of the crystalline API, even at low levels. Examples include enhancements in solubility and dissolution rates as well as melting point depressions, parameters which are critical for the controlled release of the drug. At the same time, changes in solubility of the API in the isolation solvents can have a substantial impact on the crystallization itself, especially when several polymorphs are present.

Presented in this talk is an example of this effect, in which the thermodynamic stability of the polymorphic compound anthranilic acid is fundamentally altered through the presence of low levels of the structurally similar compound salicylic acid. Anthranilic acid has three known polymorphs, Form I, II and III, which are all anhydrous. Form I is thermodynamically stable form at low temperatures, and Form III is stable at higher temperatures. As salicylic acid is increasingly entrapped in anthranilic acid, the enantiotropic transition temperature dramatically changes along a two-phase envelope in which both polymorphs exist in equilibrium with each other. The implications for drug development in general and establishing form control during process development are discussed.

212a Taylor-Couette Flow and Multiphase Operations

Georg Rudelstorfer, Rafaela Greil, Maximilian Neubauer, Annika Grafschafter, Matthaeus Siebenhofer, Susanne Lux

Continuous multiphase processes will play a major role in process development for usage of sustainable and green raw material resources. Isolation of valuable materials from residuals and low grade resources, not yet economically feasible, requires stable and flexible operation of equipment to be able to manage changing raw material qualities. With its simple and effective design, the Taylor-Couette Disc Contactor (TCDC) offers a promising apparatus design, well suited for flexible use as multiphase contactor or reactor. The unique TCDC design allows continuous and intense contact of up to four phases (liquid-liquid-gas-solid). The contact mechanism is based on formation of Taylor-Couette vortices. The vorticity, adjusted by the rotor speed, provides the flexibility in establishing a specific hold up of dispersed liquid and/or solid and/or gaseous phases. Shear forces, induced by these vortices, help in adjusting the droplet or bubble size and the phase contact area.

Several multiphase operations in the TCDC have already been investigated. The column used for multiphase experiments has an inner diameter of 50 mm and an active height of 0.6 meters. The hydraulics in liquid-liquid extraction were performed and modeled with the test system kerosene/water. Application of continuous three phase (liquid-liquid-solid) contact was demonstrated with heterogeneous esterification of aqueous (2 mol/L) acetic acid with methanol combined with solvent extraction of the reaction product. The overall conversion of acetic acid was increased from 20% without solvent extraction to 30% [1]. Gas-liquid contact was investigated and modeled with continuous chemisorption of CO_2 for neutralization of 0.1 molar sodium hydroxide solution. The neutralization reaction of NaOH with a mixture of 30% CO_2 in nitrogen was performed at three different temperatures (25°C, 40°C, 60°C). The residence time distribution and the gas phase holdup were recorded. The neutralization reaction of CO_2 with NaOH in aqueous solution was modeled with the film theory and continuously stirred tanks in cascade. Concentration trajectories of the hydroxide and carbonate species along the reactor height were modeled with the reaction rate and pH value.

These examples confirm the flexible applicability of the TCDC in multiphase operations and demonstrate a one pot engineering solution, easily adjustable to different applications and phase systems. The operation principle of the TCDC also suggests advantages in leaching and in continuous precipitation, since driving force and residence time can be adjusted independently. Appropriate test systems are investigated at present.

[1] G. Rudelstorfer, M. Neubauer, M. Siebenhofer, S. Lux, A. Grafschafter, *Chemie Ing. Tech.* **2022**. DOI: 10.1002/CITE.202100184.

212b Separation of Azeotropic Refrigerant Mixtures Using Extractive Distillation with Ionic Liquid Entrainers

Ethan A. Finberg, Mark B. Shiflett

Hydrofluorocarbons (HFCs) have been used as refrigerants globally since the 1990's and replaced chlorofluorocarbons (CFCs), which were linked to the depletion of the Earth's ozone layer. HFC refrigerants are being phased out over the next two decades due to their high global warming potential. The refrigerant industry is moving toward the production and marketing of HFOs and HFO/HFC refrigerant blends. This will require the recycling and repurposing of HFC mixtures, but many of the HFC mixtures are azeotropic or near azeotropic (i.e., R-410A, R-404A, R-407C, to name a few) making the separation of the components impossible using current methods. Extractive distillation using an ionic liquid entrainers offers a solution.

Ionic liquids (ILs) are being developed as novel entrainers for extractive distillation because of their negligible vapor pressure, thermal and chemical stability, and high and differential solubility with HFC refrigerants. Vapor liquid equilibria data for refrigerants difluoromethane (HFC-32), chlorodifluoromethane (HCFC-22), pentafluoroethane (HFC-125), 1,1,1-trifluoroethane (HFC-143a), and 1,1,1,2-tetrafluoroethane (HFC-134a) in ionic liquids 1-ethyl-3-methylimidazolium bis(trifluoromethylsulfonyl)imide ($[C_2C_1im][Tf_2N]$) and 1-butyl-3-methylimidazolium hexafluorophosphate ($[C_4C_1im][PF_6]$) were fit with the Peng–Robinson equation of state to simulate the separation of multi-component azeotropic refrigerant mixtures and to develop rate-based and equilibrium models in ASPEN Plus. Process flow diagrams (PFD) were developed and optimized based on a set of physical and chemical constraints with the goal to optimize parameters and achieve refrigerant grade (>99.5 wt%) purity. The ionic liquids were found to be effective entrainers for separating refrigerant mixtures and these PFDs are used for the initial design of a pilot scale refrigerant separation process at the UCniversity of Kansas

212c Antagonistic Role of Aqueous Complexation in the Solvent Extraction and Separation of Rare Earth Ions

Pan Sun, Erik A. Binter, Zhu Liang, M. Alex Brown, Artem V. Gelis, Ilan Benjamin, Mrinal Bera, Binhua Lin, Wei Bu, Mark Schlossman

During solvent extraction of rare earth ions, extractants in an organic solvent complex with ions in water to facilitate their transport into the organic phase. A common extractant used in the separation of rare earth ions, bis(2-ethylhexyl) phosphate (HDEHP), is partially soluble in water and hypotheses for the site of initial ion-extractant complexation include both the liquid-liquid interface and the bulk aqueous phase. To explore the role of the bulk aqueous phase, we show that Nd and Er ions that complex with HDEHP within an isolated aqueous phase exhibit behavior that appears to contradict trends observed under normal conditions for liquid-liquid solvent extraction. Interfacial tension measurements suggest that complexes of HDEHP with lighter lanthanides are enhanced at the liquid-vapor interface over those containing heavier lanthanides. Measurements of the interface with x-rays support these results in the specific cases of Nd (a light lanthanide) and Er (a heavy lanthanide) by direct detection of these elements at the interface. The greater availability of Nd-HDEHP complexes at the liquid-vapor interface suggests that contacting the aqueous solution with an organic phase would lead to more efficient extraction of Nd over Er with HDEHP. In contrast, it is known that liquid-liquid extraction into the organic phase under similar conditions is greatly favored for Er over Nd. These results reveal the antagonistic role played by complexation within the bulk aqueous phase and clarify the advantages of extractants that interact with ions only at the interface during the process of liquid-liquid extraction.

212d Continuous Millifluidic Liquid-Liquid Extraction: Digital Design and Development

Jaron Mackey, Devna Grover, Gabriella Pruneda, Zoltan Nagy

Liquid-liquid extraction is a common separation technique in chemical processing. By incorporating a continuous extraction step, impurities removal can occur without the exposure of personnel to intermediates and byproducts. The inclusion of in line separation techniques allows for the integration of unit operations. In efforts to reduce the material and personnel requirements for unit operation, digital design strategies are implemented to predict extraction platform performance and determine feasible operating space. To better understand liquid- liquid extraction at the millifluidic scale, droplet analysis and fluid flow analysis was carried out to determine the type of flow regime that was observed and other parameters to assess flow conditions.

In this work, we present and implement a workflow for solvents screening and selection, batch and flow extraction experimental design, and the determination of volumetric mass transfer coefficient and digital design for extractor design for 2-chloroethyl isocyanate, ISO, a reagent in the synthesis of Lomustine. Lomustine is an anti-cancer drug that was used to treat brain tumors and Hodgkin's Lymphoma [1]. The incorporation of the liquid-liquid extraction step in this process will allow for the continuous purification of the Lomustine intermediate prior to the second synthesis step [2].

In this study, correlations were implemented to provide predictions of the system performance beyond what was measured experimentally. Factors such as the effect of changing volume ratios on droplet size and the impact of changing residence time on volumetric mass transfer coefficient were investigated to explore the design space more fully and ultimately develop a more robust extraction process.

To our knowledge, this is the first time that this kind of analysis of the extraction parameter estimation, model selection and discrimination, and digital design for this system. The work that was conducted provide valuable tools to predict volumetric mass transfer coefficient by using dimensionless groups and the resulting extraction simulations to determine feasible operating spaces and extractor design.

1. Lee, F. Y.; Workman, P.; Roberts, J. T.; Bleehen, N. M., Clinical pharmacokinetics of oral CCNU (lomustine). *Cancer Chemother. and Pharmacol.* **1985,** *14* (2), 125-131.
2. Jaman, Z.; Sobreira, T. J. P.; Mufti, A.; Ferreira, C. R.; Cooks, R. G.; Thompson, D. H., Rapid On-Demand Synthesis of Lomustine under Continuous Flow Conditions. *Process Res. andamp; Dev.* **2019,** *23* (3), 334-341.

212e The Development of Structured Ligands and CO_2-Facilitated Hydrometallurgy to Extract Critical Elements from Unconventional Resources

Hunter Vibbert, Whai Shin Ooi, Seokyoon Moon, Ah-Hyung Alissa Park

Processes that can extract and recover critical elements from unconventional resources will be imperative to the future development of new technological innovations. Many rare earth elements (REEs) can be found in unconventional resources, like waste-to-energy ashes, waste electrical and electronic equipment, battery materials, or printed circuit boards. In conventional mineral extraction, REEs can be obtained via laborious, energy intensive processing under highly acidic conditions, high temperature roasting oxidation, and environmentally unfriendly liquid-liquid extraction processes. Here, we propose a conceptually different approach, understanding the hydrometallurgical conditions required when sustainable acids and bases (e.g., CO_2, HCl, NaOH) are used to aid extraction of critical elements. We show that molecularly designed ligand systems can be used in liquid-liquid extraction or immobilized on hybrid, magnetically recoverable supports to separate Ce from a complex background of metal ions. The application of these materials towards extraction, selectivity, and the circularity of the proposed process will be discussed.

212f Sustainable Recovery of Butyric Acid through Salt-Induced Aqueous Two-Phase Extraction Using Non-Ionic Surfactants

Kristel Gatdula, William Holmes, Emmanuel Revellame

Butyric is an important four-carbon volatile fatty acid (VFA) with broad applications in food, feed, polymer, pharmaceutical and biofuels industries. In line with the Food-Energy-Water Nexus and in response to the increasing demands for energy and chemicals while overcoming the issues associated with depleting fossil-based resources, butyric acid stimulates innovative advancement in the bio-based economy as it can be produced from urban waste streams. To simulate the enhanced recovery of butyric acid by valorization from fermented wastewater, a salt-induced aqueous system of Zigova's medium, butyric acid and non-ionic surfactant was subjected to reactive extraction. In this study, non-ionic surfactant was utilized among the other amphiphilic materials as a separating agent due to its low toxicity and high extraction effectiveness for short-chain carboxylic acids. The main goal of this research is to determine how salting-out affects the cloud point of ethoxylated non-ionic surfactant, and consequently, the butyric acid separation. The effects of varying fermentation nutrients such as $(NH_4)_2SO_4$, KH_2PO_4, $MgSO_4$, $FeSO_4$ and $MnSO_4$ on the extraction of butyric acid were determined through the acid partition coefficient and recovery ratio as the main responses. Results show that the cloud point of the system which refers to the temperature at which the mixture starts to phase separate, was sensitive to the presence of electrolytes. This caused high butyric acid partition coefficient and recovery ratio for the system with enhanced salt concentrations versus the control with conventional concentration of salts in the Zigova's medium. The salt changed the solvent structure by aggregation and formation of larger micelles. This translates into a lipophilic shift which reduced the cloud point and the surfactant hydrophilic-lipophilic balance (HLB). As this HLB was decreased, the surfactant turned to be more hydrophobic and formed better separation. Also, the type of salt influenced the interphase mass transfer. Compared with monovalent cations like NH_4^+ and K^+, the polyvalent cations such as Mg^{2+}, Fe^{2+} and Mn^{2+} were found to be more effective in decreasing the HLB thereby increasing the butyric acid partition coefficient and recovery ratio. This is because of the polyvalent cations' higher surface charge densities. The implementation of this non-energy requiring recovery of butyric acid using non-ionic surfactant induced with salt may serve as a good start-up to achieve an economical production of other VFAs, especially for upscale applications.

212g Greener Extraction of Rare Earth Elements from Coal Fly Ash

Cesar Martinez Bejarano, Yasaman Ghanbari, Kenneth Walker, April Wright, Isaiah Morones, Sergio Martinez-Monteagudo, Catherine Brewer

Rare earth elements (REE) are metallic elements that include 15 of the lanthanides, scandium, and yttrium. The global for REE increases every year as these REE's are used in many modern technological innovations from microchips, electric cars, and solar panels, to fighter jets and satellites. China controls 90-100% of the REE available on the commercial market. In recent years the U.S. has increased its efforts to secure more stable domestic sources of REE. Coal fly ash (CFA) is a very fine powdery material from the combustion of finely-ground coal in heat and power plants. CFA is one of the largest types of industrial waste generated within the U.S., producing nearly 130 million tons in 2014, and some CFA contains as much or more of the desired REE than REE ores.

Typical REE extraction processes require strong acids with the associated hazards. Researchers at Sandia National Laboratories have developed a lab-scale process to extract REE from coal fly ash under relatively mild conditions, using water, supercritical CO_2, and small organic acid chelators. The goal of this project is to determine the engineering parameters and, where possible, process simplifications to enable scale-up. Experiments have been conducted with increasing levels of complexity on two types of a locally-sourced CFA to quantify the return on investment for each increase in temperature, pressure, reaction time, and reaction components. The first experiments were conducted in batch at atmospheric pressure on a stirring hot plate with and without chelator, followed by extractions in a 1.8-L batch reactor under nitrogen then CO_2. REE were quantified using inductively coupled plasma spectroscopy and compared to the quantities extracted under full-complexity processing and under traditional strong-acid conditions. Extraction efficiencies will ultimately be combined with processing cost models to identify the most feasible REE extraction process.

OIL RECOVERY FROM PRODUCED WATER VIA SELECTIVE OIL PERMEATION

Carolyn M. Cooper, The University of Texas at Austin, Austin, TX, USA
You Wu, The University of Texas at Austin, Austin, TX, USA
Sara Alshawish, Kuwait University, Safat, Kuwait
Abdalrahman Alsulaili, Kuwait University, Safat, Kuwait
Kerry A. Kinney, The University of Texas at Austin, Austin, TX, USA
Lynn E. Katz, The University of Texas at Austin, Austin, TX, USA
A. Frank Seibert, The University of Texas at Austin, Austin, TX, USA

Abstract

High-efficiency oil-water separations are crucial to the success of treatment trains for the beneficial reuse of industrial wastewaters like produced water. While traditional membrane-based separations can achieve low minimum droplet removal and high oil removals, they often experience rapid viscous fouling of the membrane surface. Selective oil permeation is a promising alternate membrane-based approach in which oil permeates through the membrane surface instead of water. The process differs from traditional membrane-based oil-water separations by generating a high-quality oil permeate stream and minimizing viscous fouling by exploiting the preferential wetting of the membrane surface. Our previous research has observed encouraging results with synthetic oil-water emulsions over a range of operating parameters and solution characteristics. These studies have provided mechanistic insight and a fundamental understanding of the process. Yet, few studies have investigated selective oil permeation utilizing real industrial wastewaters. In this study, we apply the fundamental insights that we have gained using model systems to understand and optimize system performance with produced water. In doing so, we enhance our understanding of the opportunity space for selective oil permeation.

Introduction

Complications associated with the interplay between produced water quality,[1,2] quantity,[3] and treatment[2,4] have been discussed in numerous studies.[5,6] In the United States, these factors have resulted in the reuse of approximately 8% of produced water outside the oil and gas industry, versus the subsurface reinjection of approximately 90% of produced water.[3] Researchers have concluded that fit-for-purpose treatment that thoughtfully considers both water quality and end-use specific risks is necessary to increase the beneficial reuse of produced water. Further, to mitigate toxicity concerns in higher-order reuse (*e.g.,* streamflow augmentation, municipal reuse), advanced treatment processes (*e.g.,* advanced oxidation processes, membranes) may be required, often without confirmation that the toxicity has been reduced.[5,6] Moreover, advanced treatment often requires high-efficiency pretreatment to mitigate fouling and achieve high levels of efficacy.

Nearly all produced water management requires oil-water separations. Primary oil-water separation methods (*e.g.,* API separators) traditionally encompass gravity separation of free oil ($\geq$ 150 µm). Secondary oil-water separation approaches (*e.g.,* dissolved air filtration, hydrocyclones) target dispersed oil (> 50 µm) using chemical or centrifugal methods. Tertiary oil-water separation methods (*e.g.,* media filtration) remove emulsified oils ($\leq$ 50 µm) via either adsorption or filtration.[7,8] Researchers have also explored the application of membranes, where water permeates through the membrane, to remove emulsified oils. However, as with many membrane-based approaches, membrane-based oil-water separations often suffer from oil fouling, particularly with produced water.[9,10]

Selective oil permeation (SOP) is a promising membrane-based oil-water separations technology that permeates oil (instead of water) through a hydrophobic membrane.[11–17] SOP provides inherent advantages over traditional membrane-based oil-water separations by exploiting the preferential wetting of the membrane surface, minimizing viscous fouling, and generating a high-quality oil permeate stream. Researchers have investigated the process using various insoluble oils to determine the relationship between operating parameters, solution characteristics, and process performance.[11–13,16,17] These studies have also developed our mechanistic understanding of the process where oil flux is controlled by the approach and coalescence of the oil droplet on the membrane surface, permeation of the oil through the membrane surface, and release of the oil from the permeate side of the membrane.[11,14,17–19]

However, these fundamental studies have not addressed two factors key to produced water treatment. First, produced water, like many other wastewaters, contains relatively low oil concentrations (2–565 mg/L) that must be reduced to less than 10–35 mg/L for beneficial reuse.[2,4] Minimal work has investigated oil recovery via SOP at concentrations less than 1%.[14,17] Second, to the authors' knowledge, no studies have detailed the performance of SOP with real produced waters. Minimal work has investigated SOP performance using real wastewater. While Seibert presented successful oil recovery performance data from concentrated non-flocculated lysed algae in a microporous polypropylene membrane,[20] Leiknes *et al.,* observed water breakthrough in a microporous polyethylene membrane when recovering oil from industrial cutting fluids with high emulsifying agent concentrations (~20–30%).[15] Thus, further work is necessary to understand the applicability of SOP to produced water treatment.

This study will investigate the efficacy of SOP separations in produced water by characterizing the produced water, investigating the efficacy of SOP over a range of operating conditions, and providing a comparison of these results to the literature and an existing mass transfer model. Through this, we will provide an enhanced understanding of the broader applicability of SOP to the treatment of produced water.

Theory

Mercelat *et al.,* proposed the following model for oil flux in SOP. The model views SOP oil flux as a fraction of pure oil flux due to changes in the wetted surface area of the membrane:

$$J_{O/W} = J_{Oil} \cdot \frac{a_e}{a_t} \tag{1}$$

Where $J_{o/w}$ is oil flux from oil-water mixtures (m^3/m^2-s), J_{oil} is the pure oil flux (m^3/m^2-s), and a_e/a_t is the effective membrane surface area fraction. Pure oil flux (J_{oil}) can then be described by the Blake-Kozeny viscosity component of the Ergun equation:

$$J_{Oil} = \frac{P_T \times d_{pore}^2 \times \left(\frac{1-\varepsilon}{\varepsilon}\right)^{\frac{2}{3}} \times \varepsilon^3}{150 \times L \times \mu \times (1-\varepsilon)^2} \tag{2}$$

Where J_{oil} is the pure oil flux (m^3/m^2-s), P_T is the applied transmembrane pressure (Pa), d_{pore} is the average pore diameter (m), ε is the porosity of the membrane, L is the membrane wall thickness (m), and μ is the oil viscosity (Pa-s). The effective membrane surface area is calculated via the following semi-empirical model:

$$\frac{a_e}{a_t} = \frac{1.65\times10^{12}\cdot C_{oil}\cdot\mu\cdot P_T^{-1.6}\cdot v^{0.3}}{1 + 2.66\times10^{15}\cdot C_{oil}\cdot\mu^{1.1}\cdot P_T^{-2.1}\cdot v^{0.4}} \quad (3)$$

Where C_{oil} is the volume fraction of oil and the liquid velocity, *v* (m/s), is defined as the flow rate divided by the cross-sectional area of the module.

Methods

Solution Characteristics

This study utilized de-oiled produced water from the Eagle Ford Basin in the United States. Total Suspended Solids (TSS) and Total Dissolved Solids (TDS) were measured via ASTM D5907–18. Inductively coupled plasma mass spectrometry (Thermo Scientific ICAP TQ ICP-MS) were used to characterize cations while anions were characterized via ion chromatography conducted at the Lower Colorado River Authority Lab. Total petroleum hydrocarbons (TPH) were quantified via the Texas Natural Resource Conservation Commission TNRCC Method 1005 using gas chromatography with flame ionization detection on an Agilent 6890. Total organic carbon (TOC) was quantified via a Shimadzu TOC-L.

Isopar M served as a synthetic oil throughout the experiments. An NDJ-5S Digital Rotational Viscometer and a Mettler Toledo DE40 Density Meter characterized oil viscosity (3.5 cP) and density (0.78 g/cm^3) at 25 degrees Celsius, respectively. An inline JM Canty particle analyzer determined the emulsion droplet size distribution. The mean particle diameter was 5.7 μm. No free water was observed in the oil permeate during the study.

Membrane System and Operating Procedure

Figure 1Figure 1 shows a process flow diagram of the experimental apparatus. All reported recoveries in the following sections are from the primary microporous polypropylene hollow fiber membrane, Membrane A (3M™ Liqui-Cel™ Extra Flow Membrane Contactor: Diameter: 6.4 cm, Length: 20.3 cm, Surface Area: 1.4 m^2). The guard membrane, Membrane B (3M™ Liqui-Cel™ Extra Flow Membrane Contactor: Diameter: 10.2 cm, Length: 71.1 cm, Surface Area: 20 m^2) allowed for recycle of the system for continuous operation. Both Membrane A and Membrane B contained commercially available X50 fibers, which have a reported porosity of 40%, a thickness of 40 μm, pore diameter of 0.05 μm, and a water contact angle of 120°.[17] Membrane A was pre-conditioned to wet the membrane pores and initialize the oil-wetted surface area by flowing pure Isopar M through the membrane for 24 hours.

A peristaltic pump injected a known mass of Isopar M into the suction of a high-shear pump. The high-shear pump generated the oil-water emulsion and moved the emulsion to the bottom shell-side port of the primary membrane, Membrane A. Effluent water from the shell-side of Membrane A returned to the feed tank. Membrane B received water from the feed tank and maintained the purity of the feed tank.

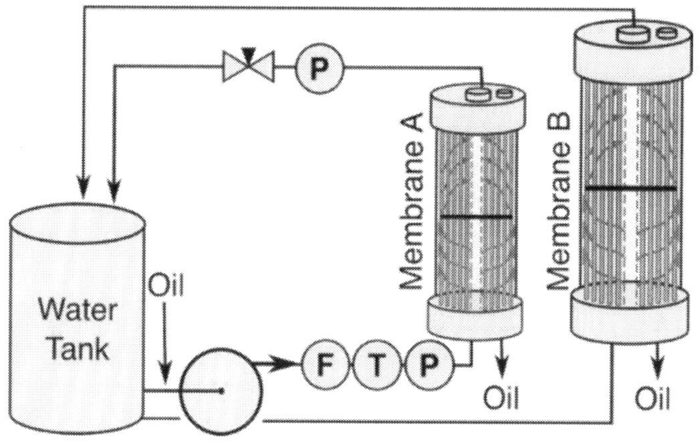

Figure 1: Process flow diagram of experimental apparatus. Abbreviations are as follows: pressure transmitter (P), temperature transmitter (T), and flow transmitter (F).

The transmembrane pressure (TMP) and influent flow rate for Membrane A were controlled via the variable speed drive on the high-shear pump and the valve downstream of Membrane A. Shell-side pressures were monitored via a transmitter while tube-side pressure remained at atmospheric pressure. The TMP was calculated by averaging the inlet and outlet shell-side pressure. Influent oil concentration, oil recovery, and oil flux were quantified gravimetrically via scale by monitoring all injected and permeated oil. Oil flux and oil recovery were calculated as a four-hour average as follows:

$$Oil\ Flux\ (\frac{m^3}{m^2-s}) = \frac{\dot{m}_{permeate}}{\rho_{oil} \times A_{membrane}} \tag{4}$$

$$Oil\ Recovery\ (\%) = \frac{\dot{m}_{permeate}}{\dot{m}_{injected}} \cdot 100\% \tag{5}$$

Where $\dot{m}_{permeate}$ was the mass flow of the permeate (g/s), $\dot{m}_{injected}$ was the mass flow of injected oil (g/s), ρ_{oil} was the density of the oil (g/cm^3), and $A_{membrane}$ was the surface area of the membrane (m^2). A DeltaV data acquisition system allowed for real-time data collection from all described instruments. The R programming environment was utilized for statistical analysis and to generate figures.

Results and Discussion

Produced Water Characterization

Measured raw produced water characteristics are presented in Table 1. The produced water characteristics are consistent with ranges observed in the literature.[1,2]

Table 1: Raw produced water characteristics. All values in mg/L unless otherwise noted.

Variable	Value	Variable	Value	Variable	Value
pH	6.9	Alkalinity	137 mg/L as $CaCO_3$	SO_4^{2-}	167
Conductivity	72 mS/cm	Na^+	12,442	F^-	<1
TDS	47,192	K^+	120	NO_2^-	<1

TSS	250	Mg^{2+}	227	NO_3^-	<1
Turbidity	334 NTU	Ca^{2+}	7,488	PO_4^{3-}	<0.1
TPH	17	Cl^-	30,500		
TOC	67	Br^-	174		

Selective Oil Permeation Process Performance

Oil recoveries of 74.3% to 99.9% were observed over sixteen experimental conditions. Some experiments achieved effluent oil concentrations of less than 1.0 mg/L, indicating the potential for SOP to regulatory requirements for beneficial reuse. Exemplar pre- and post-treatment pictures are shown in Figure 2A. Much of the visual difference between the raw produced water and Membrane A feed was achieved by the membrane system within minutes of circulating the produced water within the system.

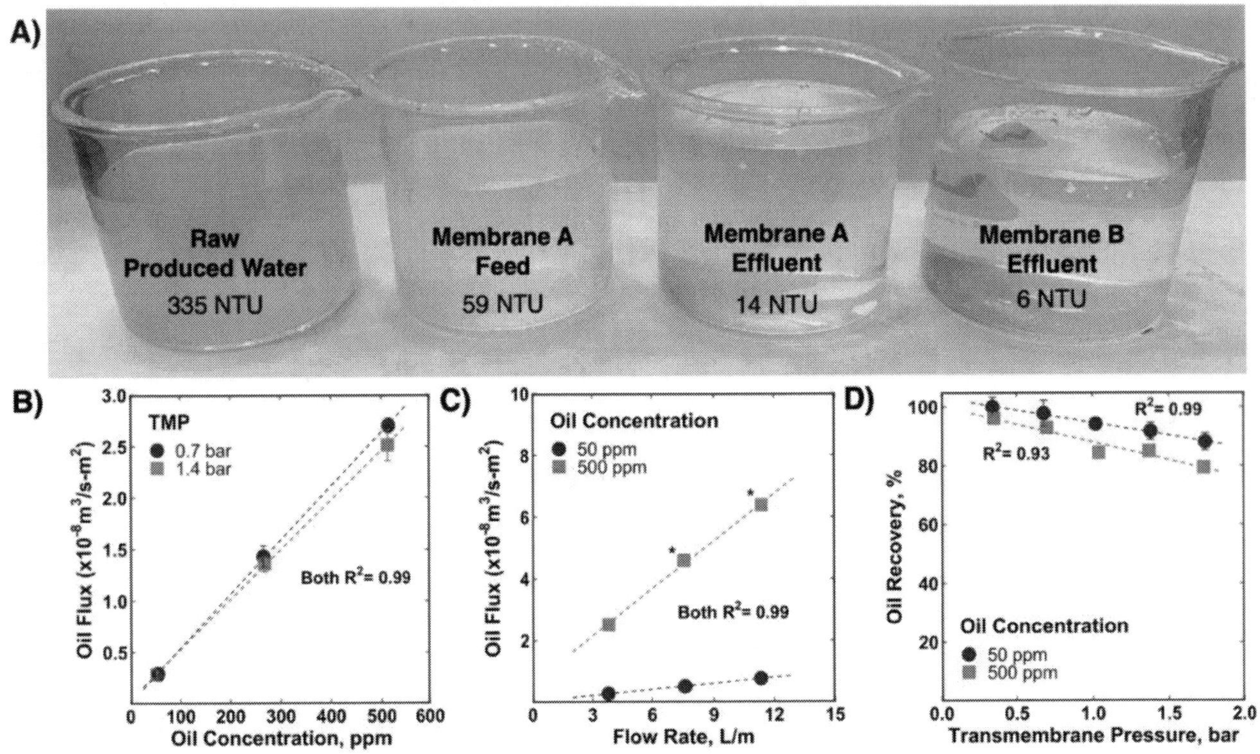

Figure 2: (A) Visual changes in produced water throughout the process at an influent Isopar M concentration of 500 ppm, TMP of 0.7 bar, and influent flow rate of 3.8 L/m. (B) The effect of oil concentration on oil flux. Experiments conducted at 3.8 L/min in Membrane A. (C) The effect of influent flow rate on oil flux. Experiments conducted at a TMP of 0.7 bar in Membrane A. Asterisks note experiments requiring two hour averaging due to system limitations. (D) The effect of TMP on oil recovery. Experiments conducted at a flow rate of 3.8 L/min in Membrane A.

Effect of Oil Concentration

Oil flux exhibited a linear relationship (R^2 = 0.99 for 0.7 and 1.4 bar) with influent concentration in this study (Figure 2B). Kong *et al.*, first observed this linear relationship between oil flux and oil concentration for oil concentrations less than 1 wt.%.[14] Mercelat *et al.*, later expanded on this work and found a saturation-style relationship between oil flux and oil concentrations ranging from 200 ppm to 100%.[17] Within the oil-disperse region, a linear relationship was observed between oil concentration and oil flux. This behavior was attributed to coalescence (instead of permeation) limiting oil flux.[17]

Oil recoveries decreased slightly at 0.7 bar (97.8% ± 4.4% at 50 ppm to 92.9% ± 1.0% at 500 ppm) and more substantially at 1.4 bar (91.7% ± 2.8% at 50 ppm to 84.9% ± 0.7% at 500 ppm) with increasing oil concentration. The decrease in oil recovery with increasing influent oil mass flow rate is also consistent with the proposed saturation-style model, where permeation becomes limiting as the oil film approaches to the maximum wetted surface area.[17]

Effect of Influent Flow Rate

A linear relationship (R^2=0.99) was observed between oil flux and influent flow rate for 50 ppm and 500 ppm (Figure 2C). Oil recovery decreased slightly for both 50 ppm (91.7% ± 2.9% at 3.8 L/m to 86.4% ± 2.0% at 11.4 L/m) and 500 ppm (84.9% ± 0.7% at 3.8 L/m to 74.3% ± 1.0% at 11.4 L/m). This result may indicate that the influent oil mass exceeds the wetted surface area of the membrane, transferring the dominant mechanism from coalescence to permeation as the influent oil mass flow rate increases. However, the relatively constant oil recovery may offer a pragmatic opportunity for SOP to serve as a high-throughput, modular treatment process.

Effect of Transmembrane Pressure

Previous researchers evaluating oil flux and recovery in oil-continuous solutions have observed relationships between oil flux and transmembrane pressure that are consistent with the Ergun equation.[11,13,17] Similarly, Kong *et al.*, found oil recovery increased with increasing TMP for influent oil concentrations <1%.[14] In contrast, this study observed an inverse relationship between TMP and oil recovery (Figures 2B and D). Magdich noted similar behavior for a 5% dodecane-water mixture, but not for experiments using a higher viscosity oil.[11] Mercelat *et al.*, also observed this behavior for influent Isopar M concentrations of 2%. Mercelat *et al.*, attributed the behavior to the rate of permeation surpassing the rate of coalescence and reducing the wetted surface area fraction (a_e/a_t) of the membrane.[17]

Increased oil recovery at lower applied TMPs offers two key advantages. First, elevated oil recovery at lower applied TMP reduces the required energy input for the process. Second, water breakthrough is hypothesized to occur when the applied TMP exceeds the critical entry pressure of the water. Thus, the operation of SOP systems at lower TMPs reduces the likelihood of water breaking through into the permeate while increasing oil recoveries.

Model Performance Comparison

Application of the existing mass transfer model to the expanded data indicated potential discrepancies (Figure 3). These discrepancies may be due to applying the initial model outside of the original parameter ranges.[17] For example, the model performs poorly for influent flow rates greater than 3.8 L/m, which may be due to the exponents associated with concentration and velocity in the original model (Equation 3) not correctly scaling with 7.6 and 11.4 L/m. Ultimately, further research is necessary to expand the original mass transfer model to adequately predict oil flux over a broader set of operating and water quality conditions.

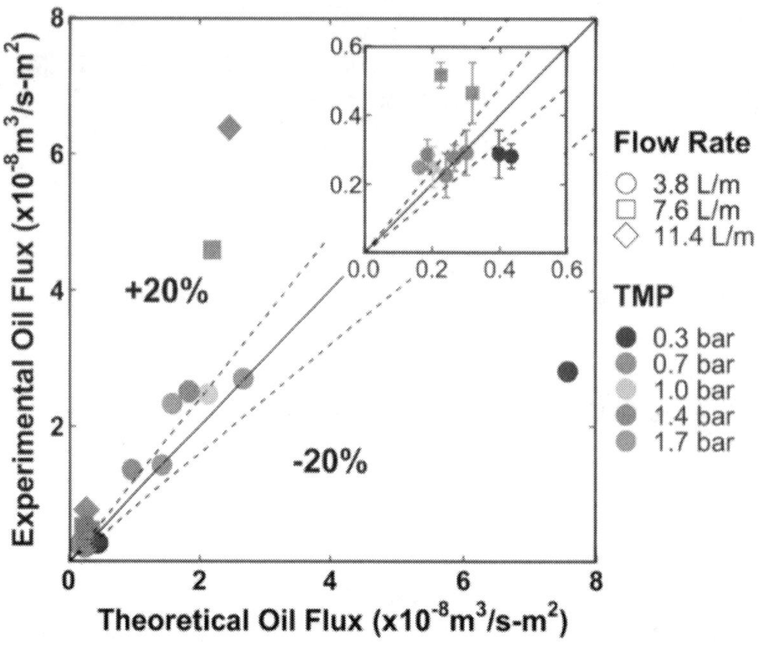

Figure 3: Parity plot between theoretical and experimental oil flux.

Conclusions

This study demonstrated the potential for SOP to achieve high oil recoveries (>90%) in oil-water separations with produced water without evidence of free water in the pure oil permeate. These oil recoveries may allow SOP to achieve oil concentrations required for beneficial reuse.

To the authors' knowledge, this study examined some of the lowest influent oil concentrations in the literature. Successful oil recovery within this concentration range demonstrates the potential for SOP to serve as a secondary or tertiary oil-water separation process. Within the examined experimental conditions, SOP process performance corresponded with competition between the hypothesized mechanisms: approach and coalescence, permeation, and release.[17] Crucially, the dominance of approach and coalescence offers advantages. For example, high oil recoveries at low transmembrane pressures also minimizes the likelihood of water breakthrough and reduces required energy throughput. The minimal relationship between oil recovery and influent flow rate suggests SOP could serve as a high-throughput, modular process at certain oil concentrations. Finally, the experimental data generally exceeded the theoretical expectations of the original model.

Broadly, these results indicate the potential for SOP to serve as a secondary or tertiary oil-water separation process for produced water treatment. While this study presents promising results for SOP, future fundamental studies are necessary in order to elucidate the effects of produced water on membrane fouling and the longevity of the membrane.

Acknowledgements

This project was funded by the Kuwait Foundation for the Advancement of Sciences under project code CN17-45EV-01. Funding for the primary author was provided through the National Science Foundation Graduate Research Fellowship under Grant No. DGE 2137420. The authors would also like to thank Benjamin A. Kienzle for help in supporting the analytical work.

References

1. Benko, K. L. & Drewes, J. E. Produced water in the Western United States: Geographical distribution, occurrence, and composition. *Environ. Eng. Sci.* **25**, 239–246 (2008).
2. Fakhru'l-Razi, A. *et al.* Review of technologies for oil and gas produced water treatment. *J. Hazard. Mater.* **170**, 530–551 (2009).
3. Veil, J. *U . S . Produced Water Volumes and Management Practices in 2017. US. Produced Water Management* (2020).
4. Chang, H. *et al.* Potential and implemented membrane-based technologies for the treatment and reuse of flowback and produced water from shale gas and oil plays: A review. *Desalination* (2019). doi:10.1016/j.desal.2019.01.001
5. Conrad, C. L. *et al.* Fit-for-purpose treatment goals for produced waters in shale oil and gas fields. *Water Res.* **173**, (2020).
6. Cooper, C. M. *et al.* Oil and Gas Produced Water Reuse: Opportunities, Treatment Needs, and Challenges. *ACS ES&T Eng.* **2**, 347–366 (2022).
7. Judd, S. *et al.* The size and performance of offshore produced water oil-removal technologies for reinjection. *Sep. Purif. Technol.* (2014). doi:10.1016/j.seppur.2014.07.037
8. Pintor, A. M. A., Vilar, V. J. P., Botelho, C. M. S. & Boaventura, R. A. R. Oil and grease removal from wastewaters: Sorption treatment as an alternative to state-of-the-art technologies. A critical review. *Chem. Eng. J.* **297**, 229–255 (2016).
9. Abbasi, M., Mirfendereski, M., Nikbakht, M., Golshenas, M. & Mohammadi, T. Performance study of mullite and mullite-alumina ceramic MF membranes for oily wastewaters treatment. *Desalination* **259**, 169–178 (2010).
10. Lee, J. M. & Frankiewicz, T. Treatment of produced water with an ultrafiltration (UF) membrane - A field trial. *SPE Annu. Tech. Conf. Proc.* (2005). doi:10.2118/95735-ms
11. Magdich, P. The Removal of Oil from Oil-Water Mixtures using Selective Oil Filtration. (The University of Minnesota, 1988).
12. Daiminger, U., Nitsch, W., Plucinski, P. & Hoffmann, S. Novel techniques for oil / water separation. **99**, 197–203 (1995).
13. Tirmizi, N. P., Raghuraman, B. & Wiencek, J. Demulsification of Water/Oil/Solid Emulsions by Hollow-Fiber Membranes. *AIChE J.* **42**, 1263–1276 (1996).
14. Kong, J. & Li., K. Oil removal from oil-in-water emulsions using PVDF membranes. *Sep. Purif. Technol.* (1999). doi:10.1016/S1383-5866(98)00114-2
15. Leiknes, T. & Semmens, M. J. Membrane filtration for preferential removal of emulsified oil from water. *Water Sci. Technol.* **41**, 101–108 (2000).
16. Ezzati, A., Gorouhi, E. & Mohammadi, T. Separation of water in oil emulsions using microfiltration. *Desalination* (2005). doi:10.1016/j.desal.2005.03.086
17. Mercelat, A. Y., Cooper, C. M., Kinney, K. A., Seibert, F. & Katz, L. E. Mechanisms for Direct Separation of Oil from Water with Hydrophobic Hollow Fiber Membrane Contactors. (2021). doi:10.1021/acsestengg.1c00055
18. Unno, H., Taguchi, K. & Akehata, T. Liquid permeation through (poly)-tetrafluoroethylene (ptfe) membrane. *J. Chem. Eng. Japan* **20**, 52–57 (1987).
19. Unno, H., Saka, H. & Akehata, T. Oil Separation From Oil-Water Mixture By A Porous Poly(Tetrafluoroethylene) (Ptfe) Membrane. *J. Chem. Eng. Japan* **19**, 281–286 (1986).
20. Seibert, F. *et al.* Non-Dispersive Process for Insoluble Oil Recovery from Aqueous Slurries. **2**, 28 (2013).

466a Light-Responsive Azobenzocrown Ethers for Enhanced Cs^+ Extraction from Contaminated Aqueous Sources

John Edward Sio, Erwin Escobar, Khino Parohinog, Wook-Jin Chung, Grace Nisola

Nuclear fission provides a cheaper and cleaner energy source. However, it also holds a highly hazardous consequence when handled improperly (e.g. Chernobyl Power Plant accident) or disrupted by natural calamities (e.g. Fukushima Daiichi Power Plant earthquake and tsunami incident). The accidental release of radioactive fission products (e.g. $^{137}Cs^+$) pose threats to the general population and to the environment by soil and water contamination. Thus, continuous monitoring and separation of these ions are essential. Liquid-liquid extraction (LLE) still serve as a good separation technology for radioactive metal ions removal from contaminated waters. LLE involves solute transport, Cs^+ in this case, from aqueous to organic solvent phase with the assistance of ionophores and other additives pre-dissolved in the organic phase. Crown ethers are known to be inert to radioactive materials and thus are good candidates for LLE. A greener approach is introduced wherein light is used as an external stimulus to enhance the removal of Cs^+ contaminants.

Herein, photo-responsive azobenzocrown ethers (a-BCEs) were synthesized and applied as Cs^+ ligands. The prepared a-BCEs and their intermediates were thoroughly characterized by FTIR, FT-NMR, and UV-vis spectroscopy. *Cis-trans* isomerization of a-BCEs was assessed by scanning 200-1100 nm wavelength range after UV or visible light irradiation. UV-vis absorption studies confirmed *cis-trans* photo-switching abilities of the compounds within 1 h of irradiation. UV-irradiated a-BCEs showed *cis*-content estimates ranging from 42% to 88% while visible light-irradiated samples showed *trans*-content estimates of more than 98%.

The LLE performance of a-BCEs for Cs^+ was evaluated by its extraction percentage (%E) and distribution coefficients (D). The a-BCEs alone showed better %E towards Cs^+ in their *cis*-forms (8-11%) compared to the *trans*-isomers (5-9%). D values for both *cis* and *trans* a-BCEs were obtained in the range of 0.05 to 0.15 where *cis* is slightly higher than *trans*. Addition of diethylhexyl phosphate (DEHPA) to a-BCEs improved the transfer of Cs^+ from aqueous to organic phase and showed higher %E in the *cis*-forms (21-22%) than the trans-isomers (15-16%) and DEHPA control (14-16%). D values also improved in the range of 0.15 to 0.30 where *cis* is slightly higher than *trans*. The findings indicate a synergistic extraction of Cs^+ by DEHPA and a-BCEs under the influence of UV light. The a-BCEs show potential application for Cs^+ removal in aqueous systems.

This study was supported by the National Research Foundation of Korea (NRF) grant funded by the Ministry of Science and ICT (2020R1A2C1003560, 2021R1A2C2093746, and 2021R1H1A2008284), Basic Science Research Program through the Ministry of Education (2020R1A6A1A03038817).

BORON EXTRACTION FROM AQUEOUS MEDIUM USING A DESIGNED DEEP EUTECTIC SOLVENT AND THE REGENERATION OF THE USED SOLVENT

Aya Ghazal [1,3], Hadil Abu Khalifeh [1], Ioannis Zuburtikudis [1], Inas M. Alnashef [2,3]

[1]Chemical Engineering Department, Abu Dhabi University, P.O. Box 59911, Abu Dhabi, United Arab Emirates.

[2]Chemical Engineering Department, Khalifa University, P.O. Box 127788, Abu Dhabi, United Arab Emirates.

[3] Center for Membranes and Advanced Water Technology (CMAT), Khalifa University, P.O. Box 127788, Abu Dhabi, United Arab Emirates.

Abstract

Boron element is one of the vital elements for human health-boosting, industrial applications, and plant growth. In fact, if the boron concentration in an aqueous stream would go beyond the permissible limit it should be removed and recovered. Therefore, boron extraction is targeted as a preventive measure to satisfy the standards of drinking water in desalination plants and protect the plant crops. Moreover, for production purposes, boron is recovered from industrial wastewater including nuclear wastewater streams to be reused in further processes. Based on previous experimental work of our group, the removal of boron is performed through liquid-liquid extraction (LLE) using deep eutectic solvents (DES), which are green solvents that have many desirable features like low volatility, simplicity of synthesis, biodegradability, and cost-efficiency. In this work, the hydrophobic deep eutectic solvent that consists of 2,2,4-Trimethyl-1,3-pentainediol (TMPD) and Decanol (Dec) is used as an extraction solvent to extract boron from aqueous medium at different operating conditions. The performance of the newly developed eutectic mixtures will be verified over various extraction operational parameters such as pH, A/O mass ratio and temperature. This will be accomplished through varying the investigated parameters simultaneously by applying the Design of Experiments (DoE) statistical methodology to observe the 'sweet spot' at which the targeted variables will have the maximum capacity of boron extraction. The optimization done takes into consideration the industrial and economical point of views in which the obtained Extraction Efficiency (EE) reached up to 93.1 ± 0.6 % and 91.4 ± 0.4 %. In addition, 97.9 ± 1.1 % of the boron loaded DES was regenerated using the back extraction method.

Introduction

Deep Eutectic Solvents (DESs) are classified as state-of-art green solvents for their worth advantages. Their favorable merits like simplicity of synthesis, low volatility, cost-efficiency, and biodegradability made them a potential extractants in distinct hydrophobic and hydrophilic applications. They are accepted to be a union of a hydrogen bond acceptor (HBA) and a hydrogen bond doner (HBD) where they are joined to each other by a hydrogen bond [1], [2]. The main feature that approves a mixture to be a DES is the deep depression in the melting point of the mixture compared to the individual constituents [3].

To satisfy the requirements of the end users or follow the environmental regulation of boron content in the water, boron removal become a targeted process in several water streams such as ground water, industrial wastewater, and desalinated water. Boron element stated to be the 10th most available element in oceanic salts [4]. Additionally, it is crucial to consider the recovery of boron as raw material.

Previously, a hydrophobic DES that consists of Decanol (Dec) and 2,2,4-Trimethyl-1,3-pentanediol (TMPD). In which the HBA is Dec and the HBD is TMPD. The optimum molar ratio of HBA to HBD (Dec:TMPD) was confirmed as 4:1 according to the supreme performance of the DES in extracting boron element from the aqueous medium where the Extraction Efficiency (EE) reached 95.44% [1].

Currently, the optimization of the multiple operational conditions will be demonstrated through the design of experiments (DoE) statistical methodology. The work will examine the significancy of the operational variables like temperature, pH and aqueous (A) to organic (O) mass ratio (A/O mass ratio) on the extraction process. These extraction operational conditions were optimized to discover the 'sweet spot' that designates the synthesized DES optimal EE. Minitab software was utilized to perform the DoE as it helped in generating the runs order as well as analyzing the observed response. In addition to the optimization, the regeneration of the boron loaded DES will be demonstrated. The regeneration of the (Dec:TMPD) DES will be implemented through the method of back extraction.

Experimental Part

Materials

The chemicals utilized to prepare the DES were Decyl alcohol (Dec) (+98%) that was purchased from Acros Organics, USA and 2,2,4-Trimethyl-1,3-pentanediol (TMPD) (97%) that was purchased from Sigma-Aldrich, USA. In addition, to synthesize the 1000 ppm boron solution, Boric Acid (analytical reagent grade) (>=99.8%) was purchased from Fisher Scientific, USA. For synthesizing the stripping solution, Sodium Hydroxide pellets (analytical reagent) purchased from Surechem Products, England. The chemicals were utilized as received.

Synthesizing and Characterizing the DES

To synthesize the hydrophobic Dec:TMPD 4:1 DES heating and mixing was crucial to obtain stable homogenous mixture. The DES constituents exposed shaking in an incubator shaker (Incubator shaker, KS 4000, IKA, Germany) at 80°C, 300 rpm for 4 hours. Prior to utilizing the mixture, it should be cooled down to room temperature (20°C). Properties such as density, viscosity, water content, water solubility, and leachability were optimized in the previous work beside the mass ratio of HBA to HBD that resulted in a desirable EE [1].

The combination of Dec and TMPD supplement each other to function as a powerful boron extractant. To explain, TMPD has a robust selective behavior towards extracting towards boron as a result of the stable complexes it creates with boric acid [5]. On the other side, Dec is responsible for maintaining a stable solvent [1].

The ongoing work characterizes the synthesized DES using the FT-IR spectra (Vertex 80v, Bruker) showing in Figure 1a and thermogravimetric analyzer (TGA 4000, Perkin Elmer, USA, heating rate of 10°C/min, nitrogen inert atmosphere) as shown in Figure 1b. On one side, in Figure 1a, the spectra of both Fresh DES (before extraction) and boron loaded DES (after extraction) screened on a range of 400-4000 cm^{-1} and 32 scans were assigned for each sample. On the other side, in Figure 1b, the decomposition temperature of the DES was investigated where the graph shows a complete decomposition of the DES at around 205°C.

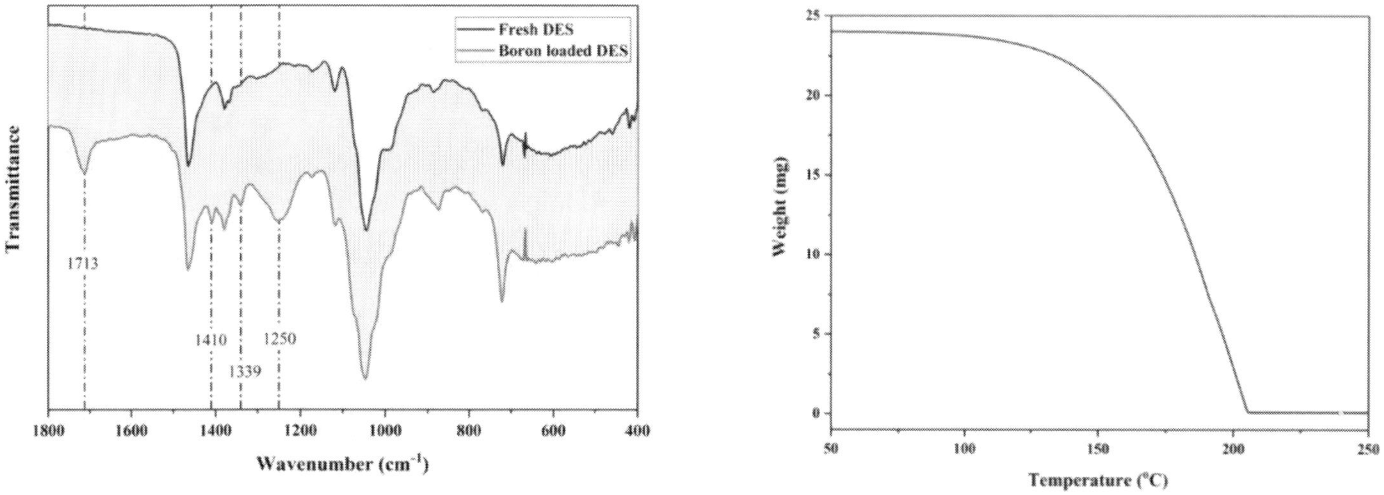

(a) (b)

Figure 1. Characterization of the Dec:TMPD (4:1) DES: (a) FT-IR spectra of the DES before and after extraction and (b) TGA thermogram of the DES.

Boron Extraction

Among the distinct ways of extraction, liquid-liquid extraction (LLE) was utilized to extract boron from the aqueous medium. The aqueous medium was synthesized from boric acid and DI water to act as a boron rich environment that has a 1000 ppm concentration of boron. The LLE process has been executed in a plastic centrifuge vial that will be placed in a mechanical incubator shaker (KS 4000, IKA®, Germany) for 3 hours at 300 rpm. The vial will comprise distinct ratios of boron solution (aqueous phase (A)) and the synthesized DES (organic phase (O)) in which upon shaking the boron molecules will be sucked by the DES leaving the water phase with much lower concentration of boron. The target of this work is to optimize the operational conditions of the LLE that will result in the optimal boron extraction from different points of view. Thus, the design of experiments methodology is in charge of determining the operational conditions that will be chosen each run. It is important to observe a complete phase separation between the aqueous and the organic phases after the LLE is done. Thus, each sample was centrifuged at 5000 rpm, 25°C for 5 minutes using (max speed 18000 rpms, Z326K, Hermle LaborTechnik, Germany). The centrifuging step results into two complete separated phases where the aqueous phase (higher density) will settle at the bottom and the DES loaded with (lower density) will remain at the top. ICP-OES (Model 5100, Agilent, USA) was utilized to observe the concentration of boron in the aqueous samples. A prior dilution for each sample is mandatory before testing samples in the ICP-OES. Therefore, a sample of around 0.5 ml was sipped from the aqueous phase within the centrifuge vial and diluted with 10 ml of DI water. The concentration of boron element was detected in the aqueous phase before and after extraction and EE was calculated through equation (1). C_f denotes the final concertation of boron in the aqueous phase related to the after-extraction sample (final concentration of boron) and C_i denotes the initial concertation of boron in the aqueous phase related to the synthesized 1000 ppm boron solution before extraction (initial concentration of boron)[5].

$$EE=\left(1-\left(\frac{C_f}{C_i}\right)\right)\times 100\% \quad (1)$$

Design of Experiments Statistical Methodology

The methodology followed for optimizing the operational conditions of the extraction process is the Design of Experiments (DoE) statistical methodology which implemented using Minitab software. The aim is to investigate the performance of synthesized DES among a wide range of operational variables. The operational variables desired to be tested, and their ranges are demonstrated in table 1. The regression analysis techniques followed was the Response Methodology (RSM) with Box-Behnken Design (BBD). The experimental runs generated consist of 17 runs related to 1 replicate with 5 center points. To decrease the influence of the uncontrolled variables the software generated experimental runs were randomized [6]. A sample of the experimental runs with the response obtained (EE) are shown in table 2.

Table 1. Operational conditions ranges of the LLE targeted for boron extraction

Operational Conditions (Constant)	Value	Continuous Factors (Manipulated)	Factors Levels
Dec:TMPD ratio	4:1	Temperature (°C)	15, 40, 65
Initial Aqueous Feed Concentration	1000 ppm	pH	2, 7, 12
		A/O mass ratio (g/g)	0.125, 4.0625, 8

Table 2. Sample of the 'Minitab' generated DoE runs, and the response (EE) observed upon performing the conditions experimentally

StdOrder	RunOrder	Temperature	A/O Mass Ratio	pH	Extraction Efficiency (EE) (%)
1	9	15	0.125	7	88.68
15	10	40	4.0625	7	82.33
13	11	40	4.0625	7	84.43
10	1	40	8	2	53.48
8	13	65	4.0625	12	1.40

Results and Discussion

After obtaining the response (EE) experimentally, all the results were inserted in 'Minitab' software to comprehensively demonstrate the model analysis. The pareto generated (Figure 2 a,b) guides to a proper identification of the variables as well as the interactions that should be considered and own a significant contribution. Initially, the pareto chart (Figure 2a) includes the impact of all the variables and interactions. To consider a variable significancy, the effect should exceed the red reference line (line 2.36, Figure 2a) [7]. As factors AB, BB, AA, A, and AC did not exceed the red line (line 2.36), a better model fit could be obtained by eliminating these variables. As a result, the temperature factor (A) was removed from the design for its insignificant single and two-way effect. Figure 2b is the modified pareto chart that indicates all significant variables that has an influence on the model. Equation (2) is a full quadratic equation that demonstrates the model summary of the data and summarizes the model validity. The determination coefficients of the model support the model validity in which R^2, R^2-adj, and R^2-pred had values higher than 90% as 97.92%, 97.23% and 94.10% respectively. The model validity confirms that it is unaffected by the elimination of the insignificant variables as the p-value of the lack-of-fit equals 0.152 ($0.152 > \alpha = 0.05$). Furthermore, the model's p-value found to be 0.000 which is a supplemental confirmation that the model is sufficiently able to describe the data [8]. The contour plot in Figure 3 demonstrates the variation of the response (EE) among wide range of pH and A/O mass ratio values. For an enhanced model fit, the contour plot (Figure 3) generated after eliminating the temperature factor for its insignificant effect in accordance with the pareto chart in Figure 3a.

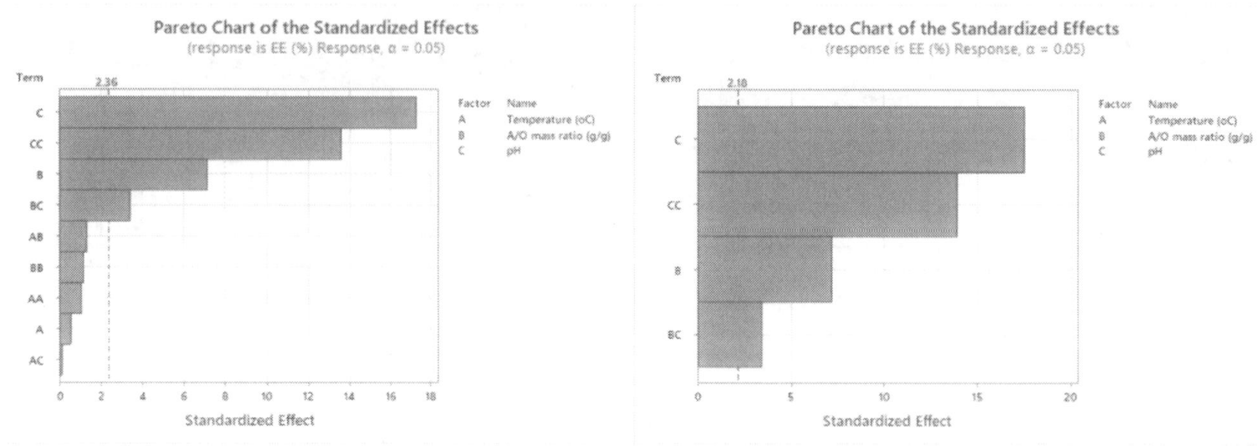

(a) (b)

Figure 2. Pareto chart of (a) showing the effect of all variables one-way and two-way interactions, and (b) after eliminating variables with insignificant impact on the model fit.

$$\text{EE (\%) Response} = 80.51 - 7.50 \text{ A/O mass ratio (g/g)} + 13.09 \text{ pH} - 1.612 \text{ pH*pH} + 0.521 \text{ A/O mass ratio (g/g)*pH} \quad (2)$$

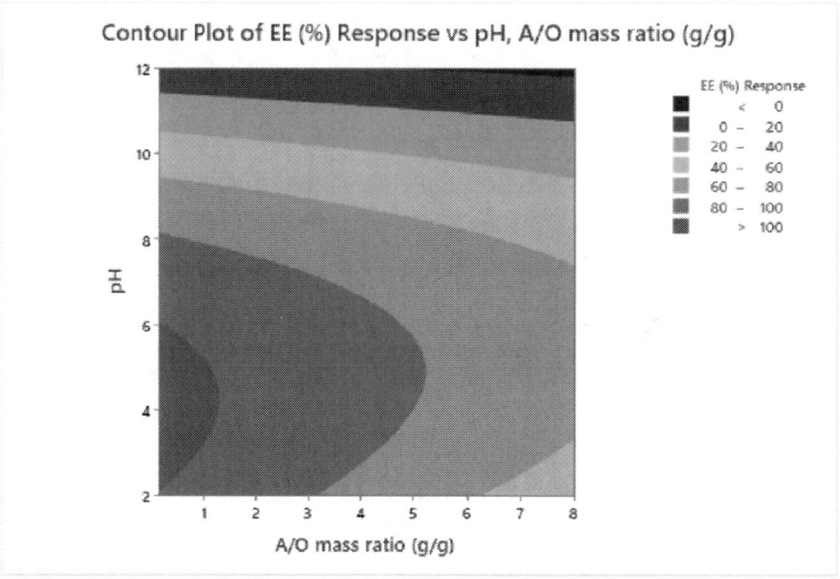

Figure 3. Contour plot to show the response (EE) over wide range of pH and A/O mass ratio

Model Validation

To validate the model generated according to the implemented DoE, it is crucial to check the validity of the predicted responses. From an industrial and economical point of view, it is convenient to treat more wastewater with lower amount of DES. As one of our aims is to meet the end user regulations of boron content in the water like those for desalination plants, the water pH should be closer to United Arab Emirates (UAE) seawater pH. A UAE seawater sample characteristics research conducted by Y Ibrahim et al. found the pH to be 7.8± 0.3 [9]. pH closer to the regional seawater makes the study more realistic. The model validation was tested for 2 different sets as indicated in table 3. Both validations were experimentally executed 3 times and their average was taken. The percentage difference between the predicted and experimental execution were less than 5% for both conditions and this suggests that the model can predict new observations sufficiently.

Table 3. Model validation for better conditions from the economical and industrial point of view.

pH	A/O Mass Ratio (g/g)	Predicted EE (%)	Average Experimental EE (%)	(%) Difference Between Predicted and Experimental EE
6	1.5	94.5	93.1 ± 0.6	1.5
5.5	1.8	95.4	91.4 ± 0.4	4.2

Regeneration of the Boron Loaded DES

One of the most notable features of DESs is being renewable and biodegradable especially for industrial applications. The biodegradability and ability to regenerate the boron loaded DES was inspected. The method followed to regenerate the used DES was the back extraction and it was calculated through finding the Stripping Efficiency (SE) equation (3) where m_W is the boron mass in the stripping agent (aqueous phase) after the stripping step and m_{DES} is the boron mass in the dirty DES before the stripping step [5]. SE was executed with the aid of a stripping agent of 0.1 M NaOH solution. When regenerating DESs the A/O mass ratio will comprise the stripping agent as the aqueous phase (A) and the boron loaded DES as the organic phase (O). The operational conditions of the back extraction process were A/O mass ratio of 1 at room temperature (25°C) for 2 hours. Single stripping cycle was adequate to regenerate 97.9 ± 1.1 % of the boron loaded DES.

$$SE = \frac{m_W}{m_{DES}} \times 100\% \quad (3)$$

Conclusion

To conclude, DoE statistical methodology was employed for the optimization of three operational conditions for the LLE process of extracting boron from aqueous media. The DES synthesized was Dec:TMPD (4:1) and the operational variables were pH, A/O mass ratio and temperature. The DoE implemented using Minitab software and by following the Response Surface Methodology. A Box-Behnken Design was used. The Design included 17 runs with 5 center points, and Minitab was used for the experimental results analysis. The analysis showed that the temperature has an insignificant contribution to the model. Thus, for a better model fit the temperature was eliminated. In addition, an optimization from industrial and economical point of view was tested where both conditions showed a robust EE of 93.1 ± 0.6 % and 91.4 ± 0.4%. The model was able to sufficiently predict the response for the new observations as, for both optimization points, the percentage difference between predicted and experimental responses was less than 5%. The regeneration of the boron loaded DES was examined using the back extraction method. 97.9 ± 1.1 % of the boron loaded DES was regenerated using 0.1 M of NaOH solution as the stripping agent. Our future plan is to expand the characterization of the Dec:TMPD 4:1 DES as well as optimizing the regeneration conditions of the used DES.

Acknowledgement

The authors are pleased to acknowledge the financial support supplied by Abu Dhabi University (Internal Fund grant no. 19300619) to Hadil AbuKhalifeh. Additionally, the authors gratefully acknowledge the support by the Center for Membrane and Advanced Water Technology at Khalifa University.

References

[1] N. Awaja, G. Almustafa, A. S. Darwish, and I. Zuburtikudis, 'Deep Eutectic Solvents for Boron Extraction', pp. 1–8.

[2] L. I. N. Tomé, V. Baião, W. da Silva, and C. M. A. Brett, 'Deep eutectic solvents for the production and application of new materials', *Applied Materials Today*, vol. 10. Elsevier Ltd, pp. 30–50, Mar. 01, 2018. doi: 10.1016/j.apmt.2017.11.005.

[3] B. B. Hansen *et al.*, 'Deep Eutectic Solvents: A Review of Fundamentals and Applications', *Chemical Reviews*, vol. 121, no. 3. American Chemical Society, pp. 1232–1285, Feb. 10, 2021. doi: 10.1021/acs.chemrev.0c00385.

[4] M. Tagliabue, A. P. Reverberi, and R. Bagatin, 'Boron removal from water: Needs, challenges and perspectives', *Journal of Cleaner Production*, vol. 77. Elsevier Ltd, pp. 56–64, Aug. 15, 2014. doi: 10.1016/j.jclepro.2013.11.040.

[5] G. Almustafa *et al.*, 'Liquification of 2,2,4-trimethyl-1,3-pentanediol into hydrophobic eutectic mixtures: A multi-criteria design for eco-efficient boron recovery', *Chemical Engineering Journal*, vol. 426, Dec. 2021, doi: 10.1016/j.cej.2021.131342.

[6] C. Mao, B. Zhang, X. Tang, H. Li, and S. He, 'Optimized preparation of zinc-inorganic antibacterial material containing samarium using response surface methodology', *Journal of Rare Earths*, vol. 32, no. 9, pp. 900–906, 2014, doi: 10.1016/S1002-0721(14)60161-7.

[7] Jiju Antony, *Design of Experiments for Engineers and Scientists*, 2nd ed. Edinburgh: Elsevier, 2014.

[8] E. Svinterikos and I. Zuburtikudis, 'Tailor-Made Electrospun Nanofibers of Biowaste Lignin/Recycled Poly(Ethylene Terephthalate)', *J Polym Environ*, vol. 25, no. 2, pp. 465–478, Jun. 2017, doi: 10.1007/s10924-016-0806-3.

[9] Y. Ibrahim, F. Banat, A. F. Yousef, D. Bahamon, L. F. Vega, and S. W. Hasan, 'Surface modification of anti-fouling novel cellulose/graphene oxide (GO) nanosheets (NS) microfiltration membranes for seawater desalination applications', *Journal of Chemical Technology and Biotechnology*, vol. 95, no. 7, pp. 1915–1925, Jul. 2020, doi: 10.1002/jctb.6341.

466d Molecular Design of Solvents for Temperature Swing Solvent Extraction

Lauren Ward, Luke Rakers, Shuai Qian, Gabriel Barbosa, C. Heath Turner, Jason Bara, Steven Weinman

Desalinating or purifying high salinity brine water presents a unique challenge. While reverse osmosis (RO) through a semi-permeable polyamide membrane has shown positive results for low salinity brine water and seawater, it is not effective to treat high salinity brine waters where the high osmotic pressure requires unreasonable operating pressures. High salinity brine is found in industrial processes and RO concentrate and is multiple times saltier than sea water. Temperature Swing Solvent Extraction (TSSE) is a recently re-discovered technique that could achieve desalination of these high salinity brines. TSSE is done by adding brine water to an amine-based or carboxylic acid-based solvent and manipulating the temperature of that mixture. The solvent's polarity is temperature dependent, such that by changing the temperature of the solution, the solvent will extract water from the brine at one temperature and then phase separate the water out at a different location when that temperature is changed. In the case of amine-based solvents, the solvent will dissolve water at a lower temperature and, by increasing the temperature, the water solubility in the solvent will decrease. The goal of this study is a rational molecular design process driven by experiments and computations by which to discover and assess optimal systematically-varied solvents that can improve the efficiency of the TSSE process. For each solvent, the efficiency was measured in terms of the amount of water the solvent extracted and released, the salinity of the brine phase after contact with the solvent, the salinity of the released product water, and the amount of solvent in the water phase at the low and high temperatures. Initial results have shown that certain amine compounds appear more favorable for solubilizing water while others appear to be more water soluble.

466e Relative Stability Complexes of Lanthanide-Ligand Complexes from Computation

Ravi O'Brien, Thomas Summers, Danil Kaliakin, David Cantu

Relative stability constants of lanthanide (Ln)-ligand complexes quantify the selectivity of a ligand toward specific lanthanide ions, which is crucial in the separation of rare earth elements from one another. Ln-ligand stability constants can be measured experimentally or calculated with binding energies. Electronic structure calculations were done to obtain binding energies of La^{3+}, Eu^{3+}, Gd^{3+}, and Lu^{3+} with ethylenediaminetetraacetic acid (EDTA) based on the structures of Ln-EDTA complexes in solution resolved with ab initio molecular dynamics simulations. The binding energies were used to calculate the relative stability constants and compared to experimentally measured ones. Ab initio molecular dynamic simulations were done at both 25 °C and 90 °C to determine how variations in temperature affect complex stability. In addition, the EDTA ligand was simulated at two protonation states corresponding to pH ~7 ($HEDTA^{3-}$) and pH ~11 ($EDTA^{4-}$) to determine the effect of solution pH has on stability of Ln-EDTA complexes. Results demonstrate that the trend of increasing complex stability along the lanthanide series was found to be correlated to decreasing structural deviations with an increase in temperature. Moreover, by protonating one amine group of the EDTA ligand, the solution structure of the Ln-ligand complex is significantly altered in a way that reduces the stability of the complex.

466f DFT Calculations for Solvent Extraction Desalination: Electrostatic Potential, Hydrogen Bonding, and Solvation Free Energy

Xiaoyang Liu, C. Heath Turner

There is a need for improved desalination technologies in order to treat naturally-occurring and industrial brines [1]. Current options generally include membrane-based processes, such as reverse osmosis (RO) [2], as well as thermal- or evaporation-based processes like multi-stage flash (MSF) and multi-effect distillation (MED) [3]. However, these relatively mature technologies are challenging to deploy in developing regions, due to a variety of limitations: the evaporative processes involve high energy consumption, both processes require advanced support infrastructure and large centralized installations [4], and the operation and maintenance of membrane systems can be prohibitively expensive [5].

Solvent extraction desalination (SED) is an alternative membrane-free desalination technology, which only requires low grade, low temperature heat, and it is ideally suited to processing high-concentration brines. However, new solvents are still needed that can improve the molecular-level water selectivity, while minimizing solvent contamination in the aqueous phase. Improvements in solvent design for SED require a fundamental understanding of the molecular interactions governing solvation behavior. Herein, quantum chemical calculations are used to delineate different fundamental interactions in water-solvent systems involving 38 different solvents, with a particular focus on the role of the electrostatic potential characteristics. These interactions are rigorously analyzed using the independent gradient model based on Hirshfeld partition (IGMH), Symmetry-Adapted Perturbation Theory (SAPT), as well as the energy decomposition analysis based on force field (EDA-FF) to compare against a standard force field model. The solvation free energies and partition coefficients are also calculated to estimate water extraction performance in a bulk solvent system.

References:

[1] D.L. Shaffer, L.H. Arias Chavez, M. Ben-Sasson, S. Romero-Vargas Castrillón, N.Y. Yip, M. Elimelech, Desalination and Reuse of High-Salinity Shale Gas Produced Water: Drivers, Technologies, and Future Directions, Environ. Sci. Technol. 47 (2013) 9569-9583.

[2] L. Malaeb, G.M. Ayoub, Reverse Osmosis Technology for Water Treatment: State of the Art Review, Desalination 267 (2011) 1-8.

[3] A.D. Khawaji, I.K. Kutubkhanah, J.-M. Wie, Advances in Seawater Desalination Technologies, Desalination 221 (2008) 47-69.

[4] G. Ni, S.H. Zandavi, S.M. Javid, S.V. Boriskina, T.A. Cooper, G. Chen, A Salt-Rejecting Floating Solar Still for Low-Cost Desalination, Energy Environ. Sci. 11 (2018) 1510-1519.

[5] S. Alotaibi, O.M. Ibrahim, S. Luo, T. Luo, Modeling of a Continuous Water Desalination Process Using Directional Solvent Extraction, Desalination 420 (2017) 114-124.

466g COSMO Screening of Hydrophobic Deep Eutectic Solvents for Platform Chemical Extraction from Water

Thomas Quaid, Toufiq Reza

Platform chemicals are value-added products from the biomass processing industries. Platform chemicals are generally produced in an aqueous solution which can be difficult to retrieve from due to their tendency for condensation at high concentrations. Using common thermal separation techniques causes polymerization among platform chemicals which makes liquid-liquid extraction favorable. In contrary, hydrophobic deep eutectic solvents (HDES) have been shown to be highly efficient at extracting platform chemicals from aqueous solution. HDES are relatively cheap, non-toxic, and environmentally benign, which make them excellent candidates for separating platform chemicals from water. Very few experimental studies have been conducted on the application of HDES in platform chemical extraction. This study looks at a comprehensive sample size of HDES (>100 solvents) and the driving forces of absorbance and selectivity for three representatives of platform chemicals: hydroxymethyl furfural, levulinic acid, and sorbitol. These three represent the three classes of platform chemicals of furan, acids, and alcohols. A conductor like screening model (COSMO) is adopted to screen suitable hydrogen bond acceptors (HBA) and hydrogen bond donors (HBD) for platform chemical extraction based on hydrogen bond properties using a triple-zeta valence polarization (TZVP) basis set. This model is evaluated compared to literature experimental studies to verify integrity of the results. It has been found that HDES with strong preference for vander Waals interactions have the highest selectivity for the platform chemicals over water. The HDES that have hydrogen bonding capabilities have a less selectivity for separating the analytes from water. These observations correspond with solvent characteristics of alkyl chain length, polarizability, and resonance structure. From the results of the binary solvents, ternary solvents were examined for synergistic effects of selectivity and absorptivity. The results of this were preferential to the acidic platform chemicals since they have a higher affinity for water than the others.

704a Polyethylene in Dead End Silica Nanopores: Forces and Mobility from Non-Equilibrium Statistical Mechanics and Exsy NMR

Ziqiu Chen, Baron Peters, Frédéric Perras

Billions of tons of plastic have been produced and only a small fraction of this has been recycled. Tennakoon et al. [Nature Catalysis 3, 893 (2020)] developed a catalyst that repeatedly cleaves C_{10}-C_{30} hydrocarbons from the end of a polyethylene chain. The reaction occurs at a Pt nanoparticle at the base of a cylindrical silica mesopore of diameter 2nm and length 100nm. ^{13}C NMR can distinguish anti, gauche, and mobile polymer configurations, and thereby reveal the amount of adsorbed polymer adsorbed (with *anti* configuration) in the narrow pores. Moreover, 2D EXSY NMR can reveal the time scale and extent of fluctuations as polymer chains move in and out of the pores. We construct a linear model for the free energy as a function of the leading polymer segment position in the pore. We then construct and solve a Fokker-Planck equation for the overdamped dynamics of the polymer position in the pore. The solutions allow us to predict the intensities of the 2D NMR peaks as a function of time. We use the solutions to extract a polymer diffusivity at each temperature and to estimate per-segment free energy, enthalpy, and entropy values for desorption.

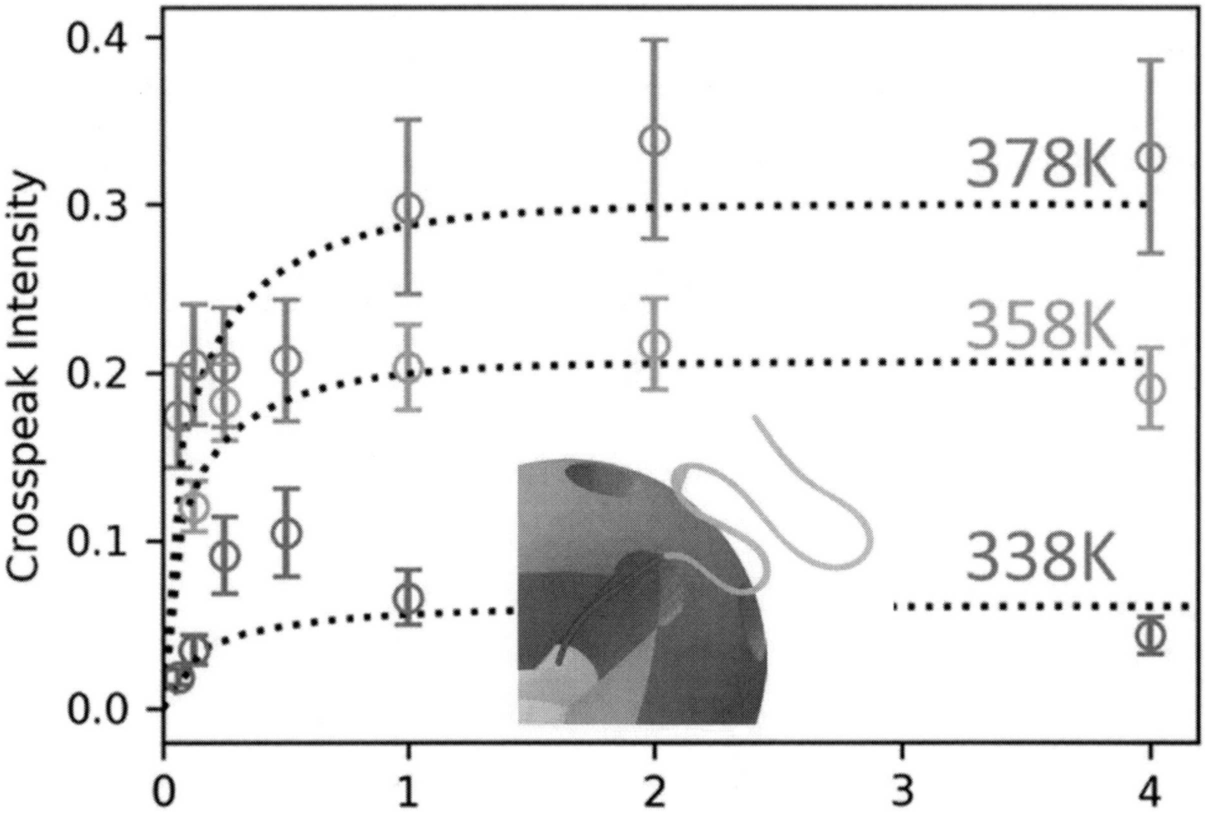

704b Analysis of the Non-Isothermal Model for Volumetric Diffusion Measurements

Stefano Brandani, JinYu Wang, Enzo Mangano, Federico Brandani, Pluton Pullumbi

Experiments carried out in a volumetric apparatus can be performed with small pressure steps to ensure linearity; nonetheless, for fast diffusing species it is difficult to achieve isothermal conditions. For this reason, a model that combines heat and mass transfer in a porous solid is developed and solved analytically to include the main effect of non-isothermality, which is the temperature dependence of the equilibrium amount. The model leads to a set of dimensionless groups and eigenvalues that can be used to understand under which limiting conditions simplified versions are applicable: isothermal; full thermal control; valve control; and regions where distinct mass or heat transfer parameters dominate the kinetics. The analysis will also show that there are ranges of parameters, where the isothermal and non-isothermal model will have the same shape. Finally, a set of recommendations will be given on how to detect the influence of the combined mass and heat transfer mechanisms.

704c The Synthesis of Hierarchical SAPO-34 Zeolite and Potential Adsorption Applications

Obaid Khan, Ninad D. Anjikar, Orhan Talu, Shaowei Yang

Zeolites are among the most mature adsorbents and catalysts. The uniform pore sizes defined the crystalline structure are advantageous for achieving molecular sieving effects for enhanced performance. On the other hand, the small sub-nanometer sized pores are also adding high transport resistance for the molecules to reach internal pores, particularly for molecules approaching the pore size. To reduce the transport resistance, the zeolite crystals could be made as either nanosized or hierarchical with larger non-zeolitic pores. However, there are typically high cost associated with the specially designed synthesis precursor or the mesopore generating structure directing agents (SDAs). In this work, we will discuss the synthesis of hierarchical SAPO-34 zeolite using a cheap secondary SDA by a dual-SDA strategy. The influence of different secondary SDAs on resultant zeolite structure and pore size distribution will be discussed. The hierarchical zeolite will be further investigated for adsorption isotherms and dynamic uptake of a few hydrocarbon molecules and compared with conventional SAPO-34 zeolite. This work will help reveal the potential applications enabled by reduced transport resistance and possible limitations of the hierarchical SAPO-34 zeolite.

704d Effect of Particle of Size on the Single Gas Diffusion of CO_2 and N_2 in 13X Zeolite Using a 100 Hz Volumetric Frequency Response System

Armin Ebner, Sarah Gustafson, Adam M. Burke, Charles E. Holland, James A. Ritter

Frequency response methods have been developed as an alternative technique for the determination of mass transfer mechanisms controlling the transport of gases into and out of microporous adsorbent materials. Recent results in the literature are showing that this technique seems to be the best for discriminating the controlling mechanism. It is also possible to reveal two or more different controlling mechanisms that may dominate in different frequency ranges. This presentation will focus on results obtained from a 100 Hz volumetric frequency response (VFR) system for exploring the diffusion of CO_2 and N_2 in 13X zeolite particles of different sizes.

It is well established that the transport of both N_2 and CO_2 in 13X zeolite is macropore limited. A newly developed volumetric frequency response analyzer that can operate up to 100 Hz is utilized to evaluate the role of the adsorbent particle size on the adsorption kinetics of these two species as single gases on 13X at temperatures of 30 and 50 °C and pressures of 100, 200 and 760 Torr. The experimental results will be evaluated against kinetic models to assess the possible transition from macropore limited diffusion to micropore limited diffusion of these species as the adsorbent particle size decreases. The latest results of this study will be presented.

704e Unraveling Water-Ion Dynamics in Reverse Osmosis Membranes with Nuclear Magnetic Resonance Spectroscopy

Marta Hatzell, Mahsa Abbaszadeh

The development of highly permeable thin-film composite (TFC) polyamide membranes with high salt rejection properties has allowed reverse osmosis (RO) to become the leading technology in desalination. The-state-of-art Reverse Osmosis (RO) membranes are being used industrially for desalinating high salinity (seawater) and lower salinity (brackish water) where unwanted ions are retained by the membrane resulting in the production of drinking water. Most commercial RO membranes possess an aromatic polyamide active layer which is made via interfacial polymerization of m-phenylene diamine (MPD) in the aqueous phase and trimesoyl chloride (TMC) in the organic phase. As the demand has grown for selective separation of single solute from water, more focus is needed on understanding the water-ion dynamics at the membrane interface. The exact nature of ion and water transport in polymeric membranes is often ambiguous and depends on the complex interplay between polymer structure and dynamics that facilitate transport. Experimentally, the Nuclear magnetic resonance (NMR) technique has a great potential to be used as an emerging technology for characterizing the structure and mobility of solutes in polymeric membranes. We have investigated the dynamics-transport of water and sodium ions in complex ionic mixtures of a polyamide polymeric membrane using solid-state NMR spectroscopy. Water and sodium ion diffusion measurements are made using pulsed-field gradient NMR diffusometry. The energy barriers for water and ion transport within the active layer of the RO membrane are assessed using magic angle spinning (MAS) and static NMR spectral line shape analysis. Through monitoring ^{13}C, ^{23}Na, and ^{1}H NMR nuclei, quantitative insights into the diffusion of water (^{1}H), sodium mobile ion (Na^{+}), and polymer membrane dynamics (^{13}C) are extrapolated. Furthermore, the spin-lattice (T_1) relaxation time of sodium and water ions is investigated to obtain quantitative insight into how sodium and water ions rotational movements occur in the absence and presence of other competing seawater cations such as potassium. Room temperature spectra and spin-lattice relaxation times indicated the presence of both mobile and rigidly held ionic species in the polyamide at different levels of relative humidity. These species are attributed to a highly immobile, 'bounded' (or aggregated) sodium/hydrogen and a significantly more mobile sodium/hydrogen for which concentration is temperature, ionic, and polymer dependent. NMR relaxation measurements detect how the water rotation is modified by sodium and potassium cations in the polyamide membrane. Results show that at low relative humidity and in the presence of competing ions, there are sodium ion resonances for crystalline sodium ions and rigidly held sodium ions on the surface. With increasing relative humidity and decreasing competing ions, the resonance suggests solution-like dynamic rotational movements of sodium ions in polyamide. The results confirm the change in activation energy barrier for sodium and hydrogen ion permeation in polyamide upon the presence of potassium ions.

704f Investigation of CO_2 Adsorbents for Direct Air Capture: Equilibrium, Kinetic, and Stability Data

May-Yin Low, David Danaci, Hassan Azzan, Lucy Victoria Barton, Camille Petit

The 2015 Paris Agreement saw 190 countries plus the European Union commit to limiting global temperature rise to below 2 °C, and preferably 1.5 °C. According to the latest reports by the Intergovernmental Panel on Climate Change, this can only be achieved with negative emission technologies (NETs). The use of solid adsorbents to remove CO_2 directly from the atmosphere, *i.e.* direct air capture (DAC), is a NET actively being investigated. However, due to the low concentration of CO_2 in the atmosphere, it remains a challenging process.

A first generation of pilot scale adsorption-based DAC units have already been commercialized. However, there is still a need for accurate performance, energy, and cost estimates for these processes. This requires the collection of critical data such as an adsorbent's working capacity in the relevant low-pressure range, mass transfer kinetics, selectivity towards other gases, and long-term stability. Today, much of this data is scarcely available.

Our work aims to address this knowledge gap by collecting such data, which is necessary to evaluate DAC adsorbents using process modelling, and eventually to assess their potential at scale. This will allow for improvements in the adsorbent design, which could then further reduce the cost and energy requirements and improve their competitiveness in comparison to other negative emission approaches or technologies.

In this contribution, we have investigated as potential DAC adsorbents: a commercial resin (adsorbent A), and TIFSIX-3-Ni, an ultramicroporous MOF first developed by Kumar et al. [1]. We have compared their performance to that of a benchmark DAC adsorbent, *i.e.* Lewatit VP OC 1065. To study the equilibrium sorption properties of these adsorbents, we have collected volumetric isotherms of CO_2 and N_2 between 15 °C and 120 °C, of H_2O at 15 °C, 25 °C, and 35 °C, and of Ar and O_2 at 25 °C. We are using this data to perform isotherm fittings and to calculate the heats of adsorption for CO_2, N_2, and H_2O. To investigate the dynamic sorption properties of these adsorbents, we are collecting kinetic data using both an in-house dynamic adsorption setup and a thermogravimetric analyzer (TGA). To assess the stability of these adsorbents, we are performing cyclic adsorption-desorption tests at 400 ppm CO_2 *via* TGA, and are investigating the CO_2 uptake performance and structural stability of TIFSIX-3-Ni after exposure to 90% relative humidity.

Our results demonstrate that adsorbent A has the highest CO_2 uptake at 400 ppm levels, but still maintains comparable N_2, Ar, O_2, and H_2O uptakes to the benchmark. TIFSIX-3-Ni is shown to have the lowest CO_2 uptake at 400 ppm levels and the highest N_2, Ar, O_2, and H_2O uptakes. Yet, preliminary investigations point to faster adsorption kinetics compared to the other two adsorbents. Based on these results thus far, both adsorbent A and TIFSIX-3-Ni could prove to be strong candidate adsorbents for DAC.

References:

1. A. Kumar, C. Hua, D. G. Madden, D. O'Nolan, K.-J. Chen, L.-A. J. Keane, J. J. Perry, and M. J. Zaworotko, "Hybrid ultramicroporous materials (HUMs) with enhanced stability and trace carbon capture performance," *Chemical Communications*, vol. 53, no. 44, pp. 5946–5949, Apr. 2017.

704g Iron-Based Metal-Organic Framework for Superfast Adsorption of Uncharged Urea from Water: Synthesis, Characterization and Mechanism

Tin Le, Milad Esfahani

Urea is widely used in fertilizer production for agricultural purposes which risks runoff into soil and water sources. An excess of urea can result in algal or toxic blooms which can poison wildlife or even humans by accumulation in food sources. Much research has been conducted on a variety of methods to remove environmental using adsorbents, but most of them lack effective removal, long (> 2 h) process time, lack re-generality, and losing materials after a few adsorption processes. Metal organic frameworks (MOFs) are the new generation of adsorbents with excellent structural and functional group tunability. In this study, we synthesized

MIL-100 (Fe), an iron-based metal-organic framework (MOF), as an efficient adsorbemt for removal of uncharged urea from water. The urea adsorption capacity of MIL-100 (Fe) was tested under varying experimental conditions such as pH (2-10), temperature (25-45°C), MOF concentration (25-400 ppm), and urea concentration (10-1000 ppm). Removal data showed the superfast adsorption (reached almost 90% removal within 1 min) of neutrally charged urea molecules. The MOF was able to reach a maximum adsorption efficiency of around 98% with a maximum uptake capacity of 4430 mg/g. MIL-100 (Fe) retained up to 88% removal efficiency after regeneration. In addition, the adsorption kinetics, mechanism and behavior of urea on Fe-MOF were studied at different temperatures and pH.

145a Optimization of the Vapor Flow Distribution in a Distillation Column Using Computational Fluid Dynamics

Yu Feng, Tony Cai

Chemical separations account for approximately half of the U.S.'s industrial energy use and 10-15% of the U.S.'s total energy consumption. More energy-efficiency chemical separation processes could save 100 million metric tons of CO_2 emissions annually. Specifically, distillation is the most widely used chemical separation technology, which accounts for approximately 40% of total energy consumption in petrochemical and chemical plants. Unfortunately, the overall efficiency is just around 11%. Therefore, it is necessary to improve the distillation efficiency, of which optimization of the spatio-temporal vapor distribution inside the column is essential. In this study, a computational fluid dynamics (CFD) model has been developed and employed to predict the vapor velocity distributions in a virtual 3-dimensional (3D) distillation column, with multiple inlet diameters (12 inches and 15 inches) and vapor flow rates (from 7929 lb/h to 27,750 lb/h) as the two key design parameters. The column height is approximately 8 feet, and the column diameter is 4 feet. Mellapak 250Y (M250Y) is used as the packing bed. Velocity contours and secondary flows at the top of the packed bed were compared to find the optimal design, with the minimum velocity variations across the plane. Coefficients of variation on the velocity distributions were also calculated and compared. Results indicate that with the increase of inlet Reynolds number, either due to the increase in inlet flow rate or the decrease in inlet diameter, the flow distributions become more uneven at the cross-section when the vapor flow exit the porous media region. Therefore, a low inlet Reynolds number is recommended for achieving a more evenly distributed vapor flow velocity.

145b Gravity Flow and Internal Piping in Distillation Columns – the Devil Is in the Details

Andrew Sloley

Internal piping and ducts are often used in inside distillation towers. Applications include routing liquid overflow from collectors, between geometry transitions (trays to packing or changes in tray passes), complex tray geometries (some types of radial flow trays and orbit trays), and mixing internal liquid flows with externally supplied liquids (feed and pumparound returns). Internal supplied liquid may enter piping in some combination of six major flow regimes. These are (a) weirlike flow, (b) full pipe flow, (c) Borda free flow with orifice control, (d) Borda full flow, (e) perfect weir flow, and (e) Borda free flow with weir control. Each of these has a different effective entrance coefficient into the piping. The entrance coefficient and flow rate determine the actual liquid head necessary inside the column. These flow regimes are examined and recommendations are made on why every effort should be made to seal piping to encourage full pipe flow.

145c Quantifying the Effect of Stripping Factor on Tray Column Efficiency Under Non-Total Reflux Conditions

Rukhsar Ahmed, Tony Cai, Kenneth McCarley, Sayeed Mohammad, Michael Miranda, Clint Aichele

Optimal design of distillation columns requires reliable estimates of column efficiencies. Several models and correlations have been developed over the years to predict column efficiencies. However, these models may not be truly representative of actual efficiencies observed in the plant as most experimental data used to develop these models were obtained from distillation tests under total reflux conditions. Since commercial scale distillation columns operate under non-total reflux, there is a need to obtain accurate experimental data on efficiencies under non-total reflux conditions. These data will be helpful in developing reliable efficiency models using data on both total reflux and non-total reflux conditions.

In this study, a modified Oldershaw column was installed and commissioned that is capable of operating under total as well as non-total reflux modes of operation. Furthermore, the Oldershaw column was validated with cyclohexane/n-heptane test system. Additional test systems were carefully selected to ensure that large VLE errors are avoided. The test systems also provided a wide variation in physical properties for the measurement of distillation column efficiencies. The effect of stripping factor on distillation column efficiency was studied by systematically varying the stripping factor for several test systems. The point efficiencies of the column under non-total reflux were compared with the results of our previous study under total reflux conditions. This paper will discuss important findings from the study.

145d Optimal Design and Operation of Ternary Hybrid Dividing Wall Columns

Dian Ning Chia, Fanyi Duanmu, Eva Sorensen

Within the chemical process industry, distillation remains unrivalled as the main method of liquid-vapor separation. However, when it comes to azeotropes, which have the same vapor and liquid compositions at a certain temperature and pressure, a single simple distillation column is insufficient as a mode of separation. For a binary azeotropic mixture, a typical way to separate azeotropes is to use an extractive distillation process, where a solvent and additional columns are needed, thus increasing the overall energy consumption and capital costs. A hybrid distillation process, which integrates a membrane with at least one distillation column, is often suggested as a more energy intensive alternative to the extractive distillation process. Due to the separation principle of membranes, which does not depend on vapor-liquid equilibrium, the separation of an azeotrope can be carried out without the addition of a solvent. Although a hybrid distillation process is a good candidate for separation of azeotropic mixtures, the application of a membrane unit can be restricted due to its low capacity (when compared to the high throughput of a chemical plant) and high capital costs (Lee et al., 2016). In order to overcome these limitations, instead of using the membrane to purify the product to meet the product specifications, the membrane can be used merely to break or cross the azeotropic point. The downside of such a structure is that another column is therefore needed to purify the permeate/retentate to the product specification. Nevertheless, the two distillation columns can be combined into a single column shell, thus forming a dividing wall column. The integration of the DWC with a membrane system will then form a hybrid dividing wall column (H-DWC) as proposed in our previous work (Chia et al., 2022).

In the previous work (Chia et al., 2022), the application of the H-DWC for the separation of minimum boiling binary azeotropic systems was explored, where the H-DWC was compared with three different hybrid distillation-pervaporation processes: a membrane system installed before a column (pervaporation-distillation, P-D structure), a membrane system installed after a column (distillation-pervaporation, D-P structure), and a membrane system installed between two columns (distillation-pervaporation-distillation, D-P-D structure). The comparison showed that D-P-D and H-DWC structures have similar performances with the least total annualized cost, compared to P-D and D-P, for the binary systems under consideration. The H-DWC is, however, superior to the other hybrid processes, including D-P-D, with increasing number of components in the feed, because with more components, more distillation columns will be required for the other hybrid processes. For the H-DWC, a single column with multiple walls is sufficient, i.e. only one column shell is needed, leading to reduced capital and/or operating cost for multicomponent systems. Energy savings can be achieved with the middle boiling component(s) freely distributed around the dividing wall(s), thus reducing the mixing effect within the column. To further explore the potential of the H-DWC, in this work, a H-DWC design for a ternary mixture with a minimum boiling binary azeotrope is proposed, and compared to the other three conventional hybrid processes (i.e. the P-D, D-P, and D-P-D structures), as well as to a standard extractive process.

The case study considered is the separation of a ternary mixture with a minimum boiling binary azeotrope. For the three conventional hybrid processes (D-P, P-D and D-P-D), an extra distillation column is required to remove either the lightest component A or the heaviest component C depending on the azeotrope. The proposed ternary H-DWC has two dividing walls and can separate all the three components in a single column shell.

Figure 1 illustrates the process when the azeotrope is between the lightest two components A and B, and where it is also assumed that the concentration of component A is lower than the composition of component B, i.e. all of component A forms the azeotrope with most of component B. In addition, the figure assumes that component B permeates through the membrane. The remaining component B is freely distributed around the first wall in the column. In the illustration, the composition of component B is built up in the middle section of the column (i.e. section between the first and second dividing wall), and therefore high-purity component B can be collected there. The heaviest component C is purified at the bottom of the H-DWC. The azeotrope is processed in the membrane unit, where most of component B permeates through the membrane and is mixed with component B from the middle of the column. The retentate stream, here containing component A and some azeotrope, is recycled back to the column. Component A can be recovered from the H-DWC at the right side of the column (i.e. opposite of the feed entry) as only the azeotrope and component A are present in that location, and because component A is heavier than the A-B minimum boiling azeotrope.

For all the hybrid structures studied (P-D, D-P, D-P-D, and H-DWC), the membrane network considered the number of membrane stages in series, as well as the number of membrane modules in each membrane stage, as part of the optimisation (Chia and Sorensen, 2022). All the design and operating conditions are optimized simultaneously, i.e. number of stages, feed location and reflux ratio etc for the columns, and for the DWC also the location of the walls, the number of stages per section, and the liquid split between the walls. All simulations are performed in gPROMS Process (Process Systems Enterprise, 2021) and all the optimization variables (i.e. variables in the membrane network and the columns) are optimized simultaneously using Genetic Algorithm (Chia et al., 2021) coded in MATLAB (The MathWorks Inc., 2019).

This work will provide a comprehensive comparison between the separation and economic performances of the H-DWC, the three other distillation-pervaporation hybrid processes (D-P, P-D, D-P-D), as well as a traditional extractive distillation process, for the separation of a ternary system with a binary minimum boiling azeotrope between either A and B, or B and C, and for different feed composition conditions. The results will show that the HDWC can save both capital and/or operating cost due to its highly integrated design, thus showing its great potential for industrial applications.

References

Chia, D. N., Duanmu, F., and Sorensen, E. (2021). Optimal Design of Distillation Columns Using a Combined Optimisation Approach. In Turkay, M. and Gani, R., editors, *31st European Symposium on Computer Aided Process Engineering*, pages 153–158. Elsevier B.V.

Chia, D. N., Duanmu, F., and Sorensen, E. (2022). Optimal Design of Hybrid Dividing Wall Columns for Azeotropic Separations, to be presented at Distillation andamp; Absorption 2022, Toulouse, September 2022.

Chia, D. N. and Sorensen, E. (2022). Optimal Design of Hybrid Distillation/Pervaporation Processes, to be presented at PSE2021+, Kyoto, June 2022.

Lee, H.-Y., Li, S.-Y., and Chen, C.-L. (2016). Evolutional Design and Control of the Equilibrium-Limited Ethyl Acetate Process via Reactive Distillation–Pervaporation Hybrid Configuration. *Industrial andamp; Engineering Chemistry Research*, 55(32):8802–8817.

Process Systems Enterprise (2021). gPROMS Process version 2.2.

The MathWorks Inc. (2019). MATLAB R2019b version 9.7.

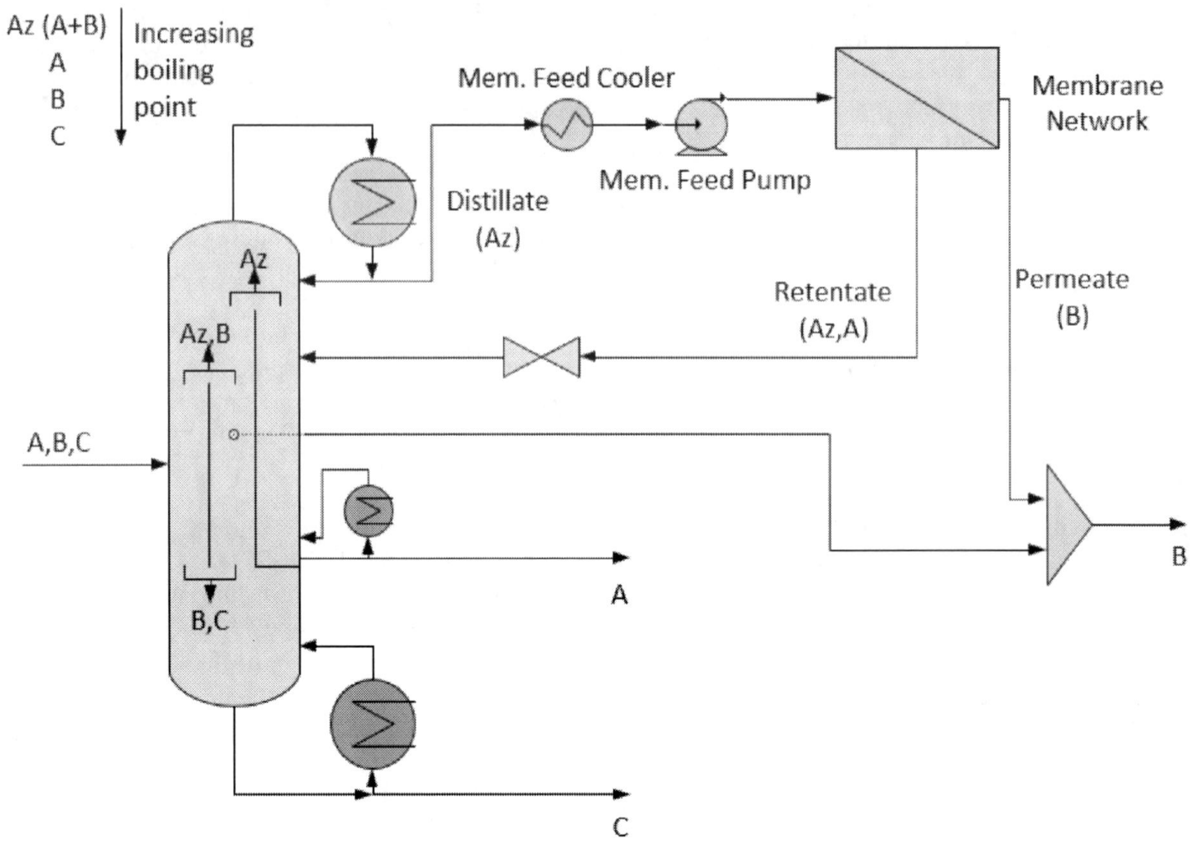

Figure 1. Hybrid dividing wall column (H-DWC) for separation of a ternary system with a binary azeotrope between components A and B, with a composition of component A lower than that of component B, and where component B permeates through the membrane.

145e Performances of Advanced Structured Packings Produced By 3D Foam Printing for Gas-Liquid Multiphase Contactors in FGD Process Intensification

Domenico Flagiello, Daniele Tammaro, Alessandro Erto, Pierluca Maffettone, Amedeo Lancia, Francesco Di Natale

Introduction

Multiphase contactors are widespread in chemical engineering processes and applied for the synthesis and purification of compounds for petrochemical, food, chemical and pharmaceutical industries, and many others targeted applications, e.g. flue-gas cleaning. Packed-bed towers are largely adopted for the process intensification of gas-liquid and liquid-liquid processes, thanks to their capacity to promote interfacial mass transfer rate. Moreover, they allow exploiting the liquid-phase chemical reactions at their highest level, and a reasonable pressure drop costs. Over the last 50 years, several studies[1-3] investigated the performance of different types of random and structured packings for different applications, correlating their performances with the geometric characteristics of commercial packings provided by major vendors, e.g. Sulzer, Koch Glitch, Raschig, etc.

In the last decade, 3D printing technology has increasingly spread in industrial applications to substitute conventional manufacturing processes and to allow faster cheaper, and in-situ production of commercial products[4]. Recently, this technology has been adopted to produce several process equipment, such as gas and liquid distributors, mixers, catalytic monoliths and packings for packed towers[5,6]. For packings, the main benefit of 3D printing is the possibility of producing plastic items in an extremely flexible manner, to meet different application needs and allow a process intensification, thanks to an easier tailor-made design for specific plant size and operating conditions. Besides, the wide range of 3D printing materials and the possibility to perform different types of printing add further flexibility to the production of packings for chemical reactors and separators.

In this paper, we report an extensive study on the performance of a Y-type corrugated packing produced by 3D foam printing. The packing was tested in a desulfurization chemical absorption tower (Wet-FGD) chosen as a model liquid-gas multiphase reactor due to its key role in clean coal technologies, for the reduction of sulfur compounds emissions. Experiments aimed at evaluating the pressure drops, mass transfer rates (i.e. the liquid and gas side mass transfer coefficients) and the SO_2 removal efficiencies of a pilot-scale scrubber equipped with the new packing operated in the typical conditions of wet-FGD processes for coal-combustion plants. The experimental results were compared with those of a commercial Y-series metallic packing, Mellapak 250Y (M250.Y) by Sulzer Chemtech. The pressure drops for Mellapak 250Y were recovered by the experiments of **Brunazzi and Paglianti**[7] who tested the M250.Y in both dry and wet-mode using an air/water system at 25°C and 1 atm. The mass transfer coefficients were calculated with the **Brunazzi and Paglianti**[3] model which is based on an experimental dataset on absorption columns equipped with M250.Y and Sulzer BX. Finally, the M250.Y desulfurization efficiencies were calculated by process simulations in ASPEN PLUS operating under the same experimental conditions of the present work. The simulations follow the same approach proposed in **Flagiello et al.**[8]: a rate-based block was used to simulate a counter-current flow absorption column, then a hydrodynamic and fluid dynamic model was implemented based on the data provided by **Brunazzi and Paglianti**[3,7]. Finally, the equilibrium model of **Flagiello et al.**[8] for SO_2 solubility in water was used.

Methodology

The experiments were performed in a fully instrumented pilot-scale scrubber[8] equipped with a DN100 column and filled with 892 mm (Z) of structured packing, designed via CAD and manufactured using 3D foam printing technology. The column operates in counter-current flow with the gas stream fed at the bottom. The packing prototype was designed based on the typical configuration of corrugated sheets, which form the classical packing structure. All geometric parameters were fixed trying to obtain a nominal area ($a_n = 240\ m^2/m^3$ without enhancement due to the printing technology and foaming) and void fraction ($\varepsilon_p = 0.96$) as close possible to that of the benchmark (Mellapak 250Y: $a_n = 250\ m^2/m^3$ and $\varepsilon_p = 0.97$). The experimental activity was divided into three sets of experiments, detailed below.

- Set A consists in measuring of specific pressure drops in dry ($\Delta P_d/Z$) and wet ($\Delta P_w/Z$) mode of the prototype using an air-water system at 25 °C. Dry mode tests were performed by varying the airflow, in the range 5 - 160 m^3/h (gas load factor $F_G = 0.20 - 6\ Pa^{0.5}$). Wet mode tests were carried in the same gas conditions with water flow rates variable between 40 and 160 L/h (liquid load factor F_L = 5.10 - 20.38 m/h).

- Set B consists in experimental tests to measure the contribution of purely physical mass-transfer coefficients of both gas and liquid-side ($k_{G,SO2}a_e$ and $k_{L,SO2}a_e$) provided by prototype. To this end, we performed SO_2 absorption experiments either with NaOH aqueous solutions (0.05 M) and distilled water acidified with HCl (0.1 M)[6]. Experimental tests (25 °C and 1 atm) with NaOH aqueous solution have been used to estimate $k_{G,SO2}a_e$ coefficient while those with HCl aqueous solution allows determination of $k_{L,SO2}a_e$ coefficient.

- Set C consists in the FGD experiments from model flue gas at 32 m^3/h ($F_G = 1.25\ Pa^{0.5}$) constant flow rate with either 500 or 1000 ppm_v SO_2 concentration, as a reference case, in order to evaluate the separation efficiency of the prototype. The liquid stream was an alkaline tap water (collected in the North-West area of Naples) at pH 7.6 and its flow rate was varied from 40 to 160 L/h (F_L = 5.10 - 20.38 m/h).

Selected Results

The experiments showed that the pressure drop (Set A) of the 3D foam printed packing increases with the gas load at similar rates as compared to the commercial Mellapak 250Y. These results were mainly due to the similar values of surface area, void fraction and hydraulic diameter. Despite the prototype channels were narrower, both the packings showed similar flooding limit and could operate up to pressure drops in the range of 6.0 - 6.5 mbar/m, hence covering gas loads between 2.8 - 3.8 $Pa^{0.5}$ corresponding to the liquid loads investigated.

The desulfurization tests (Set C) revealed very promising performances: in fact, for 500 ppm_v of inlet SO_2 concentration tested, at a high liquid load (above the F_L = 16 m/h), the separation efficiency (> 95%) was similar to the figure retrieved for the Mellapak 250Y from ASPEN PLUS simulation. For the lowest liquid loads, a marked improvement of the separation efficiency for the 3D printed packing was observed especially for the experiments at F_L = 5.10 m/h, where efficiency of about 28% more was achieved. Moreover, a 42% increase was obtained at 1000 ppm_v of inlet SO_2. This mirrors a faster SO_2 absorption rate that is mostly related to the liquid phase mass-transfer coefficient: in fact, the mass-transfer characterization (Set B) showed higher transfer rates provided by the packing prototype, i.e. for $k_{G,SO2}a_e$ up to 10% more and $k_{L,SO2}a_e$ more than 20 - 45%. This improvement likely depends on the greater specific surface transfer area generated by the surface foam packing with an enhancement of approximately 30% but also by improved wettability of the packing plastic foamed surface as assessed by dedicated tests to measure the liquid-solid contact angle for the prototype. Another relevant point is the optimization of the packing elements including the baffle rings and of its specific mechanical design, which allows improving the liquid-film dispersion reducing the liquid maldistribution on column walls and between packing elements.

Key Conclusions

The use of 3D foamed printed packings allows an intensification of desulfurization processes by exploiting specific surface properties, also assuring same pressure drops. In general, they enable higher degrees of flexibility and customization in gas-liquid multiphase reactors and separators.

Literature Cited

1. **Billet, R., and M. Schultes,** "Predicting mass transfer in packed columns," *Chem. Eng. Technol.*, **16**, pp. 1–9, (1993).
2. **Olujić, Ž., *et al.*,** "Predicting the efficiency of corrugated sheet structured packings with large specific surface area," *Chem. Biochem. Eng. Q.*, **18**, pp. 89–96 (2004).
3. **E. Brunazzi, and A. Paglianti,** "Liquid-film mass-transfer coefficient in a column equipped with structured packings," *Ind. Eng. Chem. Res.*, **36(9)**, pp. 3792–3799 (1997a).
4. **Kalsoom, U., *et al.*,**"Current and future impact of 3D printing on the separation sciences," *Trends in Analyt Chem.*, **105**, pp. 495–510 (2018).
5. **Miramontes, E., *et al.*,**"Additively manufactured packed bed device for process intensification of CO_2 absorption and other chemical processes," *Chem. Eng. J.*, **388**, pp. 124092 (2020).
6. **Parra-Cabrera, C., *et al.*,**"3D printing in chemical engineering and catalytic technology: Structured catalysts, mixers, and reactors," *Chem. Soc. Rev.*, **47**, pp. 209–230 (2018).
7. **E. Brunazzi, and A. Paglianti,** "Mechanistic pressure drop model for columns containing structured packings," *AIChE journal*, **43(2)**, pp. 317–327 (1997b).
8. **Flagiello, D., *et al.*,**"Characterization of mass transfer coefficients and pressure drops for packed towers with Mellapak 250.X," *Chem. Eng. Res. Des.*, **161**, pp. 340–356 (2020).

145f Relaxing the Constant Relative Volatilities and Constant Molar Overflow Assumptions in Distillation While Retaining Their Benefits

Tony Joseph Mathew, Mohit Tawarmalani, Rakesh Agrawal

In the previous century, shortcut models for evaluating distillation (such as the McCabe-Thiele method, the Fenske method, and the Underwood method) were commonly used due to the ease and simplicity of calculations by hand they provided. With the outstanding rise in computational processing power and the development of process simulation software, distillation columns can now be evaluated with a click of a button and, most importantly, with higher fidelity thermodynamic models. This may give the impression that such shortcut models currently serve a role only in the classroom and for toying with perfectly ideal systems in the realms of academics. Yet despite the advances in computational performance, evaluating a multicomponent distillation configuration separating four or more components with multiple columns, sloppy splits, and thermal couplings is not straightforward. Furthermore, the process engineer may want to know which among the numerous different configurations (which number to 6,128 for five-component separations) is best for their separation, and how much benefit they might expect compared to their current choice. Including optimization for each configuration, this endeavor quickly becomes unwieldy.

Shortcut models that capture essential physics adequately offer a solution to this dilemma. By optimizing such a shortcut model formulation for each configuration, the process engineer can screen the vast search space to quickly identify a handful of attractive configurations, and subsequently determine their precise benefits using a detailed simulation. But while shortcut models offer computational simplicity, they have been censured for being at the cost of real mixture behavior. The two most commonly criticized aspects of the traditional shortcut models is their usage of the constant relative volatilities (CRV) and the constant molar overflow (CMO) assumptions. CRV assumes that component relative volatilities (parameters which describe vapor-liquid equilibrium) are the same throughout a configuration. CMO assumes that the molar flowrates of vapor and liquid traffic are the same on every stage in a column section, with the underlying assumption being that the different components have the same latent heats of vaporization. Clearly, these two assumptions do not strictly hold up in the real world.

In this talk, we present simple methods to relax the CRV and CMO assumptions, to account for the real-world physics they appear to ignore, but without compromising on the simplicity they offer. For CRV, we use the well-established idea of parameter estimation to fit α in the CRV equation to VLE data, treating it as a surrogate model for the true VLE [1,2]. This allows a single choice of relative volatility values to capture the wide variations in a distillation configuration. For CMO, we employ a variable transformation to generate latent-heat variables corresponding to every molar variable [3–6]. For example, heat duty is the latent-heat analog of vapor duty. Then, by simply replacing each molar variable in a CMO model by its latent-heat analog, we obtain a new model which now accounts for the different latent heats of vaporization, and thereby varying molar traffic within a column. Most importantly, the final model after applying these two methods is mathematically equivalent to the original CMO and CRV model. Thus, the relaxed model accounts for the missing physics of the traditional model while retaining its computational benefits.

Using these relaxation methods, we have built a formulation to optimize multicomponent distillation configurations. On a four-component mixture of paraffins, we will demonstrate that these methods allow our formulation to correctly predict the ranklist of configurations compared to a sensitivity analysis in Aspen Plus. We then identify a heat-integrated energy-efficient configuration for a five-component separation of aromatics, among the 6,128 possible. Thus, we can now identify energy-efficient configurations while accounting for varying relative volatilities and the different latent heats of components.

1. Anderson, N. J., andamp; Doherty, M. F. (1984). An Approximate Model for Binary Azeotropic Distillation Design. *Chemical Engineering Science*, 39 (1), 11–19.
2. Mathew, T. J., Tumbalam Gooty, R., Tawarmalani, M., andamp; Agrawal, R. (2020). Quickly Assess Distillation Columns. *Chemical Engineering Progress*, 116 (Dec.), 27–34.
3. Mole, P. (1958). A unified theory for ideal separation processes. *Chemical Engineering Science*, 8 (3), 236–243.
4. Rev, E. (1990). The constant heat transport model and design of distillation columns with one single distributing component. *Industrial andamp; Engineering Chemistry Research*, 29 (9), 1935–1943.
5. Doherty, M. F., andamp; Malone, M. F. (2001). Conceptual Design of Distillation Systems. *McGraw-Hill chemical engineering series*. McGraw-Hill, 1st ed.
6. Ravi, R. (2008). Transformations for a generalized McCabe–Thiele method: A retrospective and a new perspective. *International Communications in Heat and Mass Transfer*, 35 (7), 860–866.

85a Membrane Materials Under Complex Conditions

Zachary Smith

Membrane-based separations can provide an attractive alternative to other separation processes because of their modularity and compact size. However, literature data rarely matches performance under realistic conditions. This presentation will discuss current challenges related to gas-separation membranes, including stability to strongly interacting penetrants and the role of gas mixtures on membrane performance. The topics will focus on microporous membrane materials, including polymers of intrinsic microporosity (PIMs), catalytic arene-norbornene annulation (CANAL) polymers, and metal–organic frameworks (MOFs). Studies will demonstrate opportunities to leverage polymer and MOF structure and chemistry to suppress plasticization effects and to leverage competitive sorption, enabling material property sets measured under complex mixtures that can actually surpass those of property sets predicted from simple pure-gas experiments.

85b Tray Design, Operation, and Troubleshooting

Mark Pilling

Unlike packings, trays have a multitude of mechanical design parameters and degrees of freedom. Given the same column operating conditions, five tray subject matter experts will present you with five different "optimum" designs. This presentation will discuss interesting design parameters and performance characteristics of trays over a wide range of applications and operating conditions. The presentation will also cover troubleshooting techniques and unexpected findings.

85c Which Separation Method Needs Lower Power: Membrane or Distillation?

Jose Adrian Chavez Velasco, Mohit Tawarmalani, Rakesh Agrawal

Reducing the energy intensity of molecular separation was Identified as one of the big challenges by the National Research Council 2006 report for achieving the development of a sustainable future chemical enterprise. Distillation has been and continues to be the most common method to separate a variety of molecular mixtures in chemical and petrochemical industries. Nevertheless, In this work, we systematically compare the power requirements of membrane and in recent years there has been an increasing interest in replacing distillation processes with membrane based processes with the claim that they have the potential to improve energy efficiency by tenfold. distillation processes for binary liquid and gaseous hydrocarbon mixtures where the desired product component is more permeable and also more volatile. We first provide a shortcut method calculation to compare the efficiency of heat pump and steam driven distillations. Then, power requirements of heat pump distillation and membrane separation are compared on an equivalent power basis. We find that both membranes and distillations have preferred energy efficient domains. Distillation generally requires lower power when either high component recoveries are needed (at all tested product purities), or high purity product streams with modest recoveries are needed. For a high purity product at modest recoveries, membranes have a potential to provide energy benefits for highly enriched feeds, especially those composed of close boiling components. Additionally, when feed concentration is moderate to high and product recovery and purity are modest, membranes are likely to show modest efficiency gain. On the other hand, distillation outperforms membranes under many situations and often the reduction in energy consumption is substantial.

85e Na^+-Gated Nanochannel Membrane for Renewable Fuel Synthesis

Miao Yu

Sustainable energy, environment, water (H_2O), and food, in a large extent, depends on acquiring/capturing/utilizing small molecules, such as H_2O, ammonia (NH_3), carbon dioxide (CO_2), methane (CH_4), ethanol, and liquid hydrocarbons, etc. Precisely designing stable, molecular-scale pores for sieving these molecules, either from the final product or during their production processes, could be an effective way of separating these molecules or promoting their production using compact and well-engineered systems. Considering the very small sizes (0.26-1.0 nm) of these molecules and tiny size difference from their contaminants/byproducts, it is a grand challenge to design these molecular-scale pores. My research interest is focused on rationally designing and preparing nanoporous structures for precisely distinguishing molecules by size/shape differences, characterizing and understanding the nanostructures, and applying them for separation and catalysis. In this talk, I will first give an overview of my research work and then present our recent research work on Na^+-gated nanochannel membrane and its application for boosting methanol and dimethyl ether (DME) synthesis.

By-product H_2O strongly inhibits the kinetics and thermodynamics of many important reactions for renewable fuel production, for example, CO_2 hydrogenation to methanol. NaA zeolitic nanochannels were found to be water-conductive and gas impeding. As a result, H_2O showed two to three orders of magnitude higher permeation rate than gases (as small as H_2) at elevated temperatures (200-250 ^oC) and pressures (21-35 bar). This is surprising because the literature in the past 20 years reported comparable water and gas permeation rates for NaA zeolite membranes prepared from all over the world. Our extensive comparative experiments suggest this is because of the drastic improvement of our NaA membrane quality, and thus the intrinsic behavior of the NaA nanochannels can be revealed. The profound impact of these nanochannels on boosting reactions, which are thermodynamically and kinetically inhibited by by-product H_2O, was demonstrated for methanol synthesis from CO_2 hydrogenation, due to the emergency of developing sustainable and renewable fuels as well as mitigating CO_2 related environmental issues. By generating a "dry" reaction environment using Na^+-gated nanochannel membrane, greatly boosted DME synthesis from CO_2 and H_2 was also demonstrated. This important discovery of Na^+-gated, water-conduction nanochannels may greatly improve energy efficiency of many important industrial processes and may also lead to novel applications utilizing special property of these water-conduction nanochannels.

608a Determining CO_2/H_2S and CO_2/CH_4 Binary Adsorption Selectivities in Metal-Organic Frameworks for Biogas Purification

Chunyi Li, Ryan P. Lively

Biogas is a promising alternative to natural gas as the global energy sources shift from fossil fuels to renewables. Biogas contains mainly CH_4 (35-70%), CO_2 (15-60%), and N_2 (15-50%), with trace impurities such as NH_3, H_2O, and H_2S. To reduce Greenhouse Gas release from biogas and prevent pipeline corrosion and catalyst poisoning in the transport and downstream process of biogas, removal of H_2S and CO_2 is critical. Besides the traditional acid gas removal methods involving aqueous amine scrubbing, less energy-intensive adsorption methods using microporous materials such as zeolites, metal-organic frameworks (MOFs), and porous organic cages (POCs) are being developed. In particular, MOFs have received much attention as they exhibit high CO_2 uptake and have tunable selectivity. Much research has focused on measuring single-component adsorption isotherms and predicting the mixture selectivity using the Ideal Adsorbed Solution Theory (IAST). As the functional groups and unsaturated metal sites often interact with quadrupolar CO_2 molecules more strongly than CH_4 or N_2, IAST predictions may not capture the nonideal adsorption behavior.[1] Mixture containing H_2S and CO_2 poses more challenges in binary adsorption selectivity prediction due to the competitive sorption of H_2S and CO_2 to active sorption sites.[2] Thus, understanding the mixture adsorption performance is valuable in adsorption process design and adsorbent selection. This work aims to measure the binary adsorption uptakes of CO_2/H_2S and CO_2/CH_4 in MIL-125-NH_2(Ti), amine-appended Mg_2(dobpdc), MIL-101(Cr), and MOF-74(Mg, Ni). A zero-length column method is used, where a small amount of MOF is first saturated with the gas mixture of interest. Then the mixture gas is desorbed using an inert gas under equilibrium conditions.[3] A mass spectrometer is used to measure the concentration of the desorption effluent gas. The experimental mixture gas results are compared to the IAST predictions, and the pre and post-adsorption MOFs are characterized by SEM, N_2 physisorption, PXRD, and XPS to understand the competitive adsorption behavior and H_2S stability in these MOFs.

1. Gharagheizi, F.; Sholl, D. S., Comprehensive Assessment of the Accuracy of the Ideal Adsorbed Solution Theory for Predicting Binary Adsorption of Gas Mixtures in Porous Materials. *Industrial andamp; Engineering Chemistry Research* **2021,** *61* (1), 727-739.
2. Joshi, J. N.; Zhu, G.; Lee, J. J.; Carter, E. A.; Jones, C. W.; Lively, R. P.; Walton, K. S., Probing Metal-Organic Framework Design for Adsorptive Natural Gas Purification. *Langmuir* **2018,** *34* (29), 8443-8450.
3. Brandani, F.; Ruthven, D., Measurement of adsorption equilibria by the zero length column (ZLC) technique part 2: Binary systems. *Industrial andamp; Engineering Chemistry Research* **2003,** *42* (7), 1462-1469.

608b Diffusion of CO_2 and N_2 in 5A Pellets Measured Using the Adsorption Differential Volumetric Apparatus

JinYu Wang, Enzo Mangano, Stefano Brandani, Federico Brandani, Pluton Pullumbi

Mass transport of pure CO_2 and N_2 in commercial 5A pellets are reported at different temperatures and pressures from 0.1-0.2 kPa to near atmospheric pressure. The experiments are performed on small sample masses (3 pellets), but even in these conditions isothermal conditions can only be achieved at the lowest pressures. The data will be used to show the transition with pressure from isothermal diffusion to combined heat and mass transfer and finally to fully heat limited cases.

Both systems are macropore diffusion controlled and for CO_2 below 293 K there is evidence of a contribution of surface diffusion.

608d Miniaturizing Dynamic Column Breakthrough Measurements to Measure Gas and Vapour Adsorption Equilibrium on Milligram Quantities of Adsorbents

Nicholas Wilkins, James Sawada, Noelie Consant, Arvind Rajendran

Dynamic column breakthrough (DCB) is a technique that can measure adsorption equilibrium, kinetics and column dynamics using a packed bed of a given adsorbent. Quantitative DCB is traditionally performed with a packed column of a pelletized adsorbent, in the gram to kilogram scale, to mitigate pressure-drop issues. Due this constraint, many newly synthesized, crystalline, and rare materials are unable to be tested in a DCB apparatus. In this presentation, a dynamic column breakthrough (DCB) apparatus is detailed that overcomes the previously mentioned shortcomings and can accurately measure single and multi-component equilibrium loadings and column dynamics of milligram-scale quantities of adsorbent rapidly. Due to the much smaller size of the DCB apparatus, breakthroughs can be performed in as little as a few seconds, and are quantifiable. The material balance is derived and presented for the DCB apparatus that accounts all possible sources of accumulation, within the solid, fluid and extra-column volumes. The apparatus was tested with dry gases on a previously studied zeolite 13X sample [1] and a commonly studied activated carbon. Single-component adsorption equilibrium data from vacuum to 1.2 bar and 30, 40 and 50°C was measured for N_2, CH_4 and CO_2 in a volumetry apparatus. This statically collected data was compared with data collected in the DCB. The equilibrium loadings for both adsorption and desorption agree with the statically collected data within 5%. Unary H_2O vapor experiments were also studied between 10 and 90% relative humidity on various adsorbents at 30°C, and agree with measurements from the literature. Afterwards, this technique was extended for multicomponent mixtures. First, mixtures of N_2 and CH_4 were tested. Since, both N_2 and CH_4 exhibit linear isotherm trends on zeolite 13X, they should not experience any competition between each other. This was verified with the DCB measurements. The collected N_2 and CH_4 equilibrium data all agree with the linear isotherm predictions. N_2/CH_4 mixtures were also studied on the activated carbon at 30, 40 and 50°C. The measured data suggests the mixture is weakly nonideal, which is agreement with the literature. Finally, mixtures of CO_2 and CH_4 were tested. CO_2 and CH_4 on zeolite 13X exhibit very nonideal adsorption competition. These findings suggest that the DCB can measure single and multicomponent gas and vapor adsorption equilibrium and column dynamics precisely and accurately.

References:

1. Wilkins, N. S. and Rajendran, A. *Adsorption*2 (2019): 115-133.

608e Direct Measurement of the Mass Transport Coefficient of Water in Silica-Gel Using the ZLC Technique

Stefano Brandani, Enzo Mangano

Adsorption processes are important in the recovery of low grade heat. Silica gel/water is one of the reference pairs – material/adsorbate – in such processes and while equilibrium measurements are well-established, the measurement of the mass transport coefficient remains a challenge in part due to the fact that it is nearly impossible to decouple mass and heat transfer effects. A new approach for the determination of the mass transfer coefficient of water in commercial silica gel is proposed and demonstrated with measurements on a single particle using the zero length column technique. The methodology requires measurements at low flowrates to acquire equilibrium isotherms with thousands of points in less than one day, thus allowing numerical interpolation to calculate the relationship between vapour concentration and equilibrium adsorbed amounts. The same column is then run at high flowrates to achieve mass transfer control. The novel approach determines directly all the components needed to calculate the mass transfer coefficient from the measured signals without the use a specific model.

Adsorption and desorption experiments were carried out at different flowrates and three different temperatures. It is possible to determine an average mass transfer coefficient that is in line with literature values. The results show also a complex behaviour with surface diffusion as the main contribution to the transport process.

Reference:

Brandani S. and Mangano E. **Direct Measurement of the Mass Transport Coefficient of Water in Silica-Gel Using the Zero Length Column Technique**, *Energy*, **2022**, *239A*, 121945 https://doi.org/10.1016/j.energy.2021.121945

608f Binary Adsorption of CO_2 and H_2o on Uio-66 and Amine-Functionalized Uio-66 Metal-Organic Frameworks

Arianjel Hernandez, Rebekah K. Impastato, Mohammad I. Hossain, Brooks Rabideau, T. Grant Glover

Metal-organic frameworks (MOFs) are being examined for a variety of different gas adsorption applications. In many of these applications adsorption occurs in a mixed gas stream, but due to the difficulty of completing mixed-gas adsorption isotherm measurements, there is only a limited amount of experimental isotherm data. The absence of binary adsorption isotherm data limits detailed understanding of complex surface adsorption behavior. Of the various gas separations that are commonly examined, the adsorption of CO_2 on porous solids is of particular interest because this process frequently takes place in a humid gas stream, and yet is commonly studied in dry gas conditions. The presence of humidity in this adsorption process is complex, and in some cases, the impact of humidity on the process is difficult to predict. With this in mind, we have experimentally measured the binary adsorption of CO_2 and H_2O on UiO-66, UiO-66-NH_2, and validated these results to molecular simulations of the binary adsorption isotherms. Then we extended the computational analysis to examine UiO-66 with other amine functional groups and determined the impact of MOF defects on the binary adsorption isotherms. The results show that defects have a significant impact on how water and CO_2 adsorb on the MOF. In some cases, adsorption is enhanced at low pressures, but in other cases these enhancements do not exist. To understand the enhancement in more detail, we utilized simulations and showed that in one MOF water can bridge OH terminated defect sites, which increases the interactions of CO_2 with the adsorbent and enhances CO_2 adsorption. The results highlight the importance of mixed gas adsorption measurements when examining materials for gas separations.

146b Enhanced Diffusiophoretic Mobility and Removal of Colloidal Water Contaminants By Tunable CO_2 Flux

Esai Lopez, Patryck Michalik, Shicheng Lyu, Alex D. Paulsen, Elizabeth Stewart, Andrew R Teixeira

As the demand for drinking water increases globally so does the need for distributed purification systems. The dissociation of carbon dioxide (CO_2) in water to hydrogen (H^+) and bicarbonate (HCO_3^-) ions has the potential to transform water purification by acting as an on-demand initiator for diffusiophoresis. Due to the faster diffusion of H^+ in water compared to HCO_3^-, 9.33×10^{-9} vs 1.18×10^{-9} m^2 s^{-1} respectively, an ion gradient develops when CO_2 is fed through water. This ion gradient is capable of moving and concentrating charged colloids. Bench scale experiments using a custom concentric tubular diffusiophoretic separator were conducted to optimize particle separation. The strength of the ion gradient, was dependent on the partial pressure of CO_2 applied through a Teflon AF 2400 semipermeable tube. A two-log reduction of 0.5 µm polystyrene spheres from lab grade water was achieved operating at a CO_2 pressure of 250 kPa. A computational fluid mechanics model was developed to understand the formation of the necessary ion gradients required for diffusiophoresis.

146c Thermodynamic Modeling of Mineral Solubility in High-Temperature, High-Pressure Petroleum Systems.

Isaac Gamwo, Derek Hall, Serguei Lvov, Hseen Baled

Over the past decades, most shallow reservoirs have been depleted of recoverable hydrocarbon energy sources due to increasing oil demand. Hence, oil and gas industry have expanded the search for petroleum in offshore deep oil and gas wells where elevated temperatures, pressures, and total dissolved solids prevail in the presence of mixed electrolytes including such elements as sodium, calcium, potassium, chloride, phosphate, magnesium, etc. Under these extreme conditions, minerals may easily precipitate to form mineral scales that build up on the production pipes leading to significant decreases in oil production. Therefore, accurate thermodynamic modeling at these extreme conditions to predict the formation of mineral scales is critical in the mitigation strategy of mineral scaling.

Here, we present our thermodynamic model to predict the solubility of various minerals found in the oil field. The contribution of major model components will be discussed. Our predicted results agree reasonably well with available solubility experimental data.

147b Nature Inspired Convective Chromatography Columns for Preparative Separations

Riccardo Onesti, Cristiana Boi

Continuous biomanufacturing is underlining the limits of the conventional chromatographic process based on packed bed columns, whose diffusion as primary transport phenomena and unsolvable bed inhomogeneity are the primary causes of performance reduction. From this perspective, open tubes and regular array columns showed a great potential, with significantly higher efficiency when compared to conventional columns. However, their application is currently confined to analytical separations of nano- and micro-samples, due to several limitations mainly related to the low surface area available for adsorption (owing to the absence of small diffusive pores) and to the manufacturing techniques.

In this work, novel supports with the potential to combine and expand the application of open tubes and regular array columns to preparative chromatography will be presented. These new stationary phases, inspired by the human intestine and partially made by means of addictive manufacturing, consist in a regular array of conduits in which the mobile phase flows. Several columns, that differ in conduits shape and dimension and fluid distributor design, have been manufactured. To overcome the limitations related to the employed material and to increase the surface area, the conduits surface was coated with nanofibrillated cellulose (NFC) dispersed in acetone by deep or spray coating, depending on the columns shape and design. Several NFC concentrations in acetone were tested, ranging from 0.1% to 2% wt/v. The morphology of the cellulose coating was verified by SEM analysis and its adhesion was tested after several chromatographic cycles at process operative conditions. The cellulose was functionalized using Cibacron Blue as a model ligand and bovine serum albumin (BSA) was used as target molecule in affinity chromatographic cycles, performed at several flow rates and BSA concentrations, to determine the highest dynamic binding capacity (DBC) of the support. In addition, the columns were completely characterized for their fluid dynamic performances, in terms of height equivalent to a theoretical plate (HETP) and asymmetry factor. The result obtained were encouraging, showing promising performance and moderate pressure drops even at high flow rates, making possible to significantly extend the column length to increase the surface area and for scaling up purposes. A critical approach will present and analyze this technology that has the potential to significantly improve the efficiency of preparative chromatographic separations.

147c A Comparative Study of Protein a Membranes for the Rapid Purification of Monoclonal Antibodies

Joshua Osuofa, Scott Husson

Chromatography is ubiquitous to therapeutic protein purification. When the protein is a monoclonal antibody (mAb), Protein A chromatography is used in the capture step to separate the product selectively from process impurities such as host cell proteins (HCPs), host cell DNA, viruses, charged variants, fragments, endotoxins and aggregates. Among the issues in resin-based Protein A chromatography is the low throughput associated with flowrate-dependent performance. In particular, diffusion limitations require resin columns to be operated with long residence times to achieve satisfactory binding capacities.

Protein A membrane chromatography is a solution. Recent innovations in Protein A membrane chromatography have led to products with binding capacities on par with resin chromatography using residence times that are shorter by an order of magnitude or more. The objectives of this study were to provide the first preliminary comparison of all commercial Protein A membranes in terms of performance metrics like dynamic binding capacity, equilibrium binding capacity, regeneration-reuse performance, HCP clearance, DNA clearance and elution peak volumes, and physical properties like pressure-flow rate dependence, pore size, surface area and dead volumes. The commercial Protein A membranes include Purilogics Purexa™-A, Gore® Protein Capture Device, Cytiva HiTrap™ Fibro Prism A, and Sartorius Sartobind® A chromatography columns.

The results presented in the study represent a summary of the breakthrough in performance of Protein A membrane adsorbers in comparison to resins. In terms of binding capacity and impurity clearance, these membranes are ready to compete and, in certain cases, outperform resins. The performance metrics and physical properties presented show how Protein A membranes can fit into bioprocessing schemes, especially at small scales.

147d Purification of Biotherapeutics By High-Productivity Nonwoven Based Membrane Adsorbers

Jinxin Fan, Ruben Carbonell, Cristiana Boi

To meet the growing demand for accessible biotherapeutics and accelerate the development of new therapeutic modalities, biopharmaceutical companies are seeking to transform biomanufacturing processes to improve productivity, flexibility, and cost-effectiveness. Nonwoven based membrane adsorbers could have an important impact on current downstream processing for biotherapeutics production by providing an inexpensive single-use adsorbent with a significantly higher productivity relative to chromatographic columns. We explored various high-capacity chromatographic nonwoven membranes that were developed by UV grafting of polyglycidyl methacrylate (polyGMA) on a poly butylenes therephthalate (PBT) nonwoven fabric and subsequent functionalization with different ligands, for purification of biomolecules including monoclonal antibody (mAb), single-chain variable fragment (ScFv) and adeno-associated virus (AAV). For example, iminodiacetic acid (IDA) cation-exchange (CEX) nonwoven membranes achieved high $DBC_{100\%}$ for capturing polyclonal human IgG (~130 mg/mL) and scFv (~150 mg/mL) from mammalian cell culture and the DBCs were fairly constant for residence times (RTs) in the range of 0.1-2.0 min. The CEX-IDA membrane could be reused for five bind-elute cycles in capturing mAbs from cell culture harvest at 1.0 min RT with high recovery of 94.2-99.5%. The obtained purity was 87.2-87.9% as some basic impurities were co-eluted, but these values are comparable to those obtained using cation exchange resins. The achieved high mAb loading capacity of 86.4 mg per mL of membrane volume at 1.0 min RT could be highly beneficial for high-throughput production. Other types of prepared membranes also exhibited high-capacity and high-efficiency separation of AAV2 or mAbs from culture fluids. These new nonwoven chromatographic membranes offer highly viable alternatives to chromatographic resins as efficient, high-productivity separation unit in the production of high-value biotherapeutics.

147e Membrane Adsorbers to Capture Cu from Mixed Metal Acidic Solutions in Support of Radiopharmaceuticals

Maura Sepesy, Benjamin Fugate, Joelle Scott, Alec Johnson, Christine Duval

Radiopharmaceuticals offer promising new approaches for both imaging and treating cancerous tumors in personalized medicine. An emerging isotope for use in radiopharmaceuticals is copper-67 (^{67}Cu) which can provide treatment (via beta decay) and imaging (via gamma rays). Currently, ^{67}Cu is purified using resin-packed columns, which exhibit (1) low column flow rates, (2) high elution volumes and (3) long purification times—all of which contribute to the bottleneck in the sustained supply of ^{67}Cu. Due to the short half-life (2.58 days) of ^{67}Cu, rapid purification from competing isotopes (^{57}Ni, ^{62}Zn, ^{65}Zn, and unreacted ^{68}Zn) is key for increasing the availability of the isotope for clinic trials and use. Membrane adsorbers are well-positioned to meet that need.

In this study, glycidyl methacrylate (GMA) brushes are grafted from commercial polyvinylidene fluoride and electrospun chloromethyl styrene membranes using AGET ATRP, a controlled polymerization technique. After grafting poly(GMA), diamine ligands (4-picolylamine, putrescine, or ethylene diamine) are attached to the poly(GMA) brushes by an epoxide ring-opening reaction. Membranes are washed with methanol/DI water, EDTA at pH 10 and HCl at pH 2 to remove residual Cu catalyst from AGET ATRP. Membranes are characterized by ATR-FTIR and x-ray photoelectron spectroscopy (XPS) to support functionalization. Permeability was calculated from pure-water flux experiments in dead-end filtration mode (unmodified PVDF: 5166 LMH/bar, GMA grafted: 4692-648 LMH/bar, and amine-functionalized: 4124-175 LMH/bar). Membrane morphology was examined using scanning electron microscopy (SEM). The binding capacity for Cu, Ni, and Zn were calculated by fitting the Langmuir isotherm to equilibrium adsorption data collected at pH 2. Binding capacity was calculated to be 1.65 and 1.59 for putrescine and ethylenediamine, respectively. Though low, this capacity is 2 orders of magnitude higher than the amount of radiocopper in solution. Dynamic binding experiments were conducted to calculate membrane productivity. The results of this research are laying the groundwork for implementing membrane absorbers as a new separation material for medical isotope purification.

147f Novel Adsorptive Membranes for mRNA Capture for Vaccine Manufacture

Riddhi Banik, Thomas Neumann, Mirco Sorci, Miral Al Sharabati, Zerui Hao, Manasa Chillara, Weibo Zhao, Todd Przybycien, James (Chip) Kilduff, Georges Belfort

The highlight of the Covid-19 pandemic, among others, has been the enormous potential of mRNA vaccine as therapeutics. Currently, vaccines for cancer, HIV, Nipah virus and therapeutics for auto-immune disorders based on mRNA technology are also under different phases of clinical trials. However, a robust downstream processing system that is fast, scalable and suitable for these sticky high molecular weight nucleic acids is required for greater accessibility of this technology. Continuous vaccine production via purification of mRNA will speed-up manufacturing and reduce costs of production. Replacing resin-based chromatography (diffusive, slow and hence long residence time), with membranes (convective, fast and hence short residence time) for purification of mRNA vaccines will (i) reduce the residence times of the recovery process, t_R, (ii) allow continuous processing, (iii) retain a higher percentage of folded mRNA due to the reduced treatment residence time, t_R, and (iv) significantly reduce the "footprint" of the equipment. This work introduces a membrane-based capture of mRNA with surface functionalized regenerated cellulose (RC) flat sheet membranes.

The work focuses on using the SET-LRP technique of surface grafting to introduce functional groups on the surface of RC membranes that can bind mRNA from a solution. A surface binding estimation is first carried out based on the hydrodynamic radii of the family of secondary structures of mRNA. This calculation determines the maximum adsorption capacity from a monolayer of surface probes. The RC membranes are first modified by addition of BIBB(α-Bromoisobutyryl bromide) to introduce Br sites. This is followed by either a one-step or a two-step grafting technique for surface modification. The different parameters influencing degree of grafting, such as, BIBB concentration, polymerization time, effect of solvent, monomer to solvent ratios are optimized with the goal of maximum binding of mRNA on the bottle brush layers of modified RC membrane. The modifications are carried out in a 96-well-plate format for a high throughput screening of characteristics. The modified surfaces are characterized by ATR-FTIR and surface fluorescence for and the density of functionalization is correlated with the reaction conditions.

For performance characteristics of the membranes, pure water and buffer flux data is recorded with a 13mm Swinney filtration setup for both unmodified and modified RC membranes. This is then compared with the filtration data obtained from mRNA capture with the same buffer conditions. The elution of the captured mRNA under different buffer conditions is also investigated and the dynamic binding capacity data is recorded. The challenge here is to adapt the surface grafting technique with hollow fiber RC membranes with much larger surface areas that would ensure a greater amount of capture per unit volume. A complete purification scheme involving both capture and polishing of mRNA is developed. Resolving the bottleneck in downstream processing in the manufacture of mRNA therapeutics with novel membranes for effective, fast purification is the ultimate aim of the project.

147g Demonstrating Mathematical Equivalence between Partial Least Squares and the Beer-Lambert Law in Estimating Protein Concentrations with Spectroscopic Data

Claire Velikonja, Ian Gough, Brandon Corbett, David Latulippe, Prashant Mhaskar

The pharmaceutical industry has made significant developments in upstream production through optimized cell culture and expression technologies utilizing feedback from process measurements. However, this optimization has been unmatched in downstream purification and analytical methods, partly due to lack of appropriate measurement techniques, resulting in significant manufacturing inefficiencies. These inefficiencies have motivated the development of "soft-sensor" based models, where real-time data is combined with multi-variate data analysis (MVDA), to allow for live estimation of product attributes, which in turn can be utilized to improve process operations. Partial Least Squares (PLS) is an MVDA technique that is commonly used with UV/Vis data to build models for the estimation of therapeutic protein concentration and impurity levels. The selection of UV/Vis data for this application is justified through the Beer-Lambert law, which assumes a linear relationship between protein mixture components and their UV/Vis absorbances. Existing literature has widely considered PLS and the Beer-Lambert Law as distinct modeling alternatives. However, this paper demonstrates that, in addition to being a well-established data analytics tool, the model formulation of PLS is identical to the multi-component Beer-Lambert Law model, providing a mechanistic underpinning to the selection of the technique for soft-sensor models, and inspiring potential advancements.

This work provides a derivation of the PLS formulation and coefficient equations and demonstrates that their structure is identical to the multi-component Beer-Lambert Law and its molar absorptivity values. To this end, artificial data was generated for a protein mixture of Immunoglobulin G (IgG), Bovine Serum Albumin (BSA), and Lysozyme over concentration ranges of 0 to 20 g/L using the Beer-Lambert model, with PLS models generated from the model-made data. The PLS model coefficients were subsequently compared to the molar absorptivity values used in the Beer-Lambert Law model, with an RMSE of 0.010 g/L demonstrating the equivalency between the two coefficient sets. Finally, experimental UV/Vis absorbance data for BSA and Lysozyme proteins using a Tecan plate reader was recorded. PLS analysis of the data and comparison of the produced model coefficients to the literature values for BSA and Lysozyme absorptivity was conducted, with a percent difference of 6.7% and 23% respectively, the latter of which was attributed to the interaction between the BSA and Lys resulting in higher absorptivity values.

The main objective of this work is to clarify how PLS and the Beer-Lambert Law are discussed and related to one another, in research of soft-sensors for biotherapeutics. The decision to use PLS for soft-sensor models is often framed as primarily a data analysis method selection issue, based on previous literature. Moreover, use of PLS versus the Beer-Lambert Law is often posited as distinct analysis methods to choose between. This work demonstrates that the Beer-Lambert Law and PLS are solving the same model formulation in the soft-sensor model applications outlined, providing a distinct mechanistic backing to the selection of PLS and further attributable meaning to the produced model coefficients.

642a Thermodynamic Gate-Opening Behavior Study on Network Nanostructured ZIF-8 to Control Gas Transport Selectivity

Hyunhee Lee, Zachary Smith

Inorganic materials such as metal–organic frameworks (MOFs) are of significant interest in achieving high-performance gas separation membranes with good stability. Unfortunately, MOF crystals are difficult to process into continuous films and expensive to make as modules. Instead, MOFs are often incorporated into polymers to form mixed-matrix membranes (MMMs), where percolation network formation between MOF fillers could play a significant role in fully taking advantage of high-performance MOFs. However, particles tend to aggregate, preventing the MOF network in the polymer matrix from forming and even resulting in non-selective defects. Thus, even at high MOF loadings, particle percolation, and hence, MOF-like transport is rarely accessible.

From our previous work, a facile method to synthesize branched ZIF-8 (BZ) nanoparticles at ambient conditions has been developed for efficient gas separations that can achieve this percolation network in the continuous polymeric matrix, resulting in significantly faster gas transport compared to its bulk counterpart traditional ZIF-8 composite. In addition, due to the small particle dimensions of BZ and the high surface-area-to-volume ratios, ZIF-8 ligand motion is suppressed when combining into the polymer–ZIF composites. This gate opening suppression results in exceptionally high selectivities for small gases (e.g., H2-based) compared to the typical propylene/propane size cut-off observed for traditional ZIF-8 control (CZ) particles. Additionally, mechanical property analysis and NMR relaxation studies indicate that the BZ morphology effectively stabilizes polymer chains, preventing plasticization. Here, this confinement effect has been further studied thermodynamically by varying temperatures to quantify the activation enthalpy for diffusion of penetrants. This systematic study allows us to finely tune the selectivity of different gas pairs whilst keeping the fast transport via percolated network.

642b The Mechanism of Light Gas Transport through Configurational Free Volume in Glassy Polymers

William Box, Zihan Huang, Ruilan Guo, Michele Galizia

Conventional glassy polymers suffer from three major issues, namely the permeability-selectivity trade-off, physical aging and plasticization, which represent a roadblock to progress in the field and hamper the design of membranes exhibiting enhanced selectivity and stability. Polymers exhibiting iptycene units, that is, configurational free volume, help overcome this limitation. The internal free volume of triptycenes is not related to the molecular conformation, such as the excess free volume in conventional glassy polymers, but to the molecular configuration. Moreover, the size of the internal volume of iptycene units is extremely regular and comparable to the size of one single molecule, which makes iptycene-based polymers highly selective. However, the fundamental mechanism of small molecules transport in polymers exhibiting configurational free volume is still poorly understood. To fill this knowledge gap, a class of polybenzoxazoles (TPBOs) with systematically varied molar amount of triptycene units was selected to perform a fundamental light gas (i.e., N_2, CH_4, CO_2) transport study and propose a possible mechanism of small molecule transport in glassy polymers exhibiting configurational free volume [1].

The dual mode sorption-mobility model [2] [3] was used to elucidate the specific effect of configurational free volume on small molecule transport. The Henry's and Langmuir's mode diffusion coefficients, D_D and D_H, were estimated and used to quantify the Henry's and Langmuir's contributions to CO_2/CH_4 diffusivity-selectivity as a function of the triptycene molar content in TPBOs. Remarkably, while the Henry's CO_2/CH_4 diffusivity-selectivity is fairly constant with increasing amount of configurational free volume, the Langmuir's CO_2/CH_4 diffusivity-selectivity increases to infinity when the molar triptycene concentration in the polymer is 75% or higher, showing that the diffusion of larger molecules (i.e., CH_4) is increasingly hindered relative to smaller molecules (i.e. CO_2) diffusion when more configurational free volume is incorporated in the polymer. In contrast, pure- and mixed-gas sorption, as well as solubility-selectivity, is essentially unaffected by the triptycene molar content. Therefore, our study provides the final proof of the superior size-sieving ability exhibited by iptycene-based polymers by unveiling, for the first time, its molecular origin.

Finally, the analysis of isosteric heats of CO_2 sorption provided some relevant information regarding the TPBOs plasticization propensity.

References

[1] S. Luo, J. Liu, H. Lin, B.A. Kazanowska, M.D. Hunckler, R.K. Roeder, R. Guo, Preparation and gas transport properties of triptycene-containing polybenzoxazole (PBO)-based polymers derived from thermal rearrangement (TR) and thermal cyclodehydration (TC) processes, J. Mater. Chem. A. 4 (2016) 17050–17062. https://doi.org/10.1039/C6TA03951K.

[2] W.J. Koros, D.R. Paul, CO2 sorption in poly(ethylene terephthalate) above and below the glass transition, J. Polym. Sci. Polym. Phys. Ed. 16 (1978) 1947–1963. https://doi.org/10.1002/pol.1978.180161105.

[3] D.R. Paul, W.J. Koros, Effect of partially immobilizing sorption on permeability and the diffusion time lag, J. Polym. Sci. Polym. Phys. Ed. 14 (1976) 675–685. https://doi.org/10.1002/pol.1976.180140409.

642c Selective H_2s/CO_2 Separation Using Sterically Hindered Amine Membranes

Shraavya Rao, Xuepeng Deng, Yang Han, Li-Chiang Lin, W.S. Winston Ho

H_2S, an extremely toxic and corrosive gas, is a common contaminant in fuel gas mixtures like syngas and natural gas. The selective removal of H_2S from such CO_2-containing fuel gases is important for enabling their industrial processing. This work describes the development of amine-based facilitated transport membranes (FTMs) with high H_2S/CO_2 selectivity at elevated temperature (> 100 °C) and explores the underlying mechanisms of acid gas transport in FTMs.

A series of primary, secondary, and tertiary amine carriers with different alkyl substituents was used to synthesize FTMs. Their H_2S/CO_2 transport performances were evaluated at 107 °C and 7 bar. Sterically hindered amines showed enhanced H_2S/CO_2 separation performance, with the H_2S/CO_2 selectivity rising from 4 to nearly 20 as the amine hindrance was increased. The best performance was obtained using a severely hindered di-*tert*-butylamine carrier, which showed a high H_2S/CO_2 selectivity of 19.6 and an H_2S permeance of 560 GPU. These results are supported by Density Functional Theory (DFT) calculations. The DFT results indicated that the presence of sterically bulky groups on the N site of the amine can restrict its reactivity to CO_2 through the carbamate pathway. Conversely, the amine-H_2S reaction was found to be relatively insensitive to steric hindrance. Finally, the carrier saturation behavior of the di-*tert*-butylamine carrier was investigated at H_2S concentrations of 0.5—30%. The carrier was found to be nearly inert to CO_2, achieving H_2S/CO_2 selectivities of 25.4 — 9 and H_2S permeances of 760—220 GPU. Overall, the newly synthesized membranes show good potential for syngas desulfurization, and the learnings offer some insight into the design of carriers for acid gas purification.

642d Microstructural Modification of Polycrystalline ZIF-8 Membranes By an Additive for Enhanced Propylene/Propane Separation Performance

Donga Kang, Hae-Kwon Jeong

Since the first ZIF-8 polycrystalline membranes showed promising propylene/propane (C3) separation in 2012,[1] there have been extensive works for the last decade to produce high quality ZIF-8 membranes.[2] Due to the nature of the polycrystalline materials, however, it is of critical importance to improve the grain boundary structures (defects) of ZIF-8 membranes to achieve high separation performance and reproducibility.[3] There are two approaches to control grain boundary structures: 1) modifying ZIF-8 synthesis methods and 2) post-synthetic processing to heal the defects.

In this talk, we plan to present how a simple addition of an additive can effectively improve the microstructures of polycrystalline ZIF-8 membranes, thereby enhancing their C3 separation performances. ZIF-8 membranes were fabricated from our previously-reported microwave-assisted seeding and secondary growth method.[4] We found that when a diol-based additive was added into the secondary growth solution, the grains of the ZIF-8 membranes showed unusual plate-like morphology. The resulting ZIF-8 membranes exhibited significantly improved C3 separation performances as compared with the neat ZIF-8 membranes synthesized without the additive, showing impressively high C3 separation factor of ~ 283 with the propylene permeability of ~ 12.4 Barrer. The separation factor of ZIF-8 membranes in this work outperformed most of previously reported high-quality polycrystalline ZIF-8 membranes. The high performance of the membranes was attributed to improved grain boundary structures resulting from plate-like ZIF-8 grains.

References

[1] Y. Pan, T. Li, G. Lestari, Z. Lai, Effective separation of propylene/propane binary mixtures by ZIF-8 membranes, Journal of Membrane Science 390 (2012) 93-98.

[2] H.K. Jeong, Metal–organic framework membranes: Unprecedented opportunities for gas separations, AIChE Journal 67(6) (2021) e17258.

[3] Y. Ren, X. Liang, H. Dou, C. Ye, Z. Guo, J. Wang, Y. Pan, H. Wu, M.D. Guiver, Z. Jiang, Membrane-Based Olefin/Paraffin Separations, Advanced Science 7(19) (2020) 2001398.

[4] H.T. Kwon, H.-K. Jeong, Highly propylene-selective supported zeolite-imidazolate framework (ZIF-8) membranes synthesized by rapid microwave-assisted seeding and secondary growth, Chemical Communications 49(37) (2013) 3854-3856.

642e Separation of HFC-32 and HFC-125 Using Amorphous Copolymers of Perfluoro(butenyl vinyl ether) (PBVE) and Perfluoro(2,2-dimethyl-1,3-dioxole) (PDD)

Abby Harders, Luke Wallisch, Mark Shiflett, Erin Sturd

Hydrofluorocarbons (HFCs) are a class of chemical compounds widely used in refrigeration and cooling systems. HFCs were developed to replace chlorofluorocarbon refrigerants that deplete our ozone layer and were regulated by the Montreal Protocol. Although HFCs have zero ozone depletion potential, many HFCs and HFC mixtures have high global warming potentials (GWP). Recent legislation such as the 2020 AIMS act seeks to regulate and phase-out HFC refrigerants by 80% by 2035. With the current legislative focus on the regulations of HFCs, sustainable methods for separating and recycling refrigerants are needed. HFC mixtures are often azeotropic, thus making their separation difficult and energy intensive. Membrane technology provides the opportunity to separate refrigerant mixtures with lower energy consumption and capital requirement compared to other separation methods. Amorphous perfluoropolymers are promising materials for this separation due to their higher affinity for fluorinated penetrants in comparison to hydrocarbon-based polymers as well as their resistance to plasticization. In order to investigate the use of perfluoropolymers for the separation of HFC mixtures, the permeability, solubility, and diffusivity of difluoromethane (HFC-32, CH_2F_2) and pentafluoroethane (HFC-125, CHF_2CF_3), which are components of the widely used refrigerant R-410A, were measured in amorphous copolymers of perfluoro(butenyl vinyl ether) (PBVE) and perfluoro(2,2-dimethyl-1,3-dioxole) (PDD) at a range of compositions. Solubility was modeled using the dual mode sorption model and diffusion coefficients were calculated by fitting kinetic sorption data to a Fickian model. Results indicate high selectivity of HFC-32/HFC-125 at a copolymer composition containing 80 mol% PDD.

642f Shape-Selective Filtration of Nanomaterials and Proteins Using Lamellar Block Copolymer-Based Slit Membranes

Maninderjeet Singh, Alamgir Karim

The growing need for highly efficient water purification and bioseparations necessitates innovations in ultra-filtration membranes. Block copolymer (BCP) membranes have emerged as a promising player for efficient ultra-filtration due to their uniform pore sizes and sharp cut-offs. Most of the work in BCP membranes has focused on using cylindrical pores. However, the biological ultrafiltration membranes can have complex architectures which can perform complex separations such as the square, oblique, or hexagonal ultrafiltration pores on the surface layers of the prokaryotic cells. In this work, we demonstrate novel slit-based membranes using lamellar block copolymers by taking insights from biological membranes. The lamellar BCPs are vertically oriented in one step film casting process by neutral solvent design. The vertically oriented BCPs are converted to slit membranes using a wet etching process. These slit-shaped membranes demonstrate sharp cut-offs for solute filtration. Furthermore, we demonstrate the first example of enhanced separation of 1-D nanomaterials (nanorods) as compared to the 0-D nanomaterials (nanoparticles) using these slit-based membranes, which is facilitated by the shape similarity of 1-D nanomaterials with the nano-slits. Additionally, we show that these slit-shaped membranes show 100 % permeability of similar-sized Lysozyme and Bovine Serum Albumin (BSA) proteins while retaining the Immunoglobulin G (IgG) antibodies up to 70-80 %, thus making them highly useful for bioseparations. The enhanced permeation of the BSA and Lysozyme is governed by their relative one-dimensional shapes as opposed to the Y-like shape of the IgG proteins. We believe that these slit-shaped membranes will be useful for shape-selective filtrations across a spectrum of applications.

642g Conjugated Polymers for Molecular Separation

Christopher Merkel, Febby Andini, Syed Ibrahim Gnani Peer Mohamed, Mona Bavarian, Siamak Nejati

The separation performance and permeance of membranes are dependent on pore size uniformity and porosity. To that end, the uniform pore size and porosity of conjugated microporous polymers (CMPs) attracted significant attention from scientific community. With recyclability and sustainability as a societal need for scientific advancement, development of selective nanofiltration membranes is now highly desirable. Along these lines, CMPs, with exceptional resistance to common solvents, are well-posed to be separation materials for solvent filtration. However, the preparation of robust nanofiltration membranes with CMPs is limited due to their brittle nature and their solvent intractability. Herein, we report the development of porphyrin-based CMP filtration membranes from the electropolymerization of 5,10,15,20-Tetrakis(4-aminophenyl)porphyrin (TAPP). To evaluate CMPs performance TAPP was electropolymerized on conductive glass, etched off using potassium hydroxide and then transferred to ultrafiltration supports. The separation performance of TAPP membranes was evaluated for a wide range of dye molecules with the molecular weight ranging from 200 to 1000 Da and in various solvents including water, acetonitrile, ethanol, and methanol. The composite membrane showed a molecular weight cut-off (MWCO) ~ 300 Da with acetonitrile permeance of 1.5 L/m^2 h bar. To improve the applicability of our membranes, next we fabricated a conductive support layer by filtration of titanium carbide based, MXene, on a nylon support. The as-fabricated MXene-coated membranes were used as conductive substrates for the electropolymerization of TAPP. The composite nanofiltration membranes demonstrated improved mechanical strength, uniform pore size and ductility. The engineered MXene/CMPs nanocomposite membrane exhibited enhanced acetonitrile permeance of 2.2 L/m^2 h bar without compromise in the solute rejection and MWCO of ~350 Da. We attribute the improved performance of the composite membrane to the as-formed MXene interlayer.

213b Snips Membranes-How It Came about

Klaus Peinemann

A SNIPS membrane is made by a non-solvent-induced phase separation of a block copolymer solution. Self-assembly of the block copolymer and phase separation happen nearly simultaneously. The very first isoporous SNIPS membrane was manufactured at the former GKSS (Germany) in 2006. I prepared this membrane myself, and I would describe this discovery as a pseudo-serendipitous event*. A concentrated polymer solution with two or more solvents was cast into a film and plunged into water. In less than 30 seconds, an asymmetric membrane was formed with a spongy substructure covered by a thin, highly porous skin with pores all of the same size in the low nm range[1]. This was done with a lab-synthesized block copolymer, and it was spectacular. However, it turned out that I could not reproduce this process with commercially available block copolymers, and this could have been the sad end of a story that had an exciting start. This was the right time to join forces with Suzana. However, we could not continue this work at GKSS** in Germany, and we searched for a new workplace. Luckily, we both received an offer from the King Abdullah University for Science and Technology (KAUST) in Saudi Arabia), where we started our work as founding members in the fall of 2009. I think we were the first researchers doing chemical experiments in the new KAUST labs. We were a special team because we had complementary talents.

As a hands-on experimenter, I loved to smell the mixture of solvents, and I liked to dive my fingers into the precipitation bath. On the other hand, Suzana dived deep into the mechanism of phase separation and the theory of morphology formation in block copolymer systems. She used the whole armada of characterization techniques available for us, including Cryo- FESEM, Cryo-TEM, Small and wide-angle X-ray scattering, focused ion beam microscopy and more. Suzana could show that micelle evolution and micelle self-assembly was vital for the membrane's morphology formation. Only ten months after our arrival at KAUST, we submitted a paper to Macromolecules which was pivotal for the further development of SNIPS membranes.[2] Suzana pushed hard, and soon afterward, we could publish more details on membrane formation by block copolymer micelle assembly[3,4]; the term SNIPS was not even coined at this time. In the Langmuir paper, a convincing correlation is shown between the micelle formation and self-assembly in solution and the final morphology of the solid membrane.

Thanks, Suzana! Without you, I would have given up this work at an early stage; with you, it developed into a fascinating project.

1.K.-V. Peinemann et al., Asymmetric superstructure formed in a block copolymer via phase separation, Nature Materials 2007, 6, 992, DOI: 10.1038/nmat2038

2.S. P. Nunes et al., Ultraporous films with uniform nanochannels by block copolymer micelles assembly, Macromolecules 2010, 43, 8079. DOI: 10.1021/ma101531k

3.S. P. Nunes et al., Switchable pH-responsive polymeric membranes prepared via block copolymer micelle assembly, ACS Nano 2011, 5, 3516. DOI: 10.1021/nn200484v

4.S. P. Nunes et al., From micelle supramolecular assemblies in selective solvents to isoporous membranes, Langmuir 2011, 27, 10184. DOI: 10.1021/la201439p

Pseudo serendipity means discovering something one was looking for in a surprising way (Diaz de Chumaceiro, The Journal of Creative Behavior 29, 143–147 (1995))

** see also my blog at https://peinemann-lifetwists.blogspot.com

213c Sustainable Production of Nanoemulsions By Membrane Emulsification for Biomedical and Bioprocessing Applications

Usman Syed, Joao Crespo, Carla Brazinha

Abstract

In the quest for replacing toxic, volatile and harsh solvents, and to abide by the principles of green and sustainable chemistry and engineering, various functional solvents have been introduced in the last two decades. The synthesis of perfluorocarbons and hydrophobic analogues of ionic liquids (ILs) and deep eutectic solvents (DESs) are some of the examples employing hydrophobic designer solvents. However, in order to potentiate bio-based applications wherein polar media is usually encountered, the low solubility of these hydrophobic solvents in water was tackled in this work by dispersing them as nanoemulsions[1].

Membrane emulsification was introduced in the end of the last century as an energy-efficient method to produce **microemulsions**. In the current study, various approaches of membrane emulsification processes were re-visited and re-established to sustainably produce **nanoemulsions**. Perfluorodecalin-in-water nanoemulsions with enhanced stability and energetics were formulated by *'direct membrane emulsification'* using isoporous polymeric membranes. The high viscosity of ILs was addressed by employing *'premix membrane emulsification'* to formulate IL-in-water nanoemulsions for enhanced CO_2 capture. Lastly, the unique observation of self-assembly of DES enabled the formulation of hydrophobic DES-in-water nanoemulsions with enhanced antimicrobial activities by *'membrane-assisted nanoemulsification'* using laser drilled microengineered metallic membranes[2].

References

1. Syed et al. (2021). Studies on the formation and stability of perfluorodecalin nanoemulsions by ultrasound emulsification using novel surfactant systems. Colloids Surf., A. 616: 126315.
2. Syed et al. (2020). Microengineered membranes for sustainable production of hydrophobic deep eutectic solvent-based nanoemulsions by membrane emulsification for enhanced antimicrobial activity. ACS Sustainable Chem. Eng. 8: 16526-16536.

Acknowledgements

Executive Agency for Education, Audiovisual andamp; Culture (EACEA) of the European Commission is acknowledged for the scholarship grant of Erasmus Mundus Doctorate in Membrane Engineering (EUDIME) program to Usman Syed. This work is also supported by the Associate Laboratory for Green Chemistry - LAQV which is financed by national funds from FCT/MCTES (UIDB/50006/2020).

213d Rational Design of DNA-Based Responsive Membranes

Thomas Schäfer

Aptamers are functional nucleic acids capable of specifically interacting with a molecular target. The use of DNA-aptamers in stimulus-responsive membranes provides a a modular system with a degree of freedom that is not achieved with bulk polymer membranes. By simply exchanging the recognizing element (DNA-aptamer) but without changing the overall physico-chemical properties of the system, the system can be easily adapted to different targets.

It has been reported previously that single DNA-aptamers can be used as so-called "gate keepers" in nanoporous membranes where the recognition of a molecular target results in a conformational change that in turn modulates membrane permeability. The gating function of the aptamer was shown to be specific, concentration dependent, and reversible. However, such systems depended on a fine-tuning between pore size and DNA-aptamer dimensions in order to be efficient.

To overcome this limitation, we studied DNA-aptamer sandwich and hybrid structures of different compositions. These structures are scalable and to a certain extent independent of the pore size such that the flexibility in designing responsive systems is significantly increased. In absence of a molecular target, the sandwich and hybrid structures remain stable; upon exposure to the target, the structures disintegrate as the recognition of the target is energetically more favourable than the hybridization between individual DNA segments.

We employed QCM-D and MP-SPR to monitor the systematic and controlled build-up of the DNA-sandwich and hybrid structures. This permitted a rational design of the responsive membranes using less DNA material and yielded a far lower polydispersity than a one-pot self-assembly proposed in previous studies. While exposure to the target and subsequent recognition resulted in the disintegration of the sandwich and hybrid structure, it was also found that this disintegration was heavily dependent on the liquid composition. As a lesson learnt from this systematic study, it will be shown that it is indispensable to take into account basic thermodynamics when designing DNA-based responsive membranes.

213e Integrating Biological Functions in Artificial Membranes

Lidietta Giorno

Some properties of artificial membranes, such as selectivity, control of fouling and self-cleaning, are still far from those of biological membranes. In biological systems, high selectivity is achieved by cascade separation mechanisms, and fouling is controlled by operating at low flux and by cleaning and regenerating the membrane *in situ*.

One of the strategies to improve the properties of artificial membranes is to engineer biohybrid materials by integrating biocomponents into synthetic nanostructured materials to mimic bio-structures, systems, functions.

The presentation will use case studies to show that in addition to the method by which a biocomponent (e.g., an enzyme, an antibody, a cell, etc.) is loaded onto a synthetic membrane and the physicochemical properties of the material, the topography and nanoarchitecture of the membrane play a critical role in preserving the structure of the biocomponent and its associated functions.

The need for nanostructured membranes specifically designed to accommodate and implement biological functions will be discussed.

213f Artificial Water Channels-Toward Biomimetic Membranes for Desalination

Mihail Barboiu

This lecture discusses the incipient development of the first artificial water channels systems. We include only systems that integrate synthetic elements in their water selective translocation unit. Therefore, we exclude peptide channels because their sequences derive from the proteins in natural channels. We review many of the natural systems involved in water and related proton transport processes. We describe how these systems can fit within our primary goal of maintaining natural function within bio-assisted artificial systems. In the last part, we present several inspiring breakthroughs from the last decade in the field of biomimetic artificial water channels. All these examples demonstrate how the novel interactive water-channels can parallel biomolecular systems. At the same time these simpler artificial water channels offer a means of understanding water structures useful to understanding many biological scenarios. Moreover, they can be used for the preparation of highly selective membranes for desalination.

[1] M. Barboiu and A. Gilles, Acc. Chem. Res. 2013, 46, 2814–2823.
[2] M. Barboiu, Angew. Chem. Int. Ed. 2012, 51, 11674-11676.
[3] Y. Le Duc, et al., Angew. Chem. Int. Ed. 2011, 50(48), 11366-11372.
[4] E. Licsandru, et al., J. Am. Chem. Soc., 2016, 138, 5403-5409.
[5] M. Barboiu, Chem. Commun., 2016, 52, 5657- 5665.
[6] I. Kocsis, et al. Science Adv. 2018, 4, eaao5603.
[7] M. Di Vincenzo, A. Tiraferri, V.-E. Musteata, S.Chisca, R.Sougrat, L.-B. Huang, S. P. Nunes, M. Barboiu, Nat. Nanotechnol. 2021, 16, 190-196.
[8] M. Di Vincenzo, et al. Proc. Natl. Acad. Sci. USA, 2021, 118(37), e2022200118
[9] L.-B. Huang, M. Di Vincenzo, M. Göktuğ Ahunbay, A. van der Lee, D. Cot, S. Cerneaux, G. Maurin, and M. Barboiu, J. Am. Chem. Soc. 2021, 143, 14386-14393

609a Preparation of Polyethersulfone Membranes on Non-Woven Fabric Supports

Joseph Danner, Blake Jarrell, Steven Weinman

Current commercial recycling processes for polyethylene terephthalate (PET) and polypropylene (PP) tend to produce inferior materials, whereas polyvinyl chloride (PVC) is hardly recycled. By developing strategies to "upcycle" these plastic wastes into non-woven fabrics, further plastic pollution can be prevented while simultaneously providing economic and societal benefits. This study investigates the interfacial science between non-woven fabrics made from recyclable material and the separation layer of the polyethersulfone (PES) membrane cast on top of these non-woven fabrics to determine how the fabrics preserve or enhance the membrane properties. Additionally, this study also investigates the effect of "green" solvents on membrane separation properties. Solvents were selected based on the evaluation of their Hansen solubility parameters, partition coefficient (LogP), and associated safety and hazard properties. The membranes were prepared by the nonsolvent-induced phase separation (NIPS) method using PES in two categories of solvents: conventional and "green". N,N-dimethylacetamide (DMAc) and N-methyl-2-pyrrolidone (NMP) were used as the conventional solvents and various glycerol-derived molecules were used as the "green" solvents. Physical characterization of each membrane's cross-section and surface morphology were accomplished via scanning electron microscopy. A dead-end stirred filtration cell was used to evaluate the pure water permeance and molecular weight cut-off of these membranes. Contact-angle measurements were taken to determine the hydrophobicity/hydrophilicity of membrane surfaces.

609b Pervaporative Alcohol Dehydration Using PVA-TEOS-CMS Hybrid Membrane

Musabbir Jahan Talukder, Ali Alshami

Polyvinyl alcohol (PVA) membranes are characterized by their inherent hydrophilicity which makes them attractive for highly permselective water separation for aqueous organic dehydration or water treatment applications. However, swelling and their compromising transport properties introduces control attempts like cross-linking, surface modification, and annealing methods. These methods offers a good solution in terms of stability but it highly compromises the transport properties. In this study, a hybrid organic–inorganic membranes were prepared by hydrolysis of poly(vinyl alcohol) (PVA) cross linked by tetraethoxysilane (TEOS) with carbon molecular sieves(CMS) derived from lignin. The negative effects introduced by the crosslinking is compensated by the additional selective water transport by the CMS. The use of hybrid membrane showed improvements in the mechanical and thermal properties. The chemical and thermal crosslinking conditions were optimized for the two-step crosslinking approach. Membrane were characterized by Fourier Transform Infrared Spectroscopy (FTIR), Thermo Gravimetric Analysis (TGA), X-ray Diffraction (XRD), Atomic Force Microscopy (AFM), SEM, water contact angle analysis as well as pervaporation performance for ethanol dehydration were investigated. The obtained results are crucial for the design and elaboration of the polymer membranes for the pervaporative alcohol dehydration as well as water treatment processes.

609c Tunable Amphiphilic Pdmaema-b-PS Block Copolymer and Ionic Liquid Composite Thin Films for Gas Separations

Charlie Knight, Paul Scovazzo, Sasan Nouranian, Adam Smith, Alexander Lopez

Ionic liquid addition to polymer membranes has improved permselectivity in gas separations through enhanced gas transport. Supported ionic liquid membranes (SILMs) are the most facile method to incorporate ILs into polymer films, and SILMs offer improvement of permselectivity vs. permeability when compared to conventional polymer films. However, SILMs cannot perform gas separations at transmembrane pressures above 5 bar due to the extrusion of the capillary force supported IL. A biphasic thin-film using an amphiphilic block copolymer to encapsulate ionic liquid in the hydrophilic domain eliminates the risk of cross membrane pressures overpowering capillary forces. This study investigated the incorporation of two ionic liquids, [EMIM][TFSI] and [EMIM][SCN], within an amphiphilic biphasic block copolymer thin film comprised of tunable poly[2-(dimethylamino)ethyl methacrylate-*b*-styrene] (PDMAEMA-b-PS) blocks. Reversible addition-fragmentation chain transfer (RAFT) polymerization controlled block copolymer tunability, and the resulting polymers were characterized via gel permeation chromatography (GPC) and ^{1}H NMR. Thin films were cast using solvent-non-solvent induced phase separation (SNIPS) and characterized by fourier-transform infrared spectroscopy (FTIR), atomic force microscopy (AFM), and scanning electron microscopy (SEM). The casting solution composition was investigated to identify its influence on the homogeneity and properties of the resulting thin films. Results indicate that [EMIM][TFSI] was incompatible in the neat casting solution due to the precipitation of the polymer. In addition, the solvents DMSO and a mixture of DMF with THF did not properly dissolve both polymer and IL in the casting solution. However, the resulting neat films cast from solutions in pure DMF proved to be rigid and stable in formulations containing > 20 wt% of the block copolymer. Lag-time and continuous gas permeation experiments at 5, 10, and 15 bar were conducted on stable neat and IL-loaded thin films using CO_2 and CH_4.

609d Ultrasonicated *in Situ* Growth of Ag-MOF in Polyamide Selective Layer of Thin Film Nanocomposite Membranes: Fabrication, Characterization, and Performance

Medha Kasula, Milad Esfahani

Thin-film composite (TFC) membranes consisting of a polyamide (PA) selective layer made by interfacial polymerization (IP) on a support layer have become a trending research area to synthesize a commercialized water treatment and desalination membrane. Functionalization of PA selective layer for enhanced permeability, selectivity, and antifouling using additive incorporation is a promoting method to achieve this goal by tuning the morphological and physicochemical properties of the PA layer. Incorporating nanomaterials into the PA selective layer provides additional degrees of freedom to tune the material properties and enhance the separation performance. Metal–organic frameworks (MOFs) are a new type of hybrid material with a synergistic combination of inorganic metal ions and organic ligands as linkers. MOFs' highly porous structure and tunable chemical functionalities make them promising candidate materials for functionalization of TFC membrane. Although the addition of MOF into the PA layer has been reported in the literature, however in most of the reported data, already-prepared MOF particles were added into the PA layer. In this study, we report the *in situ* synthesis and growth of MOF via ultrasonication method on the support layer and fabrication of PA layer in the presence of MOF. (i) The formation and characterization of Ag-MOF on the support layer and (ii) the effect of Ag-MOF on the formation of the PA layer during the IP process regarding the transport of trimesoyl chloride (TMC) and piperazine are discussed. The PA selective layer characteristics, including the degree of cross-linking, surface roughness and morphology, are studied and benched marked with a PA layer incorporated with already-prepared Ag-MOF and bare thin film composite (TFC) membrane. In addition, the performance (water permeability, selectivity and antifouling) of *in situ*-synthesized-Ag MOF are also compared with already-prepared Ag-MOF-membrane and bare TFC membrane.

609e High Flux Porous Structure Formation in Isoporous Thin Film Membranes through Selective Domain Assembly

Khadar Shaik, Ali Ammar, Kshitij Sharma, Diana Cousins, Maninderjeet Singh, Deepalekshmi Ponnamma, Mohammad Hassan, Samer Adham, Mariam Al-Maadeed, Anil Bhowmick, Alamgir Karim

The peculiar morphologies of block copolymer (BCP) thin films offer an expansive platform to design and adjust different properties of the membranes by simply tuning the experimental parameters. This is especially applicable for the diblock copolymers where their domain assembly could easily be tailored before, during or after the thin film formation to study their affects on membrane characteristics with great control. In this work, we report a highly porous structure of an ultrafiltration membrane with poly(styrene)-b-poly(4-vinlypyridine) (PS-b-P4VP) BCP derived selective layer. The desired domain assembly in the asymmetric BCP films is promoted by adding different mass ratios of an active additive which selectively migrates into the domains. The films are then captured above polyethersulfone support layer and immersed in selective swelling solution to enable modified pore formation. Such membrane composites exhibited water flux rates above 300 $Lm^{-2}h^{-1}bar^{-1}$ which indicate an increase in interconnectivity of pores. Furthermore, the porosity of the ~120 nm selective layer is enhanced by adding homopolymer and solvent blends and later using quaternization process. The resulting pore morphologies in relation to their blend concentrations are measured using atomic force microscopy. The asymmetric membranes prepared with highly tunable pores are demonstrated to be effective for ultrafiltration applications.

- This work is supported with funding from NPRP grant 10-0127-170269 of QNRF, NSF grant DMR-1905996

609f Fabricating a 'Breathable' Porous Flat Sheet Membrane Via Non Solvent Induced Phase Separation (NIPS) for Face Masks and Other Air Filtration Applications

Ebuka Ogbuoji, Isabel Escobar

The ongoing Covid 19 pandemic has led to growing demands for face masks around the world for protection against the infectious Sars-Cov-2 virus. 3M, a Personal Protective Equipment (PPE) manufacturing company, reported a global production of two billion respirators in 2020. The most common material used in producing these commercially available face masks and respirators is polypropylene (a non-biodegradable thermoplastic) because of its chemical resistance, lightweight, and low cost. A study in six countries across Asia, Europe, and North America showed that individuals between ages 16 and 65 consume about five (5) face masks per week, resulting in an estimated range of 3 to 928 million face mask waste per week in all countries studied. This could result in a polypropylene weekly waste ranging from 8 tons in lower populated countries to 2450 tons in denser populated countries. To reduce the waste caused by the excessive but essential use of face masks, studies have proposed implementing various decontamination techniques such as chemicals, detergents, heat, microwave, and ultraviolet light to ensure the safe reuse of facemasks. These techniques have been reported to reduce the mask filtration efficiency (FE) and compromise the mechanical properties of the mask material. This work involves the fabrication of a breathable high FE porous flat sheet polysulfone membrane resistant to common decontamination processes for face mask manufacture, which could potentially address the limitations of reusing nonwoven facemasks. The breathable membranes were fabricated by nonsolvent induced phase separation (NIPS) using casting solutions containing polyethylene glycol (PEG) as an additive to improve airflow and reduce pressure drop across the membrane. Fourier transform infrared (FTIR), x-ray photoelectron spectroscopy (XPS), and differential scanning calorimetry(DSC) analysis confirmed the absence of PEG in the fabricated membrane, indicating additive dissolution in the nonsolvent during phase separation. Scanning electron microscope (SEM) images of the fabricated membranes showed a large fingerlike pore structure, suggesting fast solvent exchange reported to result in membranes with increased flux. Air permeation and pressure drop tests conducted across the samples showed lower resistance to airflow and pressure drop across membranes made from solutions containing additives. Membranes made from casting solutions with PEG molecular weight of 1000, 4000, 8000, and 10000 daltons at a polymer-additive-solvent concentration of (15/10/75)%w/w resulted in airflow rates of 4.61±0.06, 5.14±0.01, 13.28±0.08, 20.85±0.98 liters per minute (LPM) respectively, which suggests an increase in pore size with PEG molecular weight. The FE of the highest airflow membrane was observed to be 99.5% for particles greater than 300μm after testing using aerosolized NaCl particles at 35LPM. The breathable membrane, N95, and surgical masks were washed with detergent for 1 hour and dried to estimate the effect of a common and simple decontamination process. We observed an insignificant effect of washing with detergent on the fabricated membrane sample FE; however, there was a 20-25% reduction in N95 and surgical mask FE. The decrease in FE was due to depleted surface charges responsible for high FE in electret filter media, which functions by electrostatic interaction mechanism.

609g Formation of Hydrogen-Selective Carbon Composite Membranes By Incomplete Stabilization of Polyacrylonitrile

H. Enis Karahan, Wataru Nakata, Behnam Ghalei

Carbon molecular sieve (CMS) membranes offer attractive separation performances and thermal resistance. Nevertheless, their fabrication is typically overly expensive, partly because of the cost of preferred polymeric precursors, like polyimides. And, affordable precursors, such as cellulose derivatives, come with film formation issues. At this point, being the most utilized precursor for carbon fiber production, polyacrylonitrile (PAN) is an appealing option for CMS development. But interestingly, not much has been done for utilizing PAN in CMS membrane fabrication, except for a few early studies and some other failed attempts. Therefore, we tested the utility of PAN for fabricating CMS membranes capable of separating hydrogen (H_2) and carbon dioxide (CO_2). As the membrane substrate, we have exploited gamma-alumina-coated alpha-alumina discs. To obtain uniform membranes with polymer layers of around 2 to 4 micrometers with minimal support infiltration, we used spin coating and a low-boiling-point solvent (hexafluoroisopropanol). And to prevent pinhole formation, we applied dual-layer coating. After achieving defect-free thin-film composite membranes, we optimized the heating protocol based on differential scanning calorimetry, thermogravimetric analysis, and X-ray diffraction studies. We found that the thermal stabilization (cyclization) step should be made prematurely to achieve attractive gas selectivities and permeabilities (exceeding Robeson's 2008 upper bound). Overall, as long as the filmmaking and stabilization steps are optimized, PAN is suitable for developing H_2-selective CMS membranes.

343a Effect of Molecular Dynamics Water Models on Flux and Diffusivity in Membranes

Suwei Liu, Sinan Keten, Richard Lueptow

Despite the broad application of polyamide membranes for water purification and treatment, the interactions between the membrane, water, and solutes are unclear at the molecular level. To better understand these interactions, molecular dynamics simulations are often used to model the filtration process. While realistic membrane models are crucial to investigating the membrane filtration process, the atomistic water model and simulation parameters also play important roles. Previous polyamide membrane simulation studies use a variety of water models along with different molecular dynamics thermostat settings. In this research, we find that the choices for the water model (SPC/E, TIP3P, or TIP4P), thermostat settings, and temperature have significant impact on water flux and diffusivity in simulations of charged nanofiltration membranes. We also compare the water dynamics results obtained through simulations with previous experimental and computational results. This work offers insights into the influence of simulation parameters on water dynamics and which water models most accurately represent real water behaviors in polyamide membrane filtration systems. Funded by NSF Grant CBET-1840816.

343b Optimization Based Modeling and Analysis of Osmotically Assisted Reverse Osmosis for Application in Shale Oil and Gas Produced Water Treatment

Elmira Mohammadi Shamlou, Radisav Vidic, Vikas Khanna

Substantial amount of water is required for hydraulic fracturing and large volume of water, known as produced water, with relatively high total dissolved solids (TDS), is produced during unconventional oil and gas production. Being costly and difficult to treat, hypersaline brines could contaminate surface freshwater resources and ground water reservoirs. Desalination technologies offer a promising way to treat the produced water and provide pure water. Having higher energy efficiency in comparison to thermally driven desalination processes, reverse osmosis (RO) is one of the most widespread desalination technologies worldwide. RO is a pressure driven separation process utilizing hydraulic pressure to force the solvent to move through a semipermeable membrane from higher solute concentration to lower concentration. However, RO is incapable of treating hypersaline produced water with high osmotic pressure, since the applied hydraulic pressure is limited by the maximum pressure that RO membranes could withstand. Osmotically assisted reverse osmosis (OARO) systems are emerging RO configurations that enable the pressure driven membranes to treat hypersaline solutions. OARO overcomes the problem of excessive hydraulic pressures by introducing a saline solution stream on the low-pressure side of the membrane leading to reduction in transmembrane osmotic pressure difference.

Systematic design and optimization of the OARO technology and operating conditions is required to assess the energetic and economic feasibility of this technology for produced water treatment. This work presents module scale optimization-based modeling and analysis of several OARO configurations. Specifically, brine-reflux OARO (BR-OARO), cascading osmotically mediated reverse osmosis (COMRO), consecutive loop OARO, and split feed counterflow RO configurations are modeled and studied in detail. The BR-OARO configuration is found to be superior to other configurations in terms of stage count, treatment cost, membrane area, and energy consumption. Additionally, sensitivity analysis is performed to determine the effect of input parameters such as feed salinity and pressure on the performance of each configuration.

343c Design Driven Trends in Transport and Stability of Biomimetic Membranes with Highly Selective Water Channels

Ritwick Kali, Erha Andini, Scott T. Milner

Peptide appended pillar[5]arenes are tubular molecules that can facilitate rapid and selective water transport. These artificial water channels can be embedded in self-assembled diblock copolymer bilayers to serve as desalination membranes. To unleash the full potential of these pore molecules, the membrane architecture must be properly chosen. We vary the block copolymer design to study its effect on water transport and channel stability, using atomistic molecular dynamics simulation. We find that for the pore molecule to be functional, the bilayer hydrophobic thickness must be no less than the height of the embedded channel. This dimensional parity is also an important factor for pore stability, which also correlates with the number of solvating water molecules and condensed counterions near the pore mouth. The hydrophilic block length is also crucial. If the block is too short, it tends to penetrate and clog the pore; if the block is too long, the water string tends to be unstable. Our simulations predict high water diffusivity in PAP[5] channels, consistent with experimentally reported values. Finally, we develop a novel technique to quantify ion rejection for these channels, which firmly supports their potential for desalination.

343d Mitigated Carrier Saturation of Facilitated Transport Membranes for Decarbonizing Dilute CO_2 Sources

Yang Han, W.S. Winston Ho

Although carbon capture and storage (CCS) from large stationary sources is a leading strategy to reduce the carbon emissions, a carbon-neutral scenario in the energy sector cannot be achieved by the current CCS technologies since 90% CO_2 recovery is commonly assumed in the design. By taking coal-derived flue gas as an example, a 90% recovery through a primary capture system results in a residual flue gas containing 1–2% CO_2. If not captured, the residual emissions must be mitigated via direct air capture (DAC), which is usually hampered by a substantial cost. The focus of this presentation is to address the membrane carbon capture from dilute, yet stationary sources in order to achieve an overall >90% CO_2 recovery and thereby alleviate burdens upon DAC. To this end, the CO_2/N_2 separation performances of facilitated transport membranes (FTMs) containing aminoacid salts as mobile carriers were characterized with a CO_2 concentration of 0.05–20%. At a reduced CO_2 partial pressure, the carrier saturation in the FTMs was mitigated, which enhanced both the CO_2 permeance and CO_2/N_2 selectivity. The best FTM containing 2-(1-piperazinyl)ethylamine sarcosinate exhibited an uprising CO_2 permeance from 1968 to 3822 GPU and an improved CO_2/N_2 selectivity from 249 to 472 with reducing CO_2 content from 1% to 0.1%. The feasibility of this FTM is exemplified by designing a two-stage enriching membrane cascade to further remove 90% of the CO_2 in a residual coal flue gas containing 1.75% CO_2. Techno-economic analysis indicates a low capture cost of $83.8/tonne. The marginal costs beyond 90% capture are also evaluated for a variety of residual flue gases, indicating that the FTM-based capture from the coal or cement plant residual flue gas is more cost effective than DAC.

343e A Model for the Separation of Complex Liquid Mixtures with Glassy Polymer Membranes: A Thermodynamic Perspective

Bennett Marshall, Joshua Allen, Ryan P. Lively

In this work, a new model is developed to predict the separation of organic solvent mixtures with glassy polymer membranes. Particular attention is paid to thermodynamic contributions such as guest solubility and guest-induced polymer swelling, and how these contributions affect the flux in multi-component mixtures. The thermodynamic contributions are complicated by the fact that the membranes are glassy polymers. The thermodynamic aspects of the diffusion problem are calculated using DGRPT (dry glass reference perturbation theory), which extends the non-equilibrium thermodynamics of glassy polymers (NETGP) such that polymer swelling is calculated self-consistently in the theory. The diffusion model for pure fluids diffusing through the membrane is obtained through Maxwell-Stefan equations where the polymer membrane is stationary. The mixture diffusion model is then constructed using a simple diffusivity averaging scheme on this pure fluid result. The model is then shown to accurately predict binary and complex mixture separations in SBAD-1, STARMEM-240, and Torlon membranes. We highlight the ability of the model accurately predict membrane based fractionation of crude oil, using only the sorption isotherms of toluene and octane as input to the model.

343f Moisture Driven Pump to Concentrate CO2 from Air

Jennifer Wade

Strong base anion exchange membranes (sb-AEM) can bind CO_2 isothermally and reversibly from air (40 Pa) and release at pressures up to 4 kPa using a change in water activity . This work presents the results of an unsteady finite-difference numerical diffusion-migration-reaction model that predicts the flux of carbon through these materials with an applied moisture gradient. CO2 reacts with hydroxide counter ions in the AEM materials forming bicarbonate species. The application of moisture shifts the equilibrium of the HCO3/CO3 distribution - the moisture-swing - creating a net current of dissolved inorganic carbon against the CO2 partial pressure gradient. The flux of CO2 is enhanced by the ionic species transport and is limited by the extent of the moisture swing response and magnitude of partially hydrated ion diffusivities. The model uses experimentally determined water transport and isotherm properties of commercial hydroxide exchanged quaternary amine AEM. In the limit of fast reaction, CO2 flux is expected to approach 20 x 10^{-6} mol m^{-2} s^{-1} under room temperature conditions. A pumping flux may be sustained up to counter pressures of 4 kPa if a sufficient moisture gradient is sustained. This presentation will report on the optimal application of a water gradient and discuss relevant material properties to best exploit the moisture driven response. This work reveals a new approach for carbon dioxide concentration that has potential to be coupled with other regeneration mechanisms for energy efficient direct air capture.

343g A Spatiotemporal Model for Dynamic RO Simulations

Mingheng Li

Dynamic reverse osmosis (RO) processes (including batch RO, closed-circuit RO and flow reversal RO) have niche applications in water desalination. In these non-traditional ROs, process variables are functions of both time and location. Traditional steady-state models are not adequate to advance the fundamental understanding of these processes. In this work, a spatiotemporal model is developed for dynamic simulations of RO desalination. The model explicitly accounts for pressure drop, concentration polarization and axial dispersion along the pressure vessels. A detailed numerical method, which is based on orthogonal collocation on finite elements in the spatial domain, is presented to solve the coupled partial differential equations (PDEs) in the model. Examples are used to illustrate the application of the developed model.

References:

[1] Li, M.; Chan, N.; Li, J. Novel Dynamic and Cyclic Designs for Ultra-High Recovery Waste and Brackish Water RO Desalination, Chemical Engineering Research and Design, 179, 473-483, 2022.
[2] Li, M. A Spatiotemporal Model for Dynamic RO Simulations, Desalination, 516, 115229, 2021.
[3] Li, M. Residence Time Distribution in RO Channel, Desalination, 506, 115000, 2021.

278a Integrated Facilitated Transport Membrane Modules for Highly Selective Syngas Purification and Carbon Capture

Yang Han, Yutong Yang, Ruizhi Pang, W.S. Winston Ho

Amine-containing facilitated transport membranes (FTMs) are of great interest for syngas purification. Herein, we report a new method to enhance the CO_2 permeance by manipulating the steric hindrance of the amine carrier. A series of α-aminoacids with different alkyl or hydroxyethyl substituents were deprotonated by 2-(1-piperazinyl)ethylamine, resulting in nonvolatile aminoacid salt carriers with different degrees of steric hindrance. In the presence of moisture, a bulkier alkyl substituent increased the steric hindrance and hence destabilized the carbamate adduct to afford bicarbonate through hydrolysis. This drastically increased the chemisorption of CO_2. The enhanced CO_2 solubility significantly mitigated the carrier saturation, and an unprecedented CO_2/H_2 selectivity greater than 125 was demonstrated at 107°C and 13.8 bar of CO_2 partial pressure (Type I). As the CO_2 partial pressure reduced to 1.1 bar, a less hindered amine yielded a higher reactive diffusivity of CO_2, resulting in a CO_2 permeance of 217 GPU with a selectivity greater than 268 (Type II). These two types of FTMs were successfully scaled up by a roll-to-roll continuous coating machine. The 14″ wide scale-up membranes were then rolled into two prototype spiral-wound modules containing 800 and 1600 cm^2 of the Type I and Type II membranes, respectively. At 90% CO_2 capture, the Type I and Type II modules achieved a CO_2 purity of ca. 95%, but the H_2 recoveries were only 98.2% and 97.1%, respectively. However, when the two modules were connected in series with the Type I membrane treating the syngas near the feed inlet (i.e., high CO_2 partial pressure) and the Type II membrane separating the gas in the proximity of the retentate outlet (i.e., low CO_2 partial pressure), the H_2 recovery was improved to 99.3%. The integrated membrane array also demonstrated 200-h stability under simulated syngas.

278b Hydrogen-Sieving Zeolitic Films By Filter Deposition on Porous Polymeric Support

Kumar Varoon Agrawal, Xuekui Duan

The synthesis of high-performance gas-sieving zeolitic film is challenging because it involves seeded secondary growth which is complex and difficult to reproduce. Additionally, expensive asymmetric inorganic porous supports are required to allow high-temperature activation. Herein, scalable preparation of H_2-sieving RUB-15 nanosheet films is reported bypassing the need for secondary growth and expensive supports. A novel, low-cost, thermally-robust, porous polybenzimidazole copolymer (PBI-AM Fumion®) support is prepared which allows deposition of thin, compact, and oriented RUB-15 nanosheets films without cracks or pinhole defects. The film hosts two transport pathways, H_2-sieving six-membered silicate rings in the nanosheets and nonselective intersheet gaps maintained by the organic guest species in the gallery spacing. The latter is eliminated either (i) low-temperature (330 °C) calcination or (ii) room-temperature extraction where the crystalline order in the nanosheets is preserved. The resulting zeolitic films on the polymeric support yield extremely promising H_2-sieving performances with H_2/CO_2 separation factor approaching 1000.

278c Supramolecular Polymer Networks of Ion-Coordinated Polybenzimidazole for High-Temperature H_2/CO_2 Separation

Leiqing Hu, Shouhong Fan, Liang Huang, Vinh Bui, Thien Tran, Sankhajit Pal, Yifu Ding, Mark Swihart, Haiqing Lin

Membrane technology with superior H_2/CO_2 separation properties at high temperatures has gained significant attention for pre-combustion CO_2 capture. Advanced membrane materials with both high H_2 permeability and H_2/CO_2 selectivity are needed for gas separation to reduce the capital and energy cost. Herein, we demonstrated facilely novel thermally-stable supramolecular polymer networks (SPNs) comprising of metal salt (palladium trifluoroacetate and nickel trifluoroacetate: $Pd(CF_3COO)_2$ and $Ni(CF_3COO)_2$) and polybenzimidazole (PBI) for H_2/CO_2 separation at high temperatures. With metal ions coordinating imidazole rings on PBI chains and trifluoroacetate anions unwinding the chain packing, the supramolecular network formed a bottleneck/microcavity bimodal pore architecture, leading to simultaneously increased H_2 permeability and H_2/CO_2 selectivity. Particularly, PBI-$Ni(CF_3COO)_2$ supramolecular network with a salt doping level of 0.17 displayed H_2 permeability of 62 Barrer at 150 °C, increased by ~ 100 % from pristine PBI with H_2/CO_2 selectivity increasing from 10 to 15. Those supramolecular network films also exhibited improved mechanical properties and stable mixed-gas H_2/CO_2 separation performance at high temperatures. This study unveiled a facile approach to design membrane materials with bottleneck/microcavity bimodal pore architectures, thereby achieving both high gas permeability and selectivity.

278d Poly(4-vinylpyridine)-Based Catalytic Membranes for Integrated CO_2 Capture and Conversion

Casey O'Brien, Renxi Jin, Hui Xu, Justin Easa, Sarah Pate

Current sorbent-based CO_2 capture processes are prohibitively expensive because they require energy intensive regeneration stages to release CO_2 and regenerate the sorbent. Additionally, they produce CO_2 streams that must be compressed and transported to sites for storage, which reduces efficiency. In this talk, I will describe a new catalytic membrane technology that we are developing to address these challenges and potentially enable commercially viable CO_2 capture by integrating CO_2 capture and CO_2 conversion in a single unit process. The membrane acts as both the CO_2 separation and conversion medium, providing an energy- and atom-efficient alternative to sorbent-based CO_2 capture, compression, transport, and storage. In principle, the membrane can couple CO_2 capture with any transformation that involves carbamate or bicarbonate intermediates including, but not limited to, cyclic carbonate synthesis from epoxides, urea synthesis from ammonia, methanol synthesis from hydrogen, and formic acid from water.

We have recently discovered that poly(4-vinylpyridine) (P4VP)-based membranes are promising for demonstrating the potential of nitrogen-functionalized membranes as an integrated CO_2 capture and conversion platform tunable for CO_2 conversion to a variety of products. In this talk, I will demonstrate that P4VP-based membranes are permeable and selective for CO_2 separation from dilute mixed gas streams and are catalytically active for cyclic carbonate synthesis at mild conditions (room temperature and pressure). I will also show that the membrane separation and catalytic performance of P4VP can be tuned and enhanced by quaternization with alkyl bromides to the corresponding quaternary ammonium salt. Finally, I will demonstrate that quaternized P4VP membranes can integrate CO_2 capture from simulated flue gas and convert it to cyclic carbonates in a separate stream at mild conditions. This work establishes the basis for integrating CO_2 capture and CO_2 conversion to valuable products in a continuous process at mild conditions, which represents a transformative shift in carbon capture technology that also mitigates the need for CO_2 compression, transport, and storage.

278e Molecularly Mixed Nanocomposite Membranes Based on High-Performance Commercial Polymer Blends for Efficientpre-Combustion CO_2 Capture

Fan Feng, Ji Wu, Can-Zeng Liang, Martin Weber, Sui Zhang, Tai-Shung Chung

Nanocomposite membrane materials with a good combination of attractive separation performance and potential commercial scalability are of pivotal role to realizing efficient pre-combustion CO_2 capture from syngas for producing clean hydrogen fuel. However, the utilization of nanoparticles in commercial polymeric membranes is still facing many incompatibility issues till now. Here, we report a novel nanocomposite membrane design based on a soluble organic macrocyclic cavitand, calix[6]arene (CA6), and polymer blends of commercial polyphenylsulfone (PPSU) and polybenzimidazole (PBI). The solubility in polymer solvents and good affinity for organic polymers allows CA6 to be able to homogeneously mix with the PPSU/PBI blends on a molecular basis. A series of PPSU/PBI blend based molecularly mixed composite membranes (MMCMs) are fabricated by the incorporating different amounts of CA6. The membranes achieve promising H_2/CO_2 separation performance owing to the effective size-sieving gas pathways through partially intruded CA6 cavities. 50%PPSU40-50%PBI-10%CA6 membrane achieved a 25% increase in H_2/CO_2 selectivity compared with the pristine 50%PPSU40-50%PBI membrane while the 50%PPSU40-50%PBI-15%CA6 membrane achieved a 35% increase in H_2/CO_2 selectivity compared with the pristine 50%PPSU40-50%PBI membrane. Also, the overall performance of both 50%PPSU40-50%PBI-10%CA6 and 50%PPSU40-50%PBI-15%CA6 can surpass the upper bound of H_2/CO_2 separation. This study could provide a new toolbox to design commercial polymer based nanocomposite membrane materials.

278f A Facilitated Transport Membrane Composed of Restricted Ionic Liquid in Graphene Oxide/Single Walled Carbon Nanotube Framework for Efficient Carbon Capture

Dinesh Behera, Bratin Sengupta, Huazheng Li, Shiguang Li, Miao Yu

Carbon Dioxide (CO_2) emissions is the primary contributor to global warming. A significant fraction of this contribution comes from power plants. Membrane processes may offer a promising solution for low cost, robust and energy efficient CO_2 capture from these plants. Facilitated transport membranes, although showing strong capability for CO_2 capture at low concentrations and under humid conditions, suffer from the loss of facilitators, such as amines, leading to decline in performance over long term operation. Ionic liquids (ILs) are chemically and thermally stable as well as have low vapor pressure, which is ideal to prevent aforementioned losses. However, mechanical stability of supported IL membranes, especially under pressure drop conditions, is a concern. Moreover, amino acid ILs were rarely studied as an efficient CO_2 facilitator. In this work, we report the use of an amino acid IL, 1-ethyl-3-methylimidazolium glycinate ([EMIM] [GLY]), as the facilitator for CO_2 capture. In order to selectively load and confine [EMIM] [GLY] in a thin selective layer, we fabricated a Single Walled Carbon Nanotubes (SWCNT) intercalated Graphene Oxide (GO) layer (thickness:~200 nm; pore size: ~10 nm). The distribution of Ionic liquid in the GO/SWCNT framework is expected to be critical for the formation of a thin selective layer. To understand and optimize this, the ratio of GO to SWCNT was varied in an attempt to adjust the interaction between the framework and [EMIM] [GLY]. The negatively charged GO (because of hydroxyl, carboxyl and epoxy groups) acts as anchoring sites for [EMIM] [GLY] molecules (due to electrostatic interaction between anions on GO and cations of ionic liquid), while the interconnected GO/SWCNT network provides numerous nanochannels for unrestricted gas transport. The influence of membrane thickness, temperature, pressure and humidity on performance of the [EMIM] [GLY] restricted in GO/SWCNT framework was evaluated. The optimized membrane showed a CO_2 permeance of >1,500 GPU and CO_2/N_2 selectivity >300 for simulated flue gas.

278g Water-Gas Shift Reaction with CO2 Capture in Ceramic-Carbonate Dual-Phase Membrane Reactor at High Temperatures and Pressures

Oscar Ovalle Encinia, Jerry Lin

Water-gas-shift (WGS) reaction is a critical step in integrated gasification combined cycles (IGCC) power plants with CO_2 capture. Membrane reactor made with a CO_2-permselective ceramic-carbonate dual-phase (CCDP) membrane offers potential to enhance hydrogen yield with simultaneous CO_2 capture for WGS reaction. The present work studies operation of WGS reaction in a tubular membrane reactor made of samarium-doped ceria infiltrated with lithium/sodium molten carbonate mixture. The WGS reaction was performed in the membrane reactor with and without a high-temperature WGS catalyst in 800-850 °C, feed pressure of 7 bar, space velocity of 150-3000 h^{-1}, and a feed gas mixture of 45.7/13.1/41.3 mol% $CO/CO_2/N_2$ with steam to carbon ratio of 4. The results show that the catalyst-free membrane reactor can convert 92% of carbon monoxide into CO_2 and H_2 and recover 29% of CO_2 at 850 °C and space velocity of 150 h^{-1}. However in the catalyst-free membrane reactor a significant amount of unwanted carbon deposition is observed. The side reactions can be minimized by reducing the operating pressure and increasing the operating temperature and space velocity, and completely avoided using a high-temperature catalyst at space velocity larger than 500 h^{-1}. The membrane reactor with a WGS catalyst achieve CO conversion of about 85%, above the equilibrium conversion, and 40% CO_2 recovery without carbon deposition at high temperature and pressure. The membrane remains in the same structure and gas-tightness after the WGS reaction tests.

469a Process Modeling of Intermittent Wellhead RO Water Treatment Operation Via Integration of Self-Organizing Maps and Long Short-Term Memory Recurrent Neural Network (RNN)

Yang Zhou, Nora Marki, Bilal Khan, Yoram Cohen

Membrane-based water treatment and desalination processes are suitable for deployment at various levels of capacity from small home-based systems to systems suitable for municipal, industrial and even agricultural water supplies. Deployment of such water systems, particularly for distributed high recovery wellhead water purification in small communities that are confronted with impaired potable groundwater supplies, where 24/7 of operator availability is infeasible, requires self-adaptive operation that responds to temporally variable environmental and water quality conditions, in addition to fluctuating water use patterns. The operation of such systems is intermittent, and the product water must meet regulatory safe drinking water requirements, whereby removal of the contaminants of concern must ensure that the product water must be well below the regulatory maximum contaminant level (MCL). In order to develop a robust operational strategy of high recovery wellhead water treatment via reverse osmosis (RO), integrated with feed pretreatment and post-treatment, process models are needed that account for actual behavior of system components (pumps, actuators, sensors, membranes) under temporally variable conditions. Accordingly, in the present work a process model was developed for a uniquely designed high recovery wellhead RO water treatment system operating in a remote agricultural community.

Model development was based on a year of operational time-series data and utilized to evaluate system performance and operational models with respect to nitrate and salt passage, as well as membrane permeability. A forecasting approach, accompanied by a comprehensive exploratory analysis was undertaken making use of long short-term memory (LSTM) recurrent neural network (RNN) architecture with and without an attention coefficient. Performance of the RO system was evaluated for different modes of system operation that included flushing, permeate production of safe product water, startup, and shutdown. A total of twenty quantitative model attributes were utilized (including, for example, pressures, flow rates, temperature, conductivity and nitrate concentrations) and their attribute significance was determined for model building. The LSTM model in which an attention coefficient was used had excellent performance for the different modes of operation (i.e., R^2 value of ~0.95 for both nitrate permeate and salt passage). In order to address the tradeoff between model complexity and robustness, a minimal set of the LSTM model attributes was identified and ranked via Spearman's correlation coefficient and feed forward feature selection (FFFS) approaches. Subsequently, a supervised learning approach based on self-organizing maps (SOM) was applied to discover similarities (or dissimilarities) among the information provided by the different model attributes and thus ensure that pertinent information is provided for model input. It is noted that the assessment of attribute significance for the LSTM models for permeate nitrate and salt passage was improved via supervised SOM analysis that incorporated class information in the input data. The above was accomplished by inserting additional class components in the input vector to indicate the class-affiliation of the data. The class component increased the clustering tendency of the samples belonging to the same class and expelled samples of foreign classes on the map; thus, stretching the SOM separation between given classes, and optimizing the filtering/extracting attributes of higher significance per class. It is stressed that in deriving the process models, an upper regulatory nitrate level was set given the concern in ensuring that the model would provide accurate prediction of permeate product quality with respect to both nitrate and salt passage. In summary, the machine learning approach developed in this study demonstrated multitude of benefits including: (a) determining deviation of nitrate from the baseline value, (b) forecasting performance of a real RO system with respect to nitrate permeate and salt passage, (c) assessing the correlation between nitrate permeate and salt passage for a real system, and (d) providing a foundation for model-based control of the treatment system for nitrate and salt removal.

469b A Study of a Multi-Step Membrane-Based Desalination Process for Resource Recovery

Hossein Dallalzadeh Atoufi, David Lampert

The exploration and production of oil and gas generate a variety of waste materials that have to be managed effectively. The most common waste product from the oil and gas industry is the produced water that is typically extracted from hydrocarbon-bearing strata during the oil and gas extraction process. This waste stream includes any chemicals that might be added to the well or added during drilling, production, or maintenance, as well as formation water and injection water. In this waste stream, we find a significant amount of solids in the form of dissolved solids, dispersed oil and grease, salts, organic compounds, heavy metals, rare earth elements, and radionuclides. Often, produced waters are managed on-site or within an oil field using evaporation ponds and seepage pits. There is also the possibility of recycling and reusing the produced waters for exploration and production operations within the oil and gas industry. Many companies, service providers, and water treatment companies offer produced water management options to their customers. According to their analysis, the costs associated with the treatment of produced water, including desalination, are much higher than the cost associated with reusing the produced water in the oil and gas field or injecting it into disposal wells. In this research project, we are targeting to develop a desalination method for reusing and recycling produced water at a reduced price by using reverse osmosis. With our reverse osmosis system with polymer membranes, we are able to achieve the highest levels of efficiency. Based on the preliminary results of our experiments, we can conclude that this multi-step membrane process reduces salinity by more than 99% and lowers the concentration of organic and inorganic compounds in produced water by more than 95%. The research also seeks to recover valuable elements from produced water like lithium and boron to transform waste into valuable materials.

469c Recovery of Organics from Humidification-Dehumidification of Hydraulic Fracturing Wastewater Using Deep Eutectic Solvents

Elnaz Nikooei, Iolar Lima, Bahman Abbasi, Nick AuYeung

Hydraulic fracturing produces over a trillion gallons of highly contaminated wastewater every year in the United States. Contaminants are highly numerous, diverse, and in some cases, even proprietary. Humidification-Dehumification (HDH) has been proposed as a method for dealing with wastewater from the hydraulic fracturing industry. Previously, we have demonstrated an HDH concept which includes a evaporation step with air to eliminate salts and less volatile compounds, which is followed by the dehumidification step which condenses out fresh water. The resulting air stream takes away many of the contaminants. As most of the compounds remaining in the vapor phase are organic, they could have calorific value if allowed to accumulate over time. Recently, new Deep Eutectic Solvents (DES) have been developed which are hydrophobic, safe, and low cost. Here we present results from our preliminary experimental work in capturing organics from the vapor stream using hydrophobic deep eutectic solvents. Based on this work, we quantify the potential value of such organic accumulation using representative wastewater compositions.

469d Multilayered MFI-Zeolite Nanosheet Plates for Tiling Membranes on Porous Polyvinylidene Fluoride Substrates for Wastewater Desalination

Zishu Cao, Landysh Iskhakova, Xinhui Sun, Zhong Tang, Junhang Dong

Desalination of high-salinity brines is a current challenge in treatment, reclamation, and recovery of valuable materials for many industrial wastewaters. The MFI-type zeolites possess ordered channels with pore openings (d_p~0.55 nm) that are slightly smaller than the kinetic diameters (d_k) of hydrated metal ions and common anions in aqueous solutions ($d_k > 0.6$ nm) but much bigger than the d_k of water molecules (~ 0.27 nm). Because the ion hydration numbers and hence the hydrated ion kinetic sizes are insensitive to the solution concentration and temperature, the MFI type zeolite membranes are well-suited for desalination of high-salinity brines by molecular sieving effects between the water molecules and dissolved salts (i.e., hydrated ions). Recently, submicron-thick, *b*-oriented MFI zeolite nanosheet laminated membranes (ZNLMs) have been fabricated on both the macroporous ceramic and polyvinylidene fluoride (PVDF) substrates. These ZNLMs were demonstrated with extraordinary water flux and nearly perfect salt rejections in pervaporation (PV) desalination of multicomponent high-salinity brines with good stability. The enabling factors for the ZNLMs' remarkable performance in PV desalination including the minimized intercrystalline spaces that preserved the ion-sieving effect and the ultrathin ZNLM layer in preferred b-orientation that reduced the diffusion length and resistance for water transport. However, the current methods for zeolite nanosheet synthesis are inefficient and prohibitively for practical purposes. In this work, a facile self-seeded secondary growth method is established for efficient reproduction of very large flower-like MFI zeolite nanosheet (ZN) assemblies in a single step. These flower-like architectures allow ZN activation without structural destruction so that large-sized ZN plates can be conveniently harvested. In this work, the thus -synthesized ZN plates are used for tiling molecular-sieve membranes on industrially desired polymer supports and the resultant ZNLMs are tested for PV desalination of high-salinity brines for lithium recovery.

469e RO Membrane Characterization and Modeling Under High Pressure

Jeffrey McCutcheon, Mi Zhang, Yara Suleiman, Danh Nguyen, Sina Shahbazmohamadi, Vimal Ramanuj, Ramanan Sankaran, Ying Li

High pressure reverse osmosis (HPRO) is an alternative to energy intensive thermal desalination technologies for brine concentration. HPRO can potentially reduce the energy consumption of brine concentration by 90% over thermal desalination systems. Operating HPRO is challenging as state of the art thin film composite (TFC) reverse osmosis (RO) membranes do not perform well at excessively high pressures. These membranes can undergo changes when exposed to excessively high pressure. These changes can be modeled and observed in the different layers of the TFC membrane. We break down our work by layer

The polyamide selective layer – This layer is ultra-thin to enable fast water transport but highly crosslinked to restrict solute transport. Under excessive pressure, this layer may undergo microstructure changes. In this work, we valuate the microstructure changes and quantify impacts on permeance and selectivity using molecular dynamics simulations.

*Porous support layer - T*he support layer is typically comprised of that is formed onto a nonwoven polyester scrim layer. Under high pressure, these layers will compress, leading to a reduction in support layer permeance. We use X-ray microscopy to quantify features of compressed membranes (porosity, tortuosity) and to measure the individual permeance of each part of the support layer.

Our findings indicate that polyamide layers do exhibit microstructure changes and a narrowing of the free volume element sizes when under extreme pressures. We also show substantial compression in an RO membrane support layer under higher pressures with more compaction occurring in the polysulfone layer than in the polyester layer. We conduct some tests in the presence of a spacer and find that compression is more severe where the permeate spacer presses against the membrane. Such findings provide critical insight into the future design of HPRO membranes, spacers, and modules.

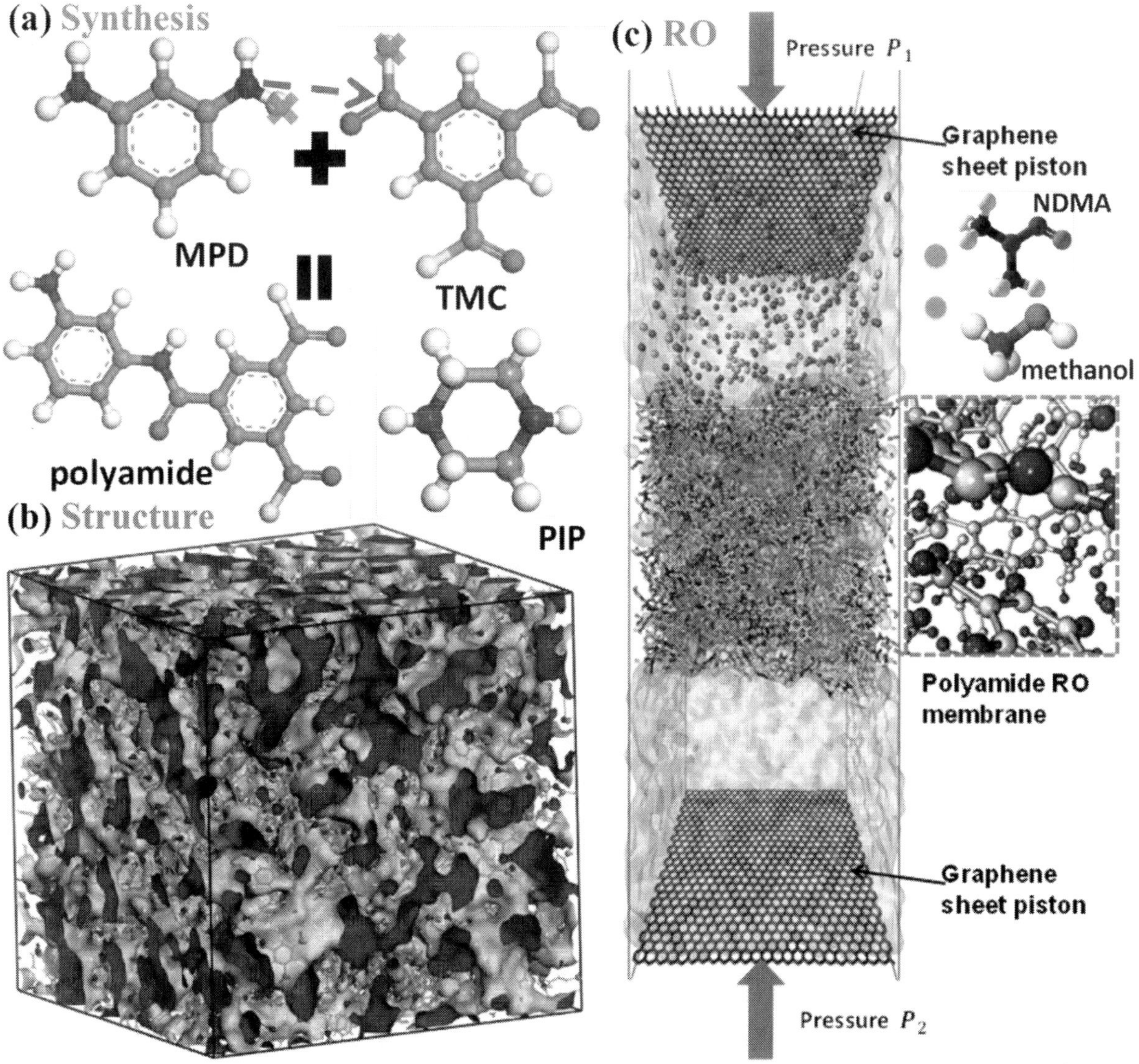
(a) Synthesis
MPD
TMC
polyamide
PIP
(b) Structure
(c) RO
Pressure P_1
Graphene sheet piston
NDMA
methanol
Polyamide RO membrane
Graphene sheet piston
Pressure P_2

469f Ultra-Strong Polymeric Hollow Fiber Membranes for Saline Dewatering and Desalination

Can-Zeng Liang, Mohammad Askari, Looh Tchuin (Simon) Choong, Tai-Shung Chung

Osmotically assisted reverse osmosis (OARO) has become an emerging membrane technology to tackle the limitations of a reverse osmosis (RO) process for water desalination. A strong membrane that can withstand a high hydraulic pressure is crucial for the OARO process. Here, we develop ultra-strong polymeric thin film composite (TFC) hollow fiber membranes with exceptionally high hydraulic burst pressures of up to 110 bar, while maintaining high pure water permeance of around 3 litre/(m^2 h bar) and a NaCl rejection of about 98%. The ultra-strong TFC hollow fiber membranes are achieved mainly by tuning the concentration of the host polymer in spinning dopes and engineering the fiber dimension and morphology. The optimal TFC membranes display promising water permeance under the RO and OARO operation modes. This work may shed new light on the fabrication of ultra-strong TFC hollow fiber membranes for water treatments and desalination.

469g Electrospun Liquid-Infused Membranes for Emulsified Oil/Water Separation

Chen Song, Gregory Rutledge

From an environmental perspective, microfiltration membranes are attractive for the separation of emulsified oils from contaminated water. However, fouling of the membrane is a major drawback of the technology. "Liquid-infused membranes" (LIM) have the potential to eliminate membrane fouling. Here we demonstrate the practical application of LIMs for the separation of oil from a stable oil-in-water emulsion, and characterize its resistance to fouling. The base membrane is an electrospun nonwoven fibrous layer of the fluorinated copolymer poly(vinylidene fluoride-co-hexafluoropropylene) (PVDF-co-HFP). The surface energy of the PVDF-co-HFP fibers was lowered by the covalent attachment of a fluorinated silane (PFOCTS), and then the membrane was infused with a perfluoropolyether (PFPE). The membrane was then challenged with model emulsions of dodecane in water in a crossflow configuration. This PFOCTS-modified LIM showed better infused liquid stability, permeation selectivity, and higher permeate flux than the unmodified LIM, and better anti-fouling properties than the bare membrane without infused liquid. We also examine the mechanism for transport of the dispersed oil phase through the liquid-infused membrane. We find a linear relationship between dodecane flux and dodecane concentration in the feed and a higher dodecane flux through the PFOCTS-modified membrane than the unmodified one, which suggests that the capture of dodecane droplets from the feed plays an important role in determining the overall rate of permeation. Other factors such as lower viscosity of the infused liquid, larger pore size, and higher operating pressure also improved the permeate flux through the LIMs. Overall, this work provides some guidelines on the design of composite membranes comprising infused liquids and the choice of operating conditions for the filtration process.

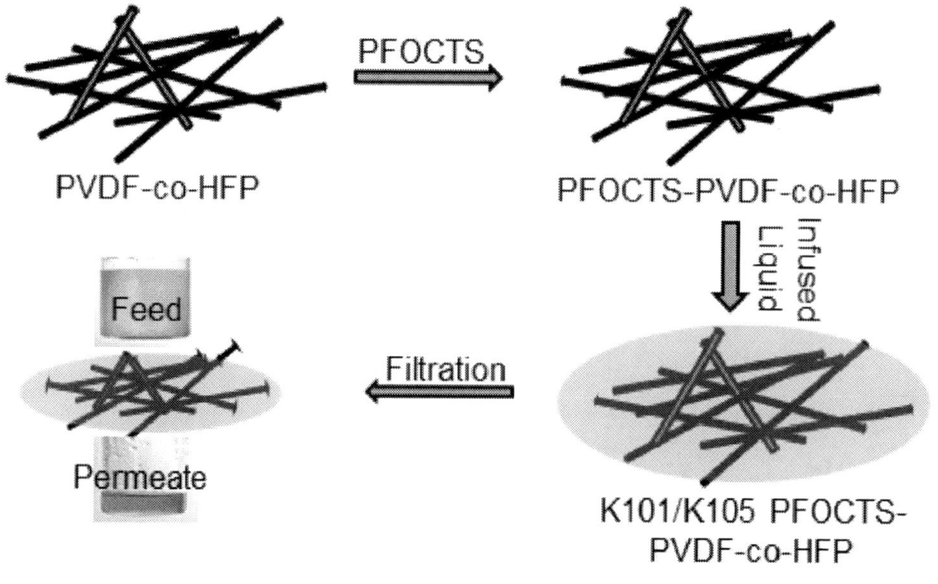

705a Synthesis and Characterization of NOVEL MIXED Matrix Membrane Comprises of Bimetallic Metal Organic Framework for Separation of Carbon Dioxide from Methane

Zartasha Azhar

Earth's population is increasing tremendously due to which the energy demand is also increasing abruptly. All this economic development and advancement in technology during the last century is based on our energy system. To fulfill demand for energy, fossil fuels are being used over a long time due to their high calorific value and availability. Upon combustion they produce large amount of carbon dioxide (CO_2) which readily enters in the environment and causes global warming and climate change.

Natural gas present in reservoirs is not in its purest form and a number of other impurities are present along with huge amount of carbon dioxide which needs to be separated to avoid corrosion in pipes and reduce compression cost. Various technologies are available for separation of CO_2 from natural gas. Among them membrane is the latest.

Mixed matrix membrane (MMM) is the modern membrane synthesis technique, the MMM is prepared by introducing different inorganic filler with superior selectivity and permeability into polymer matrix. MMM have shown huge potential for the separation performance of polymeric membrane

In this study, novel MMM was synthesized by introducing Zn/Co-ZIF composite MOF in PEBAX®1657(block copolymers made up of rigid polyamide blocks and soft polyether blocks) and have been tested for the separation of CO_2 from CH_4 and N_2. Permeation test with different loading of in polymer were investigated, to find out the performance and behavior of the membrane with the increased filler quantity. Characterization techniques like XRD, SEM-EDS and FTIR, TGA, BET were carried out to gain essential and practical knowledge about materials and to correlate properties with structure and their results will be discussed.

705c Facile Defect Engineering of Zeolitic Imidazolate Frameworks Towards Enhanced Propylene/Propane Separation Performance

Heseong An, Jong Suk Lee

Mixed matrix membranes (MMMs) are hybrid membranes consisting of molecular sieves and polymeric membranes for dual advantages of good separation performance and processability. However, several challenging issues still remain prior to commercialization of MMMs, including engineering of pore structures of molecular sieves and interfacial voids between polymer and molecular sieves. In this presentation, I will report highly processable zeolitic imidazolate framework-8 (ZIF-8) nanoparticles for enhanced propylene/propane separation performance by facile defect engineering [1]. The addition of amine modulators under the suppressed nuclei concentrations allows the formation of new Zn-alkyl amine (Zn-AA) point defects in the resultant ZIF-8 nanoparticles, but restrains the formation of typical Zn-OH point defects by fast coordination reaction. Our Zn-AA defective ZIF-8s exhibit a more restricted flipping motion of the ligands, and thus, possess a higher framework rigidity. Furthermore, partially coordinated amine modulators successfully enhance the interfacial interaction between various engineering polymers and ZIFs, enabling accommodation of high ZIF concentrations in MMMs as well as excellent propylene/propane separation performance.

[1] H. An et al., Adv. Funct. Mater., **2021**, *31*, 2105577.

705d Facile Approach to Simultaneously Reducing Physical Aging and Swelling While Enhancing Permeability in Microporous Polymer Membranes

Jing Deng, Cara M. Doherty, William Box, Laura Matesanz Niño, Angel Lozano, Cristina Álvarez, Anita J. Hill, Michele Galizia

Although exhibiting superb performance in the short time operation, microporous and ultra-microporous glassy polymers suffer from physical aging and swelling, which hampers their scale up to the industrial level in gas, vapor as well as organic solvent separations. In this study, we propose a facile approach to manipulate free volume morphology and distribution, as well as polymer chain dynamics, to reduce physical aging and swelling in microporous glassy polymers, while enhancing gas permeability. A model microporous high free volume glassy polymer, PTMSP, was blended with 5wt.% hyper cross-linked triptycene-isatin porous polymer network exhibiting configurational free volume. As a result, gas permeability increased by a factor of three compared to neat PTMSP, with a minor loss in ideal selectivity. Simultaneously, polymer swelling in the presence of high pressure CO_2 and physical aging rate were substantially reduced. The combination of PALS analysis, density measurements and free volume calculations indicate that blending with PPN provides a unique opportunity to exquisitely tune the size and architecture of free volume elements, leading to the above mentioned enhancement of permeability, physical aging and swelling resistance. The molecular mechanism by which these effects take place was elucidated in detail, and possible strategies to enhance selectivity other than permeability and long-term stability were devised.

Remarkably, although the incorporation of PAFs (Porous Aromatic Frameworks) in high free volume glassy polymers can successfully slow down the aging rate while enhancing gas permeability and retaining selectivity, their cost is very high and their synthesis pathway complicate, which hampers their industrial scale-up. In sharp contrast, the PPNs synthesis pathway is easy and one order of magnitude cheaper, which makes our approach highly scalable.

705e One-Step Fabrication of Highly Propylene-Selective Asymmetric Mixed-Matrix Membranes with *in-Situ* ZIF-8 Filler Formation

Yinying Hua, Sunghwan Park, Hae-Kwon Jeong

Separation of propylene from propane is one of the most energy-intensive separation processes due to the similarities in their physico-chemical properties[1]. Zeolitic-imidazolate framework-8 (ZIF-8), consisting of a zinc center tetrahedrally coordinated with 2-methylimidazolate linkers, is one of the most promising molecular sieve materials for propylene/propane separation. Recently, our group developed a scalable mixed-matrix membranes fabrication strategy using *in-situ* Zif-8 formation approach, named PMMOF[2]. The PMMOF enables formation of MMMs not only with much improved gas separation performances[2] but also in a scalable geometry (i.e., asymmetric hollow fibers with submicron thick selective MMM skin layers)[3]. However, the PMMOF suffers from two major challenges: 1) multi-steps involved that are relatively complicated and 2) limited choice of polymer matrices (i.e., polyimide-based polymers).

In this presentation, we plan to present a novel second-generation one-step ZIF-8-based asymmetric MMM fabrication strategy, addressing the challenges of the first-generation PMMOF process. The new process, termed "PIMOF", enable rapid formation of high ZIF-8 loaded asymmetric MMMs. The as-prepared MMMs showed unprecedentedly high C_3H_6/C_3H_8 (C3) separation performance as compared to other propylene-selective MMMs reported so far with the C_3H_6 permeance of ~ 7.5 GPU and C_3H_6/C_3H_8 separation factor of ~ 107. It is worthy of noting that our MMMs exhibit comparable C3 separation performances even with polycrystalline ZIF-8 membranes. The unexpectedly high C3 performance of the membranes was attributed to 1) the uniform distribution of the extremely small ZIF-8 nanoparticles (~ 5-10 nm), 2) unusually high ZIF-8 nanofiller loading in the selective skin layers (> 50 wt%), and 3) improved molecular sieving properties of the ZIF-8 nanofillers resulting from the hindered linker flexibility upon confined filler formation.

Reference:

1. Sholl, D.S. and R.P. Lively, *Seven chemical separations to change the world.* Nature, 2016. **532**(7600): p. 435-437.
2. Park, S., M.R. Abdul Hamid, and H.-K. Jeong, *Highly Propylene-Selective Mixed-Matrix Membranes by in Situ Metal–Organic Framework Formation Using a Polymer-Modification Strategy.* ACS Applied Materials andamp; Interfaces, 2019. **11**(29): p. 25949-25957.
3. Park, S. and H.-K. Jeong, *Transforming polymer hollow fiber membrane modules to mixed-matrix hollow fiber membrane modules for propylene/propane separation.* Journal of Membrane Science, 2020. **612**: p. 118429.

705f Tailoring Ultramicropores in Hybrid Carbon Molecular Sieve Membranes to Achieve High H_2 Selectivity

Leiqing Hu, Vinh Bui, Ashwanth Subramanian, Kim Kisslinger, Lingxiang Zhu, Shouhong Fan, Yifu Ding, Chang-Yong Nam, Haiqing Lin

Porous carbon materials with tunable pore structures have attracted substantial attention for membrane separation. Particularly, carbon molecular sieve (CMS) membranes with micropores (7 to 20 Å) that promote gas transport and ultramicropores (7 to 20 Å) derived from controlled pyrolysis of polymer precursors that confer strong size-sieving ability, have emerged for gas pair separation, including CO_2/CH_4, olefin/paraffin. However, rare CMS membranes have high selectivity for H_2/gas separation due to small molecular size of H_2. Here, we report advanced CMS membranes derived from a supramolecular mixed matrix network (SMMN) with high H_2 selectivity against other gases including CO_2, N_2, CH_4, and C_2H_6. Specifically, ZIF-8/polybenzimidazole (PBI) SMMN containing crystalline ZIF-8 and amorphous Zn/imidazole coordination on PBI chains was prepared via the in-situ ZIF-8 growth in PBI. With porous ZIF-8 promoting gas transport and Zn/imidazole coordination cross-linking PBI chains and then enhancing size-sieving ability, ZIF-8/PBI SMMN displayed both increased H_2 permeability and H_2/CO_2 selectivity. Then, the SMMN was carbonized from 400 to 900 °C. Compared to PBI/CMS prepared at same temperatures, SMMN/CMS films exhibit both higher H_2 permeability and selectivity, even far surpassing Robeson's upped bounds 2008. It can be attributed to Particularly, SMMN/CMS carbonized at 900 °C exhibited H_2/CO_2 selectivity of 220 at 70 °C, higher than those of other leading membrane materials. This study opens up a new approach to design advanced CMS materials for H_2 separation and other applications.

705g Tunable Supramolecular Cavities Molecularly Homogenized in Polymer Membranes for Ultraefficient Precombustion CO_2 Capture

Ji Wu, Can-Zeng Liang, Ali Naderi, Tai-Shung Chung

Processable molecular-sieving membranes are important materials for realizing energy-efficient precombustion CO_2 capture during industrial-scale hydrogen production. However, the promising design of mixed matrix membranes (MMMs) that aims to integrate the molecular-sieving properties of nanoporous architectures with industrial processable polymers still faces performance and fabrication issues due to the formation of segregated nanofiller domains in their polymer matrices. Here, we propose an unconventional nanocomposite membrane design using soluble organic macrocyclic cavitands (OMCs) with tunable open cavity sizes that not only mitigate the formation the discrete nanofiller phases but also deliver distinct molecular-sieving separations. The versatile organic-solvent solubility coupled with highly interactive functionalities of OMCs allows them to obtain molecularly homogeneous mixing with matrix polymers and form only one integral continuous phase crucial to the robust processability of polymers. A series of polybenzimidazole-based molecularly mixed composite membranes (MMCMs) were fabricated via the incorporation of a soluble and thermally stable OMC choice, sulfocalixarenes, with various cavity sizes. These membranes achieved outstanding high-temperature mixed-gas H_2/CO_2 separation performances comparable with several state-of-the-art molecular-sieving membranes owing to effective size-sieving gas passages through the open or partially-intruded supramolecular cavities. The broadly tunable structures and functionalities of OMCs would make their MMCMs attractive for other energy-intensive molecular separations.

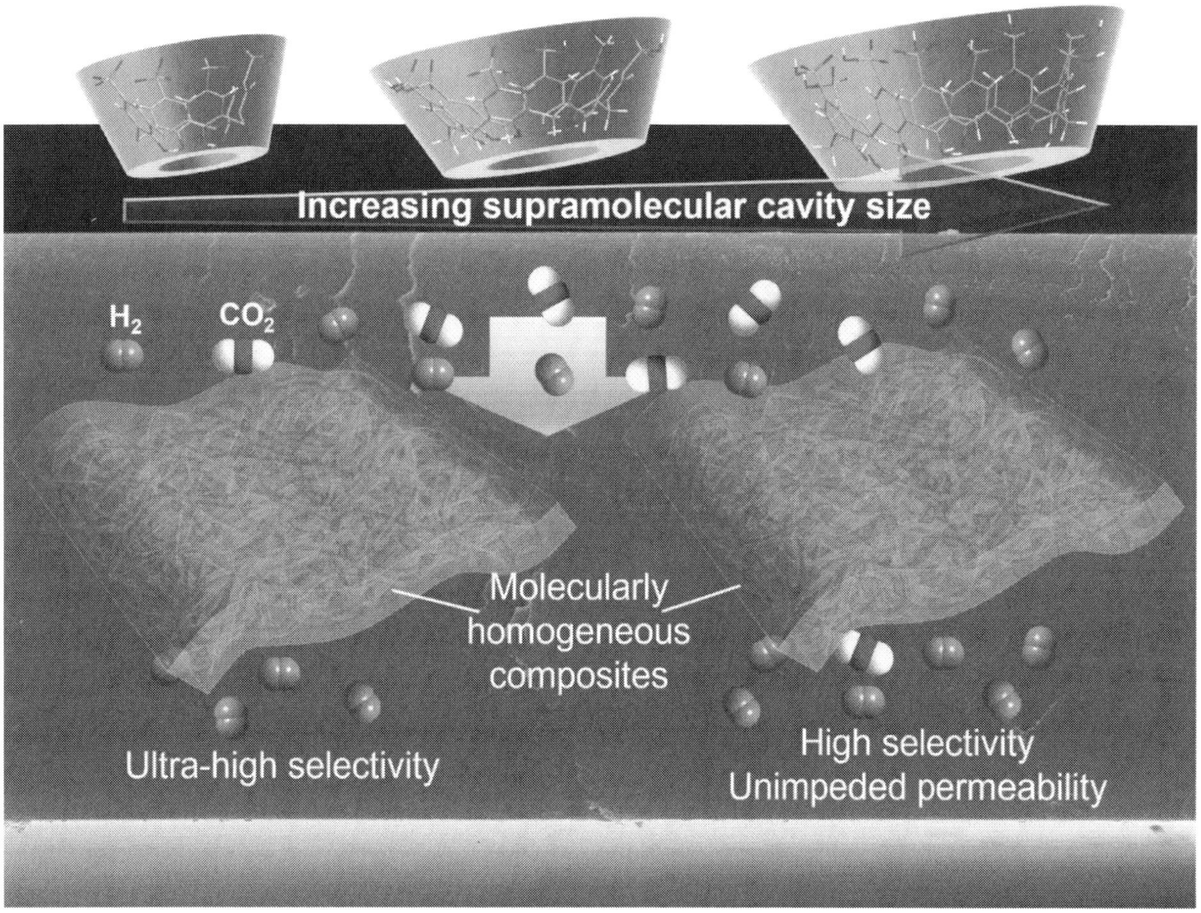

148b Developing a Novel Size-Dependent Growth Modeling Method for the Batch Crystallization of Carbamazepine from Variable Seed Crystal Size Distributions

Harrison Kraus, David A. Acevedo, Thomas O'Connor, Dongxia Liu, Adil Mohammad

Developing a crystallization model that accurately predicts crystal growth and nucleation has been an important topic in the pharmaceutical industry for the last few decades. Particularly, as the industry shifts toward continuous manufacturing (CM), crystallization modeling will both reduce the workload for experimental optimization and allow for model-based control systems that can yield more consistent quality output. In this work, a unique approach for modeling size-dependent growth was applied to a set of batch cooling crystallizations. The cooling crystallization of carbamazepine (CBZ) in ethanol was monitored using in-line Raman spectroscopy for solute concentration measurement as well as off-line laser diffraction for seed and product crystal size distribution (CSD) measurement. Modeling was performed using the MATLAB software utilizing a combined quadrature method of moments and method of characteristics (QMOM-MOCH) technique in conjunction with a modified expression for size-dependent growth. This work expands upon past work done in modeling the cooling crystallization of CBZ by evaluating the effect of variable seed CSD on crystal growth rates as well as the accuracy of the model-predicted product CSD. Using typical modeling techniques, variation in seed CSD resulted in significant error in the model CSD predictions especially for the product D10 value, thus demonstrating the necessity for novel methods for modeling size-dependent growth. This error was reduced by varying the size-dependent growth parameters as a function of the seed CSD. Simpler size-dependent equations have proven accurate in the past, particularly when seed CSD was held constant. However, this new technique provides a more accurate representation of how crystal growth rates are affected by the overall distribution of crystals present in the system. This enhanced understanding will benefit future research into continuous crystallization by enabling more accurate start-up simulations as the system CSD changes from the seed to the steady state of crystallization. It will also reduce experimental optimization workload and provide better control for perturbations in either the seed or the system CSD.

148c Digital Design of Crystallization Process: Application of a Mechanistic Morphological Crystallizer Model to Improve Powder Flowability Via Aspect Ratio Reduction

Niall Mitchell, Efty Hadjittofis, Filipe Calado, Sélim Douieb, Ugo Cocchini, Jan-Sebastiaan Uyttersprot, Jerome Mantanus, Nicolas Carly

Introduction

Crystallization is the most widely used method for separating an active pharmaceutical ingredient (API) from a reaction product mixture. It functions to control purity, crystal form, particle morphology, and particle size distribution (PSD). The application of population balance modelling (PBM) to crystallization process development has been growing as a result of improvements in process analytical tools (PAT) and numerical software packages. One such improvement has been the ability to account for high aspect ratios particle morphologies, such as needles or plates, by tracking the major and minor particle axis using a 2-dimensional (2D) framework. This approach allows for each particle axis to have independent growth kinetics, so that crystal size and shape evolution can be described over the course of a crystallization process. Particles with needle-like morphology lead, in many cases, to poor manufacturability in downstream drug product unit operations due to poor powder flowability, driving formulation and sequence of unit operations to adjust the physical properties often creating the need for redundancy for improving control and robustness of the process. Hence, reducing particle aspect ratio (length/width) can be beneficial for enhancing manufacturability of an API.

In this work, process simulation and mechanistic modelling were applied to gain further process understanding and investigate how better control over particle size and shape could be achieved, with a view to improving flowability of the crystallized material. In order to probe the crystallization behaviour of the solution system, lab scale experiments were performed at a 300mL scale using an RC1 MT system. The crystallization process considered a sequence of heating and cooling cycles in order to investigate the effect on the shape of product crystals aswell as wet milling steps (see steps 2a-2b and 4a-4b in figure 1 below), in order to break the needle-like particles, hence reducing aspect ratio of the crystals and to facilitate the dissolution of fines. Seeding and cubic cooling profiles were used at lab scale to ensure the crystallization process studied was predominantly growth driven. Solute concentration over time was tracked via PAT and offline samples were taken for particle size and shape analysis using Malvern Morphologi G4, at stages 2a, 2b, 4a, 4b and of the final product in stage 6, as outlined in figure 1. These process measurements were used for model validation purposes.

Key results

The crystallization process was modelled using a mechanistic morphological crystallizer model, which utilises a 2D framework to describe the evolution of particle size and shape along two perpendicular axes (major andamp; minor), from which shape descriptors, such as aspect ratio can be determined. Mechanisms and model discrimination was performed, yielding good predictions of the solute concentration over the cycles, as shown in figure 2. The model prediction of the evolution of particle size and shape during the process was coherent with what was qualitatively observed with the PVM images, in addition of providing a reasonable prediction of the quantitative values of volume-based PSD quantiles at the end of each pair of temperature cycles for the major and minor axes, as shown in figure 3(a) and 3(b), respectively.

The validated morphological crystallizer model of lab scale was subsequently utilised to investigate optimization strategies and potential process changes, such as heating andamp; cooling rates and profiles, number of cycles and wet milling steps and wet milling duration, which could be utilised to reduce the aspect ratio (major axis length/minor axis length) of the crystallized material and hence improve the downstream flowability of the powder.

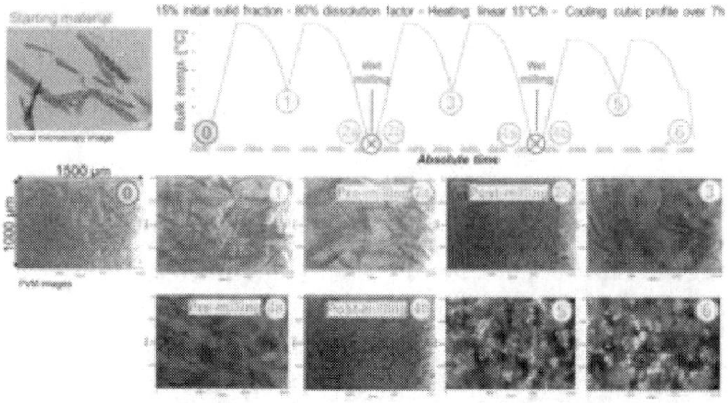

Figure 1: Process description and recipe utilised at lab scale to probe system. Wet milling via rotor-stator mills (not modelled) are utilised from stage 2a-2b and 4a-4b. PVM image shown at stages 0-6 for the time points shown in Temperature versus time plot.

(a) (b)

Figure 2: Model predictions versus experimental solute concentration during heating and cooling cycles (a) 1&2 and (b) 3&4 (after wet milling step at end of cycle 2).

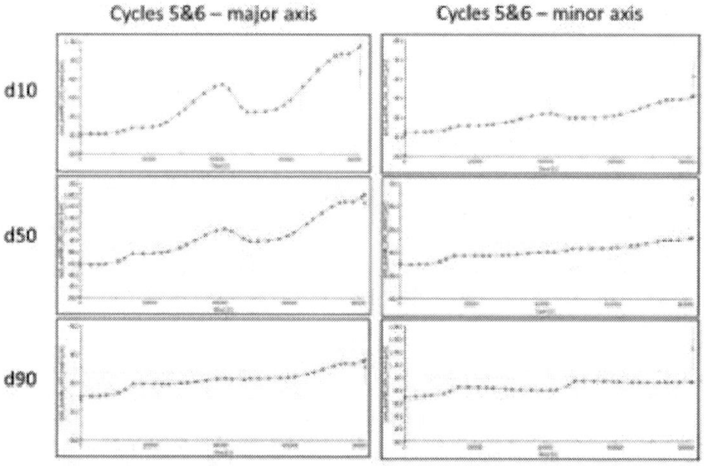

(a) (b)

Figure 3: Model predictions versus experimental volume-based PSD quantiles (d10, d50 and d90) for (a) major and (b) minor axis in temperature cycles 5&6. Model predictions are represented by the dotted line (" " ") while experimental data are represented by single data points ('). |

Scale-up of a yield-stress slurry of needle-like crystals using the cavern model

Authors: Paul Larsen*, Taylor Hudson, Steve Coats, Navraj Hanspal, Megan Cismesia, David Couling, Chris Chappelow

Acknwledgment: Richard Grenville (SPX Flow)

To minimize scale-up risk for solution crystallization processes, it is critical to establish mixing requirements based on lab studies to guide the commercial operation. Common mixing scale-up metrics include power per unit mass, turbulent shear rate (vessel-averaged or impeller-averaged), blend time, and meso-mixing time constant. The appropriate scaling parameter for a given crystallization is determined on a case-by-case basis.

This presentation reviews a case study for an agricultural active ingredient in which crystallizer scale-up based on power per mass resulted in unexpected mixing behavior at scale. The unexpected behavior was rationalized by applying the Elson cavern model, demonstrating the utility of this model for situations where the crystallization slurry is a yield-stress fluid.

[1]T.P. Elson, D.J. Cheesman, A.W. Nienow, "X-ray studies of cavern sizes and mixing performance with fluids possessing a yield stress," *Chemical Engineering Science*, 1986.

[2]A.W. Russell, L. Kahouadji, K. Mirpuri, A. Quarmby, P.M. Piccione, O.K. Matar, P.F. Luckham, C.N. Markides "Mixing viscoplastic fluids in stirred vessels over multiple scales: A combined experimental and CFD approach," *Chemical Engineering Science*, 2019.

148d Process Intensification for Continuous Manufacturing of Energetic Materials Via Model-Free Quality-By-Control Direct Design and Model-Based Digital Design Approaches

Montgomery Smith, Monika Neal, Daniel Laky, Wei-Lee Wu, Jaron Mackey, Zoltan Nagy

Crystallization is an essential process of solids manufacturing and is left unoptimized in several fields, including energetics manufacturing. Unoptimized crystallization protocols can lead to particles with undesired physical and chemical characteristics, such as particle morphology, sensitivity, detonation potential, manufacturability, and overall crystal quality. First manufactured for use in World War II, nearly a century ago, Research Department/Royal Demolition Explosive (RDX) and High Melting Explosive (HMX), have been applied to military munitions, propellants, and general explosives and continue to be two of the largest manufactured energetic materials. Although researchers have extensively studied the solubility of the common energetic materials, RDX and HMX, little work has been completed on the process intensification of RDX and HMX crystallization.

In previous work, we demonstrated the application of model-free Quality-by-Control (mf-QbC) to temperature profile development which successfully controlled the crystal size distributions/class of energetic materials. In this work, we demonstrate the combination of mf-QbC and model-based digital design to develop a robust process intensification framework for the continuous crystallization of energetic materials. Small-scale experiments completed with Crystalline showed that RDX and HMX have high solvent power (solubility) in γ-Butyrolactone and good temperature sensitivity which is desirable for crystallization control. mf-QbC allows for the reduction of experiments and exposure by utilizing feedback control strategies for desired critical quality attributes (CQAs) (Simone *et al.*, 2015). In application with *in-situ* process analytical technology (PAT) tools, the two applied direct design approaches, direct nucleation control (DNC) and super saturation control (SSC), allowed for the selection of crystallization design parameters that control the desired CQAs of RDX and HMX, including but not limited to polymorphic form and overall crystal quality, and the further optimization of the selected design parameters via model-based digital design. Intensifying the industrial crystallization of common energetic materials, such as RDX and HMX, will improve the quality of manufactured energetic materials, and future process development for the manufacturing of energetic materials. Further applying this process intensification and system information to the continuous crystallization of energetic materials reduces manufacturing time, and variability during the manufacturing of energetic materials.

1. Bakar, M. R. A., Nagy, Z. K., Saleemi, A. N., Rielly, C. D., 2009. The Impact of Direct Nucleation Control on Crystal Size Distribution in Pharmaceutical Crystallization Processes, *Crystal Growth and Design*, 9 (3): 1378–1384.
2. Saleemi, A. N., Rielly, C. D., Nagy, Z. K., 2012. Comparative Investigation of Supersaturation and Automated Direct Nucleation Control of Crystal Size Distributions Using ATR-UV/Vis Spectroscopy and FBRM, *Crystal Growth and Design*, 12 (4): 1792–1807.
3. Simone, E., Zhang, W., Nagy, Z.K., 2015. Application of PAT-based feedback control strategies to improve purity and size distribution in biopharmaceutical crystallization, *Crystal Growth and Design*, 15 (6): 2908–2919.

148e Design of the Cooling Crystallization Process Using the Machine Learning-Based Strategy

Yiming Ma

Crystallization is typical solid−liquid separation unit operation that can be employed to purify a great diversity of chemical compounds in many industries, including pharmaceuticals, fine chemicals, food, agrochemicals and various personal consumer products. In the industrial crystallization process, it is essential to precisely obtain the yield and particle size distribution (PSD) of the product under specific crystallization conditions before the industrial plant start-up because it can be used to evaluate and optimize the crystallization process comprehensively, and also dramatically affects the downstream process design such as filtration and drying.

At present, the approach adopted by researchers and industry is to predict product quality through modeling. That is, relying on the PAT tool, the relationship between concentration changes over time should be obtained through experiments, and parameters in the nucleation and growth kinetic model can be obtained by combining PSD changes before and after. Validation experiments are needed to evaluate the reliability of the obtained kinetic model. Furthermore, the PSD and process yield under unknown crystallization conditions can be obtained by solving the population balance equation (PBE) and mass balance (MB) including the kinetic models. For some systems, the contribution of particle agglomeration or breakage to the PSD should be considered as well. Although the modeling approach as mentioned earlier has been widely used in crystallization process design and process control, it still has some application limitations. First, the quantitative relationship between particle size and concentration over time in the solution system during crystallization, which is necessary for nucleation and growth kinetic models, needs to be obtained using the process analytical technology (PAT) tools. However, PAT tools, such as Raman spectroscopy, attenuated total reflectance Fourier-transform infrared(ATR-FTIR), ultraviolet−visible(UV-Vis), and focused beam reflectance measurement (FBRM), are pretty expensive. Also, these PAT tools have a lower limit of detection for spectroscopic instruments, limiting the detection of insoluble compounds at low concentrations. Secondly, the currently widely used invasive probe may also lead to fouling on the surface of the probe, which makes the detection impossible. Secondly, the accurate determination of crystallization kinetic parameters requires professionals to design experiments to isolate the processes of primary nucleation, secondary nucleation, growth, particle agglomeration and breakage, which will reduce the degree of difficulty in multi-parameters optimization. In addition, for some compounds where an empirical formula is not available, such as the protein crystallization process with long induction periods, the nucleation kinetics cannot be accurately obtained using the traditional primary nucleation modeling. Therefore, the development of robust prediction methods to reduce costs and improve accuracy is the key to breaking through the limitations of current applications.

Machine learning (ML) is a data-based approach for problem solving and information mining. It relies on self-learning ability and can generate information automatically. As the most common supervised learning algorithm in ML models, Artificial neural network (ANN) has been widely used in engineering, especially in the field of crystallization. In recent years, we have continued to focus on the cross-discipline between crystallization and ML, and have tried to use the ANN algorithm to replace complex crystallization processes or predict the special properties of products and have already made some progress[1,2]. Therefore, we believe that the optimized ML models can be used to establish the quantitative relationship between process parameters and yield and PSD during the crystallization process. Obtaining a reliable dataset is the focus of constructing ML models. When literature data is not available, it is necessary to obtain a large number of training and testing set samples through high-throughput experiments. Although a certain number of experiments is necessary, it cannot be denied that if we conduct close to or more than a hundred sets of experiments as an initial dataset for ML models, it is contrary to our desire to develop alternative methods with the advantages of efficient, accurate and fast. It will also cause materials waste and labor costs. To expand the initial dataset, we propose the CrystGAN, a generative adversarial network (GAN)-based model that learns to generate new data including crystallization conditions and product results while ensuring high generation accuracy, which could be treated as an alternative way to large-scale crystallization experiments. It is also the first time use GAN in the industrial crystallization process.

In this work, ML models for PSD and yield prediction were built based on two algorithms, named GAN and ANN, to replace traditional time-consuming and complex crystallization kinetics-based modeling. The real dataset was achieved using *CrySiV* software (developed by Zoltan Nagy, Purdue University) simulation[3]. Each group in the dataset contains initial conditions (including particle size distribution represented in D_V values, seed loading, compound concentration), temperature profile (including five values for different cooling rates from 60°C to 10°C), and results (including final PSD represented in D_V values, yield).

In order to use as little real data as possible, 20-40 real data groups are adopted as the input of the CrystGAN model, and the generated accuracy in using different real data is compared. The generated data and the real data set are combined to train the ANN model. It can be found that after using more than 35 real data groups, the prediction ability of the CrystGAN-ANN model has not been significantly improved. Therefore, we can determine that the minimum number of real data used to accomplish the PSD and yield prediction of the cooling crystallization process is 35 groups. In addition, the genetic algorithm (GA) is used to optimize the hyperparameters of the ANN model further. Finally, the root mean square error (RMSE) of the optimal yield prediction value of the proposed CrystGAN-ANN-GA model is 4.49, and the accuracy of the DV values is not higher than 10.12, which means we have achieved our preset prediction purpose.

At present, there is increasing research on applying ML to study chemical engineering. We hope to provide guidance and reference through our work for the regression and prediction of complex multi-factor systems in relevant fields.

References

[1] Y.Ma, J. Gong, et al. (2021): Design of Spherical Crystallization of Active Pharmaceutical Ingredients via a Highly Efficient Strategy: From Screening to Preparation, ACS Sustainable Chemistry andamp; Engineering 9, 27, 9018–9032.

[2] Y.Ma, J. Gong, et al. (2021): Machine learning-based solubility prediction and methodology evaluation of active pharmaceutical ingredients in industrial crystallization, Frontiers of Chemical Science and Engineering 16, pages523–535 (2022)

[3] B. Szilagyi, Z. Nagy, et al. (2021): Cross-Pharma Collaboration for the Development of a Simulation Tool for the Model-Based Digital Design of Pharmaceutical Crystallization Processes (CrySiV), *Crystal Growth andamp; Design* 21, 6448−6464.

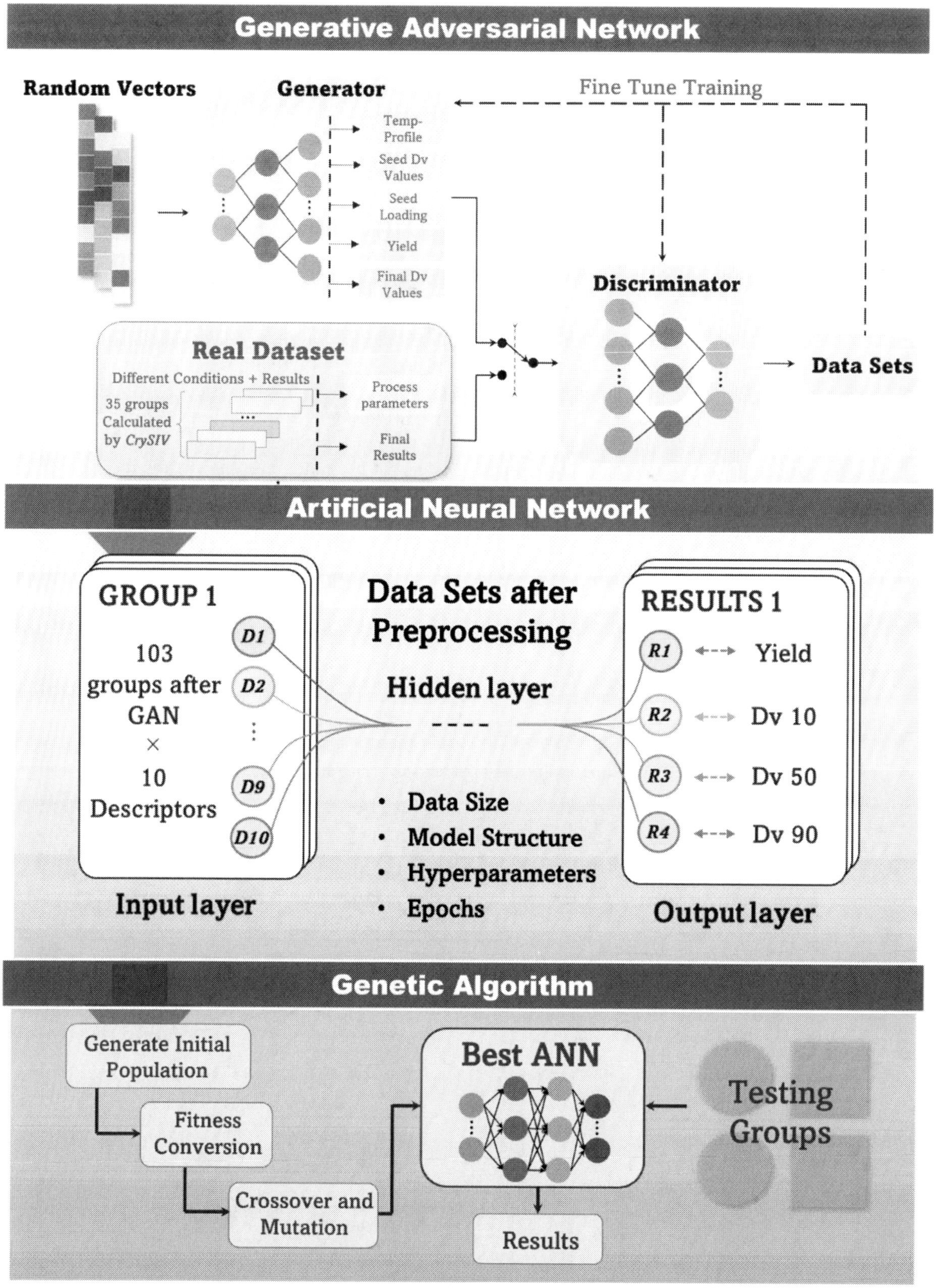
Generative Adversarial Network
Random Vectors
Generator
Fine Tune Training
Temp-Profile
Seed Dv Values
Seed Loading
Yield
Final Dv Values
Discriminator
Real Dataset
Different Conditions + Results
35 groups Calculated by CrySIV
Process parameters
Final Results
Data Sets
Artificial Neural Network
GROUP 1
103 groups after GAN × 10 Descriptors
D1
D2
D9
D10
Input layer
Data Sets after Preprocessing
Hidden layer
Data Size
Model Structure
Hyperparameters
Epochs
RESULTS 1
R1 Yield
R2 Dv 10
R3 Dv 50
R4 Dv 90
Output layer
Genetic Algorithm
Generate Initial Population
Fitness Conversion
Crossover and Mutation
Best ANN
Testing Groups
Results

572a Active Learning for Efficient Navigation of Adsorption Landscapes in MOFs

Krishnendu Mukherjee, Alexander Dowling, Yamil Colón

Machine learning (ML)-based surrogate models are becoming popular in applied sciences since they are computationally inexpensive and have great predictive ability.[1] Also, these cheap ML-based models can replace the conventional simulation methods for future predictions. However, one needs a large set of training data to build a ML-based surrogate model and often these datasets are expensive to generate. This is particularly true for adsorption where data generation is costly as they require molecular simulations which can take a few minutes to a few days to finish. This creates a bottleneck in scaling these models to add multiple features. In this prospect, we present an active learning (AL)-based protocol to predict gas adsorption in metal-organic frameworks (MOFs). AL builds the training set based on the uncertainty in its output prediction and thus it needs a smaller number of simulations to build itself.[2] We fit this AL model sequentially, starting from a prior dataset created using grand canonical monte carlo (GCMC) simulations and then sequentially build this model by sampling the next posterior point using the uncertainty in the predicted adsorption on the test set. Gaussian process regression (GPR) is used as a regression model for emulating the GCMC simulations. We continue to build the AL model until the uncertainty in adsorption for each point on the test set is lower than a threshold value (which is decided a priori). In this work, we first show how AL can be employed to build surrogate models for pure methane and carbon dioxide adsorption isotherms; pressure is the sole feature. Finally, we showcase how a GPR model can predict adsorption landscapes (varying both pressure and temperature) with an order of magnitude smaller number of simulations. In this process, we also compare different prior selection criteria and evaluate their performance. Based on our results, we believe this AL-based protocol has potential for adsorption prediction in more complex systems.

References

[1] A. Sturluson, M. T. Huynh, A. R. Kaija, C. Laird, S. Yoon, F. Hou, Z. Feng, C. E. Wilmer, Y. J. Colón, Y. G. Chung, D. W. Siderius and C. M. Simon, Molecular Simulation. 45 (2019) 1082–1121.

[2] E. Uteva, R. S. Graham, R. D. Wilkinson and R. J. Wheatley, The Journal of Chemical Physics. 149 (2018) 174114.

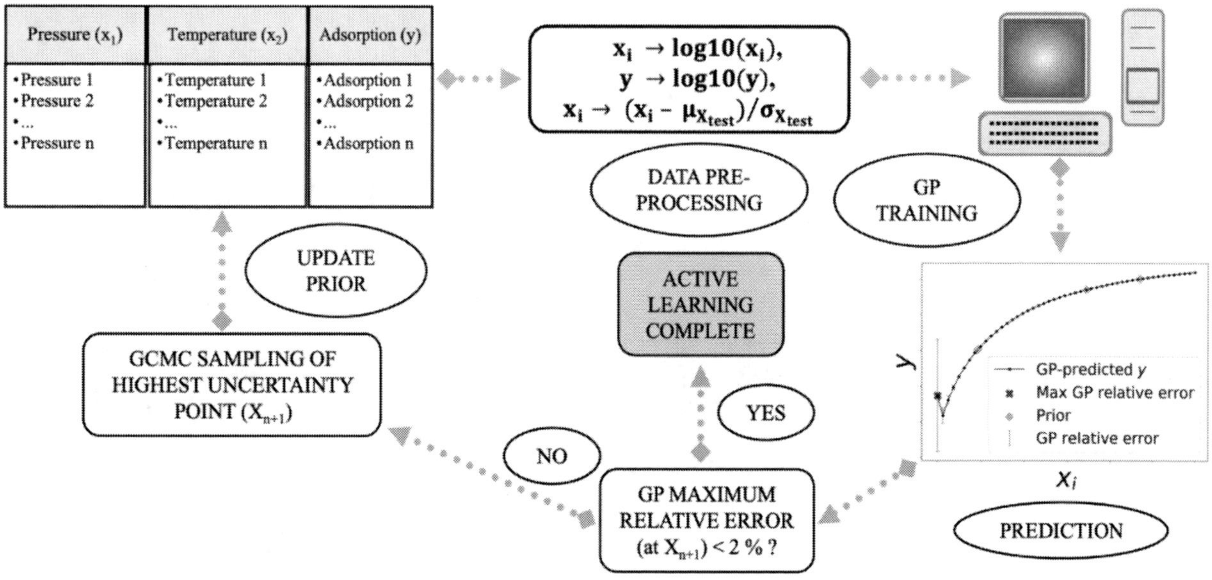

572b Adsorption Isotherms of Argon, Nitrogen, Carbon Dioxide, *n*-Butane, and Water for Pore Characterization of Chromatographic Particles

Chun-Kai Chang, Hsiao-Feng Liu, Joern Siepmann, Carlos Collados Cuadrado, Carola Schlumberger, Matthias Thommes, Mark R. Schure, Stephanie A. Schuster

Gibbs ensemble Monte Carlo simulations are carried out to assist the pore characterization of chromatographic particles via adsorption and intrusion experiments. The simulations use planar slit pore models to represent the surface chemistries of three prototypical chromatographic materials that range from hydrophilic to hydrophobic: (a) a pristine silica phase (a hydrophilic material), (b) a dense dimethyl octyldecylsilane phase with some residual silanols (C18, with a grafting density of 3.65 mol/m^2), and (c) a very hydrophobic, dimethyl octyldecylsilane phase with additional trimethylsilane endcaps (C18E, with C18 and C1 grafting densities of 3.65 and 0.54 mol/m^2, respectively). The temperatures studied are 87 K for argon, 77 K for N_2, 273 K for CO_2 and *n*-butane, and 298 K for H_2O, and the simulations cover adsorption/desorption from the gas phase and, if needed to close the hysteresis loop, also from the liquid phase. The sorption isotherms from molecular simulation are in qualitative agreement with experimental measurements that investigate superficially porous particles with more complex pore morphology. Analysis of the simulation trajectories provides information on adsorbate siting and orientational distribution along the adsorption/desorption paths and changes in the structure of the C18 ligands triggered by the wide range of temperatures and affinities for the different adsorbate molecules.

572c Mofdb: An Accessible Online Database of Computational Adsorption Data for Nanoporous Materials

Scott Bobbitt, Kaihang Shi, Benjamin Bucior, Haoyuan Chen, Nathaniel Tracy-Amoroso, Zhao Li, Yangzesheng Sun, Julia Merlin, Joern Siepmann, Daniel Siderius, Randall Snurr

MOFDB is a new freely accessible database containing simulated adsorption data from molecular simulation screening studies in nanoporous materials. The database includes about 3 million simulated adsorption data points for H_2, CH_4, CO_2, Xe, Kr, N_2, and Ar in over 160 000 MOFs and over 2000 zeolites, as well as textural properties like pore sizes and surface areas and the structure files for each material. We also include all relevant metadata such as details about Monte Carlo simulations, force field parameters, and links to original to facilitate reproducibility of the calculations. The database is easily searchable by MOF properties or adsorbate, and the adsorption data are stored in a standardized JSON format that is fully interoperable with the NIST adsorption database. Each MOF structure has a unique identifier (MOFid and MOFkey) that is searchable and amenable to cheminformatic analysis. All structure files (cifs) are provided, and the database is easily accessed using a Python API.

As an example of analysis facilitated by the database, we identified MOFs that meet high performance targets for multiple applications simultaneously, such as hydrogen and methane storage or carbon capture and Xe/Kr separation, including a MOF with predicted Xe capacity of 3.8 mol/kg and Xe/Kr selectivity of 15.6.

We have made this data publicly available in a standardized format in the hope that it will enable new machine learning analysis and lead to further discoveries from this data. We also encourage the community to adopt this JSON format as a standard way of storing and sharing isotherm data.

SNL is managed and operated by NTESS under DOE NNSA contract DE-NA0003525. SAND Number: SAND2020-3912 A

572d A Combined Deep Learning and Classical Potential Approach for Modeling Diffusion in Uio-66

Siddarth Achar, Jacob Wardzala, Leonardo Bernasconi, Linfeng Zhang, Karl Johnson

We have developed an accurate deep learning potential (DP) for UiO-66, a metal-organic framework, that is trained using data generated from density functional theory (DFT) calculations to reproduce energies and forces on atoms with close to DFT accuracy. The generation of training data was split into two steps. First, we employed an iterative data generation technique that explores compressed and expanded configurations of UiO-66 based on the DP's ability to accurately reproduce the DFT-computed equation of state. This initial step was performed to ensure accurate description of mechanical properties of UiO-66. Second, we focused on improving the description of framework flexibility by generating data from high temperature active learning iterations assisted by DFT-molecular dynamics (MD) simulations. The resultant DP achieved very low root-mean squared errors in predicting energies 1.65 10^{-3} eV/atom and atomic forces (7.8 10^{-2} eV/Å). The DP has a 0.012 % error in predicting the lattice constant and 0.1 % error in predicting the bulk modulus compared to DFT. We developed a *hybrid* DP and Lennard-Jones (DP/LJ) potential to account for adsorbent-adsorbate interactions and used this to compute diffusivities of guest fluids like Ne and Xe through UiO-66. This hybrid approach does not require refitting the DP for new adsorbates. Self-diffusion coefficients were calculated at finite Ne-loading from the DP/LJ potential and from DFT-MD. The DP/LJ results are in excellent agreement with DFT-MD simulations. Diffusion coefficients were also computed using two classical force fields for UiO-66 and were found to give similar results. Zero-loading simulations were performed using DP/LJ at three different temperatures in order to estimate the Ne diffusion barrier. The DP/LJ calculations give an activation energy for Ne diffusion of 2.67 0.18 kJ/mol. The DP/LJ was also used to perform zero-loading diffusion simulations of larger guest fluids like Xe into the MOF to demonstrate the DP's ability to model the framework flexibility of UiO-66. The self-diffusion coefficient of Xe predicted by DP/LJ was compared to classical force fields. Our hybrid DP-classical potential approach can be applied to other MOFs and other adsorbates, making it possible to use an accurate DP generated from DFT simulations of an empty MOF in concert with existing classical potentials for adsorbates to model adsorption and diffusion within the MOF. Our results show how very accurate potentials for porous adsorbents can be generated and then applied to modeling diffusion of various adsorbates without generating potentials for each adsorbent-adsorbate pair.

572e Machine Learned Disposable Force Fields for Fluid-Solid Simulations: Application to Water Transport in Carbon and Boron-Nitride Nanotubes

Fabian L. Thiemann, Cristoph Schran, Patrick Rowe, Ondrej Marsalek, Angelos Michaelides, Erich A. Muller

Simulation techniques based on accurate and efficient representations of potential energy surfaces are crucial for the understanding of adsorption at solid–liquid interfaces. Here we present a machine learning framework that enables the efficient development and validation of force fields with quantum-level accuracy with the aid of data-driven machine learning algorithms. Instead of trying to deliver a globally optimal machine learning potential, we propose to develop models applicable to specific thermodynamic state points in a simple and user-friendly process: a "disposable" potential which can be sourced with minimal effort.

Essentially, after an initial ab-initio simulation, a machine learning potential is constructed with minimum human effort through a data-driven active learning protocol. Such models can afterward be applied in exhaustive simulations to provide reliable answers for the scientific question at hand. The accuracy of this approach is benchmarked with respect to the underlying ab-initio reference and the resulting models are evaluated in detail with an automated validation protocol that includes structural and dynamical properties and the precision of the force prediction of the models.

We showcase this methodology by shedding light onto the fascinating and puzzling behaviour of water in carbon nanotubes where experimental measurements have reported ultra-fast and radius-dependent water transport which are absent in isostructural boron nitride nanotubes. Our simulations using disposable models reveal a large, radius-dependent hydrodynamic slippage on both materials with water experiencing indeed a 5 times lower friction on carbon surfaces compared to boron nitride. Analysis of the diffusion mechanisms across the two materials reveals that the fast water transport on carbon is governed by facile oxygen motion, whereas the higher friction on boron nitride arises from specific hydrogen-nitrogen interactions.

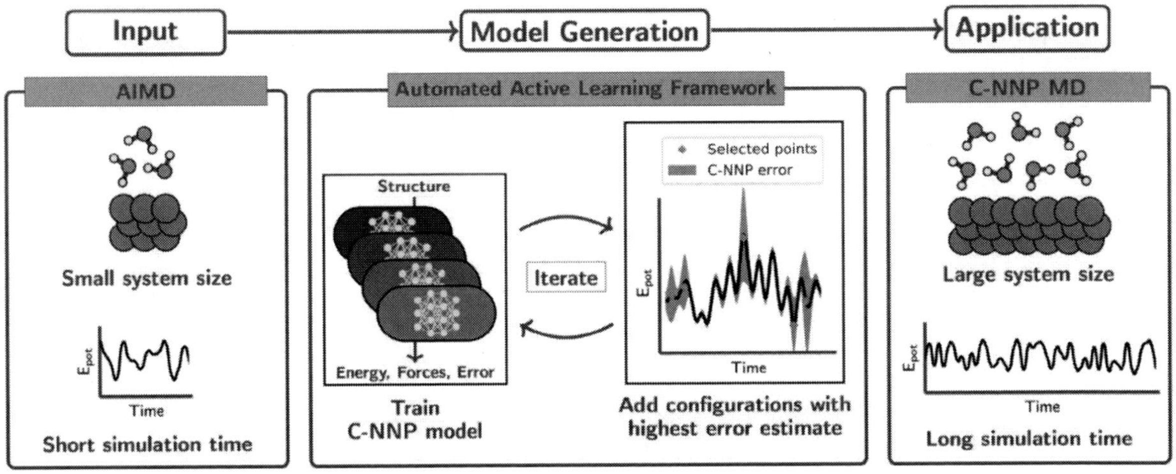

645e Molecular Simulation of Compressibility of Water in Carbon Nanopores

Gennady Gor, Jason Ogbebor, Alexei Khalizov

Compressibility of fluids in porous media determines the response of those media to mechanical loads, such as elastic waves propagation. Therefore, compressibility is one of the key parameters used in geophysical models for retrieving the properties of porous rocks. If the pores are in the nanometer size range, many thermodynamic properties of the fluid confined in such pores are altered, and fluid compressibility is not an exception.[1] The adsorption-ultrasonic experiments on mesoporous solids have shown that the compressibility of a fluid adsorbed in a nanopore is noticeably lower than in the same bulk fluid.[2] Recent molecular simulation studies of confined simple fluids, such as argon[3] and nitrogen,[4] have confirmed these observations quantitatively, also predicting a linear relation between the reciprocal compressibility and the reciprocal pore size. While the observations for nitrogen and argon fluids are of interest, it is the behavior of water under confinement that is of practical importance in many fields, including geophysics. Here we used the TIP4P/2005 water model,[5] which is known to quantitatively predict the compressibility of bulk liquid water in the wide range of pressures.[6] We calculated the compressibility of water confined in carbon pores of different geometries and sizes, using Monte Carlo and molecular dynamics simulations. Similar as in the studies of argon and nitrogen, we observed a significant deviation of compressibility of confined water from the bulk values. The pore size dependence, however, appeared to be different from the argon and nitrogen case, and we attribute this difference to the solvophobic instead of solvophilic confinement.

References:

1. C. D. Dobrzanski, B. Gurevich, and G. Y. Gor. *Appl. Phys. Rev.,* 8 (2021): 021317.
2. K. Schappert, and R. Pelster, *Europhys. Lett.* 105(5), (2014), 56001.
3. C. D. Dobrzanski, M. A. Maximov, and G. Y. Gor. *J. Chem. Phys.,* 148 (5) (2018): 054503.
4. M. A. Maximov, and G. Y. Gor. *Langmuir,* (2018): 34 (51), 15650-15657.
5. J. L. Abascal, and C. Vega, *J. Chem. Phys.*, 123(23), (2005) 234505.
6. H. L. Pi et al., *Mol. Phys.*, 107(4-6), (2009), 365-374.

572g Understanding and Virtual Design of Low-Volatility Ionic Liquid Solvents for Spacecraft CO_2 Separations

ISAAC Armstrong, Hendrik Heinz

Separating Carbon Dioxide (CO2) from other light gases for purposes such as carbon capture or removal of metabolically-produced CO2 from space vehicles, remains an exorbitantly expensive process in 2022. Liquid solvents, such as the industry standard monoethylamine (MEA), are leading contenders for energy-efficient CO2 separations. Classical molecular dynamics (MD) simulations are a valuable tool for predicting CO2-separation performance properties in liquid solvents at the atomic scale. Ionic liquids, composed of free-floating cations and anions, are exceedingly stable in the liquid phase, which is a particularly important consideration for in-space systems where solvent losses are exceedingly difficult to replace. Ionic Liquid viscosities have been drastically reduced in recent years, but little work has been done to characterize design tradeoffs between volatility and viscosity in ionic liquids versus other solvents. In this work, we developed and refined force field parameters for over 25 ionic liquids as part of the Interface Force Field (IFF), including some unusual chemistries, as well as several recently characterized CO2 separation amines such as 2-EEMPA. The IFF incorporates atomic charges that closely represent chemical bonding and non-reactive chemical bonding as well as non-bonded van der Waals and electrostatic interactions, and its wide transferability across organic and inorganic compounds enabled us to model diverse solvent molecules within the same framework. The models were validated with experimental densities, as well as vaporization enthalpies when possible, to ensure enough accuracy to observe trends in the desired properties. We then applied the models in extensive molecular dynamics simulations of bulk properties, internal molecular recognition, CO_2 adsorption, CO2 solubility, viscosity, heat capacity, and relative volatility. The developed models show that while both amines and ionic liquids can achieve lower solvent viscosities by reducing intermolecular interactions such as hydrogen bonding, ionic liquids will still exhibit lower volatility due to their split-ion nature. We will share detailed trends in molecular recognition and for a series of properties. Preliminary results indicate that, in contrast to other liquid solvents, ionic liquids can be designed for superior low volatilities while maintaining satisfactory CO2 separation performance.

572h Machine Learning Approach for Construction of Fingerprint Kernels for Pore Structure Characterization of Metal-Organic Frameworks

Shivam Parashar, Alexander Neimark

Practical Metal-Organic Frameworks (MOFs) materials are not the ideal crystals. They usually contain secondary phases, residual chemicals, and various types of defects. It is of paramount importance to have an ability to evaluate the extent of sample crystallinity and accessibility of pore compartments for hosting guest molecules. To this end, we recently suggested the method of fingerprint isotherms based on the comparison of the experimentally measured isotherms and Monte Carlo generated isotherms of molecular models of ideal crystals. The proposed method relies on the decomposition, or compartmentalization, of the total isotherm into the fingerprint isotherms corresponding to adsorption in individual pore compartment of the 3d crystal pore network. This method was demonstrated using the examples of PCN-224, ZIF-412 and Cu-BTC structures using Ar, N_2 and CO_2 at their normal boiling temperatures.[1,2]

One of the problems in the fingerprint method is how to divide the MOF pore network into individual pore compartment with invoking an arbitrary ad-hoc geometrical boundaries. In this work, we develop an unsupervised machine learning algorithm to calculate the fingerprint isotherms based on in-silico generated database of the spatial distribution of adsorbate molecules at different pressures. The algorithm learns from the distributions of adsorbate molecules at saturation conditions and then predicts which compartment a given adsorbate molecule belongs at any other pressure. To differentiate between the adsorbate molecules within different pore compartments, we use a combination of centroid and density-based clustering algorithms with cluster size constraints. Based on this procedure, we calculate the kernel of fingerprint isotherms and match it against the experimental isotherm to predict the accessibilities of pore compartments and overall non-ideality of the sample. The algorithm is applicable for calculating the kernels of fingerprint isotherms on any MOF. The proposed method is expected to be instrumental for the selection and design of novel MOF-based materials with improved properties for gas separations, energy storage, and catalysis.

1. Parashar, S.; Zhu, Q.; Dantas, S.; Neimark, A.V. Monte Carlo Simulations of Nanopore Compartmentalization Yield Fingerprint Adsorption Isotherms as a Rationale for Advanced Structure Characterization of Metal–Organic Frameworks. *ACS Applied Nano Materials* 2021 *4* (5), 5531-5540. DOI: 10.1021/acsanm.1c00937

2. Dantas, S.; Neimark, A.V. Coupling Structural and Adsorption Properties of Metal–Organic Frameworks: From Pore Size Distribution to Pore Type Distribution. *ACS Applied Materials andamp; Interfaces* 2020 *12* (13), 15595-1560. DOI: 10.1021/acsami.0c01682

572i High-Throughput Screening of Hypothetical Functionalized-Irmofs for Separation of Alkanes

Fangxi Wang, Abhishek Sose, Samrendra Singh, Sanket Deshmukh

Metal organic frameworks (MOFs) have gained wide popularity for their applications, such as gas storage and gas separation, due to their high porosity, large surface area, and tunability of pore size. The development of MOFs with enhanced selectivity and permeability for n-alkanes could result in significant energy and cost savings associated with alkane separation compared to distillation in the petrochemical industry. In this study, we performed high-throughput screening using dual force zone non-equilibrium molecular dynamics (DFZ-NEMD) simulations of hypothetical and known MOFs from existing databases for the separation of linear hydrocarbons, *e.g.*, hexane and nonane. Data generated from this screening was used to develop deep machine learning models. Then these ML models were integrated with our in-house structure generation code that systematically added functional groups on selected MOFs to predict the separation performance of functionalized MOFs. The best candidates were then validated by performing DFZ-NEMD simulations. Our design scheme and the molecular-level insights from the results of this research might help in the design of functionalized MOFs at the molecular level that could be used to separate different hydrocarbons from mixtures.

572j Efficiently Exploring the Adsorption Space of Molecules in MOFs Combining the Use of Molecular Simulations, Machine Learning, and IAST

Xiaohan Yu, Sihoon Choi, Andrew Medford, David Sholl

The combination of billions of possible molecules and thousands of reported Metal Organic Frameworks (MOFs) forms a vast chemical space. The limited variety of experimental isotherms that are available, however, make it very challenging to comprehensively assess adsorption-based separations in MOFs. To explore a much wider adsorption space, we carefully selected a large and diverse set of molecules and MOFs to generate molecular simulation data and train machine learning models. With the resulting models, accurate and efficient predictions of adsorption properties including Henry's constants and adsorption isotherms of arbitrary molecules in large libraries of MOFs are possible. By combining these predictions with Ideal Adsorbed Solution Theory (IAST) we make predictions for the adsorption properties of millions of distinct binary mixtures in thousands of MOFs. We will illustrate how this approach can be used to gain insights into the characteristics of MOFs suitable for a diverse range of challenging separations.

645a Effect of Energetical and Geometrical Heterogeneity of Kerogen on BET Surface Area Characterization and Methane Adsorption

Yingnan Wang, Wanying Pang, Zhehui Jin

As a transition fuel, natural gas plays an ever-increasingly important role to meet the global energy demand, while reaching net-zero carbon emission by 2050. On the other hand, due to the continuous depletion of conventional natural gas reservoirs, shale gas has become an important natural gas source. For example, in the United States, ~80% of total dry natural gas production is from shale formations in 2020. Unlike conventional reservoirs, surface adsorption plays a dominant role in shale gas due to the presence of a significant amount of nanosized pores in shale media. Kerogen is the main constituent of organic matter, which generates hydrocarbons via chemical decomposition. It is also the main methane storage site as methane adsorption capacity in shale rocks has shown a positive correlation with the total organic carbon (TOC) content. Therefore, the accurate characterization of surface area in kerogen nanoporous media becomes utterly important in the prediction of shale gas-in-place (GIP).

The Brunauer–Emmett–Teller (BET) method has been widely used to characterize 77 K N_2 adsorption to obtain the surface area (the so-called BET surface area, S_BET) in various porous media, such as activated carbons, metal organic frameworks (MOFs), silica, and zeolite. It is also one of the standard methods to obtain the surface area in isolated kerogen. BET theory assumes that the multilayer adsorption of ideal gas takes place on a perfectly-smooth ideal homogeneous surface. On the other hand, in contrast to the basic assumptions in BET theory, kerogen surface may not be perfectly smooth (*i.e.*, geometrical heterogeneity), while it carries energetical heterogeneity with a number of heteroatoms such as N, S, and O. While there have been a number of studies to investigate the effect of energetical and geometrical heterogeneity on S_BET, the careful analysis of their effect on kerogen S_BET is still lacking. In addition, whether S_BET can be a good indicator for methane adsorption capacity in kerogen nanopores remains unanswered.

In this work, we conduct 77 K N_2 adsorption in 13 kerogen slit mesopores to study the effect of kerogen geometrical and energetical heterogeneity on the S_BET characterization by using grand canonical Monte Carlo (GCMC) simulations. We find that within the BET pressure range, N_2 adsorption sites are mainly within concave ('basin') and valley regions on kerogen surfaces, while in convex regions its adsorption rarely takes place. In addition, while S_BET agrees well with the geometric surface area (S_GEO) in graphene mesopores, in kerogen mesopores, S_BET is generally lower than the S_GEO. The deviation becomes more significant as kerogen surface roughness becomes more significant and the fraction of convex ('ridge') surface increases. On the other hand, surface chemistry shows non-negligible impacts on N_2 adsorption when the presence of heteroatoms affects surface topology. Interestingly, S_BET correlates well with methane excess adsorption in kerogen mesopores, while it outperforms S_GEO. It is probably because both nitrogen-kerogen and methane-kerogen interactions are mainly van der Waals type, while their adsorption sites largely overlap.

This work provides some crucially important fundamental understanding about S_BET characterization in kerogen mesopores which can guide methane adsorption capacity prediction in kerogen nanoporous media and shale gas-in-place estimation.

645b Deformation of Nanoporous Carbons Induced By Multicomponent Adsorption: Insight from the SAFT-DFT Model

Nicholas Corrente, Alexander Neimark

Predicting adsorption and mechanical properties of nanoporous carbons is of paramount importance for numerous practical applications, such as mixture separations, electrodes and supercapacitors, gas storage, carbon capture, and enhanced natural gas recovery. Most industrial applications involve multi-component adsorption that is complicated in the case of compliant nanoporous carbons, which expand and contract upon adsorption and desorption of guest molecules.

In this work, we consider the deformation of nanoporous carbons during the adsorption of mixtures, focusing on the specific context of carbon dioxide displacement of natural gas from shale and coal. [1] We use the density functional theory (DFT) augmented by the perturbed chain statistical associating fluid theory (SAFT) [2] to model adsorption of fluid mixtures in carbon nanopores at geologically relevant temperatures and pressures. The SAFT-DFT model allows for calculating the adsorption stress that is the driving force of adsorption-induced deformation. [3] First, we consider the case-study system of adsorption of methane and carbon dioxide mixtures in pores of various sizes and calculate the adsorption stress as the function of the fluid composition. Next, we explore the mixtures of carbon dioxide and alkanes and determine how the hydrocarbon chain length affects the carbon dioxide co-adsorption and the stress exerted on the carbon matrix. Calculations are performed in the wide range of pressures and temperatures characteristic to geological conditions.

1. Corrente, N. J.; Zarębska, K.; Neimark, A. V., Deformation of Nanoporous Materials in the Process of Binary Adsorption: Methane Displacement by Carbon Dioxide from Coal. *J. Phys. Chem. C* **2021,** *125* (38), 21310-21316.
2. Liu, J.; Wang, L.; Xi, S.; Asthagiri, D.; Chapman, W. G., Adsorption and phase behavior of pure/mixed alkanes in nanoslit graphite pores: an iSAFT application. *Langmuir* **2017,** *33* (42), 11189-11202.
3. Ravikovitch, P. I.; Neimark, A. V., Density functional theory model of adsorption deformation. *Langmuir* **2006,** *22* (26), 10864-10868.

645j Designing Highly-Selective Sorbents for Diagnosing Respiratory Disease Via Breath Analysis

Scott Bobbitt, Michael E. Chandross, Jason Sammon, Jacob I. Deneff, Dorina F. Sava Gallis

There are numerous volatile organic compounds (VOCs) present in exhaled human breath. Pathogens and disease states produce unique profiles of some of these VOCs, providing a unique chemical fingerprint that has the potential to be used as a diagnostic. However, detecting these specific molecules in a complex mixture of air, water, CO_2, and other non-marker VOCs is a significant challenge. Metal-organic frameworks (MOFs) are porous crystalline materials consisting of inorganic nodes connected via organic linkers. MOFs offer a variety of pore sizes, pore shapes, and chemical properties, implying that they can be tailored to specific adsorption applications. We use molecular simulation and high-throughput screening to design or choose sorbents tailored to adsorb specific target VOCs with high selectivity over competing molecules (e.g. water, CO_2). We have identified and synthesized several promising MOFs with high selectivity for VOCs -- known to be indicators for COVID-19 -- and validated their adsorption capacity. We will also discuss how arrays of highly selective sorbents can be used as a diagnostic tool for disease by detecting specific profiles of VOCs .

This work was funded by the LDRD program at Sandia National Laboratories, a multi-mission laboratory managed and operated by National Technology and Engineering Solutions of Sandia, LLC, a wholly owned subsidiary of Honeywell International, Inc., for the U.S. Department of Energy's National Nuclear Security Administration under Contract No. DE- NA0003525

645d Flat-Histogram Simulations of Water in MOFs: Advanced Strategies for Overcoming Sampling Challenges

Daniel Siderius, Harold Hatch, Vincent K. Shen

Adsorption of water by porous materials has recently reemerged as an area of interest due to its possible application as a platform for harvesting water from humid air. Combined with the widespread application of computational screening of adsorbent materials for various applications, molecular simulations of water in confinement are once again an area of ongoing investigation. However, the key challenge of water simulations in porous materials remains unchanged: due to strong water-water hydrogen bonding, adsorption of water occurs primarily at high relative humidity or low temperature and typically over a narrow pressure range [1], akin to Type V isotherms in the standard IUPAC classification scheme. Hydrogen bonding networks also encourage coalescence of water in pores rather than monolayer formation [2,3]. These challenges are exacerbated in hydrophobic metal-organic frameworks (MOFs) with cage-like pores, where adsorption occurs non-sequentially, but via abrupt steps in which one cage (cavity) fills with a single cluster of water in each step [1]. These simulation challenges are manifested by slow equilibration, inefficient sampling, and large uncertainties in ensemble averages [1]. The introduction of so-called superhydrophobic materials only amplifies these simulation challenges. Recent work has explored strategies to overcome the sampling challenges by using energy look-up tables, insertion (deletion) biases, and continuous-fractional component Monte Carlo [1].

We approach the problematic sampling of water adsorption in hydrophobic materials through a similar suite of simulation tools, starting with flat-histogram Monte Carlo (MC) and adding advanced MC moves, and then parallel simulations, all focuses on improving the efficiency with which the statistical phase space is sampled. In particular, we introduce configurational-bias MC operations that are tuned for adsorption in crystalline materials and specialized MC moves that address the proclivity of water to form clusters. We implement flat-histogram MC using Wang-Landau and Transition-matrix biases, which additionally allows us to examine water cluster formation via the macrostate probability distribution, such as through identification of free-energy barriers between individual clusters. However, the water-clustering effect necessitates the introduction of new analysis tools for converting the macrostate probability distribution to ensemble averages including the adsorption isotherm. These advanced strategies are implemented using the FEASST Monte Carlo toolkit [4], which provides modularity and is accessible to other researchers. The end effect is to both present simulation results for specific materials that highlight the challenges of water adsorption and use these prototype systems to suggest specific strategies for other researchers that simulate water adsorption.

[1] Zhang and Snurr, J Chem Phys C, 121:24000 (2017)
[2] Striolo, Gubbins, Chialvo, and Cummings, Mol Phys 102:243 (2004)
[3] Sarkisov, Centineo, and Brandani, Carbon 118:127 (2017)
[4] Hatch, Mahynski, and Shen, J Res NIST, 123:123004 (2018)

572f An Efficient Featurization Scheme for Machine-Learning Predictions of Diverse Molecules in Metal-Organic Frameworks

Sihoon Choi, Xiaohan Yu, David Sholl, Andrew Medford

For the past decade, machine learning (ML) studies of adsorption in metal-organic frameworks (MOFs) have proliferated based on the numerous existing databases covering tens of thousands of structures and rapidly improving computer resources. However, many of these studies are focused on individual adsorbates and thus share a similar limitation: transferability of the resulting ML models between different adsorbates. Here, we propose a single ML model that overcomes this limitation and predicts an adsorption property, the Henry's constant, of a diverse set of > 40 adsorbates in > 400 MOFs. Host-guest interaction energies are known to be critical to determining the Henry's constant of adsorbates in MOFs. Here, we develop an efficient new descriptor scheme called Gaussian Integral (GI) descriptors in which the energy landscape of host-guest interactions is transformed into a vector with fixed size. GI descriptors are especially suitable to numerically expressing the dispersion potential landscape. With this novel scheme, we show that ML predictions of Henry's constants across many materials and adsorbates are improved and achieve mean absolute errors within chemical accuracy (1 kcal/mol) as well as an orders-of-magnitude increase in speed as compared to conventional molecular simulations.

645f Impact of Metal-Organic Framework Characteristics on Electrostatic Interactions for GCMC-Simulated Adsorption

Brandon Bout, Krista Walton

Metal-organic frameworks (MOFs) have received a great deal of attention in recent years as showing significant potential towards a wide variety of applications due to their tunable nature. One such application is that of mixture adsorption, a complicated issue with many factors involved.[1] As mixture adsorption is both challenging and time-consuming to experimentally measure, predictive methods such as the Ideal Adsorbed Solution Theory (IAST) and molecular simulations have been widely used as an easier alternative to measure mixture adsorption.[2] Grand Canonical Monte Carlo (GCMC) simulations can directly compute adsorbed loadings from a gas mixture onto a given MOF, but these results are dependent upon simulation parameters.[3] One component of these simulation parameters are the electrostatic interactions between the adsorbent and adsorbate which can be either included or excluded in the simulations. Though typically included, the comparison of adsorbate loadings when including or excluding these electrostatic interactions can provide insight into their role in adsorptive MOF applications as well as the impacts of MOF characteristics on these interactions.

Unary CO_2 adsorption was simulated using GCMC with frameworks exhibiting differing characteristics such as varied pore size, inclusion of functionalized ligands, and open-metal sites. Two sets of simulations were completed, one including electrostatic interactions between the adsorbent and adsorbate and the other excluding them, allowing investigation into framework characteristics' effects on the electrostatic interactions. Binary adsorption with mixtures of CO_2 and C_1-C_4 hydrocarbons was also simulated, both with and without electrostatic interactions, to observe the impact of electrostatics on these non-charged C_1-C_4 species in mixtures.

References:

1. Li, J. R.; Kuppler, R. J.; Zhou, H. C., Selective gas adsorption and separation in metal-organic frameworks. *Chem. Soc. Rev.* **2009,** *38* (5), 1477-1504.
2. Myers, A. L.; Prausnitz, J. M., Thermodynamics of mixed-gas adsorption. *Aiche J.* **1965,** *11* (1), 121-127.
3. Dubbeldam, D.; Torres-Knoop, A.; Walton, K. S., On the inner workings of Monte Carlo codes. *Molecular Simulation* **2013,** *39* (14-15), 1253-1292.

645h Modelling Hierarchical Carbon-Based Materials for CO_2 Capture and Separation

Daniel Bahamon, Nour Alkhatib, Mohammad Abu Zahra, Maryam Khaleel, Lourdes Vega

Among the capture and separate CO_2 (CCS) from diluted sources, such as gases emitted from fossil fuel combustion and other industrial processes, adsorption with solid materials is emerging as a viable alternative to the traditional amine absorption process due to the improvement in energy efficiency and potential cost reductions [1]. Common materials that have been extensively studied as potential adsorbents for CO_2 capture from a flue gas include zeolites, silicas, polymeric resins, Metal-Organic Frameworks (MOFs) and carbon-based structures [2]. Carbon-based materials are very attractive for this endeavor due to their versatility, as do not require moisture removal (unlike most zeolites and some MOFs), their ability to present high CO_2 capacity depending on their preparation process, while being easy to regenerate and with excellent mechanical and chemical stability [3]. However, further knowledge is still needed about the effect of surface characteristics (e.g., pore size distribution, surface area, density and species of surface groups, etc.) and the topological nature of the connected pore structures on the phase changes for the confined fluids for their optimization for final use.

The work presented in this contribution combines molecular and process modelling for developing efficient, low-cost adsorbent materials for carbon dioxide capture, aiming at using the fundamental understanding of the phenomena happening at the molecular level to optimize their performance under process conditions. Moreover, such molecular simulation methods can be used as a screening method at a very moderate computational expense to fine-tune the process needs, thus making an important contribution to the optimization of the swing adsorption process performance. We present and discuss results concerning two different carbon-based materials: zeolite-templated carbons (ZTCs) and activated carbon samples developed from seeds of date fruits (Phoenix dactylifera, AC-DS) from the United Arab Emirates [4]. Furthermore, the effect of the inclusion of hierarchical features, i.e., a collection of micro and mesopores, is analyzed in detail. We showcase the capabilities and limitations of a developed molecular simulation-based screening tool for the efficient removal of CO_2 from industrial gas streams –CO_2/N_2 and CO_2/CH_4, with and without including water traces as impurities- by adsorption at relevant gas separation process conditions. The modelling approach for the carbon-based solid sorbents, previously validated and systematically applied for other processes [4], consists of a collection of three-dimensional platelets, packed in random positions and orientations in a periodic simulation cell, with defects and oxygenated groups. Results presented include the adsorption of the pure components, the properties of both binary CO_2/N_2 and CO_2/CH_4 mixtures, and the effect of T,P,X conditions on the performance of pressure and temperature swing adsorption processes.

This work is funded by Khalifa University of Science and Technology under projects RC2-2019-007 and CIRA2018-103.

References

[1]. Rao, A.B., Rubin, E.S., Keith, D.W., Morgan, M.G. Energy Policy 34(18), 3765–3772 (2006).

[2]. Bahamon, D., Vega, L.F. Chem. Eng. J. 284, 438-447 (2016).

[3]. Rashidi, N.A., Yusup, S. J. CO_2 Util., 13, 1-16. (2016).

[4]. Bahamon, D., Ogungbenro, A.E., Khaleel, M., Abu-Zahra, M.R.M., Vega, L.F. Ind. Eng. Chem. Res., 59, 7161-7173 (2020).

645i Modeling of Adsorption and Diffusion of Chemical Agents and Simulants in UiO-66

Chinmay Mhatre, Jacob Wardzala, Meiirbek Islamov, Paul Boone, Christopher E. Wilmer, Karl Johnson

Capture and destruction of chemical warfare agents (CWAs) is an active and important field of research. The UiO-6x metal-organic framework (MOF) family is especially promising for both adsorption and chemical reaction of CWAs. Defects, particularly missing linker defects, are critical for the performance of UiO-66 for adsorbing and destroying CWAs. The pores of UiO-66 are too small to allow facile diffusion of CWAs and without defects, the zirconium nodes of the MOFs are not reactive, since they are fully coordinated. This work explores adsorption in the defective structures of MOF UiO-66 having various defect patterns, using the grand canonical Monte Carlo method. Widely studied CWA simulant dimethyl methyl phosphonate (DMMP), another simulant dimethyl phosphite (DMP), CWA sarin (GB), and soman (GD) are included in the adsorption calculations. Two different framework charge types are studied to see their effect on adsorption capacity at low and high pressures. Adsorption of simulants is lower in the case of defective MOFs than pristine UiO-66 at lower pressure, and the opposite trend is observed for high-pressure adsorption. In conjunction with the adsorption capacity of MOFs, self diffusivities of CWA and their simulants at zero loading are calculated for hydroxylated pristine UiO-66, using the dynamically corrected Transition State Theory (dcTST) approach. The parameterization of the force field to include the hydrogen bonding between the CWA/simulants and framework atoms impacts the zero loading self-diffusivities of both CWA and their simulants. The pristine MOF serves as a good starting point to study defective structures in the future.

467a Lanthanide Binding Tag Peptides for the Foam Fractionation of Rare Earth Elements

Luis Ortuno, Pan Sun, Stephen Crane, Dr. Kathleen J. Stebe, Wei Bu, Binhua Lin, Mark Schlossman, Raymond S. Tu, Charles Maldarelli

Solvent extraction is widely used for separation of rare earth elements trivalent cations (REEs or Ln^{3+}) from an aqueous phase into an organic solvent. This process is energy intensive and environmentally unfriendly, requiring large volume of organic solvent and organo-phosphate surfactants. Here, we exploit the high affinity of a surface-active Lanthanide Binding Tag (LBT) peptide (LBT1, YIDTNNDGWYEGDELLA), that coordinates selectively with Ln3+ ions for its use in a bioinspired/eco-friendly extraction processes in which the complexed LBT1-Ln^{3+} peptide adsorbs to the air/aqueous interfaces of bubbles for foam recovery. To understand the surface activity and identify the bound cationic species at the air-water interface, we characterized the surface molecular adsorption and arrangement of LBT1 and the more surface active LBT1-LLA, which has three additional hydrophobic residues (-LLA) attached to the C-terminus of LBT1. For this, we measured the relaxation dynamic tension of the each LBT in the absence and presence of Tb3+ cations. X-ray reflectivity (XRR) and X-ray fluorescence near total reflection (XFNTR) measurements on the adsorbed layer were used to compute the surface concentration of the peptide and the Ln^{+3} cation, and the thickness of the interfacial layer. The addition of three hydrophobic residues to LBT1 increased the surface density of the peptide by three-fold. Additionally, the LBT1-LLA improved the adsorption of Tb3+ ions to the air-water interface up to approximately 30% over LBT1 peptides. The interfacial density of Tb-bound LBT1-LLA increased between 50% and 100% compared with the Tb-free, unbound LBT1-LLA. We also observed a four-fold increase in the surface coverage of LBT1 in the Tb-bound state over the Tb-free state. The enhancement in the surface activity of LBTs peptides is improved by the addition of hydrophobic residues appended to the C-terminus of the peptide. This improvement in the hydrophobicity of these biomolecules could be advantageous for a green, eco-friendly separation method of REEs.

467b Eco-Friendly Detergent Combo for Viral Inactivation in Protein Manufacturing

Yuanyuan Ji, Melissa Holstein, Zhi Li, Dong Yang, Sanchayita Ghose

Protein viral contaminants can be a major concern in biologics manufacturing. Therefore, most biologics manufacturing processes need to effectively inactivate and remove viruses to comply with regulatory requirements and ensure safe administration of therapeutics to patients. Detergent-based treatment is an effective way to inactivate enveloped viruses, and a widely used commercial detergent is Triton X-100. However, due to environmental concern, the European Chemical Agency has mandated a sunset date of 2021 to replace Triton X-100. Hence, there is an industry-wide initiative ongoing to find eco-friendly Triton X-100 alternatives. The purpose of this study is to propose a novel eco-friendly detergent combo (referred as "OG/DDM") for viral inactivation in biologics manufacturing.

The OG/DDM was evaluated on three aspects, i.e., viral inactivation effectiveness, impact on product quality, and impact on process robustness. Sets of viral inactivation, detergent incubation, and detergent spike studies were conducted to obtain data on the first, second, and third aspect, respectively. The same studies were also conducted on Triton X-100 and another eco-friendly commercial detergent, i.e., EcoSurf EH9®, and associated data were used as control groups. The data associated with OG/DDM were compared with the data associated with Triton X-100 and EcoSurf EH9® to assess its suitability.

First, the OG/DDM provided satisfactory viral inactivation effects, i.e., its log reduction value for two model viruses, i.e., amphotrophic murine leukemia virus and herpes simplex virus, was above 4 under ambient temperature after 15, 30, and 60 minutes of treatment. Second, the detergent combo showed negligible negative impact on product quality. The impurity species in product pool, i.e., high molecular weight, deamidated, and oxidized species, were all within acceptable range after detergent treatment. Besides, the protein sialic acid content, which was an important species to control immune response, was also within acceptable range after detergent treatment. Third, the detergent spike study demonstrated that the detergent level in product pool was below the limit of detection after a protein A chromatography step. Therefore, the OG/DDM can be cleared by subsequent purification steps and will not disturb manufacturing process. In addition, all results associated with OG/DDM are comparable to the results associated with Triton X-100 and EcoSurf EH9®, which demonstrated that the performance of OG/DDM was comparable to conventional commercial detergents.

The proposed novel eco-friendly detergent combo, OG/DDM, demonstrated effective viral inactivation, negligible negative impact on product quality and process robustness. Besides, the performance of OG/DDM was comparable to Triton X-100 and EcoSurf EH9®. It is demonstrated that OG/DDM is a suitable eco-friendly alternative for Triton X-100 and can be used for viral inactivation in protein manufacturing.

467c Magnetic Fractionation of Red Blood Cells Based on Oxygen Saturation in Sickle Cell Disease.

Jacob Strayer, Mitchell Weigand, Xian Wu, Hyeon Choe, Jeffrey Chalmers

Anemias are a broad classification of red blood cell (RBC) disorders which effectively result in inadequate oxygen transport in the body. Sickle Cell Disease (SCD) results from a point mutation in hemoglobin, and the hemoglobin-oxygen equilibrium becomes right shifted (i.e., SCD RBCs do not oxygen saturate to the same degree as healthy RBCs). RBC exchange transfusion is a common clinical practice to increase the oxygen carrying capacity of a SCD patient - a process which the patient's own RBCs are exchanged with healthy donor RBCs. During this process, large amounts of donor RBCs are discarded along with the patient's SCD RBCs. Here we present a new RBC separation modality based on hemoglobin-oxygen equilibrium. It is well known that RBCs become magnetized as a function of the oxygen saturation of its hemoglobin. Due to the difference in hemoglobin-oxygen equilibrium for healthy and SCD RBCs we demonstrate a magnetic method of fractionating RBCs based on oxygen saturation. In our study, we use the SCD transfusion waste product and sort this sample into healthy fractions and SCD fractions using a field flow permeant magnetic separator. Under variable oxygen tension of the sample, we preformed the magnetic sorting. The results of the magnetically sorted outlets show that the separation produced populations of cells that are different based on hemoglobin-oxygen equilibrium curves. By incorporation of magnetic separation into RBC exchange processes, more efficient selection, and removal of SCD RBCs can be accomplished.

467d Ribosomal Peptides Containing Fatty Acyl Moieties Synthesized By Noncannonical Kas III Enzymes

Hengqian Ren, Chunshuai Huang, Haiyang Cui, Yuwei Pan, Huimin Zhao

Lipopeptides constitute a large group of natural products with significant medical importance. While most of the representatives are produced by nonribosomal peptide synthetases (NRPSs), few lipopeptides were characterized with ribosomal origins. Here we report the discovery of ribosomally derived, N-acylated aminovinyl-cysteine-containing lipopeptides, hereafter termed as lipoavitides. Structural elucidation of lipoavitide A, the first member in this class obtained through heterologous expression, revealed its unique combination of nonproteinogenic residues including an aminovinyl-cysteine (AviCys), a β-hydroxy-*N,N*-dimethyl-*L*-histidine (hdmHis), a glycosylated tyrosine, a *D*-alanine, as well as an unprecedented N-terminal 4- hydroxy- 2, 4- dimethyl-pentanoyl (HDMP) moiety. Genes involved in synthesizing these nonproteinogenic residues were predicted by bioinformatics and confirmed by gene deletion study. Additionally, further characterization of the biosynthetic machinery implied that the HDMP moiety is formed by a multistep reaction process initiated by a noncanonical KAS III enzyme LipC. Overall, lipoavitides represent a new class of ribosomally derived lipopeptides, and its distinct biosynthetic machinery has significant potential to be adapted to the lipopeptide engineering for drug development.

467e The Surface Rheology of Interacting Peptide Surfactants and Rare Earth Elements at Air-Water Interfaces

Stephen Crane, Jiayi Deng, Mehdi Molaei, Luis Ortuno Macias, Dr. Kathleen J. Stebe

We study peptide-based surfactants (PEPS) at air-water interfaces for selective binding of rare earth elements (REEs). PEPS comprise a hydrophilic amino acid sequence based on lanthanide binding tags that selectively bind REEs and a hydrophobic sequence that confers amphiphilicity. We envision a REE separation process based on interfacial foam fractionation (IFF) using high affinity and highly selective REE-binding peptide surfactants (PEPS) at air-water interfaces. PEPS added to REE feedstocks will associate selectively with REE cations and adsorb to air water interfaces for recovery. The proposed process would be green in comparison to solvent extraction.

We use particle tracking methods to study the interface of soluble solutions of PEPS as a function of bulk REE concentration to probe for the formation of surface domains. we also characterize the interfacial rheology using correlated displacement velocimetry (CDV). In this method, we study the flow field around Brownian colloids at the interface by measuring their correlated displacements at short lag times. The thermally induced flow can be decomposed into interfacial hydrodynamic multipoles which reveal information about the interface's compressibility, providing fundamental insights on the mechanical response of PEPS monolayers. This study guides the design of PEPS/fluid system essential to the envisioned separation processes.

467f Understanding Cerium Binding Affinity in Lanmodulin Derived Peptides

Geeta Verma, Jacob Hostert, Julie N. Renner

The re-processing of industrial waste streams and recovery of rare earth elements (REEs) is desired for a green and circular economy. There is an urgent need to identify efficient separation methods that recycle REEs from the waste streams and provide a steady, domestic source of these elements. Among several approaches that have been proposed, bio-based strategies have been given extensive attention due to several technical and environmental advantages such as high selectivity, high regeneration, fast kinetics, eco-friendly chemicals, and low cost. In the present work, we designed two peptides derived from the EF hand loop I of lanmodulin and assessed their binding affinity for the cerium (III) ion. Both experimental and simulation techniques are used to explore the binding potential of these novel peptides and investigate the underlying molecular mechanisms.

Here, we present the preliminary results from circular dichroism (CD) analysis of two peptide sequences with cerium ions at different molar ratios. The results indicate that there is a change in the conformation of the peptides upon cerium loading and thus, provide an indication for binding. The CD results are fitted to a binding model to estimate the dissociation constant (Kd). Isothermal titration calorimetry (ITC) is utilized to confirm the dissociation constant in solution and obtain thermodynamic understanding. Molecular simulations techniques have become important tools in the interpretation of the experimental results and provide insight at an atomic and molecular scale. Therefore, we used Molecular Dynamics (MD) simulation methods to perform a configurational analysis of the peptide-cerium ion complex in water. Several atomic-scale analyses are performed to corroborate experimental results including radius of gyration, radial distribution functions, water coordination in the binding site, and intermolecular interactions. Finally, the dissociation constant of surface-bound peptides is measured using quartz crystal microbalance with dissipation (QCM-D). Overall, the results from the experiment and simulation confirm that the designed peptides have a strong affinity for cerium ions and have the potential to be used in future separation technologies.

467g Monolithic Silica Microchannels Enable Thin Layer Chromatography Analysis of Single Cells

Yuli Wang, Ming Yao, Christopher Sims, Nancy Allbritton

Introduction. Thin layer chromatography (TLC) is a widely used analytical technique for a variety of applications, but its use for extremely miniature samples, such as a single mammalian cell with a volume of ~1 picoliter, has not been realized to date. A major challenge is that the dimensions of adsorbent layer in conventional TLC significantly exceeds that of single cells. The large volume of the adsorbent does not limit diffusional spread of the ultra-small volume resulting in sample dilution during the chromatographic separation step. To address this challenge, we report a new platform for TLC by creating an array of open microchannels filled with monolithic silica. The channel sidewalls and base reduce the lateral diffusion and confine the movement of compounds along the microbands, enabling TLC separation and detection at the single-cell level. The platform is termed picoscale Thin Layer Chromatography (pTLC).

Materials and Methods. The pTLC chip was fabricated by combining sol-gel chemistry and microfabrication technology. pTLC consists of a single or an array of microscale bands made from highly porous monolithic silica designed to accept picoliter-scale volume samples (Figure A). The width and depth of each band was 80 µm and 12 µm, respectively. Droplets of model fluorescent compounds with a volume of 34 pL were spotted on the microband using an ink-jet microdispenser (Microfab Technologies) and then successfully separated upon exposure of the channel inlet to 1-butanol:water:acetic acid (18:4:4 vol:vol:vol). To demonstrate single-cell analysis, K562 cells were loaded with sphingosine (2-amino-4-trans-octadecene-1,3-diol)-CY5 and single cells were then spotted on separate microbands. The cellular metabolites were extracted from the individual cells and separated using pTLC.

Results and Discussion. Model lipids, 18:1 PE CF (1,2-dioleoyl-sn-glycero-3-phosphoethanolamine-N-(carboxyfluorescein)) and Texas Red DHPE(1,2-dihexadecanoyl-sn-glycero-3-phosphoethanolamine), were successfully separated using in as little as 3 min with a separation distance as short as 360 µm. Since Texas Red DHPE has more charged groups (two tertiary amines, one sulfonate) than 18:1 PE CF (one carbonate), Texas Red DHPE had more interaction with the surface charge of the silica resulting in a slower migration rate relative to that of 18:1 PE CF.

To further demonstrate the power of pTLC, single cells were assayed for enzymatic production of sphingosine 1-phosphate (S1P) from sphingosine (Sph). S1P is a metabolic product of Sph formed by the action of sphingosine kinase (SK) as part of an important signal transduction pathway within cells. Analysis of S1P and Sph of biological samples is useful in revealing lipid signaling activity in many diseases including cancer, autoimmune and neurological diseases. S1P supports cell proliferation in tumors and thus fuels their malignant behavior so that SK is often mutated, overexpressed or otherwise upregulated in tumor cells.

To assess the separation and detection of Sph and S1P, fluorescein-labeled Sph (SphF) and S1P (S1PF) in solution was spotted at the inlet of the channels of a pTLC plate followed by separation. The average migration speed of the analytes was 170 µm for SphF and 36 µm for S1PF over the time of 1 min. The negative phosphate group of S1PF hinders its migration due to its interaction with silica. The above results demonstrate that pTLC is capable of separating micro-spotted organic compounds. One unique advantage of pTLC was that the migration was confined along the microbands without lateral diffusion. Other derivatized lipids were suitable for separation and assay by pTLC, including Bodipy FL PIP2 (Phosphatidylinositol 4,5-bisphosphate) and PIP3 (Phosphatidylinositol 3,4,5-trisphosphate), PC-CY5 and PE-CY5. Using PC-CY5 (1,2-dioleoyl-sn-glycero-3-phosphocholine-N-(Cyanine 5)) and PE-CY5 (1,2-dioleoyl-sn-glycero-3-phosphoethanolamine-N-(Cyanine 5)), the detection limit of analytes of pTLC was 3×10^{-19} mole, which is significantly better than commercial high-performance TLC (HPTLC) plates.

To demonstrate the ability to assay single-cell SK activity, a single cell loaded with Sph-CY5 was spotted on a pTLC microband. The pTLC plate was developed in 1-butanol:1-propanol:water:triethylamine

(18:2:4:1, vol:vol:vol:vol) and the fluorescence of CY5 measured within the channel. Two bands composed of Sph-CY5 and S1P-CY5 eluted from the cell, migrated along the microband, and were separated in 4 min (resolution = 1.8). Quantitative analysis of the pTLC chromatogram at 11 min showed that the S1P-CY5 peak area was 33.8% and Sph-CY5 peak area was 66.2% of the total CY5 fluorescence measured. From these data, this particular cell was shown to convert 33.8% of the loaded Sph-CY5 to S1P CY5 in 30 min via the action of sphingosine kinase. To confirm that the conversion of Sp-CY5 to S1P-CY5 was likely due to SK, cells were analyzed after incubation with the SK inhibitor PF543 (Figure B). As expected, incubation of cells with the inhibitor abrogated the S1P-CY5 peak. This result demonstrated that pTLC can be used for analyzing biologically important compounds at the single-cell level.

Conclusion. We have demonstrated a novel technology for single-cell analysis and picoliter-scale specimens. We have demonstrated the use of pTLC for the analysis of sphingolipid signaling in single cells. We envision pTLC will have potential applications in many areas where miniature specimens and high-throughput parallel analyses are needed, such as diagnosis of diseases, cell heterogeneity analysis, monitoring droplet-based reactions, detecting miniature environmental, food, forensic and pharmaceutical samples.

Reference:

1. Proctor, A.; Allbritton, N. L., 'Fix and assay': separating in-cellulo sphingolipid reactions from analytical assay in time and space using an aldehyde-based fixative. *Analyst* **2019,** *144* (3), 961-971.
2. Nakanishi, K.; Shikata, H.; Ishizuka, N.; Koheiya, N.; Soga, N., Tailoring mesopores in monolithic macroporous silica for HPLC. *HRC-J. High Resolut. Chromatogr.* **2000,** *23* (1), 106-110.
3. Jiang, Y.; Kang, X.; Gao, D.; He, A.; Guo, R.; Fan, X.; Zhai, Y.; Xia, J.; Xu, Y.; Noda, I.; Wu, J., Finding a suitable separation condition for TLC/FTIR analysis by using multiple-narrow-band TLC technique. *RSC Advances* **2015,** *5* (28), 21544-21549.
4. Castano-Alvarez, M.; Ayuso, D. F. P.; Granda, M. G.; Fernandez-Abedul, M. T.; Garcia, J. R.; Costa-Garcia, A., Critical points in the fabrication of microfluidic devices on glass substrates. *Sens. Actuator B-Chem.* **2008,** *130* (1), 436-448.
5. Fujima, T.; Futakuchi, E.; Tomita, T.; Orai, Y.; Sunaoshi, T., Hierarchical Nanoporous Glass with Antireflectivity and Superhydrophilicity by One-Pot Etching. *Langmuir* **2014,** *30* (48), 14494-14497.
6. Hara, T.; Desmet, G.; Baron, G. V.; Minakuchi, H.; Eeltink, S., Effect of polyethylene glycol on pore structure and separation efficiency of silica-based monolithic capillary columns. *J. Chromatogr. A* **2016,** *1442*, 42-52.

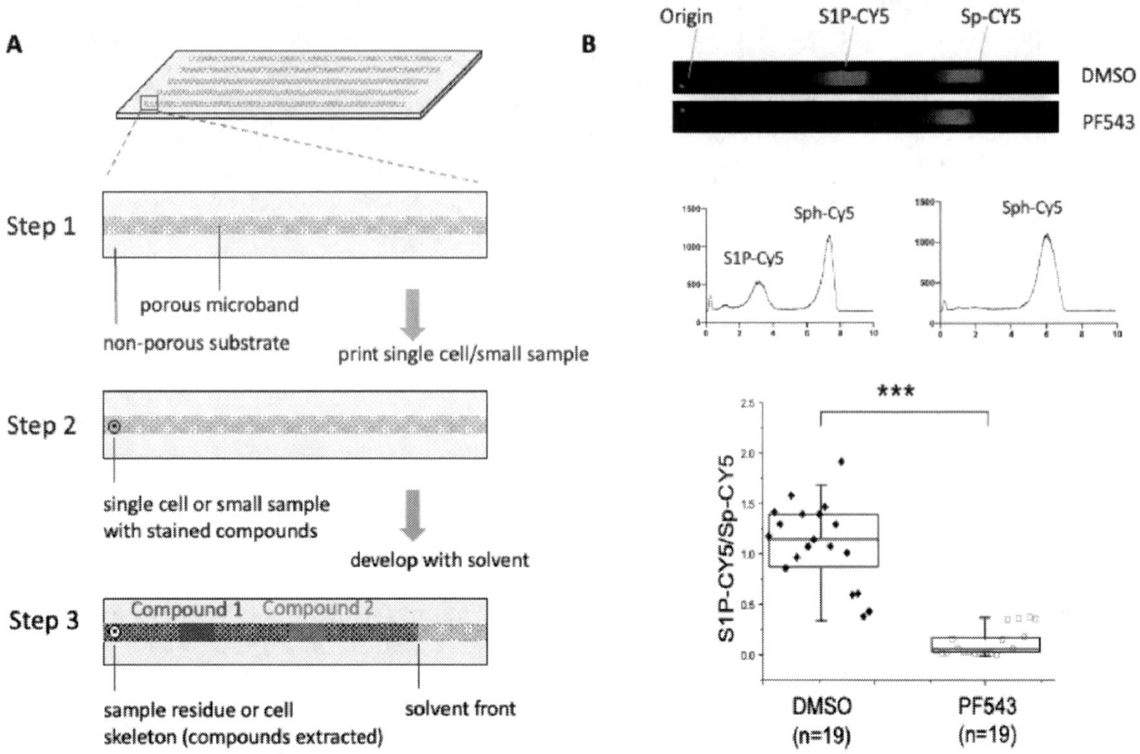
A
Step 1
porous microband
non-porous substrate
print single cell/small sample
Step 2
single cell or small sample
with stained compounds
develop with solvent
Step 3
Compound 1
Compound 2
sample residue or cell
skeleton (compounds extracted)
solvent front
B
Origin
S1P-CY5
Sp-CY5
DMSO
PF543
S1P-Cy5
Sph-Cy5
Sph-Cy5
***
S1P-CY5/Sp-CY5
DMSO
(n=19)
PF543
(n=19)

279a Non-Equilibrium Kink Density Model for Organic Molecular Crystals Containing Two Growth Units

Neha Padwal, Michael F. Doherty

In layered regimes, crystals grow via lateral progression of steps flowing across a planar surface at a constant step velocity. Under supersaturated conditions, growth units (e.g., molecule, dimer) mainly integrate into specific sites called kink sites along a step or edge. This is attributed to the fact that kink sites regenerate themselves and attachment occurs with no surface energy penalty. Hence, the step velocity depends on (1) the rate of attachment at kink sites and (2) density of kink sites. Kink density is often approximated by its equilibrium value given by the Boltzmann distribution. However, it is known that kink density depends on supersaturation and quickly deviates from the equilibrium approximation.[1] Several models have been proposed in the literature for non-equilibrium kink density calculations.[2,3] However, these approaches have been built for the Kossel crystal and do not readily extend to any realistic molecular crystal systems. Another modeling endeavor is the Simplified Steady-State Framework (SSSF) proposed by Padwal and Doherty[1], which provides non-equilibrium kink density estimation by accounting for the surface kinetics of the most-likely surface events. Such a framework differs from other approaches in that it is simple enough to be extended to more complex molecules.

In this work, SSSF is extended to the case of asymmetric organic molecular crystals with two growth units i.e., Z=2, Z'=1 or Z=2, Z'=2. Such a study is the first of its kind attempted for asymmetric molecules. The two growth units, A and B, are identical in the solution phase and integrate into the crystal lattice in different orientations. An AB crystal system comprises of faces with different growth unit patterns and constitute edges with different row configurations. Different edges consist of different types of kinks differing in densities and rates of detachment. Different kinks are formed and destroyed via different surface processes along the edge. Based on the SSSF, steady-state master equations are constructed for the most stable kink types accounting for the more-likely events that form or destroy these kinks. This is implemented using a rate model for the physical reaction of growth unit attachment/detachment and a stochastic approach for modeling densities of various types of sites along the edge. The steady-state master equations constructed are then solved numerically to yield non-equilibrium kink densities. Step velocity is then estimated as a function of non-equilibrium densities followed by crystal growth rate estimates to yield morphology predictions.[4] Inputs required are growth environment conditions such as temperature, supersaturation, solvent information; and an appropriate atom-atom forcefield. Such an analysis is applied to real organic molecular crystal systems to demonstrate its applicability to morphology prediction and control. This will serve as a blueprint to solve the problem of non-equilibrium kink density prediction in systems with more than two growth units, thereby marking the first step to accurate growth models for complex organic molecules with any number of growth units.

References:

1. Padwal, N. A.; Doherty, M. F. A Simple Accurate Non-equilibrium Step Velocity Model for Crystal Growth of Symmetric Organic Molecules. *Crystal Growth and Design* (under review).
2. Cuppen, H. M.; Meekes, H.; van Veenendaal, E.; van Enckevort, W. J. P.; Bennema, P.; Reedijk, M. F.; Arsic, J.; Vlieg, E. Kink Density and Propagation Velocity of the [010] Step on the Kossel (100) Surface. *Surface Science* **2002**, *506* (3), 183–195. https://doi.org/10.1016/S0039-6028(02)01427-9.

3. Joswiak, M. N.; Peters, B.; Doherty, M. F. Nonequilibrium Kink Density from One-Dimensional Nucleation for Step Velocity Predictions. *Crystal Growth andamp; Design* **2018**, *18* (2), 723–727. https://doi.org/10.1021/acs.cgd.7b01092.
4. Li, J.; Tilbury, C. J.; Kim, S. H.; Doherty, M. F. A Design Aid for Crystal Growth Engineering. *Progress in Materials Science* **2016**, *82*, 1–38. https://doi.org/10.1016/j.pmatsci.2016.03.003.

279b Growth Mechanism of Penta-Twinned Ag/Cu Nanowires: Multiscale Theory

Jianming Cui, Junseok Kim, Kristen Fichthorn

Penta-twinned metal nanowires are finding applications in many different technologies, ranging from health monitoring and human-machine interfaces to smart fabrics to windows that change color or tint in response to light, provide shielding from electromagnetic interference, and something about heating. These nanowires find applications in catalysis, photothermal desalination, transparent and flexible electronics applications, such as touch screens, and various elements of solar cells.

The penta-twinned nanowires can be produced with high aspect ratios from diverse face-centered cubic (fcc) metals, with {100} as sides and {111} as ends. However, the mechanism of their anisotropic growth is still poorly understood. We develop a model combining deposition and surface diffusion to predict the final aspect ratio of Ag and Cu nanowires. We describe two aspects of nanowire growth, denoted as the seed stage and the wire stage. In the seed stage, the deposition is much slower than the surface diffusion, so that the seed shape only depends on the difference between interfacet diffusion rates. In the wire stage, the nanowires continue to grow longer until the diffusion is slower than deposition.

The structure of nanowire seeds is similar to Marks decahedra. In the uncapped Ag case, the diffusion from {100} to {111} is comparable to the opposite and both inter-facet diffusion processes are faster than deposition in seed stage. The result shows that the uncapped Ag seeds could lead to three kinds of products with different strucutures as nanowires, decahedron particles and giant decahedron seeds. For the nanowire products, our predictions well match the experimental results. In the uncapped Cu case, we found out that the diffusion from {100} to {111} is slower than the opposite, which impies that the seeds would prefer to grow to large decahedron particles. In experiments, the Cu nanowires were always synthesized with some surfactants. In the case of Cu seeds with Cl on the surface, the relationship between the inter-facet diffusion processes are inversed. The diffusion from {100} to {111} is the fastest and the diffussion from {111} to {100} is even slower than the deposition, which means that the Cu seeds with Cl would prefer to grow to wire products. Based on our model, the predictions are able to represent the average sizes from experimental results.

279d A Mathematical Model to Describe the Oriented Attachment of Molecules in Crystal Growth.

Ravi Kumar Reddy Addula, Baron Peters

In the classical view, the attachment of a single molecule/atom to the nucleus leads to crystal growth. However, non-classical mechanisms are also common in crystallization literature for the past two to three decades. Irrespective of the nucleation mechanism, solute molecules first diffuse from bulk to the nucleus surface and orient themselves to be part of the growing crystal. The orientational ordering of molecules is crucial for crystal growth, for example, maintaining a hydrogen-bonded network and crystal polymorphism. Oriented attachment of molecules plays a vital role in deciding the outcome of the crystallization process. Overall, we see this process as diffusion-limited. Hence, we present a mathematical model to study the oriented attachment of disc-shaped molecules near the absorbing surface. The proposed model is rigorous and simple to describe the oriented attachment of molecules to crystal growth. This model will also help study of biomolecules' interaction with the target surfaces, which requires many direction-specific interactions.

279e Influence of Transient Nucleation Phenomena on Induction Time at Low Supersaturation

Venkateswarlu Bhamidi, Brendan P. Abolins

Experimental induction time distribution (ITD) data are usually analyzed through a survival probability analysis that assumes steady-state kinetics to be influencing the nucleation process. This analysis procedure cannot explain the 'apparent dead time in nucleation' routinely observed in experiments at low supersaturations. Under these conditions, the cumulative probability distributions of induction times exhibit a 'sigmoidal' shape rather than a sharp rise characteristic of an exponential decay of survival probability. To account for this uncharacteristic shape of ITD curves, a 'crystal growth/detection time' is often invoked. However, attempts to fit experimental data to this 'corrected process time' typically result in dubious and unphysical values for these detection/growth times. Thus, a gap exists in traditional methods used to analyze ITD data.

In this work, by considering transient nucleation phenomena, we show that the sigmoidal shape of the ITD frequently observed at low supersaturations has its origins in the time delay involved in 'equilibrating' the system to the imposed supersaturation. This time lag is significant at low supersaturations, thus resulting in a time-varying process intensity. We revisit experimental ITD data sets published in the literature and show that the proposed approach that considers transient nucleation kinetics (instead of steady-state kinetics) can explain these ITD. Our approach does not require one to assume (often trivial) growth or detection times to explain the 'dead time' of nucleation observed in ITD at low supersaturations. This analysis method also provides one with information on the transient processes involved in nucleation mechanisms and paves the way for further understanding the metastable behavior of crystallizing solutions.

646b Crystallization Kinetics and Solubility of Amoxicillin Trihydrate in the Presence of Synthesis Substrates

Patrick Harris, Hossein Salami, Ronald Rousseau, Martha Grover, Andreas Bommarius

Amoxicillin is the most consumed antibiotic in the world in terms of doses and consequently vital to global human health. Originally, the manufacturing of amoxicillin was conducted using a multi-step chemical synthetic pathway, consisting of protecting and deprotecting groups, cryogenic refrigeration units, and hazardous solvents. More recently, the use of biocatalytic processing has been investigated; the process consists of a single synthesis step and can be conducted under mild aqueous conditions, resulting in a greener process. The main drawbacks to biocatalytic amoxicillin production are the unwanted hydrolysis catalyzed by the biocatalyst. Improvements to the base biocatalytic process have been achieved by facilitating the reaction and crystallization of similar antibiotics in the same vessel, termed reactive crystallization[1]. Due to the complex nature of reactive crystallization, models describing the system are valuable in predicting optimum operating conditions, feed concentrations, residence times, reactor configurations, and other process considerations. Multi-module reactive crystallization models have been developed for similar systems, such as with cephalexin monohydrate[2]. On the other hand, neither the crystallization kinetics of amoxicillin trihydrate nor the effect of other reaction components on the solubility and crystallization kinetics of amoxicillin trihydrate also have been reported. It has been revealed that the degradation products of amoxicillin inhibit its crystallization. Also, it is known that reaction components in similar systems interact, causing inhibition of crystallization as well as changes in solubility, both phenomena greatly impact the outcomes of the process[3]. Accordingly, a comprehensive model of the reactive crystallization of amoxicillin trihydrate would only be feasible with improved knowledge of the phenomena occurring between the solid and liquid phase.

In this work, we investigated the solubility and crystallization of amoxicillin trihydrate with and without the presence of various reaction components, such as 4-hydroxyphenylglycine (4-HPG), 6-aminopenicillinoic acid (6-APA), and 4-hydroxyphenylglycine methyl ester (4-HPGME). The effects of pH, temperature, ionic strength, and presence of each component on the solubility of amoxicillin were determined. The solubility was found to increase with pH value, temperature, and ionic strength, and the three reaction components increased amoxicillin solubility, with 6-APA having the largest impact. However, ionic strength alone could not explain the increase in solubility in presence of the other reaction components, and there were possibly interactions occurring between the molecules of each reaction component and amoxicillin in the liquid phase. To confirm such interactions, single components and mixtures were analyzed via Raman and ATR-FTIR spectroscopy to explore possible interactions between the functional groups present in each compound and amoxicillin. The knowledge of amoxicillin solubility was important in determining its supersaturation under various conditions, which is the main driving force for crystallization.

We also conducted several seeded and unseeded batch amoxicillin crystallizations with and without the presence of 6-APA or 4-HPGME. We used offline microscopy and focused-beam reflectance measurement (FBRM) to monitor the solid phase, with high performance liquid chromatography (HPLC) to monitor the liquid phase. Solid and liquid phase data were fit to classical nucleation and growth models. Differences in nucleation and growth parameters with pure amoxicillin and in the presence of 6-APA or 4-HPGME were examined. A model considering the effects of 6-APA and 4-HPGME on the crystallization of amoxicillin hydrate was developed and may be incorporated into a reactive crystallization model to inform future manufacturing processes for the more efficient production of amoxicillin.

1. Encarnación-Gómez, L.G., A.S. Bommarius, and R.W. Rousseau, *Reactive crystallization of β-lactam antibiotics: strategies to enhance productivity and purity of ampicillin.* Reaction Chemistry andamp; Engineering, 2016. **1**(3): p. 321-329.
2. McDonald, M.A., et al., *Continuous reactive crystallization of β-lactam antibiotics catalyzed by penicillin G acylase. Part I: Model development.* Computers andamp; Chemical Engineering, 2019.
3. Fan, Y., Y. Li, and Q. Liu, *Enhanced Dissolution of 7-ADCA in the Presence of PGME for Enzymatic Synthesis of Cephalexin.* Applied Biochemistry and Biotechnology, 2021.

646c Application of Silane-Modified Silica of Different Morphologies Towards the Demulsification of Water in Oil Emulsions

Anirban Ghosh, Michael Miranda, Clint Aichele

The separation of oil and water poses challenges to several major industries. Industries that deal with water purification and petroleum refining routinely deal with emulsions that are difficult to separate into individual phases. The presence of natural emulsifiers in crude oils and wastewater further complicates the separation of the phases. There is evidence that silica nanoparticles can be potential demulsifiers for emulsions. This study focuses on two parameters, wettability and physical structure, that influence the ability of silica nanoparticles in their separation capacity. Fumed silica and spherical silica of two different geometries were used for the investigation. Wettability alterations were brought about by silane modifications of the silica nanoparticles. Silane chains of various lengths can be impregnated to influence the wettability of a silica particle. Silane chains are inherently hydrophobic by nature while silica is hydrophilic by nature. A longer silane chain impregnation corresponds to a decreased hydrophilicity. Hydrophilicity was measured with respect to the percentage of hydroxyl groups present in the matrix of silica particles. Contact angle measurements were made to verify the hydrophilicity of the produced silica particles. Demulsification experiments were conducted for water in oil (W/O) emulsions using the silica nanoparticles produced. Microscopic analysis was done on the emulsions before and after demulsification experiments to understand the coalescing behaviour of the emulsions in the presence of silica nanoparticles. The presentation will highlight the influence of physical structure and the wettability of demulsifiers on demulsification efficiency.

646d Segmented Flow Crystallization of Paracetamol Via Microdroplet Cooling

Jacob Crislip, Januario da Costa, Andrew R Teixeira

Efficiently controlling particle size, quality, and polymorphism challenges every branch of crystallization. Traditional top-down synthesis methods rely on cumbersome, highly sensitive processes resulting in wasted material, excessive solvents, and numerous unit operations to achieve the desired particle size distribution. Alternatively, continuous bottom-up crystallization is a precise approach to generate high-quality monodisperse products as each microdroplet experiences identical crystallization environments.

Segmented flow microfluidics offers a high throughput continuous platform to study crystal nucleation, growth, and dissolution. In situ UV-vis spectroscopy captures the liquid phase changes that occur throughout the phase change, while microscopy and laser diffraction highlight the quality of the solid products formed. Nucleation rate can be tuned according to ultrasonication energy input into the system spanning the fast generation of fines (sonics) to the slow formation of large particles (silent).

This work details the design, construction, and implementation of a continuous segmented flow crystallizer. Since solute solubility tends to be highly temperature-dependent, operating at low Reynolds and Biot numbers facilitates the isothermal step changes required to quickly span the phase diagram of cooling crystallization. Microdroplet flow crystallization is a scalable process that opens the door to directly controlling the formation of crystals from nucleation to harvesting.

646e Impurity Inclusion and Crystallization Performance in Batch and Continuous Mode

Christine Darmali, Nima Yazdanpanah, Shahnaz Mansouri, Zoltan Nagy, Meng Wai Woo

Product quality, such as purity and particle size distribution (PSD), is one of the most important process metrics. These are also defined as some of the critical quality attributes (CQAs) for formulated products in the pharmaceutical and food industries. Crystallization process development, at the Randamp;D stage, is usually carried out by utilizing pure solutions, when the emphasis is on the process development, yield optimization, and growth kinetic parameters estimation. However, in manufacturing and real-life scenarios, there are many different impurities present in the mother liquor, which impact the performance of the crystallizer, and product quality and other CQAs. In addition, the isolation step includes crystallization, filtration and drying stages, in which the filtration and drying also impact the CQAs of the product.

In this research, impurity inclusion, crystallization and filtration performance were studied for acid whey solution in batch and continuous modes. Three different impurities, and different mechanisms of impurity inclusion were studied. Yield, distribution coefficient, PSD, process performance, and process stability were evaluated and compared for batch and continuous processes in pure and impure solutions.

646a Simultaneous Lattice Incorporation and Co-Precipitation of Impurities in Solution Crystallization

Fredrik Nordstrom, Qi Jiang, Tommasina Bramante, Mitchell Paolello, Joseph Kratz, James Geier, Michael Toresco, Kayla Bensley, Gerard Capellades

Incorporation of impurities in a crystallizing suspension is a common challenge during crystallization of small organic molecules. Process development methods tend to characterize the product's purity as a single metric, representing an average value that does not inform about a specific incorporation mechanism. However, multiple impurity incorporation mechanisms can simultaneously occur during crystallization, sometimes for the same impurity, and knowledge of those mechanisms is the first step for the design of preventive and corrective actions.

In this talk, we will discuss two academic and industrial case studies presenting simultaneous lattice incorporation and impurity co-precipitation in solution crystallization. The focus is first on the diagnosis of the incorporation mechanisms, and how the contributions of each mechanism on the average powder purity were quantified using widely available technologies. The second half of the talk will focus on preventive and corrective strategies, rooted on the acquired mechanistic knowledge, for the rejection of those impurities during and after crystallization. The cost, time and resource benefits of adopting an effective diagnosis are contrasted with current standard practices in the pharmaceutical and chemical industries.

341b Nonclassical Mechanisms to Irreversibly Suppress β-Hematin Crystal Growth

Peter Vekilov, Wenchuan Ma, Victoria Balta, David Sullivan, Jeffrey Rimer

Intricate crystal architectures in nature evolve shepherded by minority solution components; industrial crystals are guided to preferred morphologies by specific modifiers. The modifiers' activities are commonly ascribed to their adsorption to specific crystal surface sites and the reversibility of adsorption inevitably predicts that growth fully recovers after the inhibitor is removed. Multiple instances of permanently poisoned crystals and terminal crystal sizes stand out of the realm of the classical inhibition mechanisms and remain elusive. Here we demonstrate irreversible inhibition of hematin crystallization, an essential element of heme detoxification of malaria parasites, due to cooperative behaviors that activate at high crystallization driving forces. The evolutions of the crystal shapes after limited-time exposure to both artemisinin- and quinoline-class antimalarials report that crystal growth remains suppressed after the drugs are purged from the solution. Treating malaria parasites with the same agents reveals that three- and six-hour drug pulses kill with efficacy comparable to that of drug exposure during the entire parasite lifetime. Time-resolved *in situ* atomic force microscopy (AFM), complemented by light scattering, reveals two molecular-level mechanisms of inhibitor action that operate at elevated hematin concentrations and prevent β-hematin growth recovery. The hematin-liganded artemisinins incite copious nucleation of nanocrystals, which invade larger growing crystals, whereas pyronaridine, a quinoline-class drug, promotes step bunches, which evolve to engender abundant dislocations. Both trapped crystals and dislocations are known to induce lattice strain, which persists and permanently impedes crystal growth. The two mechanisms of irreversible suppression of crystal growth rely on cooperativity of molecules, in crystal nucleation, or between mesoscopic objects, the steps, driven by high deviations from equilibrium. These and other cooperative behaviors can be amplified or curtailed as means to control crystal sizes, size distributions, aspect ratios, and other properties essential for the numerous fields that rely on crystalline materials.

148a Crystal Morphology Prediction: An Alternative Framework Via Kmc

Tobias Mazal, Michael F. Doherty

Accurate and rapid modelling of crystal morphologies is essential for various industrial applications, ranging from catalysis to manufacture of pharmaceuticals, OLEDs, and many others. Morphology prediction requires precise characterization of the growth conditions for determining the relative growth rates of the crystals faces and the ensuing morphology. This work strives to inform real-time morphology predictions for molecules of interest. To do so, we analyze growth rate predictions following step velocity model theory as well as those generated via Kinetic Monte Carlo (KMC) simulations and use existing visualization frameworks designed within the Doherty group to compare the predicted crystal morphologies.

In this work we investigate the potential of KMC simulation techniques to both inform step velocity models for non-ideal systems and to predict accurate crystal morphologies. We first demonstrate the validity of this approach by comparing simulation results to step velocity model predictions for the simpler systems of Kossel and centrosymmetric growth units without impurities present. We then expand to more complex systems including non-centrosymmetric growth units as well as systems containing impurities in solution. Herein, an impurity, or imposter, typically refers to a structurally similar molecule to the solute, or one which alters or inhibits the growth of, the primary crystallizing species. Our approach is demonstrated for real molecules grown from solution, both centrosymmetric (e.g. naphthalene, adipic acid) and non-centrosymmetric (e.g. paracetamol). Our results demonstrate the applicability of informing crystal morphology predictions for molecules of interest, including but not limited to active pharmaceutical ingredients. This work also discusses potential techniques utilized to reduce simulation time such as temporal coarse-graining methodologies and investigates the limitations of such approaches.

371a Guide to COF Adsorbent for Ammonia-based Green Hydrogen with Multi-scale Evaluation Approach

Nahyeon An, Seongbin Ga, Giyeol Lee, Hyungtae Cho, Junghwan Kim

Since green ammonia is holding the global spotlight as a transportation medium of carbon-free H_2, there has been active discussion on various separation technologies for H_2 purification after the NH_3 decomposition [1]. One of the H_2 purification technologies with the highest technology readiness level (TRL) is pressure swing adsorption (PSA) processes, widely used in various industries [2]. In the PSA processes, adsorbents are crucial parts where H_2 and other gases are adsorbed and desorbed selectively. Thus, the most significant improvement can be made by selecting the adsorbents with the best properties [3]. Despite the importance of adsorbent selection, the discovery of best-performing adsorbents has been barely attempted considering the circumstance where the H_2 is provided from NH_3 decomposition.

As the first attempt of the adsorbent discovery for such green H_2 separation, in this study, we explore 648 covalent organic frameworks (COFs) because COFs have high potential as adsorbents in terms of their mechanical and chemical durability [4]. In order for the practical and efficient evaluation of the adsorbents, we propose a multi-scale adsorbent screening approach, which includes both molecular and process scale evaluation. In the molecular scale evaluation, the adsorption isotherm curves of the COFs are calculated based on the grand canonical Monte Carlo (GCMC) simulations. The calculated isotherm curves, in the process scale evaluation, are employed to estimate the H_2 recovery and H_2 purity of the PSA process as performance indicators. Based on the performance indicators, we could provide a practical guide to adsorbent selection and development for green H_2 separation, where the 10 best COFs both for H_2 raffinate and N_2 raffinate cases were suggested.

Moreover, as a part of the multi-scale screening approach, we also propose a computation strategy that reduces the computational cost significantly in GCMC simulations. Based on the initial GCMC simulation results, the computation strategy determines the minimum number of GCMC simulations for adsorbent evaluation. By decreasing the number of simulations to the minimum numbers, the computation strategy could reduce at least 15 % of computational cost for our COF screening cases. For the finally selected COFs based on the performance indicators, the maximum value of normalized root mean square error (NRMSE) of all those adsorbents was smaller than 2.73 %, which barely deteriorates the entire adsorbent evaluation accuracy.

Literature cited:

[1] K.E. Lamb, M.D. Dolan, D.F. Kennedy, Ammonia for hydrogen storage; A review of catalytic ammonia decomposition and hydrogen separation and purification, International Journal of Hydrogen Energy. 44 (2019) 3580–3593. https://doi.org/10.1016/j.ijhydene.2018.12.024.

[2] A.M. Banu, D. Friedrich, S. Brandani, T. Düren, A multiscale study of MOFs as adsorbents in H2 PSA purification, Industrial and Engineering Chemistry Research. 52 (2013) 9946–9957. https://doi.org/10.1021/ie4011035.

[3] J.A. Delgado, V.I. Águeda, M.A. Uguina, P. Brea, C.A. Grande, Comparison and evaluation of agglomerated MOFs in biohydrogen purification by means of pressure swing adsorption (PSA),

Chemical Engineering Journal. 326 (2017) 117–129. https://doi.org/10.1016/j.cej.2017.05.144.

[4] G.O. Aksu, I. Erucar, Z.P. Haslak, S. Keskin, Accelerating discovery of COFs for CO2 capture and H2 purification using structurally guided computational screening, Chemical Engineering Journal. 427 (2022) 131574. https://doi.org/10.1016/j.cej.2021.131574.

371b Amine-Modified Silica Materials for Direct Carbon Dioxide Capture from the Atmosphere

Amirjavad Ahmadian Hosseini, Masoud Jahandar Lashaki

The objective of this research is to develop adsorbent materials for CO_2 capture from ambient air. There is a broad scientific consensus that Greenhouse Gases (GHGs) trap heat in the atmosphere. From 1960 to 2021, the atmospheric concentration of CO_2, the most abundant anthropogenic GHG, increased from 320 to 417 parts per million by volume (ppmv). According to NASA, this increase contributed to the 1.1°C increase in global temperature relative to the late 19th century average. In 2018, the Intergovernmental Panel on Climate Change (IPCC) urged a 45% reduction in CO_2 emissions by 2030, relative to 2010 levels, to limit global warming to 1.5°C. To curb CO_2 emissions, the Blue Map Scenario of the International Energy Agency (IEA) recommends different strategies, including switching to renewable energy sources and implementing CO_2 capture technologies. While most CO_2 capture efforts have been dedicated to large point sources such as fossil fuel power plants, CO_2 capture from the atmosphere, also known as Direct Air Capture (DAC), has been gaining momentum recently. The U.S. Department of Energy recognizes the critical role of DAC in addressing the climate crisis and achieving net-zero emissions by 2050. Owing to their superior performance, DAC applications involving cyclic adsorption-desorption of CO_2 by amine-functionalized silica materials (i.e., "aminosilicas") has attracted tremendous attention from the scientific community and industry alike. However, many relevant studies focus on near-equilibrium CO_2 uptake by aminosilicas, while treating adsorbent stability and adsorption kinetics as secondary factors, or totally disregarding them. While aminosilica stability determines its operational lifetime, fast adsorption kinetics is necessary to increase the amount of CO_2 captured per unit time. This research aims at taking an innovative approach by prioritizing adsorbent stability and fast adsorption kinetics over adsorption uptake, to develop highly efficient and durable aminosilicas for DAC applications. Aminosilicas were synthesized using (i) a commercially available mesoporous silica (G-10, Fuji Silycia), (ii) four polyamines, namely Tetraethylenepentamine (TEPA) and three branched Polyethylenimines (PEI) with different molecular weights (600, 1200, and 1800), and (iii) five levels of amine loadings (10-50 wt.%). The manipulation of synthesis conditions allowed for achieving a delicate balance between maximizing CO_2 uptake, facilitating adsorption kinetics, and maintaining material stability. All materials were evaluated using Thermogravimetric Analysis (TGA) to quantify amine loading, equilibrium CO_2 uptake, and CO_2 adsorption kinetics in the presence of *dry* CO_2 (400 ppmv, balance nitrogen) over a range of temperatures representing typical Florida conditions (e.g., 25-35°C). The performance of selected performant aminosilicas was assessed in the presence of *humid* CO_2. Finally, the long-term performance stability of the best-performing material was investigated over successive adsorption-desorption cycles.

371c Adsorption Mechanism of Ultra-Low Concentration H_2s and THT in CH_4 on Zeolites in the Presence of CO_2

Young-Ho Cho, Gina Bang, Chang-Ha Lee

Commercially available natural gas contains certain ppm of sulfur compounds such as H_2S, THT, and TBM. 5-15 ppm of sulfur content in natural gas is acceptable as fuel, while less than 1 ppm of sulfur level is required for the applications of the fuel cell. Sulfur compounds play a critical role in natural gas reforming catalysts as well as WGSR catalysts. In addition, a trace amount of sulfur compounds highly deteriorate the lifetime of cyclic adsorptive processes. The removal of high-concentration H_2S with/without other impurities in natural gas has been studied intensively. On the other hand, although the adsorption mechanism at a trace concentration from a few ppm to ppb level is practically important, few studies are conducted.

In this study, the breakthrough experiments in the bed packed with zeolites 5A and 13X pellet are carried out using a ppb level detectable analyzer at 4 ppm of H_2S or THT balanced with CH_4. In addition, since natural gas also contains a certain level of CO_2, the breakthrough experiments using the ternary mixture including CO_2 are also conducted to elucidate the effects of another impurity on the removal of sulfur compounds from natural gas. The results reveal the influence of CO_2 on sulfur removal in the ultra-low concentration region and clearly show the different adsorption behavior between two adsorbates.

371d A New Dynamic Vapor Sorption Instrument for the Investigation of Water Sorption on Porous Materials

Thomas Paschke, Frieder Dreisbach, Volodymyr Bon

Introduction

The new Discovery SA Dynamic Vapor Sorption Analyzer by TA Instruments has been designed for the reliable routine analysis of water sorption properties of a broad variety of materials in application areas including food, pharma, polymer, electronics and building materials. It can also be used to study the water sorption capacity and kinetics of porous solids including carbon materials, zeolites or MOFs with high resolution. A special feature is the possibility to measure the adsorption of humid gas mixtures such as air or variable concentrations of carbon dioxide.

Experimental

The Discovery SA instrument includes a temperature controlled symmetric microbalance, a humidity control and an autosampler. Precise humidity control is provided by two mass flow controllers that deliver one dry and one completely saturated nitrogen gas stream that can be mixed to obtain a measuring atmosphere of 0 – 98% RH. The adjustment of exact and correct humidity levels at the position of the sample can easily be verified by measuring the deliquescence of various salts according to ASTM E2551. A gas blending module can be used to provide additional reaction gases other than nitrogen in controlled concentrations.

As a first example, vapor sorption was measured on a commercial Norit activated carbon material at 15, 25 and 35°C in a range from 0 to 90% RH. While the adsorption of water is not significantly affected by the temperature change, the desorption isotherms show a larger hysteresis loop at lower temperatures that vanishes completely below 30% RH.

In another measurement, a nanoporous carbon material consisting of micro beads of 250 to 500 μm diameter was analyzed at 25°C showing adsorption and desorption of vapor only between 50% and 70% RH which indicates a quite narrow pore size distribution.

Finally, a zirconium-based DUT-67 MOF supplied by the TU Dresden was measured as example for more polar adsorbents. Because the water sorption characteristics of this DUT-67 MOF are well described in the literature [1], it is a perfect material to document the performance of the Discovery SA Dynamic Vapor Sorption instrument.

Conclusion

With three different polar and non-polar porous materials, the excellent performance of the new Dynamic Vapor Sorption instrument could be documented. Upcoming next are studies with gas mixtures of nitrogen, air and carbon dioxide at varying RH levels.

References

[1] V. Bon, I. Senkovska, J.D. Evans, M. Wöllner, M. Hölzel and S. Kaskel, Insights into the water adsorption mechanism in the chemically stable zirconium-based MOF DUT-67 – a prospective material for adsorption-driven heat transformations, J. Mater. Chem. A, 2019, 7, 12681

Fig. 1 Vapor adsorption and desorption isotherms at 15, 25 and 35°C on Norit activated carbon material

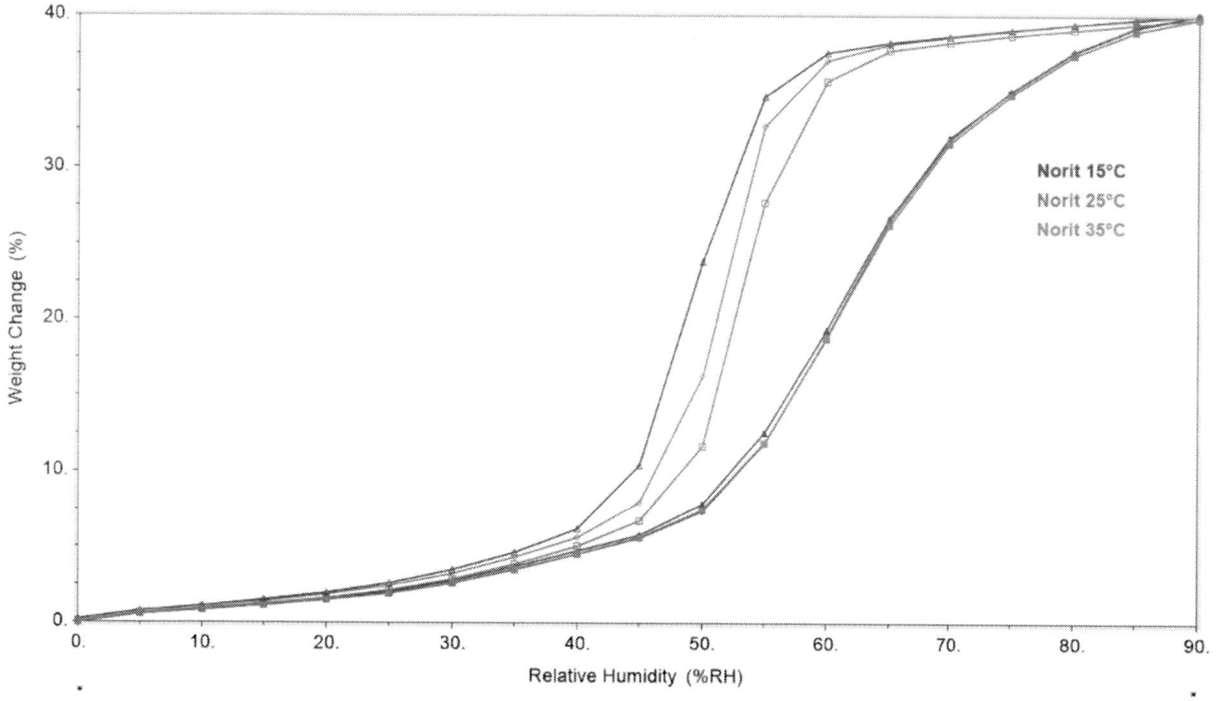

Weight Change (%)
Relative Humidity (%RH)
Norit 15°C
Norit 25°C
Norit 35°C
0.
10.
20.
30.
40.
50.
60.
70.
80.
90.

371e Adsorption Equilibria and Kinetics of CO_2, CO, CH_4, N_2, O_2, and H_2 on Silica-Based Adsorbents

Jiwon Jung, Hongjoo Do, Kyounghee Chung, Jun-Ho Kang, Moonkyung Cho, Chang-Ha Lee

Although various effluent gases have low heating values, they are used as heat sources and generate a large amount of CO_2 in industries. Therefore, the demands for carbon dioxide utilization (CCU) is rapidly increasing as a way for carbon mitigation. Especially, since an iron and steel industry, one of the highest carbon-intensive industries, accounting for 7-9 % of the worldwide CO_2 emissions, can supply various heat sources in processes, CCU applications are more suitable than other industries.

Adsorption technology is promising for CO_2 capture, but its efficiency is critically affected by a desorption energy. However, with respect to CCU in an iron and steel industry, high purity CO_2 capture does not require but CO_2 has to be concentrated to a desirable level from the effluent gases of various processes. Therefore, a weak adsorbent can be applicable for cyclic adsorption processes, considering energy-efficiency at a desorption step.

In this study, the adsorption behaviors of six components (CO_2, CO, CH_4, N_2, O_2, and H_2) on silica-based adsorbents are studied because these components are generally found in effluent gases from iron and steel industries. The adsorption isotherms of each pure gas are measured via a volumetric method at 293, 308, 323 K and pressure up to 1000 kPa. The experimental isotherms are well fitted with the Dual-Site Langmuir model and temperature-dependent Sips model. The isosteric heats of adsorption are calculated with the Clausius-Clapeyron equation. The adsorption kinetics of each component are also analyzed by a non-isothermal diffusional model. The results can contribute to designing efficient adsorption processes to concentrate CO_2 from the effluent gases in iron and steel industries.

371f Performance Evaluation and Comparison of Conventional and Emerging Adsorbents in Their Applications to a CO_2 Capture Vacuum Pressure Swing Adsorption (VPSA) Process

Yixuan Zhang, Hyungwoong Ahn

It was aimed to perform a comprehensive study to evaluate a variety of conventional and emerging porous materials having in mind their applications to a simple Vacuum Pressure Swing Adsorption (VPSA) process for CO_2 capture from a simulated flue gas consisting of 15% CO_2 and 85% N_2.

The CO_2 capture VSA runs with an adsorption cycle of five steps comprising high-pressure feed, heavy reflux, counter-current depressurization (blowdown), low-pressure purge and light product pressurization steps.

Each adsorbent candidate was assessed with respect to CO_2 product recovery that is achievable by the simple CO_2 VPSA process where the adsorption columns are packed with the adsorbent candidate. The performance of each VSA cycle was analysed by equilibrium theory that requires adsorption isotherms, bed void fraction and adsorbent density only [1].

1. Ruthven et al., Pressure Swing Adsorption, VCH, 1993.

371g Adsorbents for Selective Adsorption of VOC Biomarkers from Simulated Breath

Ojuolape Oghenetega, Krista Walton

The detection and quantification of volatile organic compounds (VOCs) biomarkers from exhaled human breath is fast becoming an acceptable technique for easy and rapid testing for several illnesses. To identify and quantify these VOC biomarkers, selective sorbent materials are of major importance. Sorbents with the desired selectivity could lead to the development of inexpensive, energy efficient and quick diagnosis of diseases using adsorption and desorption mechanism to detect and quantify these VOCs. In this work, several adsorbent materials, including activated carbon, zeolites, and metal organic frameworks (MOFs), have been tested for their VOC capacities to acquire an in-depth understanding of VOC adsorption mechanisms on these materials. The influence of VOC characteristics and properties such as polarity, boiling point, molecular diameter, etc., on their interactions with adsorbents was studied. It was observed that the adsorption loadings increased with increase in pore volume and total surface area, implying that pore filling is the prevalent adsorption mechanism. Nitrogen sites were introduced to adsorbents via amine impregnation of carbon samples and synthesis of MOFs using functionalized ligands. Adsorption loadings in functionalized adsorbents were lower than unfunctionalized adsorbents. Moreover, functionalized MIL-101 exhibited slower adsorption dynamics than the unfunctionalized MOF. CO_2 and water vapor breakthrough adsorption experiments were also carried out on the functionalized and unfunctionalized materials. Results showed both water and CO_2 breakthrough times on carbon samples with nitrogen sites increased significantly in humid streams. Therefore, this may be applied to selectively adsorb two major components of human exhaled breath prior to reaching the sorbent bed for VOC separation in a layered bed system.

371h Epoxide-Modified Amine-Based CO_2 Adsorbent for Direct Air Capture

Yao Ma, Joo-Youp Lee

A large portion of CO2 in ~5.2 gigatons (Gt) is released in relatively small quantities from distributed sources emitted each year in the U.S. Therefore, for such emissions, point source CO2 capture is not feasible and direct air capture is an indispensable part of a diversified portfolio of technologies to mitigate U.S. greenhouse gas emissions. Successful direct air capture (DAC) technologies are required to separate high purity CO_2 (e.g., >95%) with high selectivity toward CO_2, low regeneration energy requirements, minimum chemical and thermal degradation, reliability, long lifetime, etc.

In this study, epoxide-modified amino polymer-based sorbent has been studied for DAC potential. The sorbent performances were examined using epoxide-modified amino polymer-loaded silica substrates. The sorbent showed the high adsorption capacities of ~1.8-2 mmol CO_2/g of sorbent from 400 ppm CO_2 in ambient air between 25-40 °C with fast adsorption kinetics. It showed robust performances with minimum oxidative and thermal degradations over >1,000 adsorption and desorption cycles. The sorbent could also be regenerated at <100 °C that can potentially reduce the CO_2 desorption thermal energy requirement with <100 kJ/gmol CO_2.

Boosting Volumetric and Gravimetric H2 Storage Capacity of Carbon-Based Sorbents Through Tuning Surface Chemistry and Densification Approaches

Ruthra Murugavel, Fateme Rezaei

With increasing energy demand from growing population, hydrogen is considered as one of the most sustainable energy carriers for the future. Significant advancements have been made over the past few years to produce hydrogen in a sustainable manner, however, the major bottleneck currently facing the hydrogen economy lies primarily in its transportation and storage.(DOE, 2020) Adsorbed gas storage (AGS) technology offers a much safer and more economic storage option than existing cryogenic cooling or compression methods for hydrogen storage and transport. However, the current sorbents examined as H2-AGS candidates still suffer from low deliverable capacity. Therefore, to enhance the hydrogen density per volume of storage tank and to meet the DOE's set targets, extensive research on nano-porous materials and storage conditions is required. In this research, we focused our attention on exceptionally high-surface area activated carbon and graphitic carbon nanosheets to further enhance the binding energy and adsorption capacity over a pressure range of 0-100 bar and varied charge-discharge temperatures via (i) surface functionalization with light–metal dopants such as Li, Na and K, and (ii) carbon exfoliation with ionic liquids, and (ii) sorbent densification approaches. Our results indicated the effectiveness of these approaches to enhance both gravimetric and volumetric hydrogen capacity relative to the pristine sorbents.

371j Porous Carbon from Non-Recyclable Plastic Wastes

Dipendu Saha, Brian Hoffman, Joanna Weyrich, Garret Gallo

Styrofoam is the brand name forextruded polystyrene, which is one of the least recyclable plastics used all over the world. Styrofoam is used in many items, for example, packing materials, single-use cups, plates, glass, or food containers. It has the least recycling value, most of it is ultimately ended up in landfill waste. In this research, Styrofoam is converted to porous carbon after chemical modification of the Styrofoam. In the course of the research, it was observed that pristine Styrofoam has no char yield, however, prior chemical modification of Styrofoam yields a significant amount of char that can be converted to porous carbon in the subsequent steps by physical or chemical modification. In this research, different reaction parameters of Styrofoam modification and synthesis of porous carbon were investigated. The resultant porous carbon was characterized by pore textural properties, including BET surface area and pore size distribution, surface functionalities by FTIR and x-ray photoelectron spectroscopy (XPS), and shape and size by electron microscopy. The porous carbon was utilized for removing various pollutants from water.

371k Direct Ink 3D Printing of Porous Carbons

Marisa Comroe, Dipendu Saha

In modern society, porous carbon has myriad applications in numerous sectors. Currently, the majority of porous carbons are available only in powdered or briquet form which may limit their application and performance. On the other hand, 3D printing is an additive manufacturing technique that can fabricate a myriad of high-resolution and three-dimensional (3D) structures, which can be designed in a computer program. 3D printing is capable of providing excellent control over both the microstructure and component distribution of the target component in the final body thereby offering an excellent industrial production technique. The 3D printing strategy of porous carbon will create a new paradigm that may provide an unprecedented ability to control the shape, structure, and composition of porous carbon-based materials that can ultimately influence their usage. The 3D printed form of porous carbon is very limited, the majority of such forms of carbon utilized very specific polymer for lithography printing or hard-templated mesoporous carbon. In this research, commercially available activated carbon was utilized as the base material, along with resorcinol-formaldehyde/surfactant and bentonite as binding agents. Upon 3D printing, the monolithic structure was pyrolyzed to yield the final carbon that possesses the bimodal distribution of microporosity and mesoporosity. The resultant 3D printed carbon was used in different gas- and liquid-phase adsorption applications.

371l Diffusion Studies of Hydrocarbons in Ill-Crystallized or Desilicated ZSM-5 Catalysts

João Victor S. Cardoso, Laura Silva, Leandro Martins, Francisco Murilo Tavares de Luna, Celio Cavalcante Jr.

The physical and chemical properties of zeolites may be tuned by modification of their synthesis (bottom-up methods) or post-synthesis procedures (top-down). These may involve embryonic zeolites, partially formed or ill-crystallized structures, and desilication of consolidated zeolites by consuming part of the zeolitic structure in an alkaline solution. These modifications would form secondary porosity, mainly for applications as catalysts in reactions that form deactivating coke, as usually seen in the conversion of methanol to hydrocarbons (MTH), for example.

In this study amorphous, partially crystallized, and fully-grown ZSM-5 zeolites were prepared and characterized. Also, fully crystallized ZSM-5 zeolite samples were desilicated and acid leached to create different structural and morphological properties and correlate them with the accessibility to catalytic sites and their effect on catalyst deactivation. The diffusional behavior of hydrocarbons in these samples was evaluated using the Zero-Length Column (ZLC) method and compared to catalytic results in model reactions such as MTH.

371m Understanding Fluid Phase Behavior in Geometrically Disordered Mesoporous Materials

Henry R. N. B. Enninful

Mesoporous solids exhibit structural disorder which strongly influences confined fluid properties. This renders quantification of structural disorder and its correlation with physical properties of confined matter a necessary step towards their optimization in practical applications.

In this work, we present advances made in the understanding of correlations between the phase state and geometric disorder in nanoporous solids. We overview the recently developed statistical theory for phase transitions in a minimalistic model of disordered pore networks represented by the linear chains of pores with statistical disorder. Furthermore, we show that correlating its predictions with various experimental observations, the model gives notable insight into collective phenomena in phase-transition processes in disordered materials and is capable of explaining self-consistently the majority of the experimental results obtained for gas–liquid and solid–liquid equilibria in mesoporous solids. We also show how a newly-introduced interconnectivity parameter of the pore network can be assessed to describe the morphology of porous solids.

References:
[1] Schneider D.; Kondrashova D.; Valiullin R., 2017, "Phase transitions in disordered mesoporous solids", Scientific Reports, 7, 7216.
[2] Schneider, D. and Valiullin R., 2019, "Capillary condensation and evaporation in irregular channels: Sorption isotherm for serially connected pore model", Journal of Physical Chemistry C, 123, 16239.
[3] Enninful H.R.N.B., Schneider D., Hoppe A., König S., Fröba M., Enke D. and Valiullin R., 2019, "Comparative gas sorption and cryoporometry study of mesoporous glass structure: Application of the serially connected pore model", Frontiers in Chemistry, doi: 10.3389/fchem.2019.00230.
[4] Enninful H.R.N.B., Schneider D., Kohns R., Enke D. and Valiullin R., 2020, "A novel approach for advanced thermoporometry characterization of mesoporous solids: Transition kernels and the serially connected pore model", Microporous and Mesoporous Materials 309, 110534.
[5] Enninful H.R.N.B., Schneider D., Enke D. and Valiullin R., 2021, "Impact of Geometrical Disorder on Phase Equilibria of Fluids and Solids Confined in Mesoporous Materials", Langmuir, 37, 3521-3537.

371r Boosting Volumetric and Gravimetric H_2 Storage Capacity of Carbon-Based Sorbents Through Tuning Surface Chemistry and Densification Approaches

Ruthradharshini Murugavel, Fateme Rezaei

With increasing energy demand from growing population, hydrogen is considered as one of the most sustainable energy carriers for the future. Significant advancements have been made over the past few years to produce hydrogen in a sustainable manner, however, the major bottleneck currently facing the hydrogen economy lies primarily in its transportation and storage.(DOE, 2020) Adsorbed gas storage (AGS) technology offers a much safer and more economic storage option than existing cryogenic cooling or compression methods for hydrogen storage and transport. However, the current sorbents examined as H_2-AGS candidates still suffer from low deliverable capacity. Therefore, to enhance the hydrogen density per volume of storage tank and to meet the DOE's set targets, extensive research on nano-porous materials and storage conditions is required. In this research, we focused our attention on exceptionally high-surface area activated carbon and graphitic carbon nanosheets to further enhance the binding energy and adsorption capacity over a pressure range of 0-100 bar and varied charge-discharge temperatures via (i) surface functionalization with light–metal dopants such as Li, Na and K, and (ii) carbon exfoliation with ionic liquids, and (ii) sorbent densification approaches. Our results indicated the effectiveness of these approaches to enhance both gravimetric and volumetric hydrogen capacity relative to the pristine sorbents.

Keywords: Adsorption, Nanomaterials, Hydrogen storage

Reference:

DOE. (2020). Foundational Science for Carbon-Neutral Hydrogen Technologies. *DOE_BES Roundtable*, *1999*(December), 1–6.

371o Synthesis of Mussel-Inspired Polydopamine Mediated with Ionic Liquid As a Sustainable Adsorbent for the Selective Removal of Anionic Pollutants for Wastewater Applications

Rawan Abu Alwan, Botagoz Zhuman, Mahendra Kumar, Hassan Arafat, Enas Al Nashef

The extensive rise in industrial activities has led to the presence of various pollutants, including toxic synthetic dyes, in industrial wastewater effluents. From all the technologies available for treating the wastewater, adsorption is found to be the most economically feasible and easily operated technology. Different adsorbents such as activated carbon, different types of zeolites, bio-polymers, clays and various nano-composites have been used in removing impurities from water. However, there are several challenges associated with the use of many of these adsorbents. Toxicity, recyclability, scaling up and selectivity are some key challenges yet to be addressed. Therefore, development of sustainable and efficient adsorbents with minimized cross-contamination of the feed water, which have high selectivity towards certain contaminants is crucial. Recently, novel mussel-inspired material known as polydopamine (PDA) has emerged as a promising adsorbent for wastewater treatment by adsorption technology. PDA is derived from dopamine (DA) and consists of catechol, amine and imine functional moieties. Its similar structure with mussel adhesive proteins is responsible for high binding ability with different substances even on wet surfaces. However, the adsorption capacity of PDA alone is low compared to other adsorbents which meet indusial demands. Therefore, it is crucial to increase the adsorptive capacity of PDA to be useful on industrial scale and help to utilize all of its functional groups without deterioration in adsorption performance. Increasing the surface-active sites for adsorption can be achieved by modification of its surface area and charge. One possible route for developing PDA with high adsorption capacity is to modify the adsorbent via Schiff-based addition reaction with sustainable green solvent such as ionic liquids (ILs). ILs are widely used in different applications, including separation and extraction processes, as they are capable of multiple types of interactions (π–π, n–π, and hydrogen bonding). The usage of sustainable ILs is gaining more emphasis because of their high thermal and chemical stability, low volatility, environmental-friendly behavior and very high ability to dissolve a wide range of compounds.

The emerging technology of combining the unique properties of ILs along with an efficient adsorbent is still relatively not investigated, regardless of the increasing efforts in recent years to find optimal adsorbents. Adsorbents with IL functionalities have the advantages of tunability and tailor-ability according to the desired application. Due to their designability, IL-based functional adsorbents can be applied in a variety of fields. With almost endless possibilities of IL cation and anion combinations, the applications for IL-based adsorbents become vastly unlimited. In addition, the strong adhesive abilities of PDA as well as its unique functional moieties are of paramount importance in endowing the final adsorbent its chemical properties and behavior in surrounding media. Furthermore, the introduction of functional groups within the IL components may not only broaden the type of adsorbent and scope of application but may also increase the stability of the overall compound. Overall, the advent of advanced functional materials is of utmost importance due to the need for specificity and effectiveness towards a plethora of fields, whether industrial or technological. Increasing interest in IL integration with other materials finds tremendous applications in adsorption and extraction technologies for water and wastewater treatment, electrochemistry, carbon capture, desulfurization, drug delivery, material synthesis and many others.

In this work, Polydopamine-Ionic Liquid (PDA-IL) adsorbent was synthesized by a facile oxidative polymerization method and used as an efficient adsorbent for the removal of anionic dyes (Alizarin Red S, ARS) and anionic nutrients (Phosphates) from aqueous solutions. Characterizations of the synthesized PDA-IL particles were performed using multiple techniques such as scanning electron microscopy (SEM), transmission electron microscopy (TEM), Fourier-transform infrared spectroscopy (FTIR), Raman spectroscopy, thermogravimetric analysis (TGA), Zeta potential, and X-ray diffraction (XRD). Moreover, the effects of initial pH of the solution, initial concentration, contact time, adsorbent dose and temperature on the removal of ARS and phosphates using PDA-IL were investigated. Results showed that the maximum adsorption capacity at 25 °C could reach up to 234.19 mg/g for ARS removal and 93.08 mg/g for phosphate. Additionally, analysis of adsorption kinetics showed that the adsorption behavior followed the pseudo-second-order kinetic model, which indicates that adsorption is governed by chemisorption. The equilibrium adsorption data were in agreement with the Freundlich isotherm, indicating multilayer adsorption on heterogeneous surfaces. Thermodynamic analysis confirmed that the adsorption process using PDA-IL was exothermic and spontaneous, with a tendency towards increase in disorder. Furthermore, regeneration of the loaded PDA-IL adsorbent was successfully performed using methanol as an eluting agent. PDA-IL was used for 4 consecutive adsorption-desorption cycles without any deterioration in its morphology or performance. A selectivity test was also conducted to investigate the adsorption behavior of PDA-IL in dye mixtures. Lastly, a thorough understanding of the possible adsorption mechanism was proposed based on FTIR and electron-energy loss (EELS) mapping results. The high adsorption capacities obtained can be attributed to the pi–pi stacking, electrostatic interactions and the unique structural properties of both PDA and the custom-synthesized IL. Hence, it can be concluded from the results of this study that PDA-IL adsorbents could be of great potential for wastewater applications as a novel, efficient and sustainable adsorbent for removal of anionic pollutants.

371p Investigating Alkali-Activated Fly Ash Based Materials As Adsorbents for Heavy Metal Sorption

Monday Okoronkwo, Sukanta K. Mondal

Due to the high CO_2 emissions associated with the production of conventional cements, research efforts have been directed towards developing alternative low-carbon cementitious materials. One widely researched alternative cementitious material, that has also found value as adsorbent for environmental remediation, is alkali-activated cement or geopolymers. Although geopolymer material have gained wide research attention in recent decades as promising adsorbent for the removal of aqueous heavy metals, the high variability of the material and several unanswered questions have limited the development and general adoption in the industry. This study evaluates the impacts of compositional and microstructural factors on the performance of geopolymers for aqueous lead (Pb^{2+}) removal, with a view to reveal the composition-structure-property relationship thereof. The Pb sorption kinetics and efficiency of geopolymer adsorbents prepared using various fly ash precursors and activating alkaline solutions, were investigated. The investigated compositions featured high Pb removal efficiency of over 99.5%, with a slight decrease of efficiency with increasing Ca/(Si+Al) and Al/Si contents, while the sorption kinetics decreases exponentially with increasing Ca/(Si+Al) and Al/Si molar ratios. The performance of the geopolymers also shows strong correlation to the microstructure, wherein the sorption kinetic increases exponentially while the efficiency increases slightly, with increasing mass fraction of the amorphous phase in the geopolymer's phase assemblage. The results of this research indicate that by using appropriate precursor formulation and curing conditions to evoke the best microstructures, geopolymer materials can be optimized for high performance in removing heavy metals, thereby improving the chances of the material's general acceptability in the adsorbent industry.

339c Hyper-Crosslinked Tetraphenylboron (TPBx) As a Versatile Platform Material for the Development of Sorbents for Various Metal Ions

Grace Nisola, John Edward Sio, Khino Parohinog, Hern Kim, Erwin Escobar, Wook-Jin Chung

Hyper-crosslinked tetraphenylboron (TPBx) is a water insoluble, thermally stable and chemically resilient material. It is characterized by a highly porous network with hierarchical pore structure comprised mainly of micropores and mesopores, which explains its very high BET surface area (1030 $m^2\ g^{-1}$). Its chemical architecture consists mainly of benzene moieties which permit further functionalization of different moieties. Such modifications can alter its properties in liquid media and its interaction with substrates, opening TPBx to a myriad of potential applications which include cation adsorption, and establishing TPBx as a versatile platform material for sorbent development similar to carbon nanotubes and mesoporous silica. Herein, a demonstration of this versatility is presented.

Two adsorbents were prepared from TPBx through a series of simple functionalization reactions, first with sulfonation affording SO_3H-TPBx, followed by its reduction to the thia-functionalized form SH-TPBx. It is worth noting that TPBx withstands prolonged exposure to harsh reaction conditions during these functionalization steps which underscore its chemical stability --- to a strong oxidizing acid in the first step, and to reflux conditions in benzene in the second.

SO_3H-TPBx and SH-TPBx possess high BET surface areas (> 600 $m^2\ g^{-1}$) and were found to retain the morphology and hierarchical porous structure of TPBx. Both TPBx analogs exhibit remarkably enhanced wettability and dispersibility in aqueous media and effectively adsorb their target metal ions (Cs^+ and Sr^{2+} for SO_3H-TPBx, Pt^{4+} and Pd^{2+} for SH-TPBx) over a wide pH range, suggesting possible applications in metal ion decontamination and retrieval strategies in acidic and alkaline water wastes and in various types of water resources (surface and ground water). The inherent porosity and enhanced hydrophilicity of these TPBx analogs, along with their rich collection of spatially organized negative charges contributed by sulfur groups and tetraphenylboron moieties, seem to converge to create highly favorable conditions for cation adsorption resulting in very fast kinetics (equilibrium achieved in ~ 1 min), essentially complete contaminant removal even with very dilute contaminant concentrations in the feed (~ 2 mg L^{-1}), and high sorption capacities (e.g. 360 mg g^{-1} and 115 mg g^{-1} for Cs^+ and Sr^{2+}, respectively, when using SO_3H-TPBx). Moreover, the physical and chemical resilience of these TPBx analogs allows for their regeneration and repeated use. Overall, these results suggest that TPBx can serve as platform material for the development of highly efficient and practical cation sorbents.

This work was supported by the National Research Foundation of Korea (NRF) grant funded by the Ministry of Science and ICT (2020R1A2C1003560, 2021R1A2C2093746, and 2021R1H1A2008284), Basic Science Research Program through the Ministry of Education (2020R1A6A1A03038817).

143d Phosphorous-Doped Mesoporous Carbon As an Efficient Adsorbent for the Recovery of Neodymium (III) Metal Ion from an Aqueous Medium

Gebremedhn T. Gebremichael, Hiluf Tekle Fissaha, Grace Nisola, Wook-Jin Chung

Owing to their unique and excellent optical, catalytic, magnetic, chemical, and physical properties, rare earth elements (REEs) are essential resources in modern industries, such as batteries, metallurgy, medicine, turbines, electric cars engines, and communication devices. With the rapid development of high-technology industries, the demand for REEs has significantly increased throughout the world. However, the deposits of REE-containing minerals are very limited, and their supply is primarily controlled by China. Additionally, radioactive elements like Th (IV) co-exist with REE minerals causing major environmental pollution from the production of REEs from virgin ore. Hence, with the ever-increasing REE demand, lack of diversity in suppliers, and environmental challenges, efficient strategies for REE recovery (i.e., Neodymium) from REE containing End-of-life (EoL) scrap are becoming an attractive option.

Various methods have been explored to recover and separate REEs from industrial effluents and end products that contain REEs such as solvent extraction, chemical precipitation, ion exchange, and adsorption. Compared with other methods, recovery of REEs using adsorption is more viable due to its simplicity of design and operation, low environmental pollution, low cost, and high recovery efficiency. Carbon-based adsorbents are promising materials for the recovery of REEs from aqueous solutions, due to the advantage of good thermal, mechanical, and chemical stability, environmental friendliness, low cost, tunable pore size, and high surface area. On top of that, the preparation of these materials from abundant, cheap, and renewable raw materials makes them great candidates for efficient adsorption of REEs. Rare-earth phosphate precipitates are stable in an aqueous solution and most of the natural REEs ore is in the form of phosphate-bearing minerals, indicating that REEs have strong complexation abilities for phosphorus-based adsorbents.

Herein, phosphorus-doped mesoporous carbon (P-MC) was prepared by employing KIT-6 mesoporous silica as a hard template, saccharose, and phytic acid (PA) as a precursor of carbon and phosphorous source. The synthesized material was characterized using different techniques and explored for its potential application in the recovery of Neodymium ion (Nd^{3+}) from an aqueous solution. The parameters that affect the adsorption process such as solution pH, contact time, initial metal ion concentrations, and presence of interfering ions were investigated by batch adsorption experiments. The P-doped mesoporous carbon adsorption capacity was much greater than that of the pristine or un-doped mesoporous carbon, which verifies the importance of phosphorous functionalization on mesoporous carbon. Apart from adsorption kinetic and isotherm models, FTIR, XPS, and EDX analyses were performed before and after the adsorption to predict the binding mechanism Nd (III) with P-MC. Various organic and mineral acids were tested to strip the adsorbed metal ion from the adsorbent P-MC for its reusability performance. Finally, durability for adsorption capacity was assessed using adsorption/desorption runs to observe its potential for practical application.

This work was supported by NRF funded by the Ministry of Science and ICT (2020R1A2C1003560, 2021R1A2C2093746, and 2021R1H1A2008284), Ministry of Education (2020R1A6A1A03038817).

371q Demonstration of Iodine Removal Efficiency of MOF-Based Sorbents from Nuclear Waste Solutions in the Presence of Interfering Ions

Turki Alghamdi, Ali Rownaghi, Fateme Rezaei

With rapid development of nuclear industry upon ever-increasing energy demand from growing population, radioactive nuclear wastes have become an important issue. Such wastes need to be remedied urgently in an efficient and timely manner to meet stringent environmental regulations. In that regard, adsorption-based processes for capture of radioactive species from off-gas or solution streams have been demonstrated as an efficient remediation strategy. In our previous investigations[1,2] we evaluated the efficacy of aminosilane-grafted mesoporous alumina and bismuth alumina in iodine immobilization from aqueous solutions by studying the role of amine loading and bismuth content. In our previous investigations[1,2] we evaluated the efficacy of aminosilane-grafted mesoporous alumina and bismuth alumina in iodine immobilization from aqueous solutions by studying the role of amine loading and bismuth content. In this study, we aimed at investigating iodine adsorption performance of MOF-74 and UiO-66 in aqueous solution as a function of capture conditions including, feed concentration and adsorption temperature. Moreover, we aimed at evaluating the effect of interfering ions such as Cl^-, CO_3^{-2}), that may be present in the aqueous solution along with $^-I_2$, on the removal efficiency of the two MOFs. A maximum iodine removal efficiency of 97% was achieved over MOF-74 at 60 °C. MOF-74 outperformed UiO-66 in immobilizing iodine from aqueous solutions. It was also found that the presence of other interfering ions has different effects on iodine uptake over MOF-74 and UiO-66. The leach test results revealed 12 and 8% iodine leach from UiO-66 and MOF-74, respectively at 25 °C after 48 h. The kinetics and equilibrium uptake of iodine were represented by Langmuir and pseudo models.

References

(1) M. Alsalbokh, et al., *Chem.Eng. J.* **2021**, *415* (January), 128968.

(2) M. Alsalbokh, et al., *Chem.Eng. J.* **2021**, *409* (November 2020), 128277.

372b Hybrid Heat-Integrated Design Alternatives of Dimethyl Carbonate Reactive Distillation

Minyong Lee, Heecheon Lee, Chaeyeong Seo, Jeongwoo Lee, Jae Lee

Enhanced energy-efficient design alternatives for dimethyl carbonate production by combining various heat integration methods of reactive and pressure-swing distillation (PSD) columns were proposed. The design alternatives of hybrid heat-integrated sequences were based on internal and external heat integration between the columns with a vapor recompression heat pump (VRHP). The conventional PSD sequence with a reactive distillation column and the proposed hybrid heat integration sequences were compared subject to a constraint of high product purity of 99.5%. The combined sequence of internal and external heat integration resulted a cost saving of 30.28% by eliminating the low-pressure (LP) column reboiler. The external heat integration combined with VRHP fully eliminated the heat duty of the high-pressure (HP) column condenser and the LP column reboiler, which showed 38.33% energy savings and a 37.5% reduction in CO2 emissions. The hybrid sequences with external heat integration showed better performance in terms of energy, economic efficiency and CO2 emissions than the internal heat integration with VRHP, since the reflux flow rate of the HP column and the reboil flow rate of the LP column were significantly reduced.

372c A Low-Temperature Hollow Fiber Membrane Reactor for Propane Dehydrogenation

Lu Liu, Antara Bhowmick, Dongxia Liu, Chen Zhang

Propylene is a crucial petrochemical feedstock for a number of bulk chemicals and polymers. While steam cracking remains to dominate propylene production, propane dehydrogenation (PDH) has been increasingly practiced to address the gap between propylene demand and production. The propane conversion of PDH reactions is challenged by thermodynamic limitation and reaction endothermicity. High reaction temperatures (above 550 °C) are required to obtain attractive propane conversion, which leads to rapid catalyst deactivation. Membrane reactors can potentially address this challenge by integrating reaction and separation to reduce the capital cost, energy consumption, and footprint of propane dehydrogenation.

In this poster, we will present a novel hollow fiber membrane reactor for propane dehydrogenation. The membrane reactor consists of novel hollow fine fiber membranes (diameter ~ 0.3mm) for in situ hydrogen removal and platinum-based catalysts for propane activation. The hollow fine fiber membranes have high membrane packing density, which allows small membrane reactor footprint and modular propylene production. The hollow fiber membranes efficiently removed the hydrogen product from the membrane reactor and enhanced the propane conversion by 200% over the equilibrium conversion. This gave commercially attractive propane conversion (above 30%) and high propylene selectivity (above 98%) at 450 °C, which is at least 100 °C lower than commercial PDH processes. The low reaction temperature suppressed catalyst deactivation, thereby allowing the membrane reactor to provide unprecedentedly stable conversion and selectivity to outperform many PDH membrane reactors reported in literature.

372d The Technology Development in Magnesium Production and Separation

Lucas Arndt, Weiguo Xie, Richard Davis

Magnesium has taken on an increasingly significant role in our society. The global metallic Mg market In 2019 was valued at 3.5 billion dollars and is predicted to have a CAGR of 8.8% from 2020 to 2027 [1]. With that increasing demand comes an ever more critical need for clean, energy-efficient magnesium production. Currently, there are two major approaches to Mg production. The first is a variety of electrolytic routes that recover magnesium from several different materials. The second and most common is the Pidgeon process, in which dolomite undergoes calcination and silicothermic reduction [2].

In the late 1800s, an alternative method of producing Mg was introduced using carbothermic reduction. Carbon is pelletized with a source of magnesium oxide, such as dolomite. The following reaction happens when the reactants are heated to around 2,000 K at 1 atm:

$MgO + C \rightarrow Mg + CO$ (1)

Previously, it was thought that dolomite needed to undergo a calcination step before reduction, as shown in Figure 1 a. Recently the direct reduction of dolomite into magnesium and calcium has been investigated. The results show that the direct method requires less energy and releases a smaller quantity of CO_2 [3]. To further decrease the energy needed to produce the magnesium, a solar batch reactor like the one shown in Figure 2 can be used [3].

Carbothermic reduction has shown itself to be a promising alternative to current methods. It significantly decreases the quantity of energy used and the amount of CO_2 released by the process. A cradle-to-gate life cycle study found that when using solar reactors to reduce dolomite, it takes only 51.3 Mj, and results in the release of 12.7 kg CO_2 per kg of Mg produced [4]. Whereas the Pidgeon process currently used in most magnesium production facilities requires 366 Mj and releases 42 kg of CO_2 per kg of Mg produced [4]. Carbothermic reduction's significant increase in efficiency opens the possibility of once again producing magnesium domestically.

Unfortunately, the product of this method is a mixture of magnesium vapor and carbon monoxide gas, which quickly reverts the magnesium vapor into magnesium oxide once the temperature begins to decrease. Researchers have been attempting to solve this reversion reaction of equation 1 for over 100 years and have developed many methods that lessen but do not solve the problem.

The many attempts to separate magnesium from carbon monoxide can be divided into three general categories [5]. The first of these methods involves rapidly quenching the magnesium gas, resulting in the formation of small magnesium particles directly from the gas state. The second relies on rapidly condensing the magnesium onto a cold surface. Lastly, some processes use a solvent such as another metal or a hydrocarbon that cools and absorbs the magnesium. However, despite over a century of research and the development and subsequent closing of several production plants, no magnesium is currently being commercially produced via the carbothermic reduction route.

It is necessary to investigate further all the product magnesium produced by the previous separation methods. No method has been able to stop the reversion process from producing pure magnesium. An additional sublimation or distillation method has always been needed, resulting in a significant increase in the operating cost. However, there is a method that may be able to produce pure magnesium directly from the product gases. The new process being considered would separate the magnesium vapor from the carbon monoxide before the mixture can cool down from the reaction temperature. The result would be pure magnesium that had been protected from the temperature conditions that lead to oxidation. One possible way of accomplishing this task is using large beds consisting of activated carbon or other molecular sieves. Little work has been done in this area because these separations suffer an extreme loss in efficiency at high temperatures. Fortunately, even with low efficiency, the cost of running such a system would theoretically be less than having an entire distillation stage in the process. Further investigation is required to determine the efficiency of a high-temperature adsorption bed.

References:

1. "Metal Magnesium Market Size, Share andamp; Trends Analysis Report By Application (Die Casting, Aluminum Alloys, Titanium Reduction, Iron andamp; Steel Making), By Region, And Segment Forecasts, 2020 – 2027," *Grand View Research*, https://www.grandviewresearch.com/industry-analysis/metal-magnesium-market. (Accessed October 5, 2021).
2. M. Halmann, A. Frei, and A. Steinfeld, "Magnesium Production by the Pidgeon Process Involving Dolomite Calcination and MgO Silicothermic Reduction: Thermodynamic and Environmental Analyses," *Industrial andamp; Engineering Chemistry Research*, vol. 47, no. 7, pp. 2146-2154, Apr. 2008, doi: 10.1021/ie071234v.
3. A. Najafabadi, N. Ozalp, M. Epstein, and R. Davis, "Solar Carbothermic Reduction of Dolomite: Direct Method for Production of Magnesium and Calcium," Industrial andamp; Engineering Chemistry Research, vol. 59, no. 33, pp. 14717-14728, Aug. 2020, doi: 10.1021/acs.iecr.0c02329.
4. H. A. Najafabadi, N. Ozalp, M. Epstein, and R. Davis, "Solar Carbothermic Reduction of Dolime as a Promising Option To Produce Magnesium and Calcium," *Industrial andamp; Engineering Chemistry Research*, vol. 58, no. 51, pp. 23540-23548, Dec. 2019, doi: 10.1021/acs.iecr.9b04856.
5. B.A. Chubukov, S.C. Rowe, A.W. Palumbo, et al., "Investigation of continuous carbothermal reduction of magnesia by magnesium vapor condensation onto a moving bed of solid particles," *Powder Technology*, vol 365, pp. 2-11, Apr. 2020, https://doi.org/10.1016/j.powtec.2019.01.067

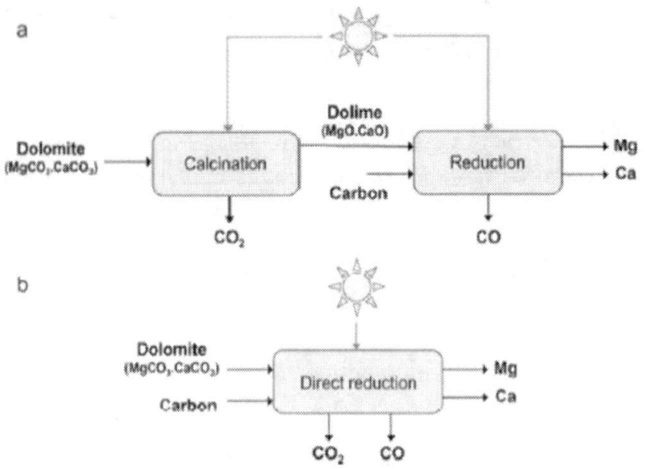

Figure 1: Solar carbothermic reduction of dolomite, (a) two-step process and (b) direct process [3].

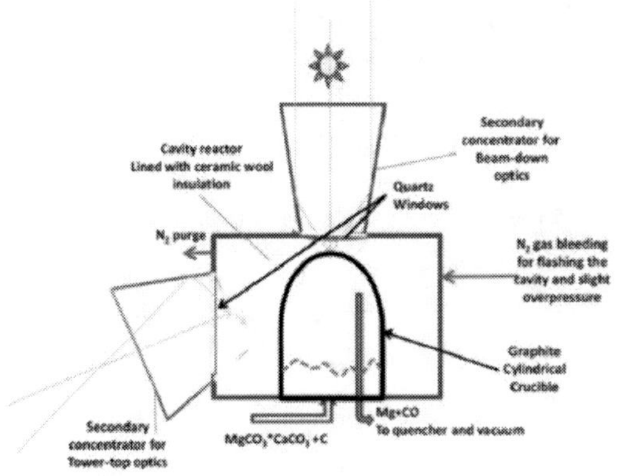

Figure 2: Conceptual illustration of a batch solar reactor for the carbothermic reduction of dolomite. [3]

372e Selective Recovery of Au^{3+} *via* Complexation-Reduction Capture and Release By Thermo-Responsive P(NIPAM-co-15TCE-4)@SiO_2 Nanoparticles.

Hiluf Tekle Fissaha, Grace Nisola, Wook-Jin Chung

Production of electrical and electronic devices with short life cycles and lack of suitable recycling technologies has led to the surging electronic waste generation worldwide. E-wastes such as spent mobile phones contain higher amount of precious metals as compared to their mineral ore sources. Hence, recovery of valuable metals like gold (Au) from waste resources has high environmental and economic significance. Many adsorbents derived from biomass, modified magnetic nanoparticles, and MOFs have been developed for the recycling/recovery of these valuable metals like gold (Au^{3+}) from e-wastes. However, most of these adsorbents are not selective towards Au^{3+} from its co-exiting metal ions such as Pd^{2+} and Pt^{4+}. Meanwhile, crown ethers (CEs) are attractive macrocyclic complexing agents as their cavity structures can be tailored to accommodate the ionic size of a precious metal and their heteroatoms (i.e. O, N, or S) can be selected to afford their affinity towards Au^{3+}. Herein, a multi-functional Au^{3+}-selective thermo-responsive adsorbent, P(NIPAM-co-15TCE4)@SiO_2, was prepared from thiacrown ether (15TCE4) with Au^{3+} recognition properties and a thermo-responsive N-isopropylacrylamide on a silica (SiO_2) support via the Activators ReGenerated by Electron Transfer; Atom Transfer Radical Polymerization (ARGET-ATRP) method. The adsorbent was successfully prepared as verified by NMR, FTIR, TGA, EA, FE-SEM, XPS, HR-TEM, and DLS. The thermo-responsiveness of P(NIPAM-co-15TCE4) @SiO_2 was observed at critical temperature $T_c \sim 34.9\ ^{\circ}C$. P(NIPAM-co-15TCE4)@SiO_2 showed Au^{3+} capture $(T > T_c)$ and release $(T < T_c)$ by varying the system temperature above and below its phase transition (LCST, T_c) point. At $T = 50\ ^{\circ}C$ $(T > Tc)$ and pH = 2, the maximum Au^{3+} capacity q_m was quickly achieved within 2 h. HR-TEM/EDS, WAXD and XPS results indicate that Au^{3+} capture involves complexation, partial reduction to Au^{+} and disproportionation to Au^{o} /Au^{3+}. Adsorption results show competitive and selective Au^{3+} uptake of P(NIPAM-co-15TCE4)@SiO_2 with K_D ~16,276 mL g^{-1} and $\alpha = 9 - 43{,}053$ for Au^{3+} against Pd^{2+}, Pt^{4+}, Cu^{2+}, Pb^{2+}, Zn^{2+} and Ni^{2+} present in a simulated mobile phone leachate. Meanwhile, desorption occurs at $T < T_c$ by the brush network movement as NIPAM component rehydrates at $T = 27\ ^{\circ}C$. Complete desorption was achieved even in a very mild eluent that eluted firmly bound Au species and regenerated the adsorbent via S = O reduction to C-S-C. The reversible behavior of P(NIPAM-co-15TCE4) @SiO_2 for Au^{3+} capture ($T > Tc$) and release ($T < Tc$) highlights the synergy between 15TCE4 and NIPAM, making it a reusable practical adsorbent with consistent performance for Au^{3+} recovery in complex feed wastes.

This study was supported by the National Research Foundation of Korea (NRF) funded by The Ministry of Science and ICT (2020R1A2C1003560, 2021R1A2C2093746, and 2021R1H1A2008284), Basic Science Research Program through the Ministry of Education (2020R1A6A1A03038817).

372g $Co_{1-X}[Fe(III)(CN)_6]/Ag$ for Electrochemical Recovery of Cobalt Ion from Spent Nickel Metal Hydride Batteries

Hana G. Zeweldi, Gebremedhn T. Gebremichael, Negasi Teklay Weldesemat, Grace Nisola, Wook-Jin Chung

Cobalt is utilized as a core element in rechargeable batteries, catalysts, magnets, inks and pigments. The rapidly growing use of rechargeable batteries in electric vehicles increases the demand for cobalt at a fast pace. Thus, exploration of feasible cobalt recovery and recycling technologies are essential to meet the global demand. In addition, cobalt is a toxic element that causes adverse effect on human health, thus recycling this element can reduce environmental pollution as well as minimize the wastage of such valuable element. Because of the uncertainty surrounding the supply of primary cobalt and its status as a critical raw material, improving the recycling and recovery of cobalt is becoming more vital. The secondary sources of cobalt include waste catalysts, magnets, super alloys, cemented carbides, rechargeable batteries, and other metallurgical waste/byproducts. The techniques used for the recovery of cobalt from these secondary sources are mainly pyrometallurgy and hydrometallurgy, which are characterized by high energy consumption, emission of pollutants, and involvement of complicated steps such as acid leaching. Thus, alternative cobalt recycling practices should be explored.

Electrochemical techniques for metal ion recovery are known to have advantages over the conventional adsorption and precipitation techniques such as high performance, easy operation, low cost, easy separation, and high rate of ion capture. Herein, cobalt hexacyanoferrate ($Co_{1-x}[Fe(III)(CN)_6]$) was utilized to recover cobalt ions from a simulated leachate of spent nickel metal hydride batteries (NiMH) batteries through electrochemical approach. $Co_{1.5}[Fe(III)(CN)_6]$ electrode delivers a high reversible capacity and a high capacity retention after several cycles. The high capacity and excellent cyclic stability of the material makes it ideal for electrochemical applications. Thus, a three-electrode assembly consisting of Co^{2+} selective intercalating electrode ($Co_{1-x}[Fe(III)(CN)_6]$), a chloride-capturing silver electrode, and Ag/AgCl reference electrode were utilized for Co^{2+} capture. The structural and morphological properties of the electrodes were thoroughly characterized. The cyclic stability of the electrodes was examined through cyclic galvanostatic charge and discharge tests. The selective Co^{2+} intercalation and de-intercalation were tested by cyclic voltammetry test and galvanostatic charge and discharge tests at different operating conditions. The overall performance of CoHCF/Ag system was examined by the Co^{2+} intercalation and de-intercalation capacity, selectivity and energy requirement.

In this study, the potential of the $Co_{1-x}[Fe(III)(CN)_6]/Ag$ electrode pair for the electrochemical Co^{2+} from a simulated spent NiMH batteries aqueous source was successfully demonstrated. The stability of the electrode pair in aqueous environment was confirmed by the material and electrochemical characterizations. Cyclic galvanostatic discharge and charge tests confirmed the stability of the material over 50 cycles. The increase in applied current and time facilitated the Co^{2+} capture. $Co_{1-x}[Fe^{(III)}(CN)_6]$ selectively captured Co^{2+} in the presence of other cations such as Ni^{2+}, Ce^{3+}, La^{3+}, Nd^{3+}, and Pr^{3+} in the simulated spent NiMH feed solution. Overall, the proposed electrochemical system is eco-friendly, effective, selective, and recyclable for Co^{2+} recovery from spent NiMH battery source.

This work was supported by NRF funded by The Ministry of Science and ICT (2021R1H1A2008284, 2020R1A2C1003560, 2021R1A2C2093746), Basic Science Research Program through the Ministry of Education (2020R1A6A1A03038817).

372i Selective Adsorption of Chlorofluorocarbons (CFC) and Hydrochlorofluorocarbons (HCFC) in Hydrofluorocarbon's (HFC) Using Adsorbent Supported Metal Oxide.

Anup Kumar Doraiswamy

Hydrofluorocarbons (HFC) are used in a variety of applications such as blowing agents, refrigerants and plasma etching gases for the semiconductor industry and are often commercially manufactured from fluorination of chlorofluorocarbons (CFC) or hydrochlorofluorocarbons (HCFC). As a result, these products may contain undesired impurities which may be toxic, have high global warming or ozone depletion potential, as well as impact the end market process use of the gas. Some of these impurities form azeotropic combinations with the matrix gas leading to difficulties in separation using conventional distillation. A novel purification method to selectively remove these undesired impurities (CFC's and HCFC's) from hydrofluorocarbons using metal oxide based adsorption bed along with specific pre-treatment processes will be discussed. Conventionally, these adsorbent beds have been employed to completely remove or scrub halocarbons from various gas streams for environmental purposes. In this application, we are taking advantage of the relatively higher affinity of chlorine containing compounds towards metals and metal oxides compared to the fluorocarbons for selective adsorption thereby producing ultrahigh purity electronic gases. Examples will be presented including 245fa (1,1,1,3,3-Pentafluoropropane) purification utilizing a metal oxide adsorption bed where some key CFC/HCFC impurities including 1,1,2 trichlorotrifluoroethane and 2,2 dichloro-1,1,1trifluoroethane were observed to be reduced from 284 ppm and 16 ppm to << 1ppm respectively when measured using GC/MS/TCD.

372j Project Earth (Environmentally Applied Research Towards Hydrofluorocarbons) Ionic Liquids Approach

Kalin R. Baca, Greta M. Olsen, Lucia Matamoros Valenciano, Madelyn Bennett, Dorothy M. Haggard, Mark B. Shiflett

Project EARTH (Environmentally Applied Research Toward Hydrofluorocarbons) is a collaborative project focused on developing sustainable processes for the selective separation of hydrofluorocarbon (HFC) refrigerant mixtures. The initial focus of the project is on R-410A, an azeotropic mixture containing 50 wt.% HFC-125 (CHF_2CF_3) and 50 wt.% HFC-32 (CH_2F_2). Project EARTH is exploring three separation approaches including, ionic liquids as entrainers for extractive distillation, polymeric membranes, and porous materials such as zeolites.

This poster will describe the work being conducted using the ionic liquids approach and will specifically focus on physical property measurements, equation of state modeling, and a lab-scale separation process. The poster will include an overview of how we use specialized equipment including a Hiden Isochema IGA gravimetric balance, a High-Pressure View Cell, a Hiden Isochema XEMIS with the Integral Mass Balance method, a VLLE apparatus, and a Setaram Calorimeter to build a comprehensive understanding of both pure HFC+IL and binary HFC+IL mixture systems.

372k Impact of Wafer Chemistry on Electrodeionization Removal and Selectivity

Leticia Santos de Souza, Jamie Hestekin, Christa N. Hestekin

Electrodeionization (EDI) removes ions using electricity as the driving force. As ions get transported from the dilute to concentrate chambers, the concentration in the dilute decreases, which increases the internal resistance for ion removal, resulting in a process limitation. To overcome that issue, a mixture so-called wafer is used in the dilute chambers. Wafers include ion exchange beads whose function is to interact with the ions in the solution, guided by the functional groups on the beads' surface. It is known that although the commercial ion exchange beads are good exchangers, their selectivity needs to be enhanced to meet the requirements of specific ion removal processes.

Previous studies have shown that material modification with various crown ether molecules has resulted in increased ionic selectivity of different ionic species based on the size of the crown molecules present in the system. For instance, 15-crown-5 ethers have been shown to complex with sodium. For this study, a multi-ion solution was run on a 1-wafer EDI using wafers produced with different cation and anion exchange beads to better understand the impact of different wafer chemistries on the ion removal process. Included in the cation exchange beads were those that had been modified with crown ethers.

The different combinations of beads had a high impact on ion removal, and the modifications of cation exchange beads resulted in an increase in the wafers' selectivity by at least 10%.

372l Estimation of Supersaturation in Pressure-Driven Supersaturated Gas-Liquid Systems

Sushobhan Pradhan, Prem Bikkina

A gas-liquid system is referred to as supersaturated when the concentration of the solute phase (gas) exceeds the equilibrium concentration in the solvent phase (liquid). A supersaturated solution can be achieved either by changing the temperature and/or pressure of the system. The gas from a supersaturated liquid escapes either through bubble nucleation that usually occurs on solid surface and/or gas diffusion through the gas-liquid interface. A pressure-driven bubble nucleation process requires a minimum threshold supersaturation to initiate the formation of gas bubble(s). In addition, a certain waiting time is required at each pressure reduction step to ensure either enough time is given for the bubbles to nucleate and grow, or to conclude that the supersaturation is not sufficient to initiate the bubble nucleation. When the supersaturation is not sufficient to cause bubble nucleation, some or all of the supersaturated gas may diffuse out from the liquid through the gas-liquid interface before the next pressure reduction step to further increase the supersaturation. In this study, a simple analytical method developed based on Fick's second law of diffusion and Henry's law to estimate the supersaturations generated in three gas-liquid systems (CO_2-water, CH_4-water, and N_2-water) with two step-down pressures (100 and 500 mbar) is presented. Characteristic times of the gas-liquid systems were estimated to validate whether the waiting times used in this study are in accordance with the semi-infinite diffusion model used to estimate the supersaturations generated. During the waiting time, some or all of the supersaturated gas may diffuse out before initiating the next step-down pressure. Therefore, it is important to know the actual supersaturation in the system as a function of time during the waiting time in order to quantify the actual supersaturation required to initiate the bubble nucleation. Therefore, in this study the influence of step-down pressure, and the waiting time between the pressure reduction steps on supersaturation is also investigated.

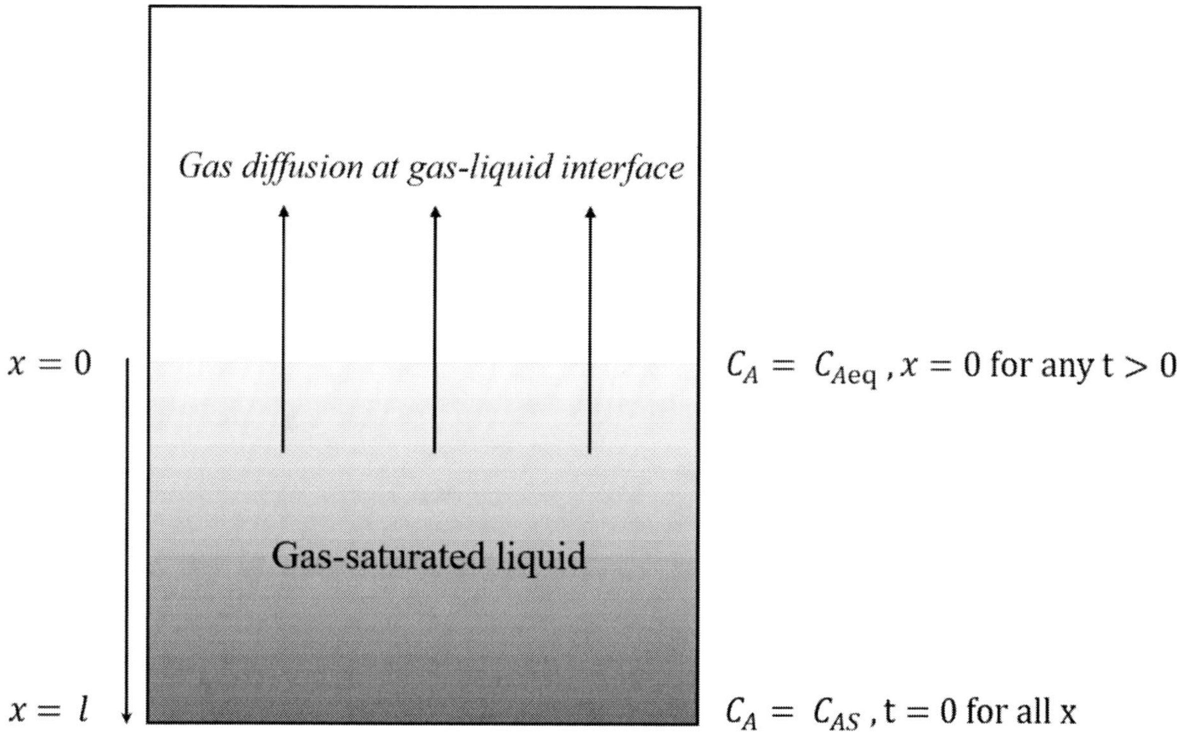

372m Identification of the Innovative Separation Technique for Sustainable Society

Akanksha Prasad, Dr. Utkarsh Maheshwari

Separation is undoubtedly a vital aspect of the chemical industry, as it is necessary to purify the components of the end product from the raw materials that are accessible to the final product that is generated.Various separation procedures exist, each utilising a distinct property such as solubility, volatility, melting point or other similar characteristics to achieve the desired separation results. Selection of separation techniques is based on the components that need to be separated. These separation procedures will be given more possibilities to decrease the amount of waste generated throughout the manufacturing process.

Nowadays, there are new technologies that are being used in order to increase the productivity of the workforce. The advancement of traditional separating methods while simultaneously reducing energy usage and improving separating operations are examples of separation engineering advancements. Innovation is essential for addressing a wide range of serious social and industrial concerns in which separation is critical, such as climate change. The current work investigates the numerous separation techniques and the improvements that have been used to ensure the long-term viability of society.

372p Phase Equilibria for HFC-32, HFC-125, and Binary Mixtures (R-410A) with a Variety of Ionic Liquids

Kalin R. Baca, Greta M. Olsen, Lucia Matamoros Valenciano, Madelyn Bennett, Dorothy M. Haggard, Mark B. Shiflett, Mark B. Shiflett

Hydrofluorocarbons (HFCs), the current generation of refrigerants, are being phased out due to high global warming potential (GWP) while hydrofluoroolefins (HFOs), the new generation of more environmentally friendly refrigerants, are being commercialized. Thus, millions of tons of HFC mixtures will need to utilize new technology to be efficiently recycled and repurposed. Project EARTH (Environmentally Applied Research Toward Hydrofluorocarbons) is focused on developing selective and efficient separation processes for HFC mixtures for this purpose. One approach is to use ionic liquids (ILs) as entrainers to selectively separate azeotropic HFC mixtures like R-410A, which is composed of 50 wt.% HFC-32 (CH_2F_2) and 50 wt.% HFC-125 (CHF_2CF_3). HFC-32 can then be blended with new HFOs to produce products and HFC-125 can be repurposed as a high-value fluorinated feedstock for production of new materials with a low-GWP.

This presentation will compare the phase equilibria data for a variety of novel ILs with HFC-32, HFC-125, and mixtures (e.g., R-410A). This presentation will discuss ILs that are both solid and liquid at 25 °C as well supporting data from a high-pressure view cell and a Calvet calorimeter. Phase equilibria data for HFC-32 and HFC-125 were measured using a Hiden Isochema IGA gravimetric balance to compare differences in gas solubility for a variety of ILs.

372q Separation of Dispersed Water from ULSD Using a Wire Mesh Electrowet Coalescer

Mohammad Assaleh, George G. Chase

Dispersed water drops in diesel fuel can cause undesirable emulsions, acting as a contaminant and can diminish the lifetime and performance of diesel engines. Therefore, separating dispersed water from diesel fuel is crucial and is of high importance. However, existing separation methods pose efficiency limitations as well as environmental and safety challenges. This work presents the performance of a compact electrical coalescer, in coalescing fine dispersed water drops in Ultra Low Sulfur Diesel (ULSD) to from larger drops that are easily separated using gravity settling. The coalescer is comprised of stainless-steel mesh electrodes separated at a relatively close distance from each other to benefit from electrode surface coalescence enhanced by electrowetting. The main objectives of this work were to select the most appropriate operating and geometrical parameters for optimum results. Laboratory tests based on a 2^n factorial design to determine the variables that have the most significant effect on the coalescence and separation performance, as well as finding the optimum conditions. The variables considered were voltage, face velocity, mesh size, and electrode gap distance, while water separation efficiency and average water drop size were the response variables. The proof of concept of the wire mesh electrowet coalescer (EWC) was conducted and showed the EWC could enlarge drops from 30 micron to about 2 mm and required minimal energy to perform. The EWC was also able to achieve a maximum of 91% separation efficiency at the the optimum conditions;150 V, 0.5mm gap distance and a 2cm/min face velocity. Having a smaller pore inlet mesh and larger pore outer mesh showed better performance.

Barriers to electrodialysis implementation: maldistribution and its impact on resistance and limiting current density

Jack Ledingham[a,1], *Kyra L. Sedransk Campbell*[a], *Ben In 't Veen*[b], *Lucas Keyzer*[b], *Alasdair N. Campbell*[a],

[a] *Department of Chemical and Biological Engineering, The University of Sheffield, Sheffield, S1 4AA, United Kingdom*

[b] *Shell Global Solutions International B.V*

This work is an extension of our recent publication: https://doi.org/10.1016/j.desal.2022.115691

1. Introduction

Electrodialysis (ED) is an emerging electromembrane technology with broad applications in electrochemical separations such as desalination, resource recovery and wastewater treatment. As the global water crisis worsens, it is more important than ever that natural freshwater resources are conserved, and new sources are established. To achieve this, significant research and development of separation technologies is required, and ED is uniquely positioned to achieve this. In ED, an electric field is applied tangentially to saline streams flowing between ion exchange membranes in a geometry analogous to that of a plate-and-frame heat exchanger (PFHEX) (Fig. 1). The electric field drives the transport of ions from one stream, through ion exchange membranes (IEMs), and into another. The membranes exist as semi-permeable barriers, theoretically allowing the transport of anions or cations only. As opposed to competing technologies such as reverse osmosis and thermal separations, in ED it is the solute, which is transported, rather than the solvent. This presents several inherent benefits such as a higher rate of separation, greater controllability, a lower susceptibility to fouling, a greater membrane lifespan, and a higher recovery ratio.

Recent developments in IEM performance and manufacturing techniques have meant that ED is, for the first time, commercially viable. However, there are several technological challenges which remain as a result of limited fundamental understanding of ED. A salient example of this is the limiting current density (LCD). Concentration gradients form adjacent to membranes as a result of concentration polarization, the magnitude of which increases as the ion flux, and thus the current density, increases. The membrane-solution interfacial concentration decreases until it vanishes, at

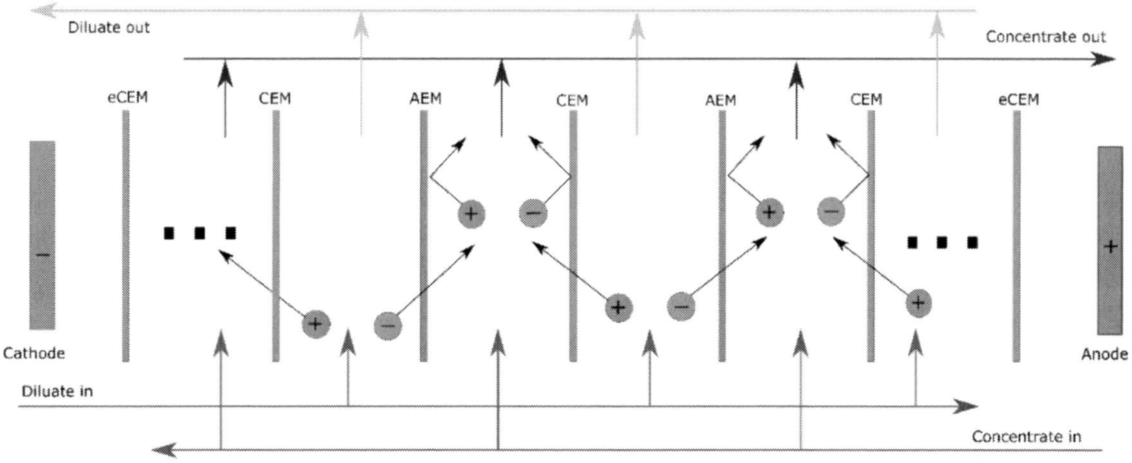

Fig. 1: A schematic of ion transport in ED, indicating the repeating pattern of cation and anion exchange membranes as well as the electrodes and end cation exchange membranes (eCEM). Blue shades show positive charges (cations, AEMs, anode) and red shades show negative charges (anions, CEMs, cathode).

which point the LCD has been reached. A substantial increase in electrical resistance and unwanted pH changes then occur due to the onset of water splitting. The LCD represents the highest possible ion flux and so has important consequences for process intensification. Complex phenomenological interactions drive the use of highly empirical power-law models as theoretical models are inadequate and over-predict the LCD. This makes models highly specific to the system they were initially built for and obscures fundamental interactions, making it difficult to identify methods to improve the LCD.

Several other prevalent assumptions about ED behavior add to the obfuscation of underlying phenomena, again inhibiting fundamental understanding. One ubiquitous assumption in ED modelling and experimental analysis is that of channel uniformity, an assumption which has never been validated. The distribution of flow within an ED stack has been given virtually no attention across published literature despite it having significant consequences for ED operation. Herein, we evaluate this assumption to its breaking point and aim to quantify the extent of maldistribution within an ED stack. Crucially, we go further to investigate what design elements affect maldistribution, as well as the effect it has on performance metrics such as the electrical resistance and the LCD.

2. Methodology

In this work, 3D computational fluid dynamics (CFD) simulations are used for the first time to generate realistic flow fields of entire lab-scale ED stacks (Fig. 2). This is a significant step forward from previous maldistribution studies which have used 2D simulations or focused on only one part of the geometry. Further, an analytical maldistribution model, originally derived for a PFHEX, was applied to the results of the CFD simulations:

$$U_c = W_0\left(\frac{A_m}{nA_c}\right)\frac{m\cosh\big(m(1-Z/L)\big)}{\sinh(m)}$$

$$m^2 = \left(\frac{nA_c}{A_m}\right)^2\frac{1}{\zeta}$$

$$\zeta = \frac{2\Delta P_{mc}}{\rho U_{mc}^2}$$

Where, U_c is the channel velocity, W_0 is the stack inlet velocity, A_m and A_c are the areas of the channel and manifold, respectively, n is the number of channels, m is a dimensionless maldistribution number, Z is the axial length along the manifold measured from the first channel. L is the manifold length from the first to last channel, ζ is a general friction coefficient, ΔP_{mc} is the mean channel pressure drop, ρ is the fluid velocity, and U_{mc} is the mean channel velocity.

Two important benefits arose from the use of this model: laborious CFD simulations can be avoided, and a dimensionless maldistribution number is provided as a suitable metric for the degree of maldistribution. The effect stack design and operation have on maldistribution was investigated by modifying the geometry and velocity boundary conditions and computing the associated maldistribution number. The specific design aspects varied were the inlet flowrate, channel width, distributor angle, number of manifolds, number of cell pairs, and the length-to-width ratio of the flow channels (Fig. 2). To determine the effect that maldistribution has on operation, a simplified 1D circuit-based model was developed and used to predict the LCD and stack resistance.

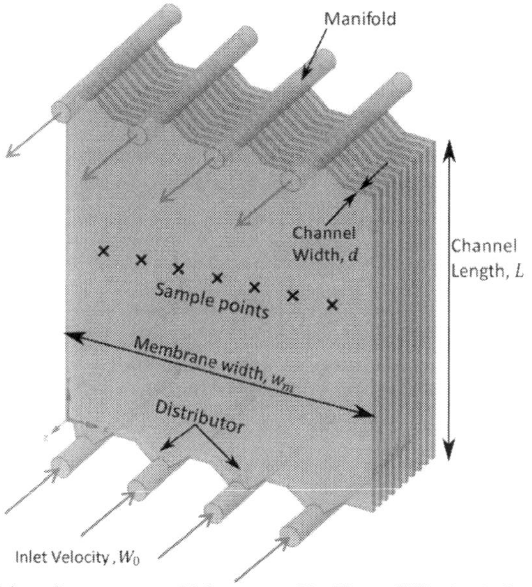

Fig 2: The geometry of a typical lab scale geometry which was used in Fluent CFD simulations. This is the flow region of one stream only, the second being rotationally symmetric to the first. Also shown are the named domains, dimensions and the locations of centreline velocity sample points.

Experimental validation of the results was conducted on a PC Cell 64004 ED stack using Fuji-12 cation and anion exchange membranes (Fujifilm). The aim of these being to validate the effect maldistribution has on the LCD. These experiments were conducted on a recirculating batch ED system (MEGA Electrodialysis) in steady state. By enacting changes to operation which only increase the maldistribution, a decrease in the LCD should be seen. Two parameters were chosen, the number of cell pairs and the volumetric flow rate. An increase in either the salt concentration or solution flowrate will increase the LCD. These effects were thus balanced out to achieve a constant salt molar flowrate while increasing the volumetric solution flowrate. Since the LCD is bound by the salt molar flowrate of the slowest channel, if no maldistribution is present then the LCD will be constant. However, if maldistribution is present and is affected by the volumetric flow rate, the LCD will change in accordance with model predictions. Likewise, the number of cell pairs was increased from 10 to 20, and the volumetric flow rate doubled. This ensures the average channel flowrate remains constant, while the maldistribution may change. Therefore the effects of maldistribution may be seen.

3. Results and Discussion

Initial CFD simulations on a standard lab-scale stack geometry revealed maldistribution to be significant (Fig. 3). The fastest channel, adjacent to the entrance and exit, was found to have twice the flowrate of the slowest, and thus half the residence time. These results were confirmed by the PIV experiments. It can therefore reasonably be concluded that maldistribution exists in lab-scale electrodialysis stacks and must affect electrodialysis experiments conducted on similar geometries. Since this geometry is commonplace to find in published research, these results are of great significance.

The analytical maldistribution model was found to fit the flow distributions very accurately with a root-mean-square error of only 0.4%. Further, the model reasonably predicted the flow distribution using common friction factor correlations. This shows the model is an invaluable tool for studying maldistribution in ED without requiring computationally demanding CFD simulations. The use of a single dimensionless number to quantify maldistribution facilitates the comparison of maldistribution between different units and is useful as a simple but objective input into phenomenological models.

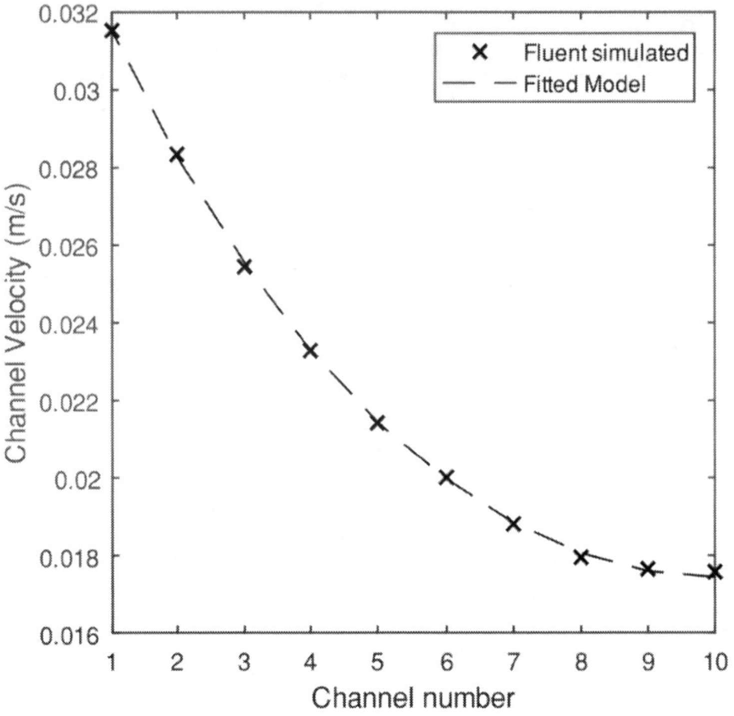

Fig. 3: Velocity distribution of the standard-case stack with the fitted analytical model shown. Channels are numbered with one being the channel closest to the inlet and outlet, and ten being the farthest

The stack design was found to have a significant effect on the maldistribution. In general, design choices which increased the channel pressure drop or reduced the channel flow rate were found to reduce maldistribution. For example, increasing the stack inlet velocity was found to increase the maldistribution (Fig 4). Additionally, the number of channels had the greatest effect on maldistribution and was found to be proportional to the maldistribution number (Fig 5). Industrial ED stacks tend to have between 50 and 200 cell pairs, and so maldistribution will likely be worse at industrial scale.

It was suggested by the 1D-modelling results that maldistribution has a detrimental effect to both LCD and electrical resistance, but that the effect on the LCD is much greater. The degree of maldistribution present in the standard lab scale stack was found to reduce the LCD by 23% but increase the resistance by only 2% (Fig 6). This may go some way to explain why there has been a distinct lack of maldistribution research present in literature. While a large and unexplained increase in resistance would be difficult to ignore, the uncertainty surrounding what affects the LCD and highly empirical nature of its modelling could easily obfuscate the impact of maldistribution.

The ED stack experiments were successful in validating the modelling results. The LCD decreased when the volumetric flowrate increased, in line with what would be expected given the predicted degree of maldistribution (Fig. 7). The LCD was also consistently lower for experiments on a 20 cell pair stack. This is strong evidence for the existence of maldistribution since the experiments at 10 and 20 cell pairs have the exact same average cell conditions. Furthermore, the difference between these lines is quantitatively inline with what the model suggested. The average LCD ratio between 10 and 20 cell pairs is 1.62. From the LCD model, this value is expected to be 1.61. However, the expected LCD measured is about 20% lower than what would be expected from the model. This shows that although absolute predictions are not captured by the simple 1D LCD model, relative behavior is captured.

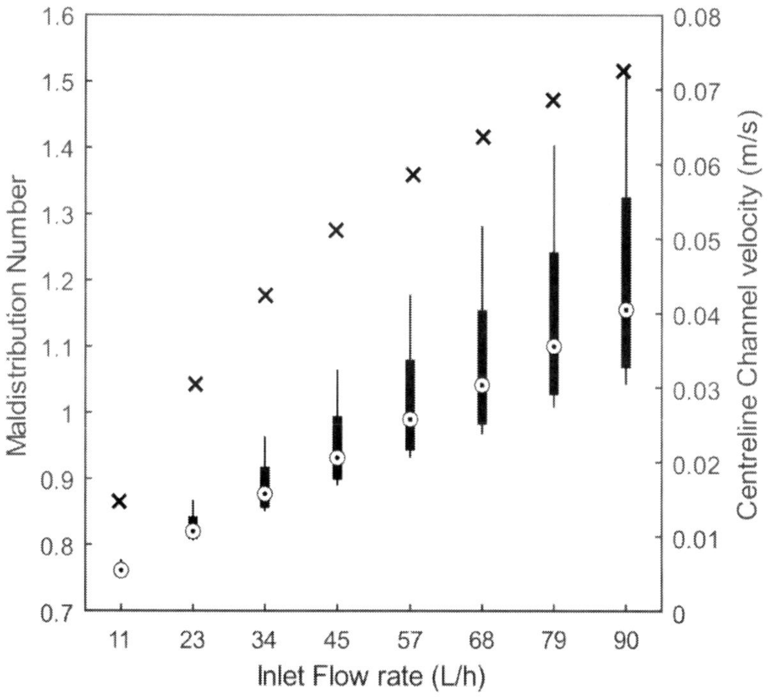

Fig. 4: Fitted values of the maldistribution number (**×**, left y-axis) and a box-and-whisker diagram showing the distribution of the velocity across the stack (right y-axis) for simulations where the inlet velocity was changed. Targets (⊙) give the median velocity, boxes represent the interquartile range, and lines show the overall range of velocity.

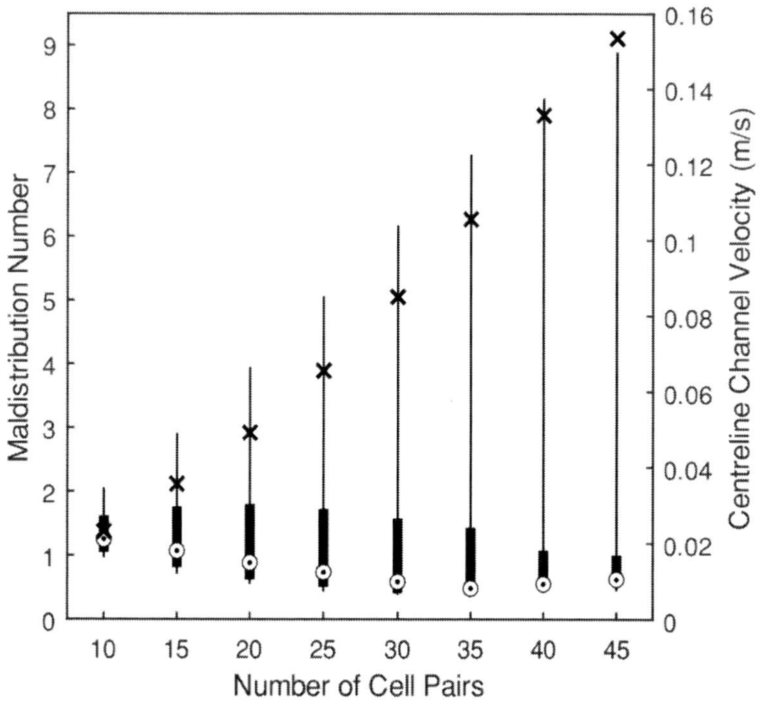

Fig. 5: Fitted values of the maldistribution number (**×**, left y-axis) and a box-and-whisker diagram showing the distribution of the velocity across the stack (right y-axis) for simulations where the simulated number of cell pairs was changed. Targets (⊙) give the median velocity, boxes represent the interquartile range, and lines show the overall range of velocity

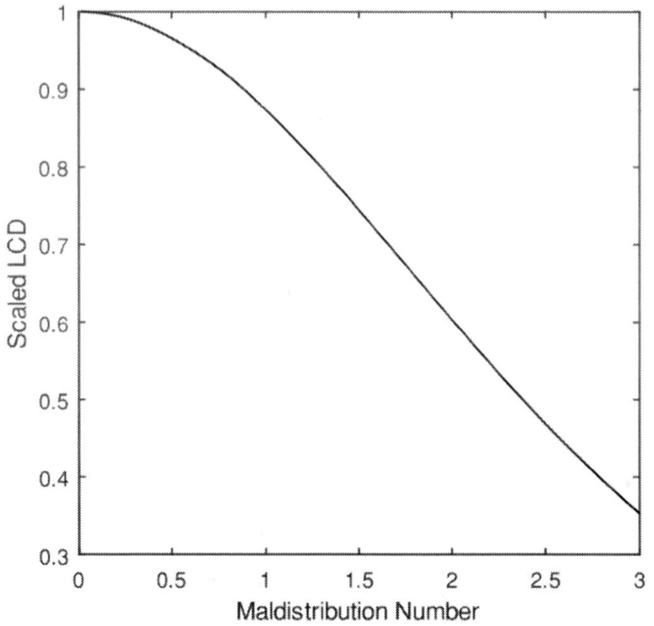

Fig. 6: A graph of the scaled limiting current density (LCD) as a function of the maldistribution number. The LCD is normalised to a case of uniform flow distribution (m = 0) to remove the effects of confounding variable such as inlet concentration and current efficiency.

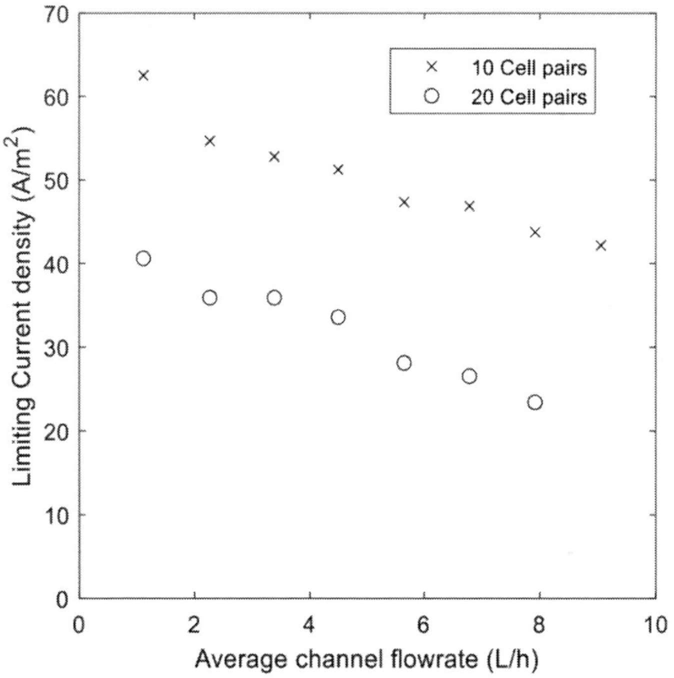

Fig.7: The experimentally measured LCD on a 10 and 20 cell pair stack while changing flow rate and salt concentration to maintain a constant average channel convective salt flux

4. Conclusion and Significance

This work has shown, for the first time, that maldistribution is prevalent in ED stacks and has a significant and detrimental impact on performance. The use of robust CFD techniques on realistic 3D geometries and experimental validation gives a high degree of confidence to these conclusions. Since maldistribution has not been considered historically, it has been unknowingly reducing performance of both industrial applications and lab experiments. This is impeding the decision-making process of industry as well as the depth of analysis that is achievable at all scales. It further highlights that our understanding of the engineering challenges of ED is far from complete and that performance may be significantly improved through fundamental research and mathematical modelling. The modelling techniques established herein can be used to develop methods to reduce or eliminate maldistribution by artificially increasing the pressure drop of channels nearer to the entrance. Subsequently, the LCD can be increased by as much as 50% for a lab-scale unit and potentially even more at industrial scale.

The fundamental and impactful nature of this research ensures that the conclusions are significant to applications beyond conventional electrodialysis. This includes bipolar membrane electrodialysis for chemical-free pH manipulation, flow batteries and reverse electrodialysis for power generation. This research can therefore aid the development and commercialization of many industrial processes which have been hampered by a less than ideal performance.

372s Utilizing Capillary Isoelectric Focusing to Identify Uremic Toxins in the Urine of Chronic Kidney Disease Patients

Haley Duncan, Christa N. Hestekin

Across America, 26 million adults suffer from chronic kidney disease (CKD). With a population of around 327 million, CKD afflicts one in ten people. The clearance of toxins in the blood directly relates with the kidney's capacity to filter blood properly. As CKD causes decrease in renal function, certain uremic toxins, otherwise known in dialysis as middle molecules, begin to gather, resulting in elevated amounts in the patient's urine. Elevated levels in urine directly correlates to elevated blood serum levels. When middle molecule levels, such as beta-2-microglobin and indoxyl-sulfate, accumulate across various systems of the body, medical complications arise caused by the addition of oxidative stress. Conventional methods fail to detect these species, as they are typically intermediates or degraded components of complicated biological processes. Capillary isoelectric focusing applies the isoelectric points (pI) of peptides and other molecular species to separate and distinguish them from each other. By the utilization of known pI markers, an unknown species may be identified from interpolation. Concentration data can be gathered and compared to normal urine. With this advancement, earlier detection of kidney disease may be possible. Early detection is crucial to avoiding the life-threatening, long-term effects of CKD.

Porous Ionic Liquids: Theoretical and Experimental Research

Hongping Li,[a] **Jinrui Zhang,**[a] **Jie Yin,**[a] **Wenshuai Zhu**[a,]*, **and Huaming Li**[a,]*

[a] *Institute for Energy Research, School of Chemistry and Chemical Engineering, Jiangsu University, Zhenjiang 212013, P. R. China*

E-mail: zhuws@ujs.edu.cn; lhm@ujs.edu.cn

Recently, based on the research of porous liquid materials, a new kind of porous ionic liquid has been reported [1]. The material combines the characteristics of ionic liquids with the advantages of porous materials, and has good performance in gas absorption and separation. In addition, we can see that the porous ionic liquid has the advantages of high specific surface area, excellent adsorption performance, high selectivity and high extraction activity, thermal stability of ionic liquid. Therefore, it has a good application prospect in the field of fuel desulfurization. This subject is divided into two aspects:

Firstly, the mechanism of absorption of CO_2, SO_2 and other molecules by silica-based porous ionic liquids is studied by using density functional theory (DFT) (Fig. 1a) [2]. It is found that the most active sites of the silica-based porous ionic liquid for absorption are hydrogen bonding, and the most active sites are around the polyethylene glycol unit. For SO_2 absorption, hydrogen bond and π-hole bonding are predicted by theoretical calculation (Fig. 1b). The strongest active sites are located at the interface between silica sphere and outer crown [3]. Secondly, a quaternary ammonium based porous ionic liquid was synthesized. The results of the test of the desulfurization performance of fuel extraction show that the desulfurization rate of the porous ionic liquid is larger than the sum of the desulfurization ratio of hollow silica ball and ionic liquid (45.7 SR%), indicating that the structure of the two parts has synergistic effect [4].

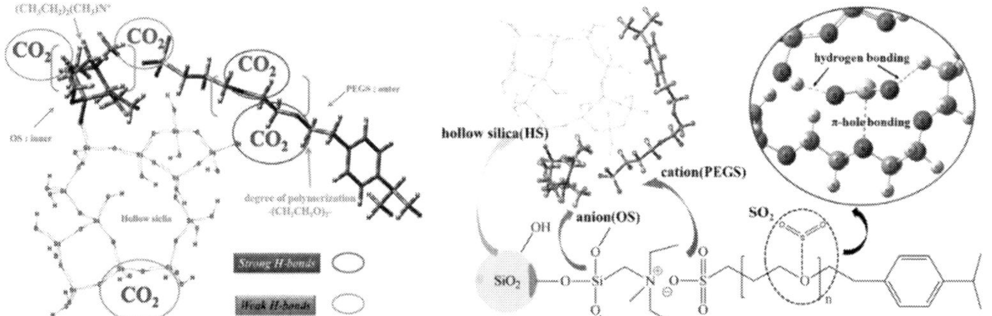

Fig. 1 silica-based porous ionic liquids (a) CO_2 absorption (b) SO_2 absorption

References

[1] (Journal) J. S. Zhang, S. H. Chai, Z. A. Qiao, et al., *Angew. Chem. Int. Ed.*, 2015, **54**: 932.

[2] (Journal) J. R. Zhang, N. X. Lv, H. P. Li. et al., *J. Mol. Graphics Modell.*, 2020, **100**, 107694.

[3] (Journal) J. Yin, J. R. Zhang, H. P. Li. et al., *J. Mol. Graphics Modell.* **2020**, *103*, 107788

[4] (Journal) J. R. Zhang, W. D. Fu, H. P. Li. et al., *Submitted.*

372u Eutectic Ionic Liquids for Efficient Gas Separation

Guokai Cui

Absorption of gases such as CO, CO_2 and SO_2 are highly important, because these gases threaten environment and people's health. Thus, new absorbent materials that can efficiently, reversibly, and economically capture CO, CO_2 and SO_2 must be developed. Eutectic ionic liquids (EILs) are considered as the alternative absorbents for gas capture due to the unique properties of ionic liquids. Compared with the conventional ILs, EILs are more attractive in CO, CO_2 and SO_2 absorption in recent years.

In this work, a series of functionalized EILs with multi-site anions were developed for efficient and reversible CO, CO_2 and SO_2 absorption through cooperative interaction. These functionalized EILs contain nitrogen and oxygen atoms as active multi-sites in their anions. It was found that EILs with Cu(I), imide anions and pyridinolate anions are efficient for CO, CO_2 and SO_2 capture under low-concentration, respectively. To study the selectivity of SO_2/CO_2, the SO_2 absorption and CO_2 absorption were performed using EIL-based task-specific deep eutectic solvents. Thus, high absorption capacities, rapid absorption kinetics, and low absorption enthalpies could be obtained through preorganization and cooperation strategy. The absorption mechanisms were investigated by using IR, NMR, and DFT calculation. This new process using multi-site anion-functionalized EILs is also applicable to the capture of other gases such as CO and NO. This work was supported by the Natural Science Foundation of China (No. 22078294).

References

- Cui, K. Jiang, H. Liu, Y. Zhou, Z. Zhang, R. Zhang, H. Lu, Highly efficient CO removal by active cuprous-based ternary deep eutectic solvents [HDEEA][Cl] + CuCl + EG, *Sep. Pruif. Technol.*, 2021, 274, 118985
- Cui, J. Liu, S. Lyu, H. Wang, Z. Li, R. Zhang, J. Wang, SO_2 absorption in highly efficient chemical solvent AChBr + Gly compared with physical solvent ChBr + Gly, *J. Mol. Liq.* 2021, 330, 115650.
- Cui, S. Lyu, H. Wang, Z. Li, R. Zhang, J. Wang Tuning the structure of pyridinolate-based functional ionic liquids for highly efficient SO_2 absorption, *Fuel*, 2021, 303, 121311.
- Cui, Quasi-chemisorption by Ionic Liquids Through Quasi-chemical Interaction, *Encyclopedia of Ionic Liquids*, 2021, DOI: 10.1007/978-981-10-6739-6_139-1
- Cui, S. Lyu, F. Zhang, H. Wang, Z. Li, Y. Li, J. Wang, Tuning Ionic Liquids with Functional Anions for SO_2 Capture through Simultaneous Cooperation of N and O Chemical Active Sites with SO_2, *Ind. Eng. Chem. Res.* 2020, 59, 21522-21529
- Cui, J. Liu, S. Lyu, H. Wang, Z. Li, J. Wang, Efficient and Reversible SO_2 Absorption by Environmentally Friendly Task-Specific Deep Eutectic Solvents of PPZBr + Gly, *ACS Sustainable Chem. Eng.* 2019, **7**, 14236-14246.
- Cui, Y. Li, J. Liu, H. Wang, Z. Li, J. Wang, Tuning Environmentally Friendly Chelate-Based Ionic Liquids for Highly Efficient and Reversible SO_2 Chemisorption, *ACS Sustainable Chem. Eng.* 2018, 6, 15292-15300.
- Huang, G. Cui, Y. Zhao, H. Wang, Z. Li, S. Dai, J. Wang, Reply to the Correspondence on "Preorganization and Cooperation for Highly Efficient and Reversible Capture of Low-Concentration CO by Ionic Liquids", *Angew. Chem. Int. Ed.* 2019, 58, 386-389.

373a Designing, Preparing and Characterizing Alkali Treated Graphitic Carbon Nitride Thin Film Composites (TFC) Membranes for Efficient Dye Salty Water Separation.

Ameya Tandel, Haiqing Lin

Ultra-high water permeance, almost zero salt rejection and ~100% dye rejection is of great interest for waste water treatment especially in textile dye-salty water separation. Herein, we fabricated alkali treated graphitic carbon nitride (GCN) ultrathin thin-film composite membranes using polydopamine (PDA) chemical crosslinking to have efficient dye separation from salts without sufficient loss in water passage. Crosslinking using PDA adhere GCN on commercial support such as polysulfone (PSF) and can fine tune surface channel sizes for effective dye water separation. First, the membranes were characterized using FTIR and water contact angle measurement to see incorporation of GCN/PDA on the surface of PSF supports. Moreover, GCN/PDA crosslinking conditions and GCN/PDA mass ratio was optimized for efficient dye-salty water separation. For e.g. 0.4 mg/0.4 mg loaded GCN/PDA on PSF20k support with crosslinking condition of 0-5°C/30 mins and 23 °C/10 mins showcases optimum separation performance (Water permeance: 345 ± 22 LMH/bar, % rejection of Congo red dye, 697 Da: 97.6 ± 0.4 and % rejection of Na_2SO_4 salt: 16 ± 0.5). Water permeance of GCN/PDA membrane is 30, 8 and 3.5 times higher than commercial NF-90 membrane, commercial ultrafiltration membranes (with molecular weight cutoff of 2000 Da) and graphene oxide (GO)/PDA crosslinked membranes, respectively. This efficient and attractive separation performance manifest application of GCN based membranes in dye-salty water separation.

373b Membrane Material Development for Spaceflight Wastewater Applications

Mary Lind, Elisabeth Thomas

Membrane processes are a vital part of the water treatment arena and a continuing area of technology development. Reverse osmosis (RO) accounts for a large portion of this technology, but is generally limited by TDS levels >100. The driving force for RO is realized via a hydraulic pressure applied to the feed side of the membrane, and utilizes dense membranes that are often polymeric. Pervaporation is a membrane process that shares some similarities with RO. In pervaporation, the membranes are typically dense and polymeric, with some other materials (such as zeolites), also being utilized. The driving force is the key feature of pervaporation, which is realized by either a sweep gas or a vacuum pull on the permeate side of the membrane, requiring a condensation/capture method downstream. Pervaporation is attractive in that it can handle highly concentrated waste streams and also be tailored for removal of water or organics from the feed stream, depending on membrane properties. Our research focuses on improving pervaporation membranes in order to enhance longevity potential of the membranes and target highly concentrated waste streams, such as those present in waste streams found on the International Space Station.

373c Effect of Ionic Liquid Cations and Anions on the Cellulose Acetate Membrane Characteristics and Performance for Water Treatment

Atta Ur Razzaq, Milad Esfahani, Jason Bara

The use of ionic liquids (ILs) as green solvents has gained a lot of interest in recent years due to their tunability and environmentally friendly compatibility as solvents. ILs have great potential as a new generation of solvents to replace the conventional organic solvents for the fabrication of membranes. In this work, the effect of three ILs, including 1-butyl-3-methylimidazolium chloride [C_4mim][Cl], 1-Butyl-3-methylimidazolium acetate [C_4mim][OA], and 1-dodecyl-3-methylimidazolium chloride [C_{12}mim][Cl] on cellulose acetate ultrafiltration membranes characteristics (pore size, pore morphology, physical integrity, hydrophilicity, etc.) and consequently on membrane performance (permeability, rejection and antifouling) were studied. A concentration of 10 wt.% of cellulose acetate was dissolved in 90 wt.% of each of these ILs at 110 rpm at 90°C for 72 hours and then degassed for 8 hours. The phase inversion method was employed for the synthesis of membranes using DI water as the nonsolvent. The scanning electron microscopy (SEM), Fourier transform infrared (FTIR), zeta potential and water contact angle of all three membranes were measured and compared to identify the effect of IL structure on membrane characteristics. Dead-end filtration was used to determine the permeability, dye rejection (methylene blue and methylene orange) and flux recovery ratio (FRR%) of all membranes.

373d 2D Assembly of Graphene Oxide Nanoribbon *Via* Slot-Die Coating and Its Application for Organic Solvent Nanofiltration

Dae Woo Kim, Ji Hoon Kim

Graphene Oxide Nanoribbon(GONR) is emerging material in various membrane research fields due to its excellent mechanical and chemical properties attributed to one dimensional(1D) structure with narrow width of single-layer graphene. When the GONR is dispersed in solvents, it can be used to build nanostructures from nanoplate to 3D scaffolds with increasing concentration. However, a high-aspect-ratio nanosheet made of GONR is hard to prepare, with 2D shape is better to use for coating and membrane fabrication. Herein, the slot-die coating method was used for GONR nanosheet fabrication by filtering 5 mg/mL concentration of GONR solution and the GONR nanosheets were used to prepare organic solvent nanofiltration(OSN) membrane. In the process of slot-die coating, nanoplate assembly of GONR was observed during injection through the die-head. When a 40 nm-thick GONR layer was coated on a porous polymer support, the membrane showed a high pure IPA permeance of 186 LMH/bar on dead-end filtration mode and 679 LMH/bar on cross-flow mode, both with 961 Da of molecular weight cutoff. The OSN performance is surpassing the previous polymeric and nanomaterials-based membranes and the fabrication method is scalable by the continuous slot-die coating method.

[REFERENCES]

1. H. Kim, *et al.*, Carbon 191, 563 (2022)

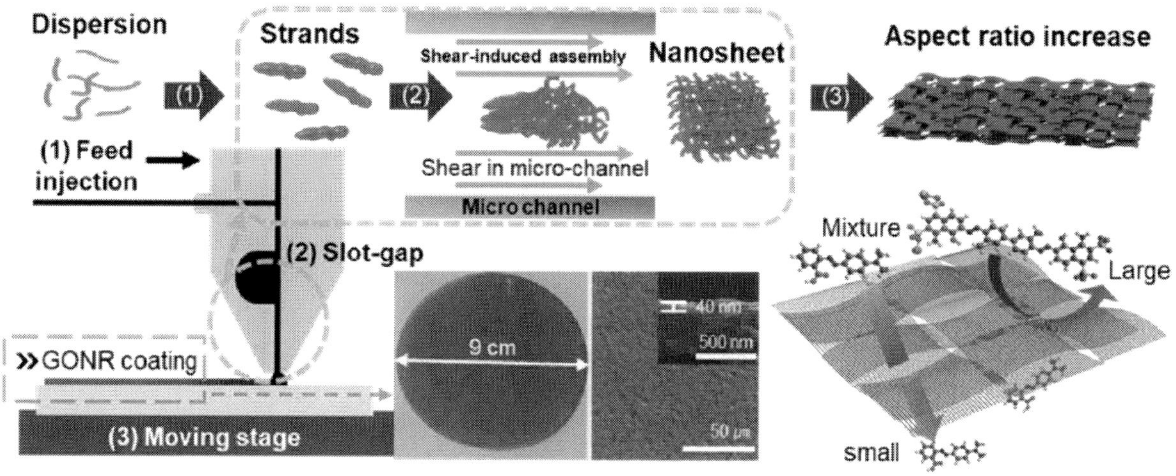

373f Microwave-Assisted Nanoporous Multilayer Graphene Membrane with Ultrafast Organic Solvent Nanofiltration

Junhyeok Kang, Dae Woo Kim

Organic solvent nanofiltration (OSN) technology has emerged due to its sustainability and cost-effectiveness in various industries for the purification and recovery of solvents. Graphene-based materials have been widely used for the OSN membrane owing to molecular sieving their interlayer structure and high stability in various solvents. However, the graphene-based membrane has suffered from low solvent flux by the narrow interlayer spacing for bulky organic solvent molecules. To enhance the solvent flux, a few approaches have been reported such as the use of nanomaterials, controlling interlayer spacing, and pore generation. Herein, we fabricated an sp^2-introduced nanoporous graphene membrane for OSN. The nanopores were activated first by using thermal treatment of graphene oxide which results in a turbostratic sp^3 carbon structure. Thereafter, the microwave treatment was used to introduce sp^2 carbons on the nanoporous graphene sheets. The prepared membrane exhibited ultrafast solvent permeance up to 13117 $Lm^{-2}h^{-1}bar^{-1}$ of acetone and high stability under the cross-flow system. In addition, the membrane performed switchable molecular sieving by differences in the swelling of graphene interlayer depending on solvents, which showed the capability of multiple chemicals separation for industries.

373g Treatment of Produced Water Using Sand Filtration and Ultrafiltration Using Modified α-Alumina Membranes

Anirban Ghosh, Diako Mahmodi, Michael Miranda, Clint Aichele, Seokjhin Kim

This study focuses on the step by step purification of produced water for obtaining clean water. The purification was achieved using a two-step purification process involving sand filtration and membrane ultrafiltration. Sand filtration was conducted to increase the longevity of the α-alumina membranes in order to separate impurities that would clog up membrane pores. Characterization was conducted of the produced water sample before and after the sand filtration step in order to understand the changes in the produced water composition. α-alumina membranes were used for the purpose of membrane ultrafiltration. Pristine α-alumina membrane although hydrophilic in nature may not be able to efficiently separate oily wastewater. In order to enhance oil-water separation, a hydrophilic modification was conducted on the α-alumina membranes the modification of α-alumina ultrafiltration (UF) membranes by surface functionalization using hydrophilic fumed silica particles. Hydrophilic modification of the membrane was conducted to enhance the fouling resistance and oil-water separation. The alumina membranes were treated with silica-nanoparticle solutions of varied silica loading from 0.025% to 3 wt.%. The impacts of various weight percentages of silica nanoparticles on the surface properties, membrane morphology, membrane flux and separation performance of the membranes were investigated. SEM (Scanning Electron Microscopy), EDX (Energy Dispersive X-ray spectroscopy), OCA (Oil contact Angle), and AFM (Atomic Force Microscopy) analyses were applied to characterize the pristine α-alumina and modified membranes. SEM, EDX and AFM served to characterize the membrane surface while OCA helped in understanding the wettability of the membranes.

373h Production of Nanofiber Membranes Using Centrifugal Spinning

Divine Kavunga, Beck Bruch, Cory Stone, Maryam Amouamouha, Travis Walker

Nanofibers that are formed into nonwomen mats can be utilized as a separation medium in membrane-based technology, considering their high separation efficiencies and relatively low costs. Additionally, their small diameter, high surface-to-volume ratio, high porosity, and unique mechanical behavior, as well as their ability to efficiently capture nanoparticles, provide promising opportunities for water and wastewater treatment. In this project, a centrifugal spinning (CFS) setup was designed, built, and operated to develop centrifugal nanofiber membranes. The designed CFS includes two nozzles that are mounted on a DC motor and a series of collector posts that can be placed at different radial distances. Different ultrahigh weight polymers, including dissolved polyethylene glycol (PEO) and polystyrene (PS 208) at different concentrations, are spun under similar conditions. The design and proper spinning parameters, including nozzle-collector distance, nozzle diameter, and rotational speed of the centrifuge, were investigated to control and investigate the structure, morphology, and strength of the nanofiber. The centrifugal fibers later follow a thermal post-treatment to increase their packing density, and graphene oxide will be used to improve the mechanical and antimicrobial properties of the centrifugally spun membranes. The performance of CFSs will be compared to commercial membranes (PES and PVDF) and electrospun membranes (ESMs). The properties and performance of CSMs were characterized using scanning electron microscopy (SEM), filtration process for particles removal and bacteria removal, FTIR spectra, and Storm 860 Molecular Imager for determination of the presence of bacteria cells on the surface of the membranes.

373i Ceramic Membrane Synthesis, Application and Characterization

Tejay Lanjewar Sr., Mahesh Varma

In today's era separation plays a very important role in the economic conservation of many systems. This membrane played a very important role. The ceramic membrane can handle a high temperature and corrosive chemical environment. Hence ceramic membrane can be effectively used for the various separation process and can also be effective for combine reactions and separation.

In separation we have used a ceramic membrane for corrosion acid separation Ceramic membrane using Silicon carbide (SiC) is fabricated and it is used for the separation of water from spent acid. As acid was used in many chemical industries, particularly in the explosive industry. SiC membrane is hydrophilic and to fill the gap between the particles of the SiC particles TiO_2 particles are used. Due to this porosity of the membrane is decreased and which helps in the separation. Various characterization of the membrane is done to check the quality of the membrane such as FTIR, FESEM, EDS and XRF. In the permeate section, a good quantity of water is collected which gives a good separation factor.

Desalination is the requirement of the future to maintain a safe and secure water supply around the globe due to water constraints. The current material for fabricating ceramic membranes is expensive. To save money on raw materials, a by-product from a thermal power plant was used to manufacture the membrane. The membranes are 2 mm thick and have a 0.9 m pore diameter on average. The membrane was grafted with poly dimethyl siloxane to give it hydrophobicity. When compared to other available ceramic materials, the membrane's thermal conductivity, which is a key parameter in desalination, was found to be very low, in the range of 0.046 W/mk. The redesigned membrane in the direct contact membrane distillation module was utilized to undertake experimental tests under varied conditions using a feed of varying molarity solutions. The selectivity of the membrane was found to be 99 %. The permeate flux values were in the range of 1-13 Kg/m^2hr which are comparative to flux achieved by polymeric membranes. The membrane's contact angle was discovered to be 119.2 degrees, and the liquid entrance pressure was measured to be 85 KPa. The temperature of the membrane wall was also computed, allowing the theoretical and experimental results to be compared.

The reaction and separation will take place on the membrane as a result of the process intensification. This will assist in lowering the cost of producing methane gas. Although hydrogen can be utilized as a fuel, storage issues emerge, and methane has a higher calorific value than hydrogen. The base material can be made of a variety of materials, however, nickel has a higher selectivity for methane and can be employed efficiently for separation. Because carbon causes a transient deactivation. It will only be on the membrane's upper surface and will be easily removed.

373k Progress in the Development of Membranes for Pressure Retarded Osmosis Process

Syed Zaidi, Haleema Saleem

Progress in the development of membranes for pressure retarded osmosis process

Haleema Saleem, Syed Javaid Zaidi

Center for Advanced Materials (CAM), Qatar University, Qatar

E-mail address: szaidi@qu.edu.qa

Abstract

Pressure retarded osmosis (PRO) has a huge possibility to develop into a viable process for converting the osmotic pressure of a salt solution into hydraulic pressure. It is an osmotically-driven membrane-based process that uses the energy of mixing between low- and high-salinity streams for producing mechanical energy. In this sustainable power generation process, water permeates across a semipermeable membrane from a feed stream of low concentration into a high concentration. Most of the proposed applications of the PRO process are established on the usage of the osmotic pressure of the brine solution obtained from the seawater reverse osmosis process. The PRO process is an effective way of brine management as it helps to dilute the brine to seawater salinity and could be directly discharged to the sea and also it produces energy at the same time. The current study analyzed the latest advances as well as utilization of this process, and possible energy that can be generated from salinity gradient sources in a multistage and single stage PRO processes. A main challenge in the PRO process is obtaining a commercially available membrane that blends the features of the reverse osmosis membrane (for withstanding higher hydraulic pressure) and forward osmosis membrane (to reduce the concentration polarization phenomenon). To address this challenge, facts regarding the commercially available PRO membranes along with the advanced lab-prepared PRO membranes are presented in the current study. The capability of the PRO process is shown by explaining the process design, salinity gradient resources, the energy required for pretreatment, the dual-stage PRO (DSPRO) process, and the PRO desalination systems. It is expected that this study could support in extensively understanding the PRO process and therefore provide significant information to simulate future research as well as development.

Keywords: Membrane, Dual-stage pressure retarded osmosis, Salinity gradient

Acknowledgement

The authors gratefully acknowledge the support of NPRP13S-0205-200263 project and the findings of the study are solely the responsibility of the author[s].

373l Separation of Olefin-Paraffin Mixtures with Optiperm™ customized Amorphous Fluoropolymer Membranes

William Charlton, Christine Parrish, Ken Loprete, Marissa Bonanno, Sudip Majumdar

Compact Membrane Systems has developed the novel Optiperm™ membrane platform for decarbonizing today's industry across three applications: olefin paraffin separation, biogas, and carbon capture. The separation of light olefins from paraffins is one of the most energy intensive in the world. The state of the art separation method is distillation, a technology that has remained unchanged over decades. Membranes hold the promise of a more modular, energy efficient and capital efficient method for separation. However, they have historically been hampered by low performance and stability.

CMS' Optiperm™ membrane it the first membrane developed to overcome the hurdles of practicing facilitated transport in the field. By utilizing a silver embedded fluoropolymer material to selectively transport olefins, the membrane is chemically stable enough to stand up to process poisons. After demonstrating over 3 years of longevity in laboratory operation, Optiperm™ membranes were trialed at the Delaware City Refining Company in over 300 days of field operations validating the robustness and capability of the technology in an operating plant environment and CMS began the process of scalingthe membranes to industrial size.

To continue the commercialization of Optiperm™ Membranes, CMS' has partnered with Braskem to jointly develop and fabricate a multistage pilot unit to test commercial scale spiral wound membranes. This pilot started up at the Braskem Marcus Hook facility in March 2022 to test commercial scale membrane modules with active areas of >100 ft2 under field operating conditions using actual process streams. Furthermore the pilot units design incorporates all of the process equipment and operations of a commercial scale unit including an advanced process control system that enables the system to simultaneously produced two high quality product streams. Scale up and pilot design considerations will be discussed. We will share the latest learnings from our pilot field testing with Braskem as well as provide updates on the scale up of our flat sheet spiral wound membrane cartridges.

373j Implications of Shear and Cation Choice on Dynamics of Nonliving Natural Organic Matter

Kathlyn Mealio, Holly A. Stretz, Martha J. M. Wells, Katherine Slamen, John Clark

Nonliving natural organic matter (NNOM) is a common material found in environmental water that presents a fouling problem in water treatment membranes. Humic acid (HA) falls under the category of NNOM and displays dynamic particle size characteristics in response to factors such as shear rate, pH, ionic strength, and temperature. These variables have been hypothesized to impact the aggregation and disaggregation of NNOM. The behavior of NNOM found in pond water as a function of shear rate was investigated under constant 0.4M $CaCl_2$ concentrations. A rheometer was used to induce shearing of the NNOM in the range of 10 to 400 sec^{-1}. The Particle Size Distribution (PSD) displayed a complex relationship between shear rate and relaxation behavior. Non-Newtonian behavior of the sample(s) was also explored. Preliminary data shows the disappearance of certain size ranges at specific shear rates and colloidal behavior.

373k Investigation of Correlation Between Membrane Characteristics and Rejection Rate of Ions

Kiwoong Kim

As a growing global population and development of industry, water consumption has shown an obvious rising trend. However, the available water resource is limited and conventional water filtration processes need not only huge amounts of energy but large-scale infrastructure. To break this, understanding of ionic behavior in the region near a charged surface is important for development of small-scale water filtration devices. In this study, formation of ion depletion zones and water filtration performance were experimentally investigated using membranes with varying surface charges and pore sizes. These membranes were fabricated using graphene oxide, poly(sodium 4-styrenesulfonate), and sodium dodecyl sulfate. The respective rejection rate of the fluorescent dyes and the ionic solutions was noticeably high as revealed by the mechanism of Donnan exclusion theory and size exclusion. In addition, the proposed membrane exhibited an antimicrobial property. These results suggested that the proposed filtration membrane can be hired as a compact and small-scale water filtration device because of its strong potential for desalinating brackish water.

*This work was supported by the National Research Foundation of Korea(NRF) grant funded by the Korea government(MSIT) (No. NRF-2020R1C1C1003813).

373m Petrified Hollow Fiber Membranes with Hierarchical Pores

Ching-En Ku, Lu Liu, Chen Zhang

Inorganic membranes often rely on porous oxide hollow fibers as substrates, which are traditionally made by high-temperature methods. For example, the phase inversion-sintering method typically occurs at 1200-1500 °C. These high-temperature fabrication processes are expensive, which limit the large-scale applications of inorganic membranes for chemical separation and filtration. In this poster, we will present the fabrication of novel porous oxide hollow fiber membranes, which we name "petrified hollow fiber membranes", via petrification of polymer hollow fiber templates at a temperature (600 °C) far below the traditional phase inversion-sintering method. Like petrified wood reproduces the geometry and cell structure of wood, petrified hollow fiber membranes reproduce the hollow fiber geometry and asymmetric macropore structure of polymer hollow fiber templates. These petrified hollow fiber membranes have a unique hierarchical micro-/meso-/macroporous structure giving both high surface area and large pore volume. Their asymmetric macropore structure allows the petrified hollow fiber membranes to efficiently reject macromolecular solutes (e.g., protein) in cross-flow filtration tests. They are also highly gas permeable, and therefore are potentially suitable as substrates of inorganic membranes for gas and vapor separations. The petrified hollow fiber membranes were extensively characterized by scanning and transmission electron microscopy, thermogravimetric analysis, nitrogen and argon physisorption, mercury intrusion porosimetry, infrared spectroscopy, and water vapor adsorption, which allowed us to propose a hypothetical formation mechanism of these novel structures.

References:

[1]Liu, L.; Ku, C.-E.; Zhang, C., Petrified Hollow Fiber Membranes with Hierarchical Pores. *ACS Materials Letters* **2022**, 938-943.

373n Improving Propylene/Propane Separation Performances of Polycrystalline ZIF-8 Membranes Via Additive-Assisted Microstructural Modification

Donga Kang, Hae-Kwon Jeong

Due to their promising propylene/propane (C3) separation, polycrystalline ZIF-8 membranes have been extensively studied for the last decade.[1] Because of the nature of the polycrystalline membranes, however, it is of critical importance to control the grain boundary structures (defects) of ZIF-8 membranes to ensure high reproducibility and separation performance.[2] There are two ways to obtain improved grain boundary structures: 1) improving synthesis methods and 2) healing defects by post-synthetic processing.

In this presentation, we plan to show how a simple addition of an additive can modify crystal growth and improve the microstructures of polycrystalline ZIF-8 membranes, thereby improving their C3 separation performances. ZIF-8 membranes were prepared using our previously-reported microwave-assisted seeding and secondary growth method.[3] We found that when a diol-based additive was added into the secondary growth solution, the grains of the ZIF-8 membranes exhibited unusual plate-like morphology. The resulting ZIF-8 membranes showed significantly improved C3 separation performances as compared with the neat membranes synthesized without the additive, showing impressively high C3 separation factor of ~ 283 with the propylene permeability of ~ 12.4 Barrer. The ZIF-8 membranes outperform most of previously reported high-quality polycrystalline ZIF-8 membranes. The high separation performance of the membranes was attributed to enhanced grain boundary structures resulting from plate-like ZIF-8 grains.

References

[1] H.K. Jeong, Metal–organic framework membranes: Unprecedented opportunities for gas separations, AIChE Journal 67(6) (2021) e17258.

[2] Y. Ren, X. Liang, H. Dou, C. Ye, Z. Guo, J. Wang, Y. Pan, H. Wu, M.D. Guiver, Z. Jiang, Membrane-Based Olefin/Paraffin Separations, Advanced Science 7(19) (2020) 2001398.

[3] H.T. Kwon, H.-K. Jeong, Highly propylene-selective supported zeolite-imidazolate framework (ZIF-8) membranes synthesized by rapid microwave-assisted seeding and secondary growth, Chemical Communications 49(37) (2013) 3854-3856.

373o Dehydrated Channel Protein-Based Biomimetic Membranes with High Breathability and Protective Capability

Hyeonji Oh, Yu-Ming Tu, Benny D. Freeman, Manish Kumar

Due to the vulnerability of human skin, protective fabrics have been developed to provide protection against hazardous substances. While current dense impermeable protective materials reject exogenous materials effectively, their low water vapor permeability makes wearers suffer from heat stress. In this work, we designed channel protein-based biomimetic membranes to offer both high breathability and protective capability by emulating the structure and function of skin. With highly packed well-defined structured pores (representing sweat pores), it shows high water vapor transport rate exceeding the performance of commercial breathable fabrics. By mimicking a 'brick and mortar' structure of human skin, we achieved high protective capacity against a range of dye molecules and a model biological agent, MS2 bacteriophage. We propose a side change rearrangement of the channel proteins to explain their unique performance change in liquid water transport upon dehydration.

373p Effect of Grafting Density and Sidechain Length on Mechanical and Gas Transport Properties of Poly(ladder) Romps

Sherrie Qian, Kayla Storme, Sharon Lin, Francesco Maria Benedetti, Timothy Swager, Zachary Smith

Chemical purification methods like distillation, drying, and evaporation have been widely used for centuries. However, these separation processes account for approximately 16 quadrillion BTUs (16 Quads) of energy in the United States annually, with distillation responsible for 53% of this energy use and roughly 16% of all CO_2 emissions. Replacing these processes with more energy efficient alternatives could save 100 million tonnes of CO_2 emissions and $4 billion in energy costs annually. Membranes are an attractive alternative to many separation processes since they do not require phase changes, moving parts, or thermal regeneration to maintain separation performance. Incorporating membrane-based technologies could lead to a 90% reduction in energy costs compared to distillation.

Membranes made from polymers like polysulfones, polyimides, polyphenylene oxides, cellulose acetates, and silicon rubber are already used in industry for hydrogen recovery, nitrogen production, vapor recovery, and natural gas treatment. Despite being in use for several decades, these membranes are limited in performance for other separation processes like CO_2 capture from flue gas. Additionally, many of these aforementioned polymers lack plasticization resistance (thereby reducing size-sieving ability) and suffer from physical aging (leading to decreased permeability over time). Polymers of Intrinsic Microporosity (PIMs), in particular PIM-1 have a ladder-like structure that limits dense packing and therefore has high free volume. PIM materials have shown high selectivity and permeabilities for CO_2 and other mixed and pure gases due to their backbone structure. However, like many other membranes, PIMs experience physical aging and are prone to plasticization.

To address these limitations, microporous bottlebrush polymers featuring rigid, pore-generating side chains on flexible backbones were synthesized through a reaction based on Ring Opening Metathesis Polymerization (ROMP). Macromonomers were produced from Diels-Alder reactions before converting into high molecular weight polymer with low polydispersity by ROMP. The first-generation poly(ladder) ROMPs showed tremendous potential for membrane-based gas separations. OMe-ROMP and CF_3-ROMP displayed permeabilities and selectivities similar to PIMs, and demonstrated stability against pure-gas CO2 plasticization with no discernable plasticization pressure point up to 51 bar. CF_3-ROMP also exhibited record-breaking CO_2 permeability, ideal for carbon capture applications.

One limitation of these poly(ladder) ROMPs is their tendency to form brittle films. Previous studies suggests that bottlebrush melts and lightly crosslinked gels can achieve higher strain at breaks by increasing the degree of polymerization of the side chains (n_{sc}) and by decreasing grafting density, resulting in super elastic materials that are 1,000 times softer than conventional elastomers.

In this study, OMe-ROMP macromonomers were either separated by molecular weight using silica gel chromatography and underwent ROMP to produce poly(ladder) ROMPs with monodispersed sidechains or copolymerized with simple norbornenes to probe the effects of increasing n_{sc} and grafting density on the elasticity and gas separation performance, respectively. Using copolymerization to alter the grafting density of poly(ladder) ROMPs resulted in more elastic films with greater selectivity, but decreased permeability and plasticization resistance. As the sidechain length increased, more elastic films were formed with improvements to gas separation performance and increased CO_2/CH_4 mixed-gas plasticization resistance. Altering grafting density and length of the macromonomers provided greater insight into the design of poly(ladder) ROMPs that gives rise to greater mechanical elasticity, unprecedented plasticization resistance, and remarkable gas separation performance.

373q Unexpectedly High Propylene-Selective Mixed-Matrix Membranes with Facile *in-Situ* ZIF-8 Filler Formation Process

Yinying Hua, Sunghwan Park, Hae-Kwon Jeong

In-situ MMM fabrication is an emerging strategy that might potentially overcome the limitations of the conventional blending methods. In this regard, metal-ogranic frameworks (MOFs) offer unique opportunity for *in-situ* MMM fabrication due to their labile chemistry and organic moieties that are compatible with polymers. Zeolitic-imidazolate framework-8 (ZIF-8), consisting of a zinc center tetrahedrally coordinated with 2-methylimidazolate linkers, is one of the most promising molecular sieve materials for propylene/propane separation. Recently, our group developed a scalable mixed-matrix membranes fabrication strategy using *in-situ* Zif-8 formation approach, named PMMOF[1], which enables formation of MMMs not only with improved gas separation performances but also in a scalable geometry (i.e., asymmetric hollow fibers with submicron thick selective MMM skin layers) [2]. However, the PMMOF suffers from two major challenges such as relatively complicated fabrication steps and limited choice of polymer matrices (i.e., polyimide-based polymers).

In this presentation, a novel second-generation PMMOF is promoted to address the challenges of the first-generation. The new process, termed "PIMOF", enable rapid formation of high ZIF-8 loaded (> 50 wt%) asymmetric MMMs. The as-prepared MMMs showed unprecedentedly high C_3H_6/C_3H_8 (C3) separation performance as compared to other propylene-selective MMMs reported so far with the C_3H_6 permeance of ~ 7.5 GPU and C_3H_6/C_3H_8 separation factor of ~ 107, comparable even with polycrystalline ZIF-8 membranes. The unexpectedly high C3 performance of the membranes was attributed to the uniform distribution of the extremely small ZIF-8 nanoparticles (~ 5-10 nm) in the selective skin layers and improved molecular sieving properties of the ZIF-8 nanofillers resulting from the hindered linker flexibility upon confined filler formation.

Reference:

1. Park, S., M.R. Abdul Hamid, and H.-K. Jeong, *Highly Propylene-Selective Mixed-Matrix Membranes by in Situ Metal–Organic Framework Formation Using a Polymer-Modification Strategy*. ACS Applied Materials andamp; Interfaces, 2019. **11**(29): p. 25949-25957.
2. Park, S. and H.-K. Jeong, *Transforming polymer hollow fiber membrane modules to mixed-matrix hollow fiber membrane modules for propylene/propane separation.* Journal of Membrane Science, 2020. **612**: p. 118429.

373r Carbon Molecular Sieve Gas Separation Membranes Pyrolyzed from Aromatic Barrier Polymers

Gaurav Iyer, Chen Zhang

Polymer membranes have excellent scalability and good separation performance for many gas pairs. However, owing to their permeability-selectivity tradeoff, their large-scale gas separation application has yet to evolve beyond air separation and natural gas purification. Carbon molecular sieve (CMS) membranes are derived by controlled pyrolysis of polymer membranes to provide a good balance of permeability, selectivity, and scalability. Formation of CMS membranes often relies on pyrolysis of advanced and expensive polymer precursors, such as polymers of intrinsic microporosity and fluorinated polyimides. Barrier polymers are much less expensive and used for beverage packaging at large scale. Barrier polymers, however, have been rarely considered for membrane-based gas separations due to their ultra-low gas permeabilities.

In this poster, we will present the transformation of impermeable barrier polymers to high-performance CMS membranes. An aromatic barrier polymer precursor was synthesized by condensation polymerization, which showed ultra-low CO_2 permeability (<0.002 Barrer) at room temperature. Pyrolysis of the barrier polymer precursor provided a novel CMS membrane that had over 100,000 times higher CO_2 permeability than the barrier polymer precursor. The barrier polymer-derived CMS membrane showed attractive selectivities, which put them above the polymer upper bounds for several commercially important gas pairs.

373s Separation of HFC-32, HFC-125, HFC-134a, HCFC-22, and HFO-1234yf Using Copolymers of Perfluoro(butenyl vinyl ether) and Perfluoro(2,2-dimethyl-1,3-dioxole)

Abby Harders, Erin Sturd, Luke Wallisch, Mark Shiflett

Recent legislation has called for the phase-out of greenhouse gases that trap heat in our atmosphere. One specific class of compounds targeted for phaseouts are hydrofluorocarbons (HFCs). HFCs are fluorinated gases commonly used in refrigeration and cooling systems. With an increased legislative focus on reducing the use of high GWP refrigerants, methods for separating and repurposing low GWP refrigerants from high GWP mixtures are imperative. Many refrigerant mixtures are azeotropic, making their separation difficult and energy intensive. Membranes provide the opportunity to separate refrigerant mixtures and achieve a high selectivity while being less energy intensive in comparison to other separation technologies. In this work, amorphous perfluorinated copolymers of perfluoro(butenyl vinyl ether) (PBVE) and perfluoro(2,2-dimethyl-1,3-dioxole) (PDD) were used to investigate the permeability, solubility, and diffusivity of 5 different refrigerants: difluoromethane (HFC-32), pentafluoroethane (HFC-125, CHF_2CF_3), 1,1,1,2-tetrafluoroethane (HFC-134a, CH_2FCF_3), chlorodifluoromethane (HCFC-22, $CHClF_2$), and 2,3,3,3-tetrafluoropropene (HFO-1234yf, $CH_2{=}CFCF_3$). Results can be used to inform the separation of a variety of commonly used refrigerant mixtures, such as R-410A (50 wt% HFC-32 and 50 wt% HFC-125) and R-407C (23 wt% HFC-32, 25 wt% HFC-125, and 52 wt% HFC-134a).

373t Iptycene-Based Polybenzimidazole Membranes for H_2/CO_2 Separation

Mengdi Liu, Ruilan Guo

H_2/CO_2 separation relatable to pre-combustion CO_2 capture from syngas after water-gas shift reactions demands robust polymer membrane materials with high thermal stability and strong size-sieving capability at high temperatures. Polybenzimidazole (PBI), such as *m*-PBI (Celazole®), has been the focus of increasing studies due to its good H_2/CO_2 separation properties and high thermal stability. However, *m*-PBI suffers from extremely low permeability. Although many previous studies were attempted to improve *m*-PBI's permeability by incorporating bulky moieties or large substituent groups, these strategies typically resulted in compromised size-sieving capability, particularly at high temperatures. On the other hand, modified PBI-based membranes, such as crosslinked PBIs, that were able to achieve ultrahigh selectivity often exhibited a significant loss of H_2 permeability. To address this issue, we developed a novel macromolecular design to incorporate shape-persistent iptycene (i.e., triptycene and pentiptycene) units into PBI structure to instill configurational free volume elements. The non-collapsible and well-defined microcavities originating from intrinsic molecular configuration of iptycene units are able to significantly enhance gas permeability as well as size sieving property. Moreover, by systematically varying types of iptycene units and pendent groups, structure-property relationships of these novel iptycene-based PBIs were investigated. In this work, a triptycene-based PBI (TPBI) and a pentiptycene-based PBI (PPBI) have been produced, and gas transport properties have been characterized. We demonstrated that by incorporating triptycene units, H_2 permeability was effectively improved by 300% relative to *m*-PBI. The bulkier pentiptycene units could disrupt chain packing more efficiently, exhibiting 30 times higher H_2 permeability than that of *m*-PBI. Post treatments including thermal treatment and crosslink via acid doping were performed to further tailor microporosity, leading to versatile and significantly improved separation performance. We found that thermal treatment at 350 ^{o}C favored rearrangement of TPBI chains to unblock more microcavities, resulting in further enhanced H_2 permeability by 200% relative to the untreated sample, which indicates the good temperature tolerance of configurational free volume. After crosslinked by doping with phosphoric acid, the packing of polymer chains was tightened due to proton transfer and hydrogen bonding interactions. As a result, the acid-doped TPBI exhibited largely improved H_2/CO_2 selectivity by 100%, outperforming the 2008 upper bound for H_2/CO_2 separation. This talk will discuss the preparation and characterization of novel iptycene-based PBIs membranes as well as the influence of configurational free volume, thermal treatment, and acid doping on their gas transport properties.

373w Application of 2^6-Factorial Design Optimization for PVA Based Membrane Synthesis for CO_2/N_2 and CO_2/CH_4 Separation

Iris Samputu, Bhavya Bhat, F Handan Tezel

The consumption of hydrocarbons and therefore the rise in global temperatures is predicted to continue increasing with the increase in globalization and the industrialization of third world countries. Therefore, separation of carbon dioxide (CO_2), the main greenhouse gas causing global warming, is of great interest. Some example separation applications are: CO_2 separation from biogas so that it can be upgraded as a renewable source of natural gas, as well as CO_2 separation from other flue gases. This research focuses on the separation of CO_2 from methane (CH_4) and nitrogen (N_2) gases using membranes. Amine-based Poly vinyl alcohol (PVA) polymeric membranes that had shown good potential for gas separation results were adapted for use in this research. The adapted membrane was then optimized using a 2^6 factorial design to optimize the membranes' performance with respect to CO_2/N_2 and CO_2/CH_4 selectivity values when tested at near atmospheric conditions. It was established that the highest selectivity, 2.45 was achieved with the CO_2/CH_4 pair where the combination of PVA, formaldehyde and Poly (allylamine hydroxide) most positively affected CO_2 selectivity over CH_4. This was a 145% increase from the base PVA membrane, which had a CO_2/CH_4 selectivity of 1.00. The most optimized CO_2/N_2 selectivity obtained was lower at a maximum selectivity of 2.22 with PVA being the sole component that most positively affected CO_2 selectivity over N_2. Furthermore, the maximum permeability value was achieved by CO_2 at 521,935 Barrers followed by N_2 and CH_4 at 521,880 Barrers and 512,708 Barrers, respectively.

373x Feasibility Study of Membrane Reactor for Reverse Water Gas Shift By ZSM-5 Zeolite Membrane

Motomu Sakai, Masahiko Matsukata, Kyoka Tanaka

Reverse water gas shift (RWGS) is one of the promised techniques for carbon utilization. To increase the level of conversion of RWGS, higher reaction temperature is required because of the large endotherm of this reaction. When water is selectively removed from reaction system by using membrane, high conversion would be obtained at lower temperature. In this study, we developed a membrane reactor for RWGS using ZSM-5 zeolite as water-selective membrane and investigated the effect of dehydration on improvement of the level of conversion.

ZSM-5 membrane was synthesized on the outer surface of porous tubular a-alumina support by a secondary growth method. The outer surface of support was seeded by a dip-coating method using a slurry of ZSM-5 seed crystals. A seeded support was hydrothermal treated at 453 K for 12 h. After hydrothermal treatment, the membrane was washed with boiling water and dried at 383 K for 12 h, and then we finally obtained a ZSM-5 membrane.

RWGS was taken place using the mixture of CO_2/H_2 (= 1/3, 24 mL(STP) min^{-1}) as feed gas with 3.2 g of Cu/Zn/g-Al_2O_3 (CuO/ZnO/Al_2O_3 = 40/40/20wt%). as a catalyst prepared by polymerizable complex method. Reaction temperature was changed in the range of 513 – 610 K. The membrane was placed in reactor by graphite o-rings and catalyst located around membrane. For comparison, RWGS without ZSM-5 membrane was also carried out using non-porous glass tube instead of membrane.

Here, we compared the yield of CO in RWGS with and without ZSM-5 membrane. Unsurprisingly, the yield of CO without membrane were almost the same as equilibrium conversion. In contrast, CO yield in membrane reactor exceeded that without membrane and thermodynamic equilibrium; i.e. the yield of CO with and without membrane at 586 K were 28.9 and 25.1%. About 20% of water generated in reaction system was removed by ZSM-5 membrane.

By selective removal of water, equilibrium conversion was shifted according to *Le Chatelier*'s law, and then CO yield improved over thermodynamic equilibrium constraint. In addition, it is expected that CO yield further increase with increasing the removal ratio of water from reaction system.

373y Integrated Module Array with Facilitated Transport Membranes for Enhanced H_2 Recovery from Syngas

Yang Han, Yutong Yang, Ruizhi Pang, W.S. Winston Ho

Amine-containing facilitated transport membranes (FTMs) are of great interest for syngas purification. Herein, we report a new method to enhance the CO_2 permeance by manipulating the steric hindrance of the amine carrier. A series of α-aminoacids with different alkyl or hydroxyethyl substituents were deprotonated by 2-(1-piperazinyl)ethylamine, resulting in nonvolatile aminoacid salt carriers with different degrees of steric hindrance. In the presence of moisture, a bulkier alkyl substituent increased the steric hindrance and hence destabilized the carbamate adduct to afford bicarbonate through hydrolysis. This drastically increased the chemisorption of CO_2. The enhanced CO_2 solubility significantly mitigated the carrier saturation, and an unprecedented CO_2/H_2 selectivity greater than 125 was demonstrated at 107°C and 13.8 bar of CO_2 partial pressure (Type I). As the CO_2 partial pressure reduced to 1.1 bar, a less hindered amine yielded a higher reactive diffusivity of CO_2, resulting in a CO_2 permeance of 217 GPU with a selectivity greater than 268 (Type II). These two types of FTMs were successfully scaled up by a roll-to-roll continuous coating machine. The 14″ wide scale-up membranes were then rolled into two prototype spiral-wound modules containing 800 and 1600 cm^2 of the Type I and Type II membranes, respectively. At 90% CO_2 capture, the Type I and Type II modules achieved a CO_2 purity of ca. 95%, but the H_2 recoveries were only 98.2% and 97.1%, respectively. However, when the two modules were connected in series with the Type I membrane treating the syngas near the feed inlet (i.e., high CO_2 partial pressure) and the Type II membrane separating the gas in the proximity of the retentate outlet (i.e., low CO_2 partial pressure), the H_2 recovery was improved to 99.3%. The integrated membrane array also demonstrated 200-h stability under simulated syngas.

373z Process Design and Optimization of Membrane-Cryogenic CO_2 Capture Process for Industrial CO_2 Emissions

Mun-Gi Jang, Jin-Kuk KIM

In order to achieve carbon neutrality by 2050, countries around the world are preparing various measures to reduce CO_2 emissions. Also, the attention of CO_2 capture technology is increasing even in industrial sites that emit large amounts of CO_2, such as cement, steel, and oil refining. Accordingly, demand for and research on CO_2 capture technology is expected to increase continuously.

Among the CO_2 capture technologies, the membrane is one of promising and attractive alternatives due to small footprint, no requirement of thermal energy, and environmentally friendly. In addition, it has been reported that a membrane-based hybrid process such as membrane-cryogenic distillation process has the potential to achieve high energy-efficiency and cost-effectiveness for treating high CO_2 contents of flue gas.

In this study, process design of membrane-cryogenic CO_2 capture process was carried out for industrial CO_2 emission sources. The optimization framework with the objective function of CO_2 capture cost can determine optimal configuration and operating conditions for hybrid capture systems, in which design interactions between the membrane and cryogenic CO_2 capture process are systematically evaluated.

Acknowledgement

This work was supported by the National Research Foundation of Korea (NRF) grant funded by the Korea government (MSIT) (No. 2019R1A2C2002263) and by the Korea Institute of Energy Technology Evaluation and Planning (KETEP) and the Ministry of Trade, Industry andamp; Energy (MOTIE) of the Republic of Korea (No. 20214710100060).

373aa Elucidating Key Factors Dominating Natural Gas Sweetening Efficiency of Membranes

Yang Liu

The world is in the middle of an energy transition period. Although renewable energy sources are expected to take an increasing share of the energy mix, natural gas, followed by oil, will still become the world's primary energy source towards 2050. Indeed, natural gas can play a central role in supporting energy security alongside variable renewables during the transition. Moreover, combusting natural gas generates ~50% less carbon than burning an equivalent amount of coal. Natural gas production will continuously increase in the next 30 years; however, producing "clean" natural gas products is still expensive and energy-intensive by using thermally-driven amine absorption methods for treating undesirable acid impurities, e.g. hydrogen sulfide (H_2S) and carbon dioxide (CO_2). By now, roughly 40% of current global natural gas resources are contaminated with high concentrations of H_2S and CO_2, and H_2S concentrations in the Middle East can reach 30 mol.% along with significant amount of CO_2. Applying membrane methods can reduce the high cost, whereas it is still challenging to achieve high H_2S and CO_2 simultaneous removal efficiency using a single membrane device. This presentation will introduce our recent progress in membrane-based natural gas sweetening, i.e. (H_2S+CO_2) simultaneous removal. Specifically, we will discuss the key factors, e.g. feed gas composition, temperature, pressure, and gas sorption, diffusion, competition, polymer plasticization etc., that dominate the overall membrane separation performance. The findings being delivered in this presentation are important for guiding new membrane developments as well as evaluating existing membranes toward realistic applications.

373ab Dynamic Modeling and Simulation of a Membrane Contactor for CO_2 Absorption

Wanderson Passos, Arioston Araújo de Morais Jr

It's known that in recent decades researchers have paid attention on the trial to reduce CO_2 emissions into the atmosphere, which has led to the development of technologies with the ability of removing the pollutant from gaseous streams. These technologies are commonly called CCS (Carbon Capture and Storage) technologies. Many methods for carbon capture were studied and optimized, although the conventional ones still present some disadvantages, such as the separation on absorption column methods. Therefore, as the membrane separation methods provide some advantages if compared to many others, this manuscript presents an approach for the modeling of CO_2 absorption carried out on a hollow fiber membrane contactor (HFMC) using an aqueous monoethanolamine solution as the solvent. The model consists on partial differential molar balances for the gas and liquid phases, so for the membrane and a global molar balance on the gas phase to give the velocity equation. Kinetics and thermal effects were not taken into account for this approach, as also the effects of pressure drop. Each concentration depends on the time and the axial coordinate, giving rise to a partial differential equations system that were discretized by the past finite difference method, which has turned it into an ordinary differential equation (ODE) system solved by an ODE solver on Python software. The geometry parameters and volumetric flow rates of the HFMC were configured to obtain a CO_2 capture ratio of 90%, so the process transient behavior could be observed under certain initial conditions. The results obtained were good and consistent with the expected behavior. The concentration of the solute on the gas phase decreases with the time and the z axis, as shown on Figure 1 (a), which means that there's CO_2 migrating from the gas to the liquid phase, as shown on Figure 2 (b). Also the maximum mass transfer can be observed by the liquid phase behavior on time equals 161 seconds. The system achieves the steady-state in about 7000 seconds and thereafter the capture ratio remains constant.

Figure 1 - HFMC behavior

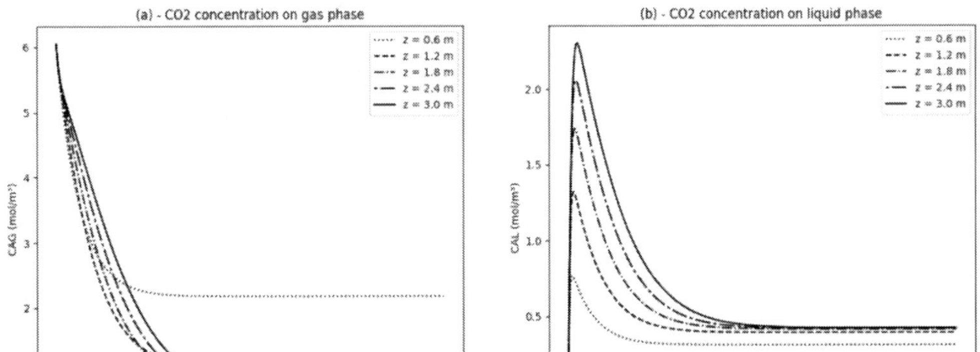

373ac Respirator Cartridge Performance Testing for Chemicals of Potential Concern (COPC) at Higher Chemical Vapor Concentrations

Satish Nune, Michael Minette, Stephen Davidson, Angela Melville, Thomas Brouns, Eugene Morrey, Michael Zabel, Parker Jones, Zachary Way

Washington River Protection System (WRPS) identified the need to test air purifying respirator (APR) and powered air purifying respirator (PAPR) chemical cartridges commonly used at Hanford tank farms in Richland, WA-USA. The main aim of this program is to determine the vapors breakthrough the cartridge when exposed to a mixture of Chemicals of Potential Concern (COPC) from vapors emanating from the headspace of tanks or exiting the exhausters from Hanford waste storage tanks. We will present our recent results on the performance of the MSA OptiFilter TL multipurpose respirator cartridge when tested against a gas mixture containing higher chemical vapor concentrations of signature vapors (Ammonia, Acetaldehyde, Furan(s), Pyridine or Acetonitrile) in controlled laboratory conditions with gas flow rate through the cartridge is 102.5 L/min +/- 5%, temperature of gas is 20°C +/- 2°C and 70% +/- 5% relative humidity.

374c Synthesis and Characterization of CO_2-Selective Membranes with Poly(ethylene formamidine)

Jingying Hu, Yang Han, Winston Ho

Amidine species was perceived as better alternatives for amine species in the facilitated transport membranes for having faster CO_2 reaction kinetics. The synthesis of poly(ethylene formamidine) (PFA) was carried out via the polycondensation of ethylenediamine and triethyl orthoformate catalyzed by acetic acid. The synthesized PFA possesses the highest amidine functional group density owing to the case using the shortest-chain diamine, ethylenediamine. Moreover, in comparison with mobile carriers, it can have a much higher boiling point, which practically eliminates the possibility of carrier volatility and leaching out. In order to obtain the preliminary gas permeation results, PFA (10 kDa MW) was blended with polyvinylalcohol (PVA, 1.1 MDa MW) and then coated onto nanoporous polyethersulfone (PES) substrates to form 170-nm thin membranes. Increasing the content of PFA from 36 wt.% to 66 wt.% in these membranes led to a substantial increase in membrane performance, especially for CO_2 permeance. Further increasing the PFA content to 83 wt.% still enhanced the CO_2 permeance but showed a moderate drop in CO_2/N_2 selectivity. Under an extremely high PFA content of 90 wt.%, the membrane showed a low CO_2 permeance of 1019 GPU but a high CO_2/N_2 selectivity of 236, indicating severe pore penetration. The membrane containing 83 wt.% PFA exhibited the best results with 2807 GPU CO_2 permeance and 72 CO_2/N_2 selectivity at 77°C. In view of the superior transport performance, it is necessary to increase the MW of PFA to make it capable of forming the membrane matrix alone to enable the incorporation of CO_2-reactive mobile carriers, thereby improving the overall CO_2 separation performance. By adopting the design of experiment technique, the synthesis conditions of PFA were improved in terms of the temperature profile, catalyst amount, and monomer concentrations, through which the MW was successfully increased to 3.3 MDa.

374d Responsive Membranes for Enhanced Capture and Detoxification of Water and Air Pollutants

Rollie Mills, J. Todd Hastings, Lindell Ormsbee, Thomas Dziubla, Dibakar Bhattacharyya

This research focuses on the development of responsive functionalized membranes that can exhibit enhanced capture of water and air pollutants, while simultaneously incorporating the detoxification of such contaminants. Functionalized membranes with zero-valent catalytic nanoparticles have shown immense promise for enhanced removal and degradation of water-based polychlorinated biphenyls (PCBs), yet reaction efficiency is hindered by mass transfer limitations between the particles and water contaminants. In recent years, the incorporation of a thermo-responsive polymer, poly(N-isopropylacrylamide) (PNIPAm), has shown to mitigate such limitations in preliminary flat-sheet membranes. In this study, enhanced degradative efficiency with PNIPAm functionalization of catalytic hollow fiber membranes (HFMs-higher particle packing density) was investigated for superior environmental remediation. Additionally, to further overcome these limitations and enhance reaction rates, transparent membranes were functionalized with plasmonic catalytic nanoparticles and PNIPAm, and their efficiency to degrade polychlorinated pollutants, such as trichloroethylene (TCE), with targeted surface heating of catalytic plasmonic particles (via Near-IR illumination) was investigated for cost-effective remediation applications. Furthermore, with respect to viral air-borne contaminants, various membrane functionalization materials, such as salts, enzymes, and silver nanoparticles, have exhibited antiviral properties to reduce or eliminate the infectivity of SARS-CoV-2. In this research, the denaturation of the SARS-CoV-2 spike glycoprotein (facilitates host cell infection) by iron-palladium-functionalized membranes was investigated for the development of enclosed-environment responsive filters. Overall, this study advances the field of separations, membrane science, catalysis, and environmental remediation with responsive materials for air and pollutant detoxification. This research is supported by the NSF RAPID Grant Program, NSF GRFP and NIEHS/SRP Program.

374e Optimal MEA/Dipa/Water Blending Ratio for Minimizing Regeneration Energy in Absorption Based Carbon Capture Process: Experimental Solubility and Thermodynamic Modeling

Mun Jihun, BeomJu Shin, Kyung-Min Kim, Jong-Ho Moon

A new CO_2 absorption solution was developed by blending monoethanolamine (MEA) and diisopropanolamine (DIPA) in H_2O. The solubility of CO_2 in MEA/DIPA aqueous solutions with various blending ratios was measured at a temperature range of 323.15–383.15 K under CO_2 partial pressures of up to 400 kPa. The successive substitution method was introduced to calculate the molar fractions of all 12 species, and the electrolyte–UNIQUAC model was applied to consider non-idealities. Then, the optimal blending ratio was obtained for four different targets: 1) max Δ_2 (total amine-based CO_2 cyclic capacity), 2) max Δ_2 (total solution-based CO_2 cyclic capacity), 3) min $-\Delta h_{abs}$ (heat of CO_2 absorption under stripping condition), and 4) min Q_{regen} (regeneration energy for the carbon capture process). Here, Δ_2, Δ_2, and $-\Delta h_{abs}$ are the properties of solution, while Q_{regen} is a performance parameter of the process. The Q_{regen}, the sum of sensible heat, latent heat, and heat of reaction, was calculated according to the amine blending ratio using the "shortcut method" [1]. In the process, besides the cyclic capacity and heat of absorption, the effect of sensible heat and latent heat must be considered simultaneously. Therefore, the most reliable method to determine the amine blending ratio is Q_{regen} minimization. The obtained minimum Q_{regen} of the process was 3.44 GJ/t CO_2 using MEA/DIPA/H_2O solutions of 4.01/25.99/70 (w/w/w), which was 8.7% and 2.6% lower than those of the single MEA and DIPA aqueous solutions due to the synergistic effect of MEA and DIPA.

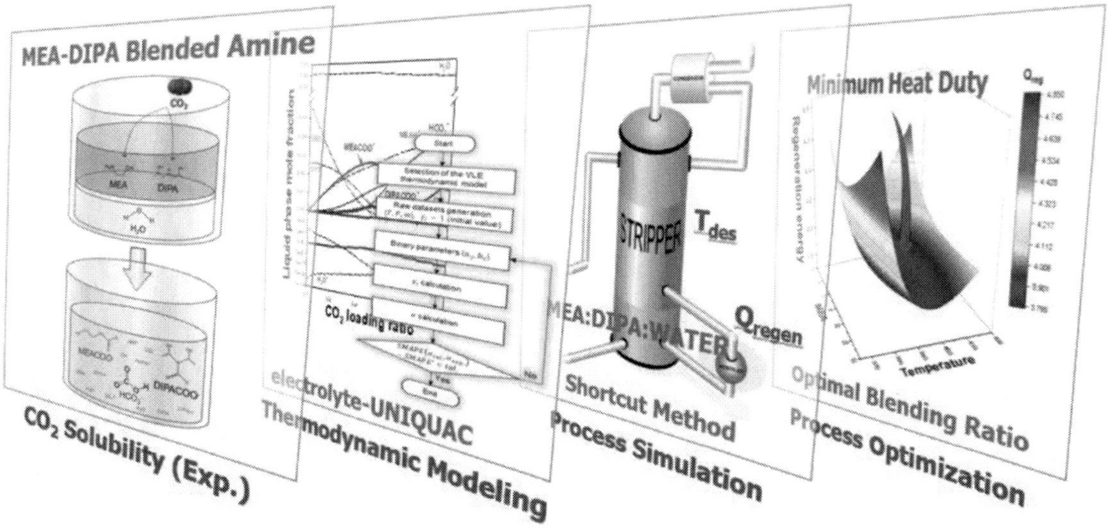

374f Precise Control of Pore Size of Clinoptilolite and Separation of N2 from CH4

Hiroshi Okaniwa, Yuki Jonoo, Yuki Oba, Satoshi Yoshida

Clinoptilolite is a zeolite with HEU structure, and natural clinoptilolite has long been used in gas separation.

We ion-exchanged synthetic clinoptilolite with Sr and Na, and evaluated the correlation between the exchange ion composition and the adsorption rate of CH4 and N2.

When the Na exchange ratio exceeded 40%, the adsorption rate of CH4 decreased markedly and the adsorption of N2 decreased. It was considered that the ion-exchanged Na inhibited diffusion in the 10-membered ring pores of clinoptilolite, resulting in a decrease in the adsorption rate of CH4.

We have prepared a clinoptilolite CH4/N2 separating agent (1.5mmφ, cylinder shape) with a suppressed adsorption rate of CH4 without decreasing the adsorption amount of N2, and confirmed that it can dynamically separate CH4 and N2, by the difference in adsorption rate between CH4 and N2.

In this work, we will also discuss about the mechanism of control of CH4 adsorption rate, and other gas separation applications.

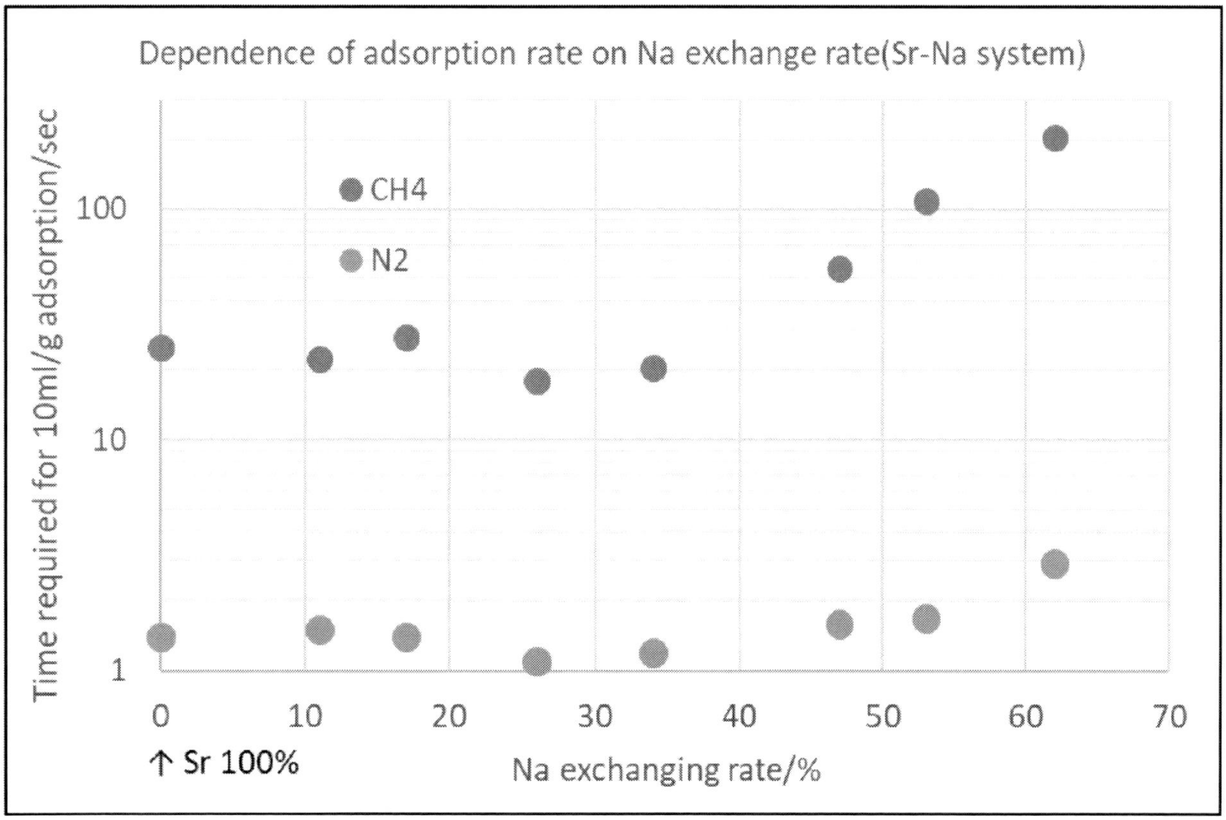

374g Performance Analysis of a Multi-Bed PSA Process Using Silica-Based Adsorbents for CO_2 Capture from Off-Gas in Iron and Steel Industry

Hongjoo Do, Jiwon Jung, Jun-Ho Kang, Moonkyung Cho

The iron and steel industry is one of the most carbon-emitting industries, constituting about 8 percent of the entire carbon dioxide emission amount in the world. Therefore, the demands for the development of carbon capture and utilization (CCU) technology in iron and steel industries are urgently increasing.

Generally, in CCS, 90% capture rate with 95 % CO_2 is recommended for emission gases. However, In viewpoint of CCU, each industry requires a specific concentration of CO_2 to meet its requirement of desired products and operating conditions. Especially, the iron and steel industry needs only the concentrated CO_2 from effluent gases for CCU, instead of 95 % CO_2. Then, the recovery rate of at least 90% becomes an important factor in process efficiency. In this perspective, the pressure swing adsorption using weak adsorbents is highly advantageous to achieve the desirable purity and recovery of a target gas.

In this study, the gas mixture (CO_2, H_2, N_2, and CO) is applied as a feed gas, which matches the composition of the real steel process off-gas. The mathematical dynamic model using experimentally obtained isotherms and kinetics is confirmed with experimental breakthrough curves using silica-based adsorbents. Then, a multi-bed PSA process using layered beds is developed. Sequential and multiple adsorption steps and purge steps are applied to design the cyclic configuration. Then, the sensitivity analysis of various conditions such as step time or feed pressure is conducted using the developed dynamic model.

374h Electrospun Polymer Fibers for Carbon Capture of CO_2

Edward Huang, Jennifer Weiser, Amanda Simson

Carbon capture in ambient air, known as direct air capture (DAC), is increasingly required to meet global climate goals. To improve DAC systems, more cheap and effective adsorbents need to be designed [1]. Branched polyethylenimine (PEI) is a networked polymer with primary, secondary, and tertiary amine groups known to be a good carbon capture material [2]. We hypothesize that electrospinning PEI and polystyrene (PS) will improve CO_2 adsorption capacity and rate, as electrospinning increases the surface area of the polymer scaffold [3]. PS is added to improve the strength and overall fiber morphology when co-spun with PEI. For a PEI/ PS mixture, the best electrospinning parameters were at 15 cm needle-to-drum distance, 1 mL/hour ejection rate, 1.5 V drum rotation voltage, and 15 kV for primary electrospinning voltage. CO_2 adsorption performance was measured with isothermal thermogravimetric analysis (TGA) over temperatures ranging between 30°C to 90°C. Varying weight percentages of PEI/PS from 0 wt% to 30 wt% PEI:PS had the highest adsorption capacity when spun at 20 wt% PEI:PS. At 20 wt% PEI:PS, the adsorption capacity was determined to be 0.97 mmol/g at 50°C with a maximum adsorption rate of 0.57 mmol/g/min. The maximum desorption rate of 0.14 mmol/g/min was found to be 4 times slower that adsorption. The results of this study will help improve the development of energy efficient and cheaper DAC technologies.

[1] International Energy Agency IEA, Direct Air Capture. 2021, Paris https://www.iea.org/reports/direct-air-capture

[2] X. Shen, H. Du, R. H. Mullins, R. R. Kommalapati, *Polyethylenimine Applications in Carbon Dioxide Capture and Separation: From Theoretical Study to Experimental Work.* Energy Technol. **2017**(*5)* 822. doi:10.1002/ente.201600694

[3] C. Huang, P. Wang, Y. Li *Fabrication of electrospun CO2 adsorption membrane for zinc-air battery application*, Chem Eng J. **2020** 395. doi:10.1016/j.cej.2020.125031

374i Carbon Capture from Residual Emissions Enabled By Facilitated Transport Membranes

Yang Han, W.S. Winston Ho

Although carbon capture and storage (CCS) from large stationary sources is a leading strategy to reduce the carbon emissions, a carbon-neutral scenario in the energy sector cannot be achieved by the current CCS technologies since 90% CO_2 recovery is commonly assumed in the design. By taking coal-derived flue gas as an example, a 90% recovery through a primary capture system results in a residual flue gas containing 1–2% CO_2. If not captured, the residual emissions must be mitigated via direct air capture (DAC), which is usually hampered by a substantial cost. The focus of this presentation is to address the membrane carbon capture from dilute, yet stationary sources in order to achieve an overall >90% CO_2 recovery and thereby alleviate burdens upon DAC. To this end, the CO_2/N_2 separation performances of facilitated transport membranes (FTMs) containing aminoacid salts as mobile carriers were characterized with a CO_2 concentration of 0.05–20%. At a reduced CO_2 partial pressure, the carrier saturation in the FTMs was mitigated, which enhanced both the CO_2 permeance and CO_2/N_2 selectivity. The best FTM containing 2-(1-piperazinyl)ethylamine sarcosinate exhibited an uprising CO_2 permeance from 1968 to 3822 GPU and an improved CO_2/N_2 selectivity from 249 to 472 with reducing CO_2 content from 1% to 0.1%. The feasibility of this FTM is exemplified by designing a two-stage enriching membrane cascade to further remove 90% of the CO_2 in a residual coal flue gas containing 1.75% CO_2. Techno-economic analysis indicates a low capture cost of $83.8/tonne. The marginal costs beyond 90% capture are also evaluated for a variety of residual flue gases, indicating that the FTM-based capture from the coal or cement plant residual flue gas is more cost effective than DAC.

374j Application of Calcium Oxide-Based Bifunctional Catalytic Sorbent to Hydrogen Production

Pilseok Kim, Ki Bong Lee

Global warming is a result of the continuous emission of carbon dioxide from the combustion of fossil fuels. Carbon capture and storage (CCS) is proposed as a mid-term solution for this global warming issue. However, nearly 80% of the total operating cost of CCS occurs in the carbon dioxide capture/separation step. Developing a cost-efficient carbon dioxide capture technology is necessary to commercialize the CCS process. Among various technologies for the carbon dioxide collection process, calcium-looping (CaL) technology that utilizes calcium oxide-based adsorbents has emerged as an effective way to capture carbon dioxide in high-temperature conditions. While calcium oxide-based adsorbents have a high theoretical carbon dioxide uptake capacity of 17.8 mol kg^{-1}, it significantly decreases as the adsorption/regeneration process repeats due to their low thermal stability. To solve this cyclic instability, it has been known that incorporating thermally stable and inert materials improve the structural stability of the calcium oxide-based adsorbents. This study developed bifunctional catalytic sorbents by adding nickel to calcium oxide-based adsorbents and applied them to the sorption-enhanced steam methane reforming (SE-SMR) reaction for high-purity hydrogen production. Carbon dioxide adsorption uptake and hydrogen production efficiency of bifunctional catalyst were measured.

374k Ca-Al Hydrotalcite-Based Mixed-Metal-Oxide for High-Temperature Carbon Dioxide Adsorption

Kuei Tan Lee, Ki Bong Lee

Carbon dioxide (CO_2) is a detrimental gas that takes a large proportion of greenhouse gases emitted from fossil fuels and one of the major causes of global warming. For the reduction of CO_2 to achieve sustainable development, worldwide cooperation between international societies has been required. Among numerous countermeasures for its reduction, carbon capture and storage (CCS), a technology related to the separation, transportation, and storage of CO_2 gas produced from several emission sources, has been promising. Especially a separation process, the most cost-consuming process, has received attention to enhance its effectiveness so that the total cost of CCS can be reduced. CO_2 separation by applying adsorbents is a feasible method to capture CO_2 because of its lower energy consumption and simple regeneration process. In specific, hydrotalcite is a CO_2 adsorbent applicable to the temperature range between 200 and 500 °C, while having a relatively low adsorption capacity. Additionally, hydrotalcite is a material of which structure is easily altered by preparation variations, leading to the change of CO_2 adsorption trend. In this study, to enhance the adsorption capacity of hydrotalcite, calcium (Ca) precursor was applied in the synthesis of hydrotalcite instead of magnesium (Mg) precursor. Moreover, variations such as calcination temperature, Ca to Al ratio, and anion types in hydrotalcite were applied to confirm their influences on CO_2 adsorption of the adsorbent.

374l Selective Removal of Cs^+ in Acidic Nuclear Water Waste Using Epoxy Resins

Erwin Escobar, John Edward Sio, Rey Eliseo Torrejos, Khino Parohinog, Hern Kim, Wook-Jin Chung, Grace Nisola

The sequestration of Cs^+ from aqueous media via adsorption had been the focus of many studies in the past few years because it complements the waste vitrification step before concentrated radioactive wastes are sent for long-term storage. Meanwhile, most studies have dealt with the treatment of contaminated surface waters and very few have reported on the adsorptive sequestration of Cs^+ from nuclear fuel processing wastes or high level liquid wastes (HLLW). HLLW contains a variety of interfering metal ions which undermines the selective sequestration of Cs^+. Also, HLLW contains high concentrations of oxidizing acid (~3 M HNO_3) which adversely affects the sequestration capacity of most adsorbents. As such, very few adsorbents have been employed successfully at removing Cs^+ from HLLW, and so far these sorbents have been prepared only with calixarenes or its derivatives via impregnation in macroporous silica or polymeric resin, or by grafting onto mesoporous carbon.

In this work, a new class of adsorbents for the sequestration of Cs^+ in HLLW is presented. Three types of low molecular weight oxygen- and sulfur-bearing bis-epoxides (BE) as binding sites for Cs^+ were reacted with aliphatic diamine curing agents to obtain highly crosslinked epoxy resins as Cs^+ adsorbents. The effects of diamine linker length, BE structure, and complexing atoms were navigated and correlated with the adsorption performance of the epoxy resins, particularly in terms of their capacity and selectivity for Cs^+ in simulated HLLW.

Results reveal that the most Cs^+-selective epoxy resin is prepared from a bis-epoxide bearing a benzene moiety having oxygen and sulfur donor atoms in *ortho* position and cured with diaminopropane (resin BE2-Ep). BE2-Ep is a yellow glassy resin that is thermally stable and mechanically tough in acidic medium. It has the right dimension of channel spaces which permit Cs^+ permeation through its matrix. BE2-Ep registers a maximum capacity of 1.004 mmol g^{-1} (133.5 mg g^{-1}) which is higher than that of other known Cs^+ sorbents used in highly acidic media. BE2-Ep is most selective towards Cs^+, rejecting other competing metal ions in simulated HLLW. This is the first report on epoxy resins with innate selectivity and capacity for Cs^+.

This study was supported by the National Research Foundation of Korea (NRF) grant funded by the Ministry of Science and ICT (2020R1A2C1003560, 2021R1A2C2093746, and 2021R1H1A2008284), Basic Science Research Program through the Ministry of Education (2020R1A6A1A03038817).

374m Tunable H_2s/CO_2 Separation Using Sterically Hindered Amine Membranes

Shraavya Rao, Xuepeng Deng, Yang Han, Li-Chiang Lin, W.S. Winston Ho

H_2S, an extremely toxic and corrosive gas, is a common contaminant in fuel gas mixtures like syngas and natural gas. The selective removal of H_2S from such CO_2-containing fuel gases is important for enabling their industrial processing. This work describes the development of amine-based facilitated transport membranes (FTMs) with high H_2S/CO_2 selectivity at elevated temperature (> 100 °C) and explores the underlying mechanisms of acid gas transport in FTMs.

A series of primary, secondary, and tertiary amine carriers with different alkyl substituents was used to synthesize FTMs. Their H_2S/CO_2 transport performances were evaluated at 107 °C and 7 bar. Sterically hindered amines showed enhanced H_2S/CO_2 separation performance, with the H_2S/CO_2 selectivity rising from 4 to nearly 20 as the amine hindrance was increased. The best performance was obtained using a severely hindered di-*tert*-butylamine carrier, which showed a high H_2S/CO_2 selectivity of 19.6 and an H_2S permeance of 560 GPU. These results are supported by Density Functional Theory (DFT) calculations. The DFT results indicated that the presence of sterically bulky groups on the N site of the amine can restrict its reactivity to CO_2 through the carbamate pathway. Conversely, the amine-H_2S reaction was found to be relatively insensitive to steric hindrance. Finally, the carrier saturation behavior of the di-*tert*-butylamine carrier was investigated at H_2S concentrations of 0.5—30%. The carrier was found to be nearly inert to CO_2, achieving H_2S/CO_2 selectivities of 25.4 — 9 and H_2S permeances of 760—220 GPU. Overall, the newly synthesized membranes show good potential for syngas desulfurization, and the learnings offer some insight into the design of carriers for acid gas purification.

374n Separation of Refrigerant R-410A Using Porous Materials: Thermodynamic Modeling and Breakthrough Experiments

Andrew Yancey, David R. Corbin, Mark Shiflett

According to a report published by the United Nations in Summer 2021, the presence of atmospheric greenhouse gases (GHGs) has perpetuated anthropogenic global climate change. Global temperatures have risen 1.0-2.0 K throughout the 20th century, and if GHG emissions are not reduced, this could result in an additional increase of 1.5-2.0 K throughout the 21st century, further perpetuating sporadic and destructive weather patterns. Hydrofluorocarbons (HFCs) are among the GHGs cited as contributing to global climate change. The high global warming potentials (GWPs) of HFCs was noticed in the early 21st century, which has led to worldwide legislation restricting production and use. The most recent is the U.S American Innovation and Manufacturing Act in December 2020.

There is currently an estimated 2,800 ktons of HFCs in global circulation. As HFCs are phased out and replaced with next-generation fluorocarbons, hydrofluoroolefins (HFOs), responsible action will be required to sustainably deal with the unused HFCs. Rather than venting or incinerating the HFCs, these unused refrigerants should ideally be reclaimed, recycled, and reused; however, such goals are difficult since many refrigerants currently in use are azeotropic or near-azeotropic HFC refrigerant blends. Before the refrigerant mixtures can be recycled, the HFCs must first be separated so that each species can be handled separately and efficiently. Traditional distillation techniques cannot effectively separate azeotropic mixtures; therefore, other separation methods must be developed so that HFC refrigerants can effectively be reclaimed and recycled. Adsorption-based separation has been proven to work for zeotropic, azeotropic, and isomeric fluorocarbon separations.

The following presentation will focus on using zeolites and carbons for the separation of refrigerant R-410A (50/50 wt% HFC-32/HFC-125), which is commonly used in commercial and domestic refrigerators and air conditioning units. Pure gas isotherm data was measured using a Hiden Isochema XEMIS gravimetric microbalance for basic and acidic zeolites, as well as activated carbons, and fit to isotherm models. Binary sorption of HFC-125 and HFC-32 on the same sorbents was additionally measured using a separate Hiden Isochema gravimetric microbalance that uses the integral mass balance (IMB) method. Sorbed phase activity coefficients were calculated from experimental data and fit to vapor-adsorption equilibrium (VAE) activity coefficient models. Both pure gas models and activity coefficient models were used in Real Adsorbed Solution Theory (RAST) to develop accurate thermodynamic models for binary sorption of HFC-125 and HFC-32. Breakthrough experiments have additionally been performed for various sorbents. Both thermodynamic modeling and breakthrough experiment results will be presented and discussed to assess the potential of the sorbents for separating R-410A.

374o Evaluating Transport Factors to Understand Electrochemical Nutrient Removal and Recovery from Synthetic Animal Wastewater

Sana Heydarian, Francesca Carney, Damilola Daramola

Background

Concentrated Animal Feeding Operations (CAFOs) are one of the main sources of introducing excessive nutrients, mainly phosphorus and nitrogen, to the environment via manure use as fertilizer. [1] These excessive nutrients are key factors in eutrophication of water bodies, and it is crucial to treat the CAFO effluents before disposal to water bodies, in order to reduce environmental impacts. [2] Furthermore, limited natural resources of phosphorus increases the need for an alternative resource for this nutrient. Simultaneous removal and recovery of phosphorus and nitrogen from wastewaters via an electrochemical approach, as a precipitate called $MgNH_4PO_4{\bullet}6H_2O$, i.e., struvite, addresses both excessive amount of nutrients in these effluents and limited resources of natural phosphorus. Struvite also has economic benefit as a slow-release fertilizer. [3] A predictive model of the effects of chemical composition and solution conditions on phosphorus removal from wastewater and recovery as struvite has been introduced in literature. [4] This model requires additional experimental data to validate (or modify) the anticipated effects of temperature and ion concentration on solid phosphorus recovery.

Methodology

In the current study, the effect of ion concentration, temperature, solution agitation, and pH on nutrient removal efficiency and the type of P-containing precipitant recovered will be investigated. This investigation will occur in a bench scale setup with well-defined mass and thermal transport. This study will help with (1) providing a fundamental background knowledge for electrochemical nutrient recovery and removal, (2) enhancing the actual efficiency of electrochemical nutrient recovery by separating the recovery due solely to chemical transport and (3) assisting with validating/modifying the aforementioned literature model.

Results

The initial experiments of this study showed higher phosphorus removal from wastewater in pH of 9 in comparison to pH of 7.6, in agreement with the model prediction. [4] Further investigation of the collected solid phase samples will be used to assess the effect of pH on the type of recovered precipitate. Additional results of solid and liquid phase analyses will be presented at the conference including the effect of other transport factors on the recovered nutrient and the delineation between literature model prediction and experimental results.

References

[1] Z. Belarbi, D. A. Daramola and J. P. Trembly, "Bench-Scale Demonstration and Thermodynamic Simulations of Electrochemical Nutrient Reduction in Wastewater via Recovery as Struvite," J. Electrochem. Soc., vol. 167, no. 4, Nov. 2020, doi: 10.1149/1945-7111/abc58f.

[2] L. Kekedy-Nagy, *et. Al*, "Electrochemical Nutrient Removal from Natural Wastewater Sources and its Impact on Water Quality,"

[3] Z. Belarbi and J. P. Trembly, "Electrochemical Processing to Capture Phosphorus from Simulated Concentrated Animal Feeding Operations Waste," J. Electrochem. Soc., vol. 165, no. 13, pp. E685–E693, Jan. 2018, doi: 10.1149/2.0891813jes.

[4] G. P. Pindine, J. P. Trembly and D. A. Daramola, "Equilibrium-Based Temperature-Dependent Economic Analysis of the Recovery of Phosphorus from Different Wastewater Streams via Chemical Precipitation" ACS EST Water, 2021, **1**, 2318–2326.

374p Intensified Water Purification Via Closed Loop Carbon Dioxide-Mediated Diffusiophoresis

Esai Lopez, Patryck Michalik, Shicheng Lyu, Alex D. Paulsen, Elizabeth Stewart, Andrew R Teixeira

Traditional water purification systems often require large energy inputs to achieve particle separation. To address the expanding need for access to clean drinking water, less energy intensive separations methods must be developed. A novel concentric tube-in-tube-in-tube separator was developed to remove colloidal water contaminants. The system utilizes carbon dioxide (CO_2) to generate an ion gradient across the inner most annulus and induce diffusiophoretic migration of charged colloids to the outer edge of the annulus. Clean water is withdrawn from the center of the inner most annulus using a capillary. An investigation was conducted into system parameters such as split ratio between clean and concentrate streams, recoverable CO_2, and capillary position to maximize clean water production. The effect on performance and throughput were evaluated across a multidimensional parameter space including operating pressure (150–250 kPa) and residence time (10–1000 s). Additionally, a computational fluid mechanics model was developed and validated against experimental flux measurements and used to evaluate key performance parameters. The results revealed a viable pathway for removal of colloidal contaminants while recovering CO_2 in a closed loop system.

374q Effect of Co-Ions on O_2 Gas Transport Properties of Cross-Linked Poly(ethylene oxide)

Taliehsadat Alebrahim, Narjes Esmaeili, Thien Tran, Haiqing Lin

Cobalt (Co)-based complexes are considered as attractive O_2-binding elements, which can reversibly and selectively interact with O_2 and affect O_2 gas transport properties. Herein, we investigate the effect of Co ions dissociated in amorphous cross-linked poly(ethylene oxide) (XLPEO) on physical properties and gas transport characteristics. Polymeric films containing Co salts were prepared by photo-polymerization of solutions containing poly(ethylene glycol) diacrylate (PEGDA), poly(ethylene glycol) methyl ether acrylate (PEGMEA), and different salts such as $Co(ClO_4)_2$, and $Co(BF_4)_2$. The salts can be fully dissociated, and the Co^{2+} ions form conjugations with the ethylene oxide in XLPEO, as validated by FTIR and XRD. Increasing the Co content increases the density (following the additive model) and glass transition temperature of XLPEO. For instance, adding 17 wt.% $Co(ClO_4)_2$ to XLPEO increases the T_g and density from -65 °C and 1.149 g/cm^3 to -26 °C and 1.304 g/cm^3, respectively. The O_2 and N_2 permeability are significantly impacted by adding high concentration of Co salts. Specifically, O_2 and N_2 permeability decreases from 26 and 10 to 0.24 and 0.14 Barrer, respectively, by adding 17 wt.% of $Co(BF_4)_2$. Interestingly, the CO_2/O_2 selectivity increases from 20 to 60 by addition of 17 wt.% of $Co(BF_4)_2$ presumably due to the retarded O_2 diffusion. Additionally, we found that O_2 solubility increases with increasing $Co(BF_4)_2$ content due to high affinity between Co ions and oxygen. For example, O_2 solubility increases from 0.2 to 6.4 cm^3(STP)/cm^3 atm at 35°C by addition of 6.4 wt.% $Co(BF_4)_2$ to XLPEO. Understanding the effect of Co ions-O_2 interactions on gas transport properties may provide guidance in designing advanced polymeric membranes with high CO_2/O_2 selectivity

374r Synthesis and Characterization of Multi-Ionic Block Copolymers for Chemical and Biological Protective Clothing Applications

Karen Barrios Tarazona, Gilberto Ramos Rivera, David Suleiman

Geopolitical instabilities in the world have created increased awareness of potential chemical and biological threats. One proactive approach to this situation has been the development of multi-ionic block copolymers as selective materials for chemical and biological protective clothing (CBPC) applications. This study discusses the synthesis of sulfonated amine block copolymers and the effect of multiple ionic domains on the morphology and transport properties of polymeric membranes for chemical protective clothing applications. The monomers 2-(tert-Butylamino) ethyl methacrylate (TBAM), 2-ethoxy ethyl methacrylate (EEM), and styrene were used to prepare the block copolymers by atom transfer radical polymerization (ATRP). The copolymers were then sulfonated by chemical grafting with pendants of sulfobutyl groups onto the polymer structure. Properties of the resulting membranes were determined as a function of block composition and incorporation of sulfonic groups, using a series of materials characterization techniques (*e.g.*, FT-IR, TGA, AFM, SAXS). The results indicate that the ion exchange capacity and water absorption of the membranes had a similar pattern after the incorporation of each block in the structure, where the increase of these properties were proportional to the amine block composition. In addition, the results show that the interaction between multi-ionic domains improves the thermal stability of the polymeric structure. This interaction also caused morphological changes, which affected the transport properties of the membranes and their performance for use in CBPC applications. For example, selectivities of up-to-three orders of magnitude for a simulant of Agent Sarin has been obtained, while maintaining low breathability (water vapor transport rate < 1500 gm^2 day^{-1}).

374s Separation of Refrigerant Mixtures Using Extractive Distillation with Ionic Liquid Entrainers

Ethan A. Finberg, Mark B. Shiflett

Hydrofluorocarbons (HFCs) have been used as refrigerants globally since the 1990's and replaced chlorofluorocarbons (CFCs), which were linked to the depletion of the Earth's ozone layer. HFC refrigerants are being phased out over the next two decades due to their high global warming potential. The refrigerant industry is moving toward the production and marketing of HFOs and HFO/HFC refrigerant blends. This will require the recycling and repurposing of HFC mixtures, but many of the HFC mixtures are azeotropic or near azeotropic (i.e., R-410A, R-404A, R-407C, to name a few) making the separation of the components impossible using current methods. Extractive distillation using an ionic liquid entrainers offers a solution.

Ionic liquids (ILs) are being developed as novel entrainers for extractive distillation because of their negligible vapor pressure, thermal and chemical stability, and high and differential solubility with HFC refrigerants. Vapor liquid equilibria data for refrigerants difluoromethane (HFC-32), chlorodifluoromethane (HCFC-22), pentafluoroethane (HFC-125), 1,1,1-trifluoroethane (HFC-143a), and 1,1,1,2-tetrafluoroethane (HFC-134a) in ionic liquids 1-ethyl-3-methylimidazolium bis(trifluoromethylsulfonyl)imide ($[C_2C_1im][Tf_2N]$) and 1-butyl-3-methylimidazolium hexafluorophosphate ($[C_4C_1im][PF_6]$) were fit with the Peng–Robinson equation of state to simulate the separation of multi-component azeotropic refrigerant mixtures and to develop rate-based and equilibrium models in ASPEN Plus. Process flow diagrams (PFD) were developed and optimized based on a set of physical and chemical constraints with the goal to optimize parameters and achieve refrigerant grade (>99.5 wt%) purity. The ionic liquids were found to be effective entrainers for separating refrigerant mixtures.

374u Energy Efficient Advanced Separation Process of Solvent Mixture (Xylene, Butyl Acetate and Methyl Ethyl Ketone) Recovery/Re-Processing for Paint Industry

Abdur Rehman

Most of the OEM (Automotive sector) paints are solvent base, so the use of different types of solvents, resins, pigments and additives is essential for paint manufacturing. In manufacturing process, companies have to manufacture all types of color paints and in order to restart the next batch process they have to wash the machine and equipment every time to avoid color contamination of previous batch. This requires large quantity of solvents for cleaning purpose. The average consumption of these solvents is around 5-10 tons/month, majorly containing Xylene (C_8H_{10}), (MEK) Methyl Ethyl Ketone (C_4H_8O) and Butyl Acetate ($C_6H_{12}O_2$). (60%, 20% andamp; 20 % weight respectively). In this study a novel method is investigated for the recycling and reuse of these solvents to reduce significant amount of cost and their hazardous effect on environment. Initially, Liquid- Liquid extraction (LLE) process is applied as the C_8H_{10} is insoluble in water so its separation will be done by LLE using water as the pure solvent. Butyl acetate and MEK are slightly soluble in water so no. of stages of extractor and type of extractor are calculated.

Secondly the process involves membrane separation. Polysulfone (PS) membrane with different pore sizes were investigated to check the separation efficiencies. Finally, the binary distillation process to separate two component mixture $C_6H_{12}O_2$and C_4H_8O both solvents possibly forms azeotropic mixture. 85% to 88% efficiencies of this combined process were calculated, the mentioned process is most viable and cost effective process for the separation of above mention solvent solutions.

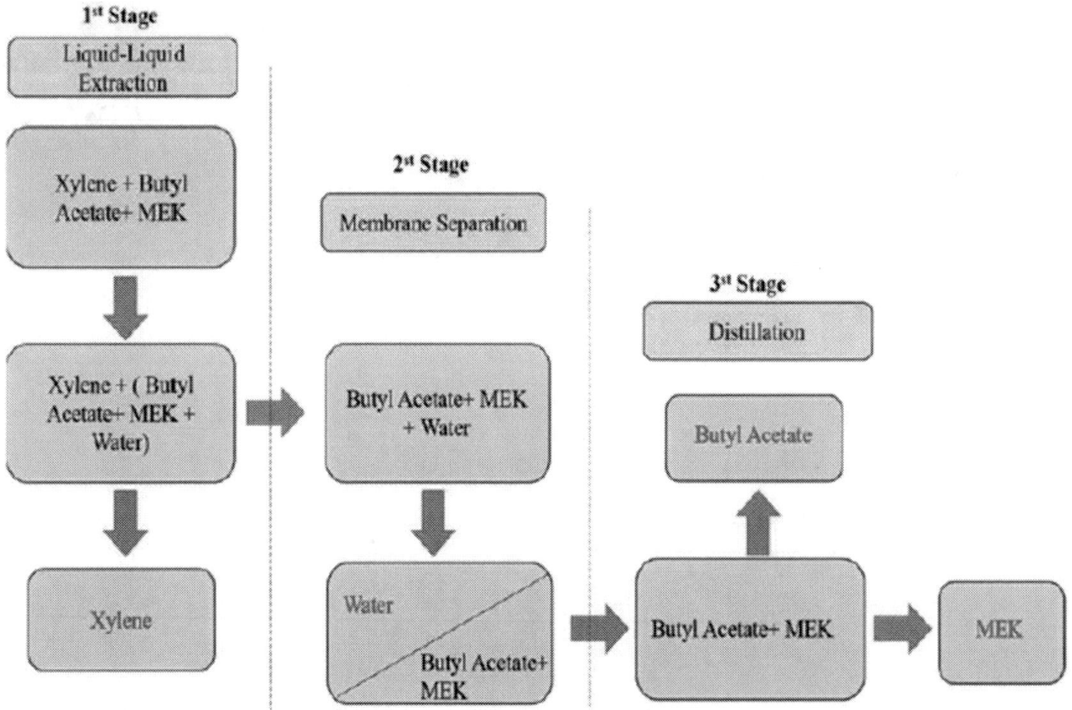

Scheme of combine separation process for solvent recovery

374v CFD-PBM Simulation of Nickel-Manganese-Cobalt Hydroxide CO-Precipitation in CSTR

Mohsen Shiea, Andrea Querio, Antonio Buffo, Gianluca Boccardo, Daniele Marchisio

Summary:

Lithium-ion batteries based on $Li(Ni_{1-x-y}Mn_xCo_y)O_2$ cathode materials have become one of the preferred battery chemistries due to their advantages such as high energy density. This fact has been driving numerous investigations in improving characteristics of these materials, which, in fact, depend on the properties of their precursor material, i.e., $Ni_{1-x-y}Mn_xCo_y(OH)_2$. As a result, the literature offers many experimental works that studied the synthesis of Ni-Mn-Co hydroxide precursors in continuous stirred-tank reactors (CSTR), which is the common system for the industrial-scale production. In particular, a lot of experimental effort has been devoted to determine the effect of operating conditions on important properties of synthesized particles, e.g., particle size distribution (PSD). However, only few modelling attempts have been made despite the fact that a predictive model can be very useful in the process scale-up and optimization. In this regard, we have recently proposed a modelling framework for the co-precipitation of Ni-Mn-Co hydroxide, which was adopted to simulate an experimental setup [1]. Here, we assess the framework by simulating the co-precipitation of $Ni_{0.8}Mn_{0.1}Co_{0.1}(OH)_2$ in a pilot-scale CSTR.

Introduction:

The synthesis of Ni-Mn-Co hydroxide is usually carried out in a CSTR by mixing metal sulphates with sodium hydroxide in the presence of ammonia as a chelating agent. The reactions involved are summarized in the work by Van Bommel and Dahn [2]. The case study is a pilot-scale CSTR with three inlets and a Rushton turbine mixer. The volume of the reactor is three litres and the stirring rate 200 rpm. The concentration of the metal feed is 2M with 80% $NiSO_4$, 10% $MnSO_4$ and 10% $CoSO_4$. The feed solutions of NaOH and NH3 are injected with the concentration of 5M and 10M, respectively. The flow rate of the inlets are calculated by assuming the residence time of 1 hr and the concentration ratio [M2+]:[NH3]:[NaOH]=1:1:2 for the mixed solution.

Modelling and simulation:

The modelling framework consists of several pieces: 1) the CFD to predict the flow fields; 2) chemical equilibria to calculate the local supersaturation generated by transported total concentrations; 3) a population balance equation (PBE) to describe the formation and evolution of particles due to the nucleation, growth, aggregation and breakage; 4) the quadrature method of moments for the solution of PBE; 5) total concentration transport equations to determine the local composition; 6) a micromixing model to consider the segregation of feed solutions. More details on the governing equations and employed models are found in our recent work [1].

The modelling framework is implemented in OpenFOAM v8.0 by modifying the "pisoFoam" built-in solver, which is suitable for the simulation of single-phase transient incompressible turbulent flows. The reason for using a single-phase solver is the fact that particles are sufficiently small (Stokes number less than 1), and therefore, their inertia has a negligible effect on the liquid flow field. Furthermore, the turbulence is solved by employing the k-ε model. Moreover, the rotation of the mixer is considered by adopting the multi-reference frame method. It should be noted that the transport equations for the total concentrations and moments are solved by the operator-splitting approach, hence the necessity of using a transient solver. This means that the solver advances in time until obtaining the steady-state solution. With the operator-splitting approach, the transport (i.e., convection and diffusion) part of the governing equations is solved first to obtain an intermediate solution at each time-step. Then, the source term of governing equations (e.g., nucleation and growth) is integrated over the time-step, taking the intermediate solution as the initial condition. The integration of the source term is done by the CVODE solver of

SUNDIALS for the stiff problems, which uses backward differentiation formulas.

Results and discussion:

Some preliminary simulations are done to assess the performance of the implementation with a particular attention to the required computational time. Considering the operator splitting method, it is expected that the computational time is determined by the time step used for the transport and the stiffness of source-terms. The former is related to the CFL condition imposed by the flow field, while the latter is mainly due to rate of particle nucleation and growth. Therefore, in these preliminary simulations, only nucleation and growth of particles are included. Once the viability of the approach is verified, the addition of the aggregation and breakage of particles should not pose a serious problem.

Simulation results show that the steady-state is achieved after more than five times the residence time, which requires more than 180000 time-steps of, for instance, 0.1 (s). In this simulation, each time-step of 0.1 s takes around 7 seconds by using 44 cores on a machine with two Intel(R) Xeon(R) Gold 6248R processors, which corresponds to a total computational time of two weeks. This amount of time will increase in the case of a larger reactor with more extreme conditions such as higher agitation rate and longer residence time. Therefore, it is important to find remedies for decreasing the total computational time. It should be mentioned that decreasing the total computational time by taking larger time-steps (i.e., reducing the total number of time-steps) is not as effective as expected, because the load of the integration over each time-step increases. In fact, there is a trade-off between the total number of time-steps (in other words, the length of the time-step) and the integration load over each time-step. Moreover, too much increase of the time-step can cause simulation instabilities and a larger splitting error. A possible solution to decrease the total computational time is to reduce the calculation time of each time-step by increasing the number of processors. Unfortunately, the conducted trials show low scalability of the source integration part of the operator-splitting method when the number of processors increases. It is unexpected since the integration depends only on the variables of each cell, and therefore, it should be highly scalable. The reason lies in the fact that the integration load differs from one computational cell to another as the nucleation and growth rates can change significantly over the domain. It eventually causes an imbalance of the integration cost across processors when the common strategies for the domain decomposition in parallel CFD codes are used. These strategies usually aim at assigning the same number of cells to each processor while minimizing the communication between them. It is obvious that changing the criteria for the domain decomposition, e.g., based on the integration load, is not a promising solution, as it can cause similar load imbalance in the solution of transport equations. Therefore, we are currently investigating possible algorithms to distribute equally the integration load over processors, without changing the common practice for the domain decomposition.

Acknowledgement:

The research reported in this work was funded by European Union, Horizon 2020 Programme, SimDOME Project, Grant Agreement No 814492. The views and opinions expressed in this publication are the sole responsibility of the authors and do not necessarily reflect the views of the European Commission/Research Executive Agency.

Rerferences:

[1] M. Shiea, A. Querio, A. Buffo, G. Boccardo, and D. Marchisio, "CFD-PBE modelling of continuous Ni-Mn-Co hydroxide co-precipitation for Li-ion batteries," Chem. Eng. Res. Des., vol. 177, pp. 461–472, 2022.

[2] A. Van Bommel and J. Dahn, "Analysis of the growth mechanism of coprecipitated spherical and dense nickel, manganese, and cobalt-containing hydroxides in the presence of aqueous ammonia," Chem. Mater., vol. 21, no. 8, pp. 1500–1503, 2009.

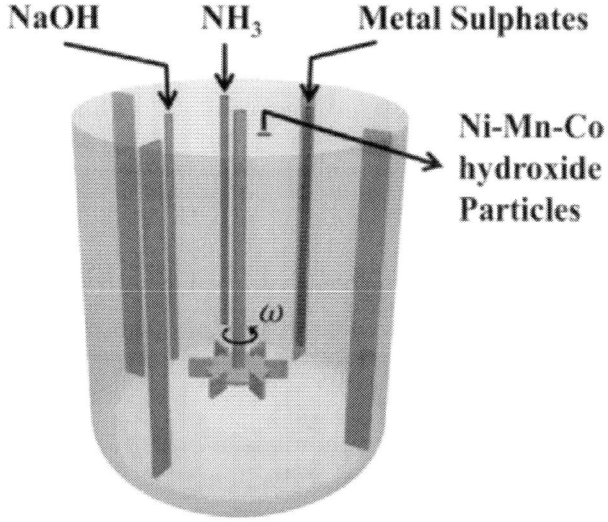

Figure 1: The geometry of the simulated CSTR

647a Practically Achievable Process Limits for a Temperature Swing Adsorption Process for CO_2 Capture from Ngcc Flue Gas

Yogashree Bharath, Kasturi Nagesh Pai, Phuc-Tien Thierry, Samuel Lethier, Philip Llewellyn, Cecile Pereira, Veronique Pugnet, Arvind Rajendran

Climate change is being tackled in many ways. Most prominent pathway includes a combination of replacing intensive fuels such as coal with natural gas and carbon capture technology. Owing to the dilute nature of flue gas emitted from natural gas fired plants (NGCC), absorption-based capture has found to be energy intensive . As an alternative, temperature swing adsorption (TSA) processes are found to be a widely acknowledged separation technology associated with dilute fuel sources. Scientists have also been involved in development of appropriate materials such as zeolites, carbons, and metal-organic frameworks (MOFs) for effective capture of CO_2. Notably, metal-organic frameworks (MOFs), that have long eluded large-scale applications have been commercialized for industrial CO_2 capture (Lin, 2021.). Conventional selection of adsorbents was based on process metrics evaluated from equilibrium measurements. Recent studies have clearly articulated the need to combine process design and optimization techniques to effectively evaluate the true potential for various adsorbents (Pai, 2021). Zeroing in on suitable material-process combination can be obtained via screening or by process inversion. This work addresses the limits that can be achieved for maximum possible performance indicators such as productivity (Pr), recovery (Re), purity (Pu), and minimum possible energy (En) for a 3-step TSA process separating CO_2 from NGCC flue gas. The pure component CO_2 and N_2 adsorption isotherms were described using single site Langmuir equations. The competition was described using extended Langmuirian model. A 3-step steam assisted TSA process was considered. Several case studies were designed and evaluated for the key performance indicators Pu, Re, En, and Pr. The case studies were curated aptly to recognize process limits as well as adsorbent limits. To overcome the cumbersome task of a conventional process optimization, a Machine assisted Adsorption Process Learner and Emulator (MAPLE) framework was employed (Pai, 2021). This technique trains surrogate models, based on artificial neural networks (ANN), to predict large-scale process performance for a given adsorbent. The ANN models were tweaked appropriately with training-testing ratio, number of neurons, number of hidden layers, and activation functions to emulate the detailed model with proximity. For each of the case studies that were considered, a large design space of MOFS and process conditions were searched with a genetic algorithm whose fitness was scored based on the MAPLE process models. Results of this study will be discussed at the meeting.

Keywords:, carbon capture, temperature swing adsorption, MAPLE, process inversion

References

Pai, K.N. et.al. (2021). Practically Achievable Process Performance Limits for Pressure- Vacuum Swing Adsorption-Based Postcombustion CO2 Capture. *ACS Sustainable Chemistry andamp; Engineering*, 3838-3849.

Lin, J. et.al. (2021.). A scalable metal-organic framework as a durable physisorbent for carbon dioxide capture. *Science*, 1464-1469.

647b Temperature-Vacuum Swing Adsorption Process for Direct Air Capture of CO_2

James A. Ritter, Marjorie A. Nicholson, Armin Ebner

The removal of carbon dioxide (CO_2) from gas streams has traditionally been utilized to purify other components of the source gas stream, e.g., hydrogen (H_2) in steam methane reforming, methane in natural gas purification, etc. The removal of metabolic CO_2 from air has also been of interest for many years, particularly with the advent of human space flight and underwater exploration. For instance, considerable effort has been put forth by NASA to remove metabolic CO_2 from spacecraft cabin air.

Removal of CO_2 from gas streams combined with the capture of the isolated CO_2 has become of increasing interest in recent years as CO_2 released from transportation, industrial processes and the electric power industry has been implicated and widely accepted as being a cause of climate change. Moreover, as improved methods for utilization of high purity CO_2 have been developed, for instance in green technologies as a fuel source, as an extraction solvent, and in algae and plant cultivation, the gas separations industry has shown great interest in improving methodologies for separating and capturing CO_2 from CO_2-containing gases, including direct capture from the air.

Adsorption technology, especially temperature-vacuum swing adsorption (TVSA), has been proposed and studied extensively for CO_2 capture. Gas separation by TVSA is achieved by coordinated pressure and temperature cycling and flow reversals over an adsorbent bed that preferentially adsorbs a component of the mixture, e.g., CO_2. Traditionally, TVSA has been utilized to supply high oxygen content streams, e.g., medical oxygen. TVSA has also been utilized in the removal of CO_2 as the final step in hydrogen synthesis for use in oil refineries, in the production of ammonia (NH_3), in the removal of hydrogen sulfide (H_2S) from hydrogen feed and recycle streams of hydrotreating and hydrocracking units, as well as in the separation of CO_2 from biogas so as to increase the methane (CH_4) ratio. Unfortunately, conventional TVSA processes make inefficient use of applied energy, and as such have remained economically unfeasible, particularly when considering separation of CO_2 from a hydrated gas stream, such as air.

What are needed are systems that can provide for improved separation of CO_2 from a gas stream, and, in particular, from a gas stream that also includes water vapor. Methods that can provide high purity CO_2 from a hydrated feed stream, for instance directly from air, could be particularly beneficial in green technologies such as algae production. This presentation will discuss a TVSA system for capturing CO_2 from a hydrated gaseous stream and, in particular, for direct air capture (DAC) of CO_2. The system utilizes a low energy TVSA process and includes two desiccant beds for the removal of water vapor coupled together with two CO_2 capture beds for the capture and concentration of CO_2. This system can produce a very pure CO_2 stream or a stream containing only a few vol% of CO_2, depending on the application. Specific details about this 4-bed TVSA system will be provided during this presentation, along with performance results obtained via process simulation using the UofSC dynamic adsorption process simulator (DAPS). One application of interest is to produce a stream containing a few vol% CO_2 for use in algae production.

647c Efficient Surrogate Optimization of Pressure-Vacuum Swing Adsorption Processes for Post-Combustion CO_2 Capture

Adam Ward, Ronny Pini

Pressure-vacuum swing adsorption (PVSA) processes are a promising candidate technology for the removal of CO_2 from post-combustion flue gases. PVSA processes operate in a series of distinct steps, in each of which the operating pressure is varied to separate CO_2. There are several important operating parameters in these systems which significantly impact the extent of the separation, including the operating pressure and duration of each cycle step, as well as the feed rate of flue gas to the process. The relationship between these operating parameters is highly non-linear, making these processes inherently challenging to design. To search for optimal operating conditions, we must typically consider the multi-objective optimization of two conflicting properties of the system, the productivity and the energy usage. The productivity specifies the amount of CO_2 extracted in the PVSA process per unit bed volume and per unit time. The total energy usage is the amount of energy required by compressors and vacuum pumps to achieve the pressure changes necessary in each cycle step. Because these two objectives conflict with one another, the solution of the process optimization problem is a Pareto front of optimal solutions – each of which represents the best available trade-off between the two objectives. The multi-objective optimization is also constrained by regulatory requirements on the purity and recovery of CO_2 extracted from the flue gas. This frames the design of PVSA processes as a constrained multi-objective optimization problem with a relatively large number of decision variables. When we consider that the evaluation of the process performance indicators for a single operating point using detailed process modelling is very time consuming, we can understand that the optimization workflow is very computationally expensive to evaluate. In our own experience, and with reference to the extensive literature available on the subject, solution for the Pareto front can take on the order of a few days.

There are several emerging trends in the literature of adsorption process design which will only serve to increase the computational complexity of this workflow further. Firstly, there is adsorbent screening, wherein many adsorbent materials need to have their optimal process performance evaluated to search for the best adsorbent material for a given application. Secondly, there are process design problems which incorporate process superstructures. In these problems, the process cycle is tailored as part of the optimization problem to best suit a given adsorbent material to find the ideal material-process combination. Finally, there are applications of alternative adsorption process configurations, such as layered bed adsorbers or beds incorporating structured contactors. In all these important applications, the number of decision variables and/or detailed models which need to be evaluated is much larger than that in the current status-quo, in which we optimize the performance of a fixed cycle with a given adsorbent material. Yet, the optimization problem within the status-quo is still, to date, prohibitively computationally expensive to readily permit an extension to any of the design problems mentioned above. One of the available options to resolve this conflict is to improve the efficiency of the optimization routine with which the detailed model is coupled.

In this work, we demonstrate the application of a so-called surrogate optimization approach for the design of a pressure-vacuum swing adsorption process for post-combustion CO_2 capture. Specifically, we consider the Thompson Sampling Efficient Multi-Objective Optimization (TSEMO), which has previously been successfully applied for the design of adsorption processes (*Hao et al., 2021*). Here, we extend previous applications of the algorithm by considering the important case of purity/recovery constrained productivity-energy usage optimization. We have carried out a post-combustion CO_2 capture process design problem to evaluate the performance of a promising metal organic framework material, ZIF-36-FRL, for separating a binary mixture of CO_2/N_2. Within the process design framework, we employ the TSEMO algorithm to optimize the operating conditions (operating pressures, cycle step durations, feed rate of flue gas) of a four-step PVSA cycle with light product pressurization to obtain a Pareto front of maximum productivity and minimum energy usage. The Pareto front is constrained by requiring both purity and recovery of the extracted CO_2 to be in excess of 90%. We have also evaluated the same design problem using the widely adopted non-dominated sorting genetic algorithm II (NSGA-II) as a benchmark.

As can be seen in Fig. 1, ZIF-36-FRL is able to achieve a lower energy penalty than zeolite 13X, a benchmark material for post-combustion capture, when applied in a post-combustion setting to a representative 1,000 MW coal-fired power plant. We have found that in the case of the constrained productivity-energy usage problem, TSEMO can obtain an essentially identical Pareto front to NSGA-II, while reducing the required single-core CPU time by a factor of 15. This result is powerful and exciting, with wide-reaching implications in the adsorption process design literature for both the status-quo of process design – as well as having applicability for the extension to emerging, challenging and novel process design tasks.

References:

Hao et al. (2021): "Efficient hybrid multiobjective optimization of pressure swing adsorption". Chemical Engineering Journal (423). https://doi.org/10.1016/j.cej.2021.130248

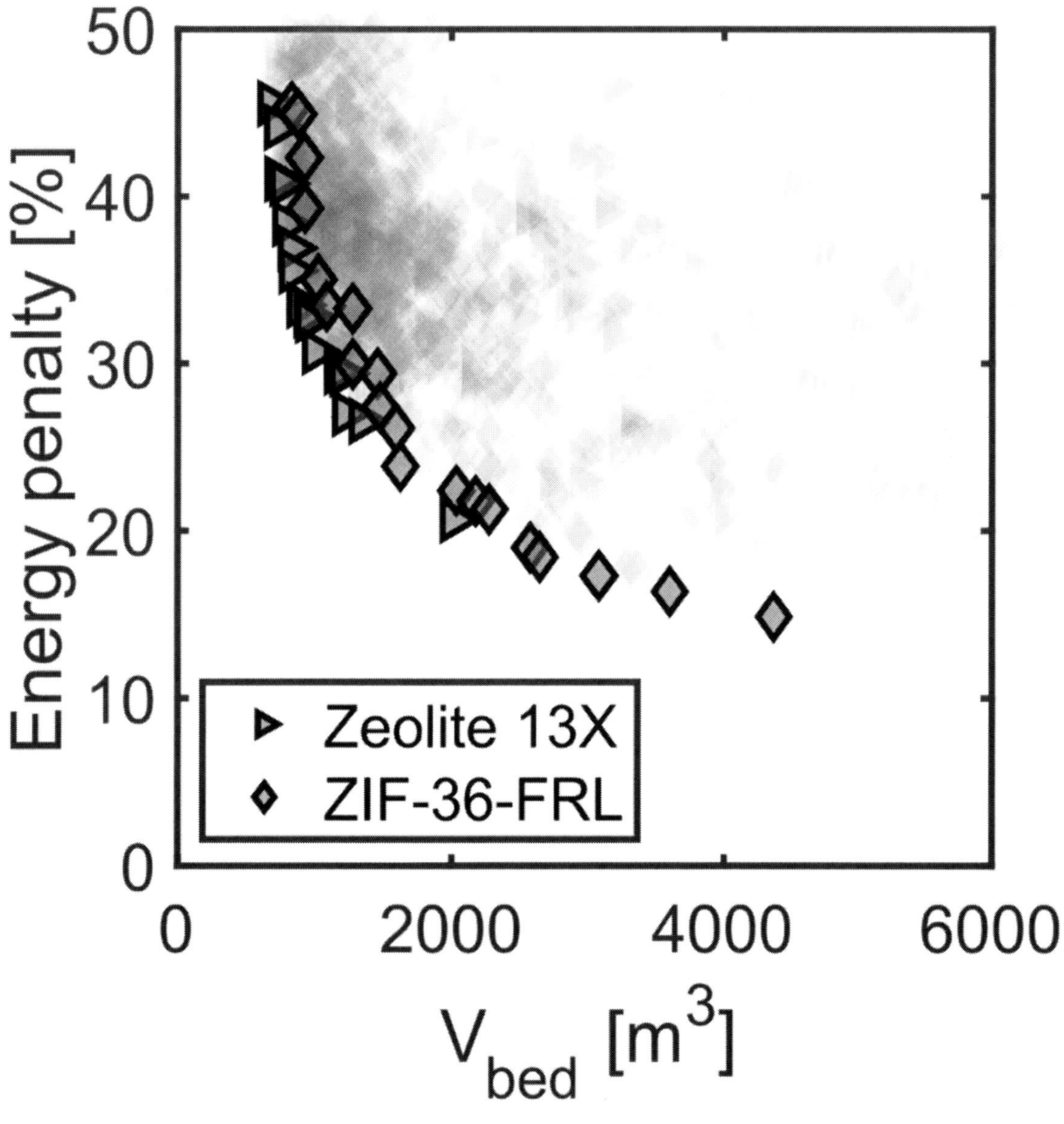

647d On the Use of Absolute and Excess Isotherms in Adsorption Processes: H_2 PSA As a Case Study.

Riccardo Rea, Hyungwoong Ahn, Stefano Brandani, Enzo Mangano

The vast majority of adsorption isotherms in literature are reported as excess isotherms. This is generally due to the fact that the skeletal solid volume required to calculate the excess isotherms can be measured relative to helium, leading often to define excess adsorption as the only directly measurable quantity[1]. Molecular simulations, process modelling and any thermodynamically consistent model, on the other hand, require the use of absolute isotherms[2,3].

The perceived advantage in measuring and reporting excess isotherms has driven the development of adsorption dynamics models where the mass balance is adapted to allow the use the parameters resulting from the direct fit of the excess isotherms.

This study explores the implications of using this approach compared to the rigorous one where a thermodynamically consistent model based on the absolute adsorbed amount is used. We apply both methods to the case study of a H_2 PSA from steam methane reforming as this is one of the separations where the distinction between absolute and excess adsorbed amounts is significant[2]. Production of ultrapure H_2 (>99.9% mol) is in fact a very challenging separation where model accuracy is key for the reliability of process predictions[4].

All fundamental adsorbent quantities (porosities, densities) as well as equilibrium isotherms were measured experimentally in our laboratory to minimize the uncertainty of the model comparisons. High pressure isotherms for H_2, CO_2, CH_4 and N_2 on the NIST NaY reference material RM8850[5,6] were measured using our custom-made Adsorption Differential Volumetric Apparatus (ADVA). The resulting excess and absolute adsorption isotherms at different temperatures, fitted using a Langmuir model, were used in the dynamic simulation of the H_2 PSA process. In addition, high-pressure binary H_2/CO_2 breakthrough experiments were carried out to validate the accuracy of the model predictions. Finally, full process simulations were implemented to compare the two methodologies in terms of the predicted purity and recovery for the separation.

[1] S. Sircar, Gibbsian Surface Excess for Gas Adsorption - Revisited, Ind. Eng. Chem. Res. 38 (1999) 3670–3682. https://doi.org/10.1021/ie9900871.

[2] S. Brandani, et al., Net, excess and absolute adsorption and adsorption of helium, Adsorption. 22 (2016) 261–276. https://doi.org/10.1007/s10450-016-9766-0.

[3] A.L. Myers, P.A. Monson, Physical adsorption of gases: the case for absolute adsorption as the basis for thermodynamic analysis, Adsorption. 20 (2014) 591–622. https://doi.org/10.1007/S10450-014-9604-1.

[4] M. Luberti, et al., Design of a H_2 PSA for cogeneration of ultrapure hydrogen and power at an advanced integrated gasification combined cycle with pre-combustion capture, Adsorption. 20 (2014) 511–524. https://doi.org/10.1007/S10450-013-9598-0.

[5] H. Giang, et al., Porosity, Powder X-Ray Diffraction Patterns, Skeletal Density, and Thermal

Stability of NIST Zeolitic Reference Materials RM 8850, RM 8851, and RM 8852, 126 (2021). https://doi.org/10.6028/jres.126.047.

[6] H.G.T. Nguyen, et al., A reference high-pressure CH_4 adsorption isotherm for zeolite Y: results of an interlaboratory study, Adsorption. 26 (2020) 1253–1266. https://doi.org/10.1007/s10450-020-00253-0.

647e Blue H_2 Production of 12-Bed PSA Process Integrated with CO_2 Absorption Process from Smr Syngas

Jun-Ho Kang, Hyun-Taek Oh, Hyeon-Hui Lee, Chang-Ha Lee

Since H_2 is produced from mainly naphtha and natural gas reforming[1], and coal gasification[2] in chemical industries, an enormous amount of CO_2 emission is inevitable at present. To meet carbon neutralization, the H_2 production strategy is essential to shift from grey H_2 to blue H_2. In the process integration for simultaneous H_2 production and CO_2 capture, H_2 PSA processes currently in operation are essentially re-optimized because feed conditions are highly changed.

Herein, as a first step, the dynamic model of 12-bed H_2 PSA process is developed, which is widely used in petrochemical industries to produce H_2 from a reforming gas. The 12-bed PSA process using the layered beds with activated carbon and zeolite is operated at a 24-step cyclic sequence. The developed dynamic model is validated with industrial operating data. When dry-based reforming gas ($H_2/CO/CH_4/CO_2$; 72.25/2.62/6.71/18.42%) is used at real operating pressure and flowrate, H_2 purity of 99.99+% with 88+% recovery can be achieved, which agrees well with the real operating result. The role of each adsorptive layer was analyzed as a bulk separator or a purifier.

Then, a highly efficient blend-amine process, which is installed before the PSA process as a pre-combustion CO_2 capture, is optimized to achieve 90-99 % CO_2 capture with 99+ % purity from SMR syngas. Finally, the 12-bed PSA process integrated with the CO_2 capture process using blend-amine is studied. The optimal operating condition for the present 12-bed PSA process is affected by CO_2 capture rate in the range of 90 to 99% because feed composition and flowrate are significantly changed. Therefore, depending on the capture rate, the performance and dynamic behavior are analyzed to gain the insights of performance and dynamic behavior by using the verified dynamic model. Due to the complexity of the integrated process optimization by multi-operating variables, a machine learning technology is applied to find the optimum operating conditions at each CO_2 capture rate.

References:

1. Y. Park., J.H. Kang., D.K. Moon., Y.S. Jo, C.H. Lee., *Chem. Eng. J.* 408 (2021) 127299.
2. D.K. Moon, D.G. Lee, C.H. Lee, *Appl. Energy.* 186 (2016) 760.

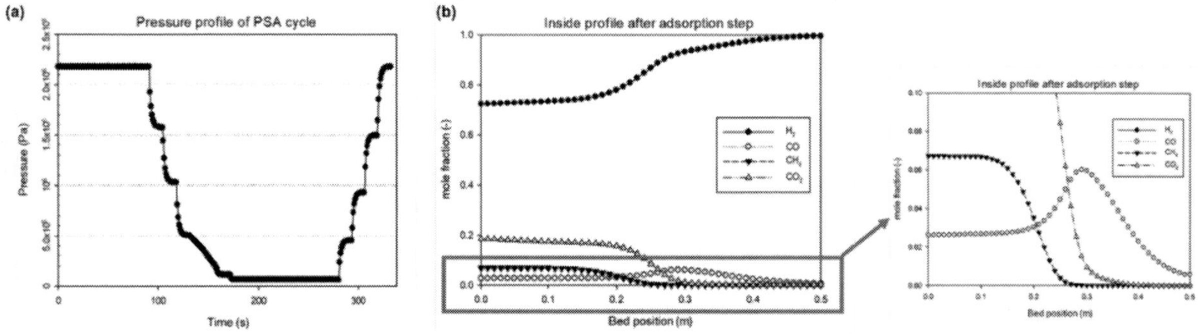

647f Evaluation of Adsorbent Performance By Optimal Temperature Swing Adsorption Process

Yuchi Sugiura, Tomoyuki Yajima, Yoshiaki Kawajiri

Adsorptive separation is one of the most promising methods for industrial gas purification. Since process performance, quantified by metrics such as purity, recovery, energy consumption, and productivity, is highly dependent on the combination of adsorbent used and operation method, analysis using a mathematical model is necessary to design a process that achieves the target performance. For this challenge, past studies have conducted predictions of process performance using process models and searched for efficient operating conditions by optimization[1,2]. In addition, various new porous materials such as zeolites and Metal-Organic Frameworks (MOFs) have been developed in recent years, and their evaluation and comparison remains another challenge, which is tackled by many researchers. Also for this problem, process modeling and model-based optimization to evaluate novel adsorbents can be a promising approach[3,4].

However, comprehensive comparison considering a variety of operating strategies optimized for each adsorbent has been rarely performed. Previous studies have focused on optimization of operating conditions, such as cycle time and operating temperature, where a certain operating cycle is assumed and fixed. In such approaches, the optimal cycle, which can be found from a wide variety of TSA operating methods, may not be found for each adsorbent.

The purpose of this study is to identify differences in optimal operating methods for each adsorbent. The operating cycle and operating conditions were optimized for multiple adsorbents, and the differences in performance and operating methods were analyzed. In this study, a rigorous mathematical model based on mass and heat balances was used to the analysis. For optimization, Pyomo, a python-based modeling package, was used to discretize the model both in time and in space, and the interior point method was employed to perform deterministic optimization considering multiple performance criteria such as CO_2 purity, recovery, productivity, and energy efficiency.

[1] Ko, D., Siriwardane, R. andamp; Bieggler, L. T. (2005). Industrial and Engineering Chemistry Research, 44(21), 8084–8094.

[2] Agarwal, A. Biegler, L. T. andamp; Zitney, S. E. (2010). AIChE J., 56, 1813-1828.

[3] Shafeeyan, M. S., Daud, W. M. A. W. andamp; Shamiri, A. (2014). Chemical Engineering Research and Design, 92, 961-988.

[4] Burns, T. D., Pai, K. N., Subraveti, S. G., Collins, S. P., Krykunov, M., Pajendran, A. andamp; Woo, T. K. (2020). Environmental Science and Technology, 54, 4536-4544.

647g Superstructure Optimization of Pressure Vacuum Swing Adsorption Processes

Kasturi Nagesh Pai, Reza Haghpanah, William Edsall

The performance of a pressure vacuum swing adsorption (PVSA) process depends on two factors, the adsorbent media and the PVSA process cycle. With the advent of metal organic frameworks, the potential pool of adsorbents has exponentially grown over the past decade. They provide the process designers with a diverse set of equilibria and other tunable properties to choose from[1]. Detailed dynamic models required to accurately capture the adsorption behavior exist but are computationally expensive to solve. Each PVSA cycle configuration has a large set of decision variables such as the pressure envelop, the feed flowrate, etc. There are potentially multiple cycle configurations for any adsorbent that one can choose to meet the desired separation objectives[2]. To achieve the true potential of a given adsorbent, the operating conditions of each cycle configuration must be optimized. Optimizing the PVSA cycle configuration and the operating conditions is not computationally scalable. Thus typically, the performance of a small set of user-defined process cycle configuration are chosen for optimization.

This work presents a superstructure model of a PVSA process as a method to design the best cycle configuration. To highlight the potential of such an approach, the case study of post-combustion CO_2 capture is presented in this work. The superstructure model presented in the work covers over a dozen possible PVSA cycle configurations. The scale of the proposed model leads to a potentially large input search region. A novel optimization strategy is also proposed to optimize the set of decision variables involved. Multiple case studies are evaluated to identify the best configurations for different processes and operational constraints.

[1]Yamil J. Colón and Randall Q. Snurr, High-throughput computational screening of metal-organic frameworks, Chem. Soc. Rev., 2014,43, 5735-5749.

[2]Shivaji Sircar, Basic Research Needs for Design of Adsorptive Gas Separation Processes, Ind. Eng. Chem. Res. 2006, 45, 16, 5435–5448.

647i A Strategy for Optimal Process Design and Operation of a Pressure Swing Adsorption (PSA) System through Pressure Equalisation and Co-Current Depressurisation

Yan Chen, Hyungwoong Ahn

A Pressure Swing Adsorption (PSA) system is known for its unrivalled capability in various industrial applications for gas separation and purification, e.g. H_2 purification and air drying. Accordingly a PSA has been actively researched to enhance its performance and adaptability, expanding its horizons for emerging gas separations, e.g. CO_2 capture.

Design of a PSA process involves its associated cycle scheduling table that is directive of how the cyclic batch system is to be operated. It is very common to see the step sequence include Pressure Equalisation (PE) step or Co-current Depressurisation (CoD) step aiming to improve the light product recovery. But it is also very hard to find a reliable guideline on how many PE steps to include and what is the optimal column pressure at the end of CoD step.

In this study, the Equilibrium Theory model for analysing a PSA system [1,2] was developed further so that it could be applied to a system with complex step configurations and non-linear isotherms, enabling a systematic analysis of the effects of PE or CoD steps on PSA performance.

It was found that in case of a feed gas containing a high fraction of the light component adding PE or CoD steps could get the product recovery improved notably, and there exists an optimum number of PE steps or CoD pressure for maximizing the product recovery. On the other hand, adding even one PE cycle to the step configuration may ruin the product recovery if the light component fraction is low in the feed. Alternatively, having CoD step instead could improve the product recovery on the ground that the CoD step pressure can be controlled more flexibly than the PE step pressure. This work did not study only the effect of feed gas composition, but it also looked into the effects of other parameters such as adsorption pressure, the extent of purging, etc.

1. Chiang, A. S. T. An Analytical Solution to Equilibrium PSA Cycles. *Chemical Engineering Science* **51**, 207-216 (1996).

2. Ahn and Lee, Equilibrium Theory Analysis of Vacuum Swing Adsorption for Separation of Ethanol from CO_2 in a Beverage Dealcoholization Process, *Industrial andamp; Engineering Chemistry Research* **59(50)**, 21948 - 21956 (2020).

570b New Amine-Containing Facilitated Transport Membrane and Process for CO_2 Capture from Flue Gas

Yang Han, Kai Chen, Witopo Salim, Dongzhu Wu, W.S. Winston Ho

Application of membrane for CO_2 capture from flue gas has been limited by the trade-off between CO_2 permeance and CO_2/N_2 selectivity for most polymeric membrane materials. To overcome this limitation, we have studied the facilitated transport membrane (FTM) possessing both high permeance and selectivity and designed a membrane process that can capitalize on high-selectivity offered by FTM. We have synthesized a new FTM containing polyvinylamine as fixed-site carrier and an aminoacid salt, 2-(1-piperazinyl)ethylamine sarcosinate, as mobile carrier. The membrane was used to fabricate 1.4-m^2 spiral-wound modules, which exhibited a CO_2 permeance of 1450 GPU and a CO_2/N_2 selectivity of 185 at 67°C with actual flue gas at the National Carbon Capture Center in Wilsonville, AL, USA. Furthermore, significant increase in CO_2 permeance has been achieved upon the bulk removal of CO_2 for the range of CO_2 partial pressure relevant to carbon capture from coal-derived flue gas. For instance, the CO_2 permeance increased from 1464 to 1918 GPU when the CO_2 partial pressure reduced from 74.1 to 3.9 kPa at 67°C. Such a carrier saturation phenomenon has been modeled and incorporated into a novel two-stage membrane process featuring partial retentate recycle. As the bulk CO_2 removal reduces the CO_2 partial pressure gradually in the membrane module, this feature mitigates the carrier saturation and results in an uprising CO_2 permeance upon the CO_2 removal. For the membrane performance at 67°C, an attractive capture cost of $41.75/tonne can be achieved for 90% CO_2 capture in addition to offering the reduced system footprint. 70% CO_2 partial capture can further reduce the cost to about $32/tonne.

570g Modifying Synthetic Membranes: A Personal View

Georges Belfort

Since our discovery nearly 30 years ago that poly(aryl sulfone) surfaces form radicals on exposure to light (UV) radiation, we have together with collaborators at RPI and MIT developed a vigorous membrane surface modification program that continues to be active today. The seminal reasons are that (a) modifying previously cast commercial synthetic membranes such as poly(ether sulfone), polyimide (PI), poly(vinylidene fluoride) (PVDF) and regenerated cellulose (RC) are widely available, (b) PI and PES are too apolar for effective use in the health, biotechnology, food and beverage industries, and (c) modifying commercially available membranes with desirable properties is less expensive and simpler than developing and processing new polymer materials as membranes. It is widely known that industrial membrane companies modify base PES and PVDF membranes to form composite membranes so as to render their surfaces polar with cross-linked hydroxyalkyl acrylate and RC[1, 2]

The Belfort group first utilized surface irradiation methods, like photooxidation and atmospheric plasma of PES for ultrafiltration, to hydrophilize the surfaces and reduce protein and organic fouling[3-5]. We developed the first truly high throughput platform to screen, select and synthesize winners from a 66-member monomer library for a series of different feeds (i.e., proteins, bacteria, mammalian cells)[6-9]. We then used solution methods such as activators regenerated by electron transfer-atom transfer radical polymerization (ARGET-ATRP)[10, 11] and single electron transfer- living radical polymerization (SET-LRP)[12] that rely on copper (Cu(1) and Cu(0) to catalyze the free radical polymerization. To better understand both processes, we implemented and solved their respective reaction schemes using MATLAB and fit the models to literature data and determined the rate constants and other parameters[10, 13]. Significant findings from our work were as follows: (a) The use of our high throughput platform to seek chemistries with desired behavior for a particular application in a very short period. GE Global used the platform to develop a DNA detection product in 2 months. (b) The discovery of surface chemistries such as specific polyethylene glycols and zwitterions that exhibit low protein fouling using our high throughput platform. (c) The discovery of amines and the natural molecule NSP1 (FG residue-rich sequence)[14] from the nucleopore complex that exhibit reduced protein fouling. (d) Demonstrating that ARGET-ATRP and SET-LRP are able to produce a new class of membranes based on grafting bottle brushes that have high surface areas and can be chemically and physically designed *a priori* for a particular application. New focus is on replacing distillation with organic nanofiltration[12], selectively purifying mRNA for vaccines, and purifying IgG using adsorptive bottle brushes.

Acknowledgement: To collaborators Chip Kilduff, Bob Langer and Dan Anderson and to my lab manager of 15.5 years, Mirco Sorci, and all the wonderful graduate students who worked on surface modifications.

References

1. Tuccelli, R.; Karnakis, A. T. Cellulosic ultrafiltration membranes. 1994.
2. Steuck, M. J. Porous membrane having hydrophilic surface and process. 1986.
3. Crivello, J. H. Y., H. and Belfort, G. Low Fouling Ultrafiltration and Microfiltration Aryl Polysulfone. 1995.

4. Pieracci, J.; Crivello, J. V.; Belfort, G., Photochemical modification of 10kDa polyethersulfone ultrafiltration membranes for reduction of biofouling. *Journal of Membrane Science* **1999,** *156* (2), 223-240.
5. Yamagishi, H.; Crivello, J. V.; Belfort, G., Evaluation of photochemically modified poly (arylsulfone) ultrafiltration membranes. *Journal of Membrane Science* **1995,** *105* (3), 249-259.
6. Zhou, M., Liu, H., Venkiteshwaran, A., Kilduff, J. C., Anderson, D. G., Langer, R. Belfort G. , High throughput discovery of new fouling-resistant surfaces. *Journal of Materials Chemistry* **2011,** *21*, 11.
7. Zhou, M.; Liu, H.; Kilduff, J. E.; Langer, R.; Anderson, D. G.; Belfort, G., High-throughput membrane surface modification to control NOM fouling. *Environmental science andamp; technology* **2009,** *43* (10), 3865-71.
8. Zhou, M.; Liu, M.; Venkiteshwaran, A.; Kilduff, J.; Anderson, D. G.; Langer, R.; Belfort, G., High throughput discovery of new fouling-resistant surfaces,. *J. Materials. Chem* **2011,** *21*, 12.
9. Zhou, M. Y.; Liu, H. W.; Kilduff, J. E.; Langer, R.; Anderson, D. G.; Belfort, G., High-Throughput Membrane Surface Modification to Control NOM Fouling. *Environmental Science andamp; Technology* **2009,** *43* (10), 3865-3871.
10. Keating, J. J.; Lee, A.; Belfort, G., Predictive Tool for Design and Analysis of ARGET ATRP Grafting Reactions. *Macromolecules* **2017,** *50* (20), 7930-7939.
11. Keating, J. J. I.; Sorci, M.; Kocsis, I.; Setaro, A.; Barboiu, M.; Underhill, P.; Belfort, G., Atmospheric Pressure Plasma - ARGET ATRP Modification of Poly(ether sulfone) Membranes: A Combination Attack. *J. Membr. Sci.* **2017,** *546*, 6.
12. Ramesh, P.; Xu, W., L.; Sorci, M.; Trant, C.; Lee, S.; Kilduff, J.; Yu, M.; Belfort, G., Organic solvent filtration by brush membranes: Permeability, selectivity and fouling correlate with degree of SET-LRP grafting. *J. Membr. Sci.* **2020,** *618*, 118699.
13. Gunter, K.; Sorci, M.; Belfort, G., Predictive Tool for Design and Analysis of SET-LRP Polymer Grafting Reactions. *ACS Applied Polymer Materials,* **2020,** *2* (11), 4924=4935.
14. Hayama, R.; Sorci, M.; Keating Iv, J. J.; Hecht, L. M.; Plawsky, J. L.; Belfort, G.; Chait, B. T.; Rout, M. P., Interactions of nuclear transport factors and surface-conjugated FG nucleoporins: Insights and limitations. *PLoS One* **2019,** *14* (6), e0217897.

570d High Purification of Binary Protein Mixtures Having Close Molecular Weights By Ultrafiltration

Lixin Feng, Yufeng Song, Sagnik Basuray, Kamalesh Sirkar, Solomon Isu, Ranil Wickramasinghe

Mixtures of proteins/biopharmaceuticals having close molecular weight ratios (MWRs) are generally separated by chromatography. Conventional ultrafiltration (UF) can separate binary protein mixtures if the MWR of the two proteins is at least ~7–10. For lower MWRs, novel countercurrent cascades of 3-5 stages can yield relatively pure proteins. To separate similarly sized proteins in one device, High Performance Tangential Flow Filtration (HPTFF) technique was developed. HPTFF achieves high selectivity but does not yield almost pure protein in the permeate. The Internally Staged Ultrafiltration (ISUF) technique was proposed to yield almost pure protein in the permeate. The ISUF technique uses a stack of 3 UF membranes without any gaskets/spacers in-between. Permeate from membrane 1 with selective skin facing the feed mixture becomes the feed to the next membrane 2 located below and so on. We have investigated separations of a number of binary protein mixtures using the ISUF configuration in a stirred cell: (1) hemoglobin (MW 64,677 Da) and BSA (MW 66,430 Da) having a MWR of 1.03; (2) α-lactalbumin (14,400, MW) and β-lactoglobulin (18,400, MW; present naturally as a dimer, 36,800) mixture having a MWR of 2.55 and encountered in whey protein concentrate separation. In each case, we have investigated the effect of pH, ionic strength and feed pressure using as required 100kDa or 30kDa MWCO Ultracel® and Omega® membranes having a particular charge. We have investigated the effect of diavolumes on the purity of the retained protein as well as other performance indicators. We have also investigated the purification of a protein from a virus using a stacked set of UF membranes. Virus filtration for validation of virus clearance is difficult when the size of the mAb approaches that of small virus particles.

570e Applying the Method of Moments to Design Gradient Elution Chromatography

Andreas Seidel-Morgenstern, Hossam Abdelghani, Xinghai An, Shamsul Qamar

Besides the widespread application of chromatography as a powerful and flexible analytical technique, there is a large interest in the pharmaceutical industry and in biotechnology to isolate and purify value added products using preparative liquid chromatography. This led in the last decades to the development of advanced operating concepts, including regimes which exploit forced modulations of certain parameters during the chromatographic process. In contrast to classical isocratic elution the application of such gradient concepts offers significant potential to improve the separation performance. However, rational design and optimization of gradient chromatography requires knowledge regarding numerous thermodynamic and kinetic parameters [1-3].

Recently an analytical solution was derived to describe the effect of linear solvent gradients on the shape of elution profiles. This solution allows generating instructive equations for the first three moments of pulse responses [4]. Considering a case study, we will apply these moment expressions to estimate jointly column efficiencies and adsorption isotherm parameters as a function of the solvent composition. The method requires only a small number of chromatograms, which should be measured for different start and end times of linear gradients.

Finally, we will demonstrate the application of a parametrized column model to identify suitable gradient shapes.

[1] Guiochon G., Felinger A., Shirazi D.G., Katti A.M.,

Fundamentals of Preparative and Nonlinear Chromatography, 2nd Edition,

Academic Press, New York, 2006

[2] Carta G., Jungbauer A.;

Protein Chromatography: Process Development and Scale-Up, 2nd Edition,

Wiley-VCH, 2020

[3] Schmidt-Traub H., Schulte M., Seidel-Morgenstern A. (Eds.),

Preparative Chromatography, 3rd Edition,

Wiley-VCH, Weinheim, 2020

[4] Qamar, S., Rehman, N., Carta, G., Seidel-Morgenstern, A.,

Analysis of gradient elution chromatography using the transport model,

Chemical Engineering Science, 2020, 225,115809

Besides the widespread application of chromatography as a powerful and flexible analytical technique, there is a large interest in the pharmaceutical industry and in biotechnology to isolate and purify value added products using preparative liquid chromatography. This led in the last decades to the development of advanced operating concepts, including regimes which exploit forced modulations of certain parameters during the chromatographic process. In contrast to classical isocratic elution the application of such gradient concepts offers significant potential to improve the separation performance. However, rational design and optimization of gradient chromatography requires knowledge regarding numerous thermodynamic and kinetic parameters [1-3].

Recently an analytical solution was derived to describe the effect of linear solvent gradients on the shape of elution profiles. This solution allows generating instructive equations for the first three moments of pulse responses [4]. Considering a case study, we will apply these moment expressions to estimate jointly column efficiencies and adsorption isotherm parameters as a function of the solvent composition. The method requires only a small number of chromatograms, which should be measured for different start and end times of linear gradients.

Finally, we will demonstrate the application of a parametrized column model to identify suitable gradient shapes.

[1] Guiochon G., Felinger A., Shirazi D.G., Katti A.M.,

Fundamentals of Preparative and Nonlinear Chromatography, 2nd Edition,

Academic Press, New York, 2006

[2] Carta G., Jungbauer A.;

Protein Chromatography: Process Development and Scale-Up, 2nd Edition,

Wiley-VCH, 2020

[3] Schmidt-Traub H., Schulte M., Seidel-Morgenstern A. (Eds.),

Preparative Chromatography, 3rd Edition,

Wiley-VCH, Weinheim, 2020

[4] Qamar, S., Rehman, N., Carta, G., Seidel-Morgenstern, A.,

Analysis of gradient elution chromatography using the transport model,

Chemical Engineering Science, 2020, 225,115809

570f Remotely Controlled Magneto-Responsive Polymeric Membranes for Controlled Release of Biomacromolecules

Mathias Ulbricht, Mohamed Elleithy

Between 2010 and 2020, FDA has approved 97 new therapeutic proteins with subcutaneous injection as the main route of administration. Stimuli-responsive membranes are an attractive field of research [1], and such membranes as barrier for a drug reservoir could offer an alternative route with high reproducibility and dosing flexibility to increase patient's compliance. For that purpose, a remotely controlled stimulation of the opening and closing of the barrier membrane's pores would be the preferred option. Previous own work had demonstrated magneto-responsive membranes with excellent switchable rejection for dextran 2000 kDa [2]. The present work aimed at developing membranes with switchable sieving for much smaller biomacromolecules such as insulin (5.8 kDa).

In the first stage of the project, thermo-responsive pore-filled membranes were developed through a two-steps approach. Polyethersulfone (PES) was used to create via casting of polymer solution and non-solvent vapor induced phase separation (VIPS) a porous support membrane, which was subsequently pore-filled via in situ cross-linking copolymerization with poly(*N*-isopropylacrylamide) (PNIPAAm) hydrogel that acts as the sieving medium. Membranes demonstrated reversible increase in water permeability by about 380 times in response to increased temperature (from 25 to 40°C). Furthermore, rejection of dextran (4 kDa, which has a comparable Stoke's diameter to that of insulin) under ultrafiltration conditions decreased from 91% to 10%; diffusion rate increased by a factor of 21. This increase in diffusion rate opens up the possibility of pulsatile biomacromolecules release from a reservoir. As an example, a membrane with a radius of 2 cm (12.6 cm^2) could deliver from a reservoir the average insulin daily requirement in 2 hours of membrane stimulation.

In the second stage of the project, to add remote control functionality, iron oxide nanoparticles as "nanoheaters" were incorporated via adapted casting and VIPS conditions into the porous PES-based support to create nanohybrid membranes. In response to an external alternating magnetic field (AMF), nanohybrid membrane temperature increases very quickly from 25°C to values far beyond the hydrogel volume phase transition temperature (33°C). Subsequent functionalization under conditions adapted to the modified structure of the base membrane resulted in PNIPAAm pore-filled nanohybrid membranes that demonstrated reversible thermo-responsive changes in water permeability as well as dextran ultrafiltration and diffusion rate, very similar to the only PES-based pore-filled membranes. Furthermore, by AMF activation of the nanoheaters in the PNIPAAm pore-filled nanohybrid membranes, the remotely controlled efficient switching of the membrane from a "closed" to an "open" state for could be shown for both dextran (4 kDa) and insulin.

Overall, highly responsive remotely controlled membranes for controlled release of biomacromolecules as small as insulin have been established. To the best of our knowledge, for the first time in literature such performance is to be reported.

[1] D. Wandera, S. R. Wickramasinghe, S, M. Husson, Stimuli-responsive membranes, Journal of Membrane Science, 2010, 357, 6-35.

[2] X. Lin, R. Huang, M. Ulbricht, Novel magneto-responsive membrane for remote control switchable molecular sieving. *Journal of Materials Chemistry B*, 2016, 4, 867-879.

147a New Design for Hollow Fiber Membrane Module: Feasibility Study Based on Computational Fluid Dynamic Simulations

Yating Xu, Raja Ghosh

Hollow fiber membrane modules are extensively used for hemodialysis, which is a medical procedure for the removal of uremic toxins such as urea and creatinine from the blood of patients with kidney disorders. Currently used hollow fiber membrane modules used for hemodialysis consist of a cylindrical shell within which a bundle of hollow fiber membranes is housed. The influent fluid is distributed to the fiber bundle through an inlet header (also called an arterial port), while the effluent fluid is collected from the fiber bundle through an effluent header (also called a venous port). One of the main problems with cylindrical hollow fiber membrane modules is flow non-uniformity, which could exist in the inlet and outlet headers, within the fibers, as well as on the shell side of the module. Flow non-uniformity can lead to several problems such as poor clearance of toxins, flow jetting into fibers close to the center of the device, high convective flow (or ultrafiltration) leading to membrane fouling, creation of zones of high shear rates where blood cells could by lysed, and creation of stagnation zones where blood clots could form. Some of these challenges could be addressed through design improvements aimed at improving flow uniformity. In this presentation, we discuss a new design for a hollow fiber membrane module which incorporates advanced flow distribution and collection features. The feasibility of using this module for hemodialysis is analysed by performing computational fluid dynamic (CFD) simulations. These simulations are used to examine the effects of design improvements on tracer dispersion, flow uniformity, velocity distribution and shear rate profiles. While the membrane module discussed in this presentation was primarily designed for use in hemodialysis, its features could be utilized to design efficient membrane modules for other applications such as ultrafiltration, microfiltration, membrane distillation, pervaporation, and membrane crystallization.

706a 3D Printed Hybrid Monolith for CO2 Capture

Ana Pereira, Alexandre Ferreira, Alírio E. Rodrigues, Ana M. Ribeiro, Maria João Regufe

The mitigation of CO_2 emissions is imperative to stop climate changes. One of the leading producers of CO_2 is power plants. To reduce the emitted CO_2 from these plants, carbon capture and storage (CCS) can be used. Adsorption-based techniques can be used for this purpose.

To ensure an efficient adsorption process, the properties of the used adsorbents must be optimized. Nowadays, traditional packed beds with pellets/beads are still used in industry. However, this type of structured adsorbent possess some drawbacks, like a high-pressure drop and attrition. This way, additive manufacturing (AM), also known as 3D-printing, emerges as an alternative to develop structured adsorbents suitable for large-scale applications. The shaping of adsorbent materials using AM technologies has still a long path of exploration.

In this work, an adsorbent monolith consisting of 50 %wt. of zeolite 13X and 50 %wt. of activated carbon was developed using direct ink writing. To evaluate the suitability of this material for large-scale applications, the impact of shaping by direct ink writing of a structured adsorbent material for CO_2 capture on the characteristics of the material and mechanical strength were analyzed. The material was characterized by employing N_2 physisorption at 77 K, CO_2 adsorption at 273 K, mercury porosimetry, and crushing tests.

Additionally, pure CO_2 and N_2 adsorption equilibrium isotherms were measured at 303, 333, and 373 K and in a pressure range of 0 to 1.5 bar and 0 to 5 bar, respectively – Figure 1. The isotherms were fitted and well described with the Dual-Site Langmuir model. CO_2 is the gas with a higher affinity towards the 3D-printed monolith. Considering a CO_2 molar fraction of 0.15, at 1 bar and 303 K, the selectivity of the monolith is 30.7.

Dynamic adsorption tests were also performed with and without the assistance of electric current to regenerate the material. The CO_2/N_2 binary breakthrough curves for these two runs are displayed in Figure 2. A mathematical model was utilized to describe the behavior observed in the breakthrough curves, and the simulation outcomes can also be examined in Figure 2. A compressive CO_2 mass front is seen in the adsorption step and a dispersive mass front in the regeneration step, which corroborates with the behavior of the measured isotherms – Figure 1. The desorption is faster when the electrification of the adsorbent material is performed (compare Figure 2 a) and b)).

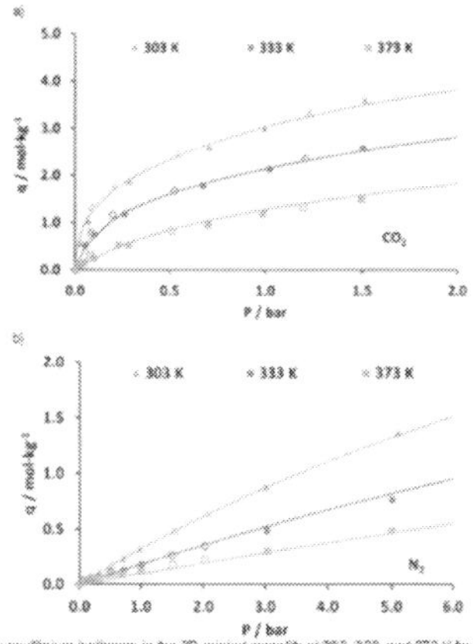

Figure 1 - Adsorption equilibrium isotherms in the 3D-printed monolith at 303, 333, and 373 K for a) CO_2, and b) N_2. The symbols correspond to experimental points and the line to the Dual-Site Langmuir fitting. Closed symbols – adsorption; open symbols – desorption.

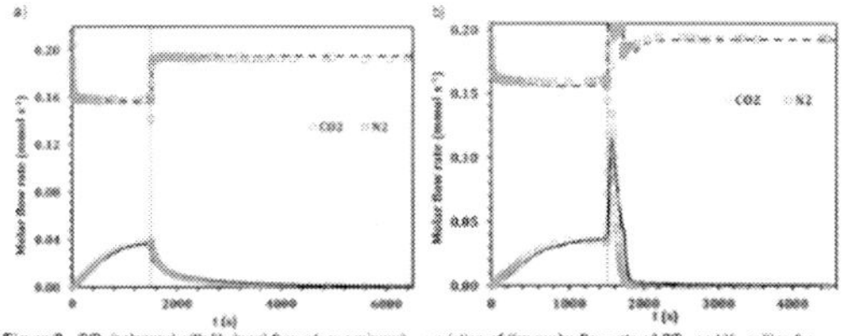

Figure 2 - CO_2 balanced with N_2 breakthrough experiments - variation of the molar flow rate of CO_2 and N_2 exiting the column during a) experiment with no electrification, b) experiment with electrification. The symbols are the experimental points, and the line is simulated by the mathematical model.

706b Development of an Expanded PTFE Structured Adsorbent Containing Activated Carbon

Sulaimon Adegunju, Ryan Sanders, Behnam F. Kisomi, Charles E. Holland, Armin Ebner, Guo Shiou Foo, Bob Grasso, Steve K. Stark, Jeff Knopf, Joe W. Henderson, James A. Ritter

The overall objective of this work is to develop a novel structured adsorbent from expanded PTFE (e-PTFE) containing activated carbon within the e-PTFE framework. Structured columns are being fabricated with various activated carbons and being compared to identical columns fabricated with the corresponding commercial granular or pelleted counterparts for testing under identical conditions. Details about the fabrication technique will be discussed along with representative results from a variety of different experimental tests being conducted. These experimental tests might include single gas adsorption breakthrough curve experiments, volumetric equilibrium adsorption isotherm experiments for column adsorbent mass determination with N_2, 1-bed pressure swing adsorption (PSA) cycling experiments, column cycling resiliency tests, single gas gravimetric uptake tests for equilibrium and kinetics, single gas 100 Hz volumetric frequency response experiments, and column axial pressure drop experiments. The most recent and compelling results will be presented.

706c Membrane Adsorbents Comprising Self-Assembled Inorganic Nanocages (SINCs) for Direct Air Capture

Thien Tran, Shweta Singh, Amanda Grass, Timothy R. Cook, Haiqing Lin

There is a significant interest in solid sorbents with high CO_2 sorption capacity for economically viable direct air capture (DAC) of CO_2. Although current solid sorbents such as polyethylenimine (PEI)-supported silica demonstrate high CO_2-sorption capacity, its kinetics is slow partially because CO_2 has to diffuse into the highly filled pores. In this work, we design highly porous membrane adsorbents comprising PEI and self-assembled inorganic nanocages (SINCs) supported by porous membranes such as Solupor membrane with a porosity of 86%, a high-density polyethylene porous support. PEI is a leading amine-based material for DAC with high CO_2-sorption capacity. SINCs have small sizes (< 10 nm), which obviates the need for CO_2 to diffuse throughout large crystallites (such as MOFs), reducing adsorption and desorption times. Zirconium-carboxylate cages (Zr-SINCs) were synthesized and further functionalized with amines to enhance the CO_2 sorption capacity. The membrane-sorbent demonstrates high PEI content up to 48 wt.% while still retain a high porosity of 75%, leading to high CO_2 permeance (~80,000 GPU, 1 GPU = 10^{-6} cc(STP)/cm^2/s/cmHg) and hence lowering pressure drop. A membrane sorbent containing 45 wt.% (PEI+SINCs) exhibits CO_2 gas sorption of 0.61 *mmol CO_2/g sorbent* at 400ppm CO_2 and 25°C under dry condition. However at 30% RH, the CO_2 capacity increases to 1.33 mmol CO_2/g. The effect of PEI and SINCs composition in the membrane-sorbents, RH, and temperature on CO_2 sorption capacity and kinetic will be discussed.

706d Sub-Ambient Direct Air Capture of CO_2 Using MIL-101(Cr) Monoliths

Yuxiang Wang, Guanhe Rim, Mingyu Song, Christopher W. Jones, Ryan P. Lively

Current carbon drawdown scenarios require deployment of direct air capture (DAC) technologies to reduce atmospheric CO_2 concentration. Supported amine materials based on zeolite, silica, and metal-organic frameworks (MOFs) have been demonstrated as promising adsorbents for DAC, but the shaping and structuring of these materials into sorbent modules for practical processes have been inadequately investigated compared to the extensive research on material development.

In this work, we demonstrate the successful fabrication of adsorbent monoliths composed of cellulose acetate and the highly porous chromium MOF, MIL-101, by solution based additive manufacturing (SBAM). These monoliths feature macroporous interpenetrated polymeric frameworks where nano MIL-101 crystals are evenly distributed. Branched polyethylenimine was successfully immobilized in these monoliths to impart CO_2 uptakes of 1 mmol $g(monolith)^{-1}$ at both ambient (25 °C) and sub-ambient (-20 °C) conditions. Temperature controlled desorption (TPD) experiments reveal two CO_2 sorption mechanisms depending on the amine loading in the monoliths and CO_2 sorption conditions. The DAC performance under both ambient and sub-ambient conditions of these monoliths were further evaluated by dynamic breakthrough experiments, and the system pressure drops and mass transfer kinetics are benchmarked with systems where adsorbents of other forms like powder and pellets are deployed. In summary, our work showcases the excellent CO_2 capture performance of PEI-loaded MIL-101 monoliths under both ambient and sub-ambient conditions, suggesting the versatility of these monoliths in DAC applications regardless of climate conditions.

706e Modelling and Experimentation on a Rotary Adsorber Applied to Direct Air Capture

Lucy Victoria Barton, Camille Petit, Ronny Pini

Direct Air Capture (DAC) is a promising technology to aid the removal of CO_2 from the atmosphere and achieve net-negative emissions. However, it produces unique challenges due to the ultra-dilute nature of the ambient air feed (0.04 mol%). For comparison, CO_2 composition in post combustion gas streams ranges from 4-25 mol% [1] which is 100 to 600 times more concentrated than CO_2 in air. Therefore, capturing 1 kg of CO_2 in a DAC process requires at least 1,300 m^3 of air compared to less than 13 m^3 of post-combustion gas (at standard temperature and pressure). Although DAC proves more challenging, unlike post-combustion carbon capture, it can achieve net-negative emissions as well as offsets for non-point source emissions such as vehicles and planes. The requirements of high volumetric air flow and net-negative emissions dictate that feasible DAC separation processes must have minimal pressure drop and low specific energy. Today, research focused on process and contactor design for DAC in the field of adsorption remains limited yet crucial for the advancement of adsorption-based DAC. Beyond process design, the development of process models and experiments allows testing of promising sorbent materials for adsorption-based DAC.

In this work, we aim to assess the design and optimisation of a rotary adsorber by applying mathematical modelling and experimental validation. The rotary adsorber is a continually rotating thin cylindrical bed with a separated adsorption and desorption section, allowing the unit to both adsorb and desorb simultaneously. The pancake style design minimises pressure drop, while simultaneous adsorption and desorption allows a continuous process with one vessel. A rotary adsorber has been considered through equilibrium modelling for post-combustion carbon capture [2]. However, in the case of DAC, a kinetic model is necessary to account for the poor adsorption kinetics in ultra-dilute conditions and the dynamic nature of a rotating process contactor. We have considered temperature swing adsorption (TSA) operation, as adequate seals for pressure or vacuum swing cannot be achieved in the rotary unit. Initially TSA with an air purge has been applied, however, this will be extended to also assess steam and inert purge TSA.

To assess the rotary adsorber, we have created a one-dimensional dynamic adsorption model that describes the spatial and temporal evolution of gas and adsorbed phase concentration profiles in the unit. Initial results demonstrate that the model can assess the impact of the wheel speed, process flow rates, desorption conditions, material properties, adsorption kinetics, division of the wheel, and wheel dimensions. The model can be used to calculate key process performance indicators including CO_2 productivity, purity, working capacity and specific energy. We have commenced the design of a laboratory scale rotary unit connected to a continuous gas analyser for model validation. We aim to extend the experimental validation to assessing the impact of optimisation parameters through adjusting the operating conditions and layout of the rotary unit. The initial results highlight that that design and optimisation of a rotary adsorber is complicated due to the interconnected nature and significant number of parameters - ultimately requiring multi-variable optimisation.

References:

1. 1. E.S. Sanz-Perez et al. Chem rev. 5 (2020) 11840-11876
2. L. Herraiz at al., Front. Energy Res. 8 (2020) 482708

706f Immobilization of Zeolite Crystals on Coc-Based Microfluidic Devices for Separation Processes

Fadi Dawaymeh, Nahla Alamoodi, Anas Alazzam, Maryam Khaleel

Immobilization of zeolite nanoparticles has attracted great interest in various potential applications such as catalysis and membrane separations[1, 2], chemical sensors[3], and microfluidic devices[4]. Self-assembly method can be used to immobilize zeolite crystal grains on different substrates through chemical linkages including the ionic, covalent and intermolecular bonds between the zeolite crystals and the substrate to form a film with good uniformity. In the present work, zeolite crystals were immobilized on a cyclic olefin copolymer (COC) using a simple wet-chemical surface treatment. During the chemical surface treatment, the hydrophilic silane 3-aminopropyl) triethoxy (APTES) was used to form covalent bonds between the zeolite crystals and the polymer surface, resulting in the formation of zeolite films with good uniformity. Three types of zeolites including zeolite LTA, zeolite β and SAPO-5 were used as model building blocks. Different characterization techniques such as FTIR, SEM and contact angle measurements were used to investigate the characteristics of COC-coated zeolites. The results confirmed the deposition of zeolite layers consisting of highly packed crystals on the COC surface, demonstrating good immobilization ability.

References:

[1] H. Jiang, B. Zhang, Y. S. Lin, and Y. Li, 'Synthesis of zeolite membranes,' *Chinese Science Bulletin,* vol. 49, no. 24, p. 2547, 2004/12/01 2004, doi: 10.1360/03wb0146.

[2] B. M. L. Dioos, B. F. Sels, and P. A. Jacobs, 'Chapter 25 - Catalyst Immobilization on Inorganic Supports,' in *Studies in Surface Science and Catalysis*, vol. 168, J. Čejka, H. van Bekkum, A. Corma, and F. Schüth Eds.: Elsevier, 2007, pp. 915-XXVI.

[3] F. Almazán *et al.*, 'Zeolite based microconcentrators for volatile organic compounds sensing at trace-level: Fabrication and performance,' *Journal of Micromechanics and Microengineering,* vol. 26, p. 084010, 08/01 2016, doi: 10.1088/0960-1317/26/8/084010.

[4] X. Huang, G. Zhang, L. Zhang, and Q. Zhang, 'Continuous Flow Synthesis of a ZSM-5 Film in Capillary Microchannel for Efficient Production of Solketal,' *ACS Omega,* vol. 5, no. 33, pp. 20784-20791, 2020/08/25 2020, doi:

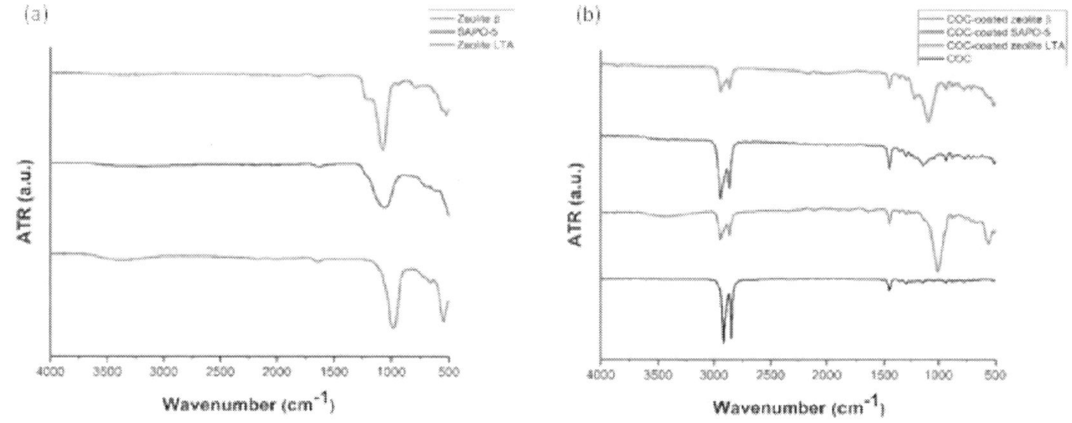
(a)
Zeolite β
SAPO-5
Zeolite LTA
ATR (a.u.)
4000
3500
3000
2500
2000
1500
1000
500
Wavenumber (cm^{-1})
(b)
COC-coated zeolite β
COC-coated SAPO-5
COC-coated zeolite LTA
COC
ATR (a.u.)
4000
3500
3000
2500
2000
1500
1000
500
Wavenumber (cm^{-1})

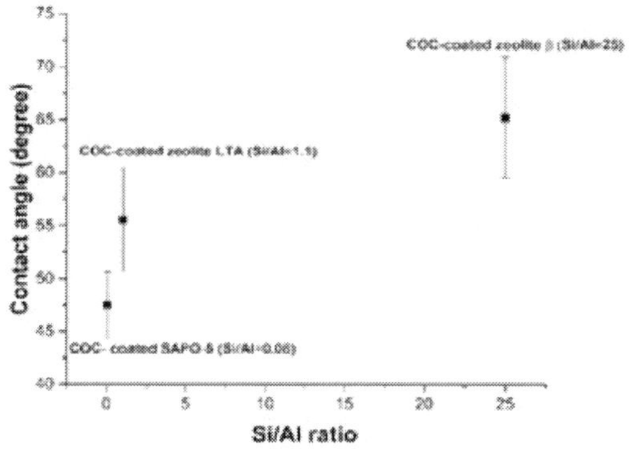
COC-coated zeolite β (Si/Al=25)
COC-coated zeolite LTA (Si/Al=1.1)
COC- coated SAPO-5 (Si/Al=0.06)
Contact angle (degree)
75
70
65
60
55
50
45
40
0
5
10
15
20
25
Si/Al ratio

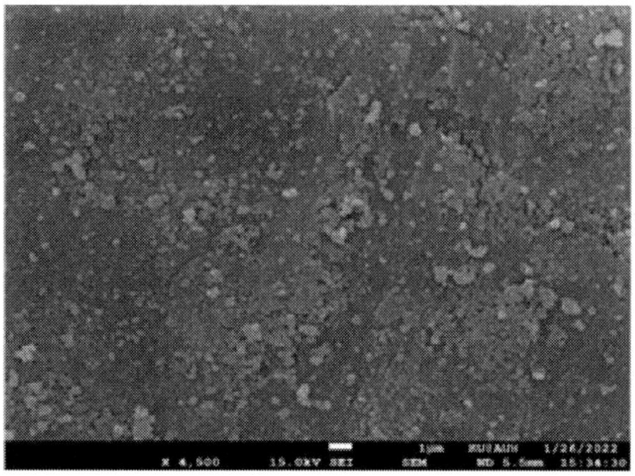
X 4,500
15.0kV SEI
SEM
1μm
WD 5.5mm
1/26/2022
15:34:30

706g Encapsulating Nanoscale Organic Hybrid Materials (NOHMs) within Polymeric Nanofibers Via Electrospinning for CO2 Capture

Jeffrey Xu, Ga Hyun Lee, Kyle Kersey, Dongjae Kim, Michelle Kidder, Yong Joo, Ah-Hyung Alissa Park

As fossil fuels continue to dominate the global energy portfolio, there is an urgent need for carbon negative technologies to reverse rising CO_2 concentration in the atmosphere. Recently, liquid-like Nanoscale Organic Hybrid Materials (NOHMs), composed of an amine polymer canopy tethered to a silica nanoparticle core, have been demonstrated as a promising CO_2 sorbent with low vapor pressure, high CO_2 selectivity, and high thermal-oxidative stability. Here, we demonstrate the incorporation of NOHMs into gas permeable polymers via nanofiber electrospinning. Several encapsulation materials were investigated; Polymers with Intrinsic Microporosity (PIM-1), polyacrylonitrile (PAN), and organic polysilazane (OPSZ), all of which are gas permeable and can selectively reject water. CO_2 capture capacity and kinetics were quantified via thermogravimetric and gas sorption analysis. Other key material properties e.g., thermal oxidative stability, fiber morphology and hydrophobicity were correlated to electrospinning conditions. Overall, these findings increase the technological feasibility of using NOHMs as a next-generation sorbent for CO2 capture.

678a RO Membranes with Surface Tethered Poly(acrylic acid) Chains for Low Fouling, Ease of Cleaning, Selective Removal and Hydraulic Permeability and Salt Rejection Tunability

Yoram Cohen, Yian Chen

Surface nano-structuring of reverse osmosis (RO) membranes has advanced in recent years demonstrating the potential application of polymeric layers tethered onto on base polyamide (PA) thin-film composite (TFC) membranes. Such layers can serve to repair imperfections in the base RO membrane, reduce fouling and mineral scaling propensity, improve cleaning efficacy, and even allow one to overcome the typical permeability-selectivity tradeoff. Increased rejection of certain solutes has also been reported. The ability of tethered layer to provide the above beneficial improvements in membrane performance and responsiveness to environmental conditions (e.g., pH, temperature and salinity) is governed by both the chemistry of the tethered polymer chains, as well as the surface number density and length of these chains. In the present work, based on compilation of multiple experiments, it is shown that tethering of stimuli-responsive poly(acrylic acid) (PAA) chains onto a PA-TCF base membrane, with a target permeability and salt rejection, can be accomplished via PA surface activation via atmospheric pressure plasma, followed by graft polymerization to attain a high density brush layer. The resulting surface nano-structured (SNA) membranes, depending on the starting properties of the base RO membranes, can achieve highly tunable performance, and have remarkable low fouling/scaling propensity, and can be easily cleaned with low salinity water. Such membranes can also improve the rejection of difficult to remove nitrate and boron. Moreover, scale-up of the above approach is feasible and can be used to prepare highly uniform (with respect to separation performance) large surface area membrane sheets that were then utilized to fabricate small spiral-wound element. The above approach will be reviewed, and performance data, based on laboratory tests with both small membrane coupon and spiral-wound membrane elements, will be presented along with technoeconomic analysis of the competitive cost of the SNS-PAA-PA membrane fabrication.

678e Electrocoagulation Combined with Electrospun Membranes for Membrane Distillation

Yuhe Cao, Ranil Wickramasinghe

Development of new cost effective unit operations for recovery, recycle and reuse of wastewater is critical as the demand for water for beneficial uses increases. Unit operations that maximize water recovery are particularly appealing. Membrane distillation is an emerging unit operation that has the potential to maximize water recovery. However, membrane fouling and pore wetting are major challenges that limit the viability of this process. Thus, development of integrated unit operations for treatment of highly impaired wastewaters will be essential. Here we focus on treating oily wastewaters from hydraulic fracturing operations as they are particularly challenging to treat given the range of contaminants present. Here we focus on combining electrocoagulation as a pretreatment step with novel electrospun membranes for membranes distillation that are shown to be fouling resistant.

We have developed a novel PVDF based electrospun membrane to simultaneously solve fouling and pore wetting. Initially a PVDF blended with etyltrimethylammonium bromide electrospun membrane was prepared. Next a second layer was electrospun consisting of PVDF. Negatively charged silica nanoparticles were attached to the PVDF blended with etyltrimethylammonium bromide electrospun membrane by electrostatic adsorption. This creates an amphiphobic surface. Finally, the unmodified PVDF fibers were modified by grafting poly(glycidyl methacrylate-sulfobetaine methacrylate) (zwitterions) to the surface of the PVDF layer. The novel bilayer electrospun membrane was tested using both synthetic and real produced water feed streams. In an alternative method, we have developed a trilayer composite membrane consisting of a hydrophilic membrane, a supporting nonwoven sheet, and a hydrophobic electrospun membrane.

Our novel membranes showed excellent fouling resistance and salt rejection when compared to base PVDF membrane. Further when combined with electrocoagulation we show excellent water recovery form highly impaired produced wasters.

678c Designing, Preparing and Characterizing Alkali Treated Graphitic Carbon Nitride Thin-Film Composites Membranes for Efficient Dye Salty Water Separation.

Ameya Tandel, Haiqing Lin

Ultra-high water permeance, almost zero salt rejection and ~100% dye rejection is of great interest for waste water treatment especially in textile dye-salty water separation. Herein, we fabricated alkali treated graphitic carbon nitride (GCN) ultrathin thin-film composite membranes using polydopamine (PDA) chemical crosslinking to have efficient dye separation from salts without sufficient loss in water passage. Crosslinking using PDA adhere GCN on commercial support such as polysulfone (PSF) and can fine tune surface channel sizes for effective dye water separation. First, the membranes were characterized using FTIR and water contact angle measurement to see incorporation of GCN/PDA on the surface of PSF supports. Moreover, GCN/PDA crosslinking conditions and GCN/PDA mass ratio was optimized for efficient dye-salty water separation. For e.g. 0.4 mg/0.4 mg loaded GCN/PDA on PSF20k support with crosslinking condition of 0-5°C/30 mins and 23 °C/10 mins showcases optimum separation performance (Water permeance: 345 ± 22 LMH/bar, % rejection of Congo red dye, 697 Da: 97.6 ± 0.4 and % rejection of Na_2SO_4 salt: 16 ± 0.5). Water permeance of GCN/PDA membrane is 30, 8 and 3.5 times higher than commercial NF-90 membrane, commercial ultrafiltration membranes (with molecular weight cutoff of 2000 Da) and graphene oxide (GO)/PDA crosslinked membranes, respectively. This efficient and attractive separation performance manifest application of GCN based membranes in dye-salty water separation.

678d Thermo-Responsive Hollow Fiber Membranes for Enhanced Detoxification of Water and Air Pollutants

Rollie Mills, Thomas Dziubla, J. Todd Hastings, Lindell Ormsbee, Dibakar Bhattacharyya

This research is aimed at developing functionalized materials that can induce infectivity loss of coronaviruses and detoxify polychlorinated compounds, with the general goal of environmental remediation of air and water sources. Functionalized membranes with zero-valent catalytic nanoparticles have shown immense promise for enhanced removal and degradation of polychlorinated biphenyls (PCBs), yet reaction efficiency is hindered by mass transfer limitations between the particles and water contaminants. In recent years, the incorporation of a thermo-responsive polymer, poly(N-isopropylacrylamide) (PNIPAm), has shown to mitigate such limitations in preliminary flat-sheet membranes. In this study, enhanced degradative efficiency with PNIPAm functionalization of catalytic hollow fiber membranes (HFMs-higher particle packing density) was investigated for superior environmental remediation. Additionally, to further overcome these limitations and enhance reaction rates, transparent membranes were functionalized with plasmonic catalytic nanoparticles and PNIPAm, and their efficiency to degrade polychlorinated pollutants, such as trichloroethylene (TCE), with targeted surface heating of catalytic plasmonic particles (via Near-IR illumination) was investigated for cost-effective remediation applications. Furthermore, various membrane functionalization materials, such as salts, enzymes, and silver nanoparticles, have exhibited antiviral properties to reduce or eliminate the infectivity of SARS-CoV-2. In this research, the denaturation of the SARS-CoV-2 spike glycoprotein (facilitates host cell infection) by iron-palladium-functionalized membranes was investigated for the development of enclosed-environment responsive filters. Overall, this study advances the field of separations, membrane science, catalysis, and environmental remediation with responsive materials for air and pollutant detoxification. This research is supported by the NSF RAPID Grant Program, NSF GRFP and NIEHS/SRP Program.

678f Enhancing PFOS Removal and Membrane Permeability By Functionalizing Polyamide Thin-Film Nanocomposite Hollow Fiber Membranes with Mxene Nanosheets

Tin Le, Elnaz Jamshidi, Majid Beidaghi, Milad Esfahani

The prolific use of per- and polyfluoroakyl substances (PFAS) in household product manufacturing has resulted in the contamination of many waterways around the United States. Due to their negative health effects and high chemical stability, PFAS removal is an emerging research area in water purification. In this work, we use MXene nanosheets to functionalize thin-film composite (TFC) hollow fiber nanofiltration (HFN) membranes during the interfacial polymerization (IP) process to enhance the removal of Perfluorooctane sulfonic acid (PFOS) from water. A MXene-polyamide nanofiltration selective layer was fabricated on top of polysulfone hollow fiber support via IP of trimesoyl chloride (TMC) and a mixture of piperazine (PIP) and MXene nanosheets. Incorporating MXene nanosheets during the IP process tuned the morphology and negative surface charge of the selective layer, resulting in an increase in PFOS rejection from 72% (bare TFC) to more than 96% (0.025% (wt.)-MXene-TFC) while the water permeability was also increased from 13.19 (bare TFC) to 29.26 LMH/bar (0.025% (wt.)-MXene-TFC). Our results demonstrate that electrostatic interaction and size exclusion are the two main factors governing PFOS rejection and both are determined by PA selective layer structural and chemical properties. In addition, our results showed that MXene nanosheets tuned the morphology and crosslinking of PA layer by affecting the transport of PIP during the IP process, which resulted in enhanced water permeability and PFAS removal without sacrificing the salt rejection of the TFN membrane.

678g Gold Nanoclusters Immobilised Onto Polymer Brushes As Stable Anti-Microbial Surfaces

Nidhi Kapil, Halan Mohamed, Shanom Ali, Marc-Olivier Coppens

Bacteria have built resistance against antibiotics due to their abusive usage, which will increasingly affect the treatment of certain infectious diseases and impact modern surgery, chemotherapy, and organ transplantations. According to the latest WHO report, at least 700,000 people die each year due to drug-resistant diseases, and this could cause 10 million deaths each year by 2050 if no actions are taken.[1] While medical scientists are working at a genetic level to combat anti-microbial resistance (AMR), it is essential to develop a methodology that can kill or prevent bacteria from colonialising surfaces in the first place (especially in healthcare) before spreading germs and causing infections. Advancements in nanotechnology have helped tackle major challenges in medicine, such as targeted drug delivery and cancer treatments.[2] In this context, atomically precise gold clusters have gained immense attention because of their high ratio of low-coordinated gold atoms on the surface with unique structural geometry.[3] Ultra-small particles of gold are known to exhibit anti-microbial activity because they can control the production of reactive oxygen species (ROS), which ultimately kill the bacterial cells.[4] Additionally, surface properties such as charge and hydrophilicity are known to exhibit bacteriostatic properties to reduce the adhesion of bacterial cells.[5] In this regard, bactericidal properties of gold and bacteriostatic properties of polymers can be combined to create antimicrobial surfaces.

This study reports the synthesis and immobilisation of atomically precise gold nanoclusters (0.8 nm in diameter)[6] onto polymer brushes grafted from the membrane surface to develop stable anti-microbial surfaces with minimum leaching of the nanoparticles. The physical and chemical properties of these surfaces are studied using XPS, AFM, SEM and UV-VIS spectroscopy. These surfaces were further tested against *Escherichia coli (E. coli)* and *Staphylococcus aureus (S. aureus)* bacteria, which are rapidly developing resistance against antibiotics and are a common cause of infection in healthcare facilities. The synthesised surfaces exhibit 90% and 96% reduction in *S. aureus* and *E. coli,* respectively, within one hour of contact. These findings suggest that these materials could be potentially used in healthcare facilities (such as fabrication of catheter surfaces or surgical coatings) to kill the bacterial cells effectively.

References

1. No time to wait: Securing the future from drug-resistant infections; Report To The Secretary-General of The United Nations. **2019**.
2. Peer, D.; Karp, J. M.; Hong, S.; Farokhzad, O. C.; Margalit, R.; Langer, R., Nanocarriers as an emerging platform for cancer therapy. *Nature nanotechnology* **2007,** *2* (12), 751-760.
3. Du, Y.; Sheng, H.; Astruc, D.; Zhu, M., Atomically Precise Noble Metal Nanoclusters as Efficient Catalysts: A Bridge between Structure and Properties. *Chem. Rev.* **2020,** *120* (2), 526-622.
4. Zheng, K.; Setyawati, M. I.; Leong, D. T.; Xie, J., Antimicrobial gold nanoclusters. *ACS nano* **2017,** *11* (7), 6904-6910.
5. Mohamed, H.; Hudziak, S.; Arumuganathan, V.; Meng, Z.; Coppens, M.-O., Effects of charge and hydrophilicity on the anti-fouling properties of kidney-inspired, polyester membranes. *Molecular Systems Design andamp; Engineering* **2020,** *5* (7), 1219-1229.
6. Kapil, N.; Weissenberger, T.; Cardinale, F.; Trogadas, P.; Nijhuis, T. A.; Nigra, M. M.; Coppens, M.-O., Precisely Engineered Supported Gold Clusters as a Stable Catalyst for Propylene Epoxidation. *Angew. Chem. Int. Ed.* **2021,** *60* (33), 18185-18193.

280a Rapidly Ordered Block Copolymer Membranes with Tunable Pore Sizes for Wastewater Treatment

Kshitij Sharma, Khadar Basha Shaik, Maninderjeet Singh, Chenhui Zhu, Mohammad Hassan, Alamgir Karim

The present surface water resources will soon be insufficient to meet the needs of the next generations. One of the solutions is to reuse water after purification. Wastewater is now contaminated with oil and other organic compounds due to the rapid rise in oil andamp; gas, petrochemical, pharmaceutical, and food processing industries. Membranes present an easy and energy-efficient solution for the removal of both particulate and oily matter from wastewater. Here we present a methodology to rapidly order low χ (Flory Huggins interaction parameter) block copolymer (Poly(styrene-b-methyl methacrylate)) thin films with well-defined through-thickness channels having minimal tortuosity. The technique involves solvent casting of asymmetric BCP films from solution mixtures doped with plasticizing additives that selectively segregate into one of the domains of the BCP and modify domain sizes. Owing to the selectivity and plasticization capability of the additive and the preferential solvation of BCP components in the solvent mixture, the film is fully ordered in the casting process. With careful selection of the casting environment, a completely perpendicular domain morphology can be achieved. These films are supported by commercial membranes like PES and treated to selectively crosslink one and etch the other block to open the pores. The tradeoff between size selectivity and permeability can be controlled through additive concentration in the system. Overall, the membranes show high water fluxes due to the lower pore tortuosity. The film microstructure is characterized via atomic force microscopy and x-ray scattering. At the same time, the membrane performance is tested using a dead end and a cross-flow fluid cell and UV-vis spectroscopy.

280b Effect of Dehydration on RO Performance and Antifouling in Fully Aromatic Polyamide Membranes Containing Functionalized Cellulose Nanocrystals (CNCs)

Connor Farrell, Ethan D. Smith, Keith Hendren, Earl J. Foster, Stephen Martin

Fully aromatic polyamide thin film composite membranes (TFCs), consisting of a thin selective layer and thick porous support, are commonly used materials for desalination by reverse osmosis (RO). Fully aromatic polyamide TFCs must be kept hydrated in order to maintain the pore structure needed for water transport. In real applications biofouling often occurs on the surface of TFCs as proteins accumulate on the surface inhibiting transport, and decreasing the efficiency of the membranes. Thin film nanocomposite membranes (TFNs) incorporate nanoparticles in the selective layer to improve upon traditional TFC performance. In this study we created TFNs by incorporating cellulose nanocrystals (CNCs) and studied the effects of CNC functionality, loading level, and dehydration on RO performance. CNCs are of interest due to their high aspect ratio, low cost, availability, sustainability, and for the possibility for future surface modifications allowing for functionalized derivatives affecting particle-polymer interactions. 2,2,6,6-Tetramethylpiperidine-1-oxyl (TEMPO)-oxidized cellulose nanocrystals (TOCNs) or zwitterionic functionalized cellulose nanocrystals (zCNCs) were incorporated within fully aromatic polyamide TFNs synthesized via interfacial polymerization of m-phenylenediamine (MPD) and 1,3,5-benzenetricarbonyl trichloride (TMC).

The flux and salt rejection were investigated for TFNs prepared with various amounts of TOCNs or zCNCs. For membranes containing TOCNs these membranes were stored in deionized water or under vacuum prior to testing. zCNCs were only stored in water. The presence of nanoparticles in the final TFNs was confirmed using ATR-FTIR. Water transport increases as pathways for molecular transport develop at the interface between the polymer matrix and nanoparticle surface. Water transport is negatively impacted by membrane dehydration as the polymer pore structure collapse, but the impact is diminished by the presence CNC nanofillers in the membrane. Antifouling behavior was studied using a test solution containing bovine serum albumin (BSA) under conditions preferable for protein deposition on the membrane surfaces. The reduction in flux due to BSA fouling was examined over several hours. TFNs containing zCNCs exhibited slower decreases in performance relative to both TFNs containing TOCNs and TFCs without any nanoparticles.

280c Engineering Reduced Graphene Oxide Membrane Via in-Situ Peracetic Acid Etching to Achieve Excellent Water Permeance for Highly Efficient Dye Recovery

Erda Deng, Kai Chen, Haiqing Lin

Graphene oxide (GO) membranes have been extensively investigated for molecular-sieving membranes for dye removal from textile wastewater thanks to the adjustable channel size and featured water transport path. Owing to the excellent dispersibility of GO in water, they are often reduced or cross-linked to improve their stability on the membrane. However, reduced GO (rGO) leads to dramatically low water permeance. This study demonstrates an effective approach of in-situ chemical etching using peracetic acid (PAA) on the rGO membrane, enhancing water permeance while retaining the dye rejection. Specifically, rGO nanosheets were prepared using hydrozine in a solvation state and vacuumed filtrated onto a nylon substrate. Then different content of PAA was deposited on top of the rGO membrane surface for different amount time to get an etched rGO (ErGO) membrane. A typical 50 nm-thick rGO membrane shows water permeance of 20 LMH/bar and congo red (CR) rejection of 99.8%, while the ErGO-10min with the same thickness exhibits water permeance of 56 LMH/bar (a 1.8 times increase) with similar CR rejection of 99.7%. ErGO membranes with water permeance of 50–80 LMH/bar and multiple dye rejection over 99.0% are superior compared to state-of-the-art commercial NF membranes, and surpassing GO membranes reported in the literature. Moreover, the stability of the ErGO membrane was corroborated in a long-term continuous filtration test and the performance remained stable. Given the good resistance to aggressive chemical washing and its excellent performance, the ErGO membranes demonstrate its potential for practical dye removal applications.

280d MIL-100 (Fe) Functionalized Thin-Film Nanocomposite Membranes for Enhanced Removal of Uncharged Urea from Water

Tin Le, Milad Esfahani

Urea is widely used in fertilizer production for agricultural purposes which risks runoff into soil and water sources. This poses a concern due to the algal blooms that may form which can poison the environment and even humans due to food source contamination. Thin-film composite (TFC) membranes are among the best performing membranes for water filtration; however, they do not show acceptable rejection for small and uncharged molecules such as urea. Adsorptive membranes combine the power of membrane filtration with the adsorption process and provide enhanced selective removal specifically for small and uncharged molecules. Metal-organic frameworks (MOF) are a new generation of effective adsorbents with excellent tuneability in structure and compatibility with polymeric membranes. We synthesized MIL-100 (Fe), an iron-based MOF using heatless method, and tested for urea adsorption. The urea adsorption capacity of MIL-100 (Fe) was tested under varying experimental conditions such as pH (2-10), temperature (25-45°C), MOF concentration (25-400 ppm), and urea concentration (10-1000 ppm). The results revealed the superfast adsorption (reached almost 90% removal within 1 min) of neutrally charged urea molecules. The MOF was able to reach a maximum adsorption efficiency of around 98% with a maximum uptake capacity of 4430 mg/g. In this project, we (i) immobilized the MIL-100 (Fe) on the surface of the polyamide layer of thin-film composite membranes and (ii) incorporated the MIL(Fe) into the polyamide selective layer of TFC for the fabrication of two different types of adsorptive membranes for enhanced removal of urea from water. The urea rejection, water permeability, and antifouling of these two functionalized membranes will be examined and compared with the bare TFC membrane. In addition, the effect of Fe-MOF on the formation of polyamide selective layer structure and morphology will be discussed regarding the impact of Fe-MOF on the interfacial polymerization process.

280e Self-Regulating Behavior of a Pilot Scale Forward Osmosis-Reverse Osmosis Hybrid System

Jeffrey McCutcheon, Noah Ferguson, Colin Fitzsimonds, Nicole Beauregard, Maqsud R. Chowdhury, Mayur Ostwal, Ranjan Srivastava

Although reverse osmosis (RO) is an industrially mature and widespread desalination technology, its susceptibility to fouling makes it challenging to operate using more difficult feedwaters. Pretreatment is one approach to reducing fouling in RO membranes and such systems often comprise a majority of a desalination plant's total footprint. Hybridization of forward osmosis (FO) with RO has been considered as an alternative configuration to more conventional pretreatment schema, but the lack of understanding of how these systems might operate and the potential for complex control schema to balance fluxes between the FO and RO systems create uncertainty in system design. In this work, we demonstrate self-regulating behavior of FO-RO hybrid systems that can lead to reduced complexity of these systems at scale. We show this behavior using a module-scale test bed that can mimic the behavior of larger scale systems. The system shows permeate flow rate near-convergence between the FO and RO modules after startup or when perturbed by a change in RO module pressure. Such a result indicates that FO-RO hybrid system do not require complex control schemes to ensure even flows between the two processes.

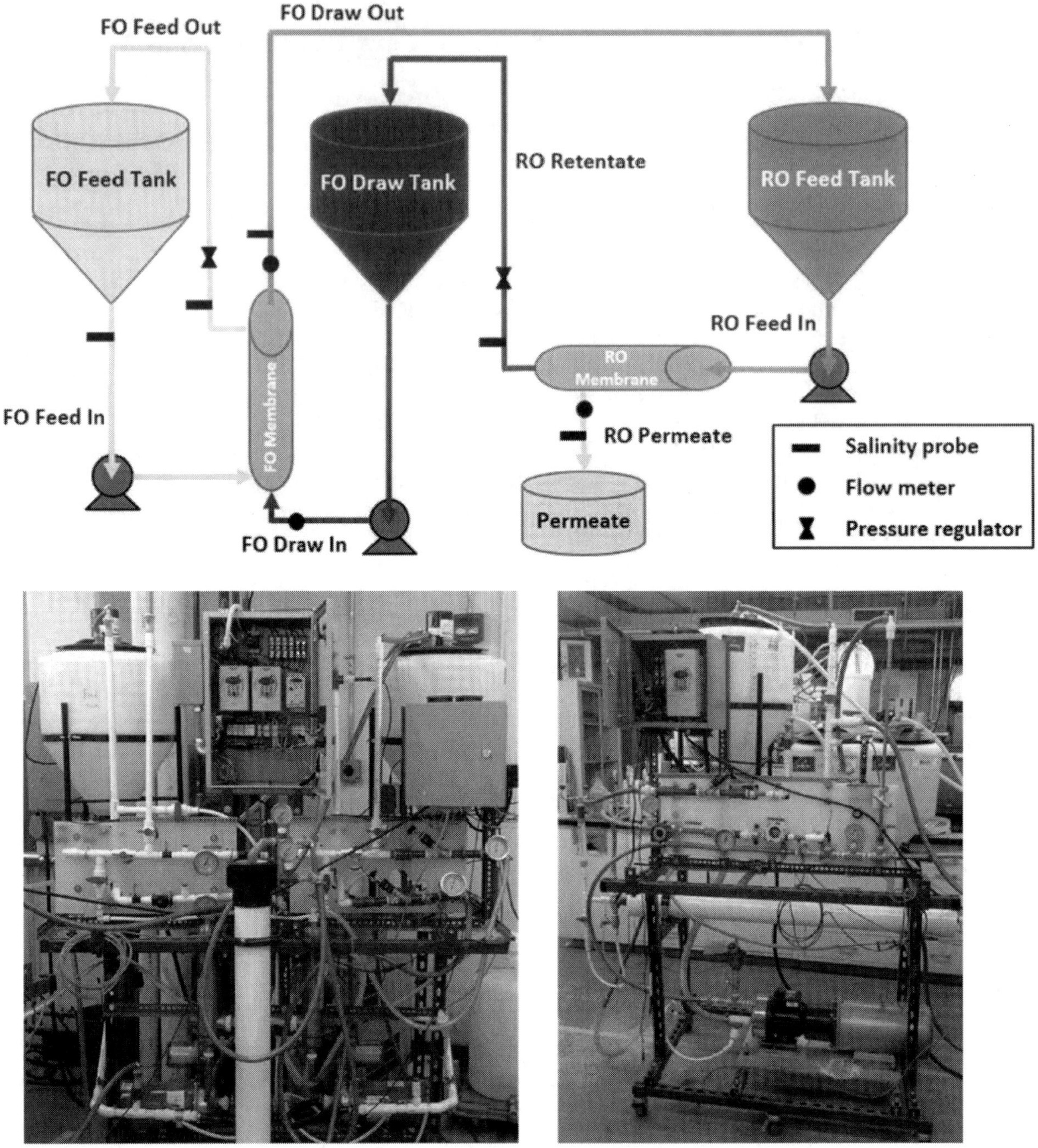
FO Feed Out
FO Draw Out
FO Feed Tank
FO Draw Tank
RO Retentate
RO Feed Tank
RO Feed In
FO Membrane
RO Membrane
FO Feed In
RO Permeate
Permeate
FO Draw In
Salinity probe
Flow meter
Pressure regulator

280f Multilayered Covalent Organic Framework Membranes Made By Sequential Interfacial Polymerization for Desalination

Xiaoli Ma, Miguel Angel Jaimes, Ho Kuan Yu, Zhiqin Qiang, Weiling Xia, Rachel Clark

Covalent organic frameworks (COFs) possess a range of unique properties including ordered pore structure, tunable pore size and functionality, and excellent stability that make them promising materials for membrane applications. However, most COF membranes reported in the literature exhibited low salt rejections due to their relatively large nanoscale pore diameters (>1 nm). Here, we report the creation of multilayered COF structures in a single membrane with the goal to enhance salt rejections and desalination performance. Multilayered COF membranes with thicknesses below 500 nm were synthesized directly on polymer supports in a matter of seconds via the scalable interfacial polymerization method. Although the membranes formed appear to be amorphous, we have demonstrated that their salt rejections can be tuned and enhanced by engineering the COF-COF interface in the membrane. The misalignment between the distinct 2D COF structures containing different functional group along their pore channels has possibly reduced the effective membrane pore size. As a result, the multilayered COF membranes exhibited salt rejections that are more than 100% higher than those of single-layer COF membranes. Detailed structure characterizations will be combined with effective pore size estimation using neutral solute permeation measurement to understand the mechanism for enhancement in salt rejections.

280g Post-Synthesis Modification of Polyamide Reverse Osmosis Membranes Via Diamines for Enhanced Urea Rejection

Shahriar Habib, Madison Wilkins, Steven Weinman

Reverse osmosis (RO) membranes are one of the most efficient membrane filtration technologies because of their strong separation capabilities and great potential for the treatment of water worldwide. Even though RO membranes exhibit excellent performance in salt rejection, they do not reject small, neutral, and uncharged molecules (SNUs), such as urea, to a level to produce potable water. Because of the rapid and uncontrolled nature of the interfacial polymerization reaction, the polyamide separation layer contains both network and aggregate free volume holes (pores). As SNU rejection is governed mainly by the size exclusion mechanism, decreasing the free volume to reduce the passage of the urea through the membrane is required. In this regard, the modification of RO membranes to increase the degree of cross-linking and/or decrease the free volume hole size is an ideal approach. We hypothesize that if the polyamide layer can be modified with a diamine, then the urea rejection will be increased. In this work, a method was developed for combining chemical modification and heat treatment of commercial DuPont XLE membranes to enhance the rejection of urea. For the chemical modification, carbodiimide chemistry was used followed by the coupling of various diamine. The modification of the commercial membrane was performed at room temperature (~22 °C) and 63 °C. The modified membranes were characterized using ATR-FTIR, XPS, SEM, contact angle goniometry, and electrokinetic analyzer. Membranes were performance tested for water permeance, NaCl rejection, and urea rejection using a dead-end stirred cell. Results show that both heat treatment and amine modification increase the urea rejection of the membranes.

644a Thin-Film Covalent Organic Framework Membranes Made By Interfacial Synthesis for Organic Solvent Nanofiltration

Xiaoli Ma, Zhiqin Qiang, Rahul Sampat Khandge

Organic solvent nanofiltration (OSN) relies on the development of robust membranes that are capable of separating organic liquid mixtures and scalable for industrial application. Covalent organic frameworks (COFs) are a new class of porous materials that possess ordered two-dimensional (2D) or three-dimensional (3D) structures, tunable pore size and functionality, and excellent chemical stability, making them promising materials for OSN membranes. Herein, we report the synthesis of β-ketoenamine linked 2D COF and imine-linked 3D COF membranes via a scalable, on-surface interfacial synthesis approach. Continuous thin-film COF membranes with controllable thickness can be directly formed on a variety of supports (ceramic or polymeric) in a matter of minutes or even seconds. Although the resultant membranes possess amorphous structure, they have demonstrated interesting OSN performances, including solvent permeances above 50 L/(m2.h.bar), molecular weight cutoff in the range of 500-1000 g/mol, and good solvent stability. The diffusion and reaction kinetics of the monomers were found to be important factors influencing the membrane structures. Detailed membrane structure characteristics, their correlations with key synthetic parameters, solvent transport and solute rejection mechanisms of these 2D and 3D COF membranes will also be discussed.

644b Quantifying Self-Diffusion of Mixed and Pure Organic Liquids in ZIF-Based Mixed Matrix Membranes By NMR

Amineh Baniani, Matthew Rivera, Ryan P. Lively, Sergey Vasenkov

Mixed matrix membranes (MMMs) are formed by dispersing filler particles, such as zeolitic imidazolate frameworks (ZIFs), in polymer films. ZIFs represent a sub-class of metal organic frameworks (MOFs). In addition to gas separations, ZIFs have shown promising performance in organic liquids separations. The diffusion process through MMMs is rather complex because diffusivities are usually different in different MMM components (i.e. polymer and filler). As a result, microscopic diffusion measurements performed selectively for a particular MMM component are needed in addition to conventional macroscopic measurements through the entire membrane for a better quantification and understanding of the MMM diffusion process. In particular, quantifying diffusion inside a ZIF component of MMMs can shed light on an influence of polymer/ZIF interfaces and possible changes of intra-ZIF transport due to a ZIF confinement in a polymer.

In this work, C-13 pulsed field gradient (PFG) NMR was used to quantify self-diffusion of pure and mixed xylene mixture inside zeolitic imidazolate framework-71 (ZIF-71) crystals dispersed in a Torlon polymer to form a ZIF-71/Torlon MMM. For the studied mixtures, selective C-13 isotopic labelling of each xylene isomer was used to ensure that the self-diffusivity of each isomer in the mixture is measured independently and correctly. The diffusion measurements were performed at different diffusion times. The observed time dependencies of intra-ZIF self-diffusivities were explained by an influence of the external crystal surface on the studied diffusion process. Analysis of these time dependencies measured for each xylene isomer allowed obtaining intra-ZIF self-diffusivities not perturbed by crystal surface effects as well as the corresponding diffusion selectivity for the studied sorbate mixture. These results will be presented and discussed in detail. The reported selectivity demonstrates promising separation behavior of the considered membrane type.

644c Tuning Polyimide Thin Film Composite Membranes for Organic Solvent Reverse Osmosis Separations Via Boc Protected Amine Solid-State Crosslinking

Yacine Feliachi, Ryan P. Lively, M.G. Finn

Organic solvent separations are energy intensive yet ubiquitous in the chemical, petrochemical, and pharmaceutical industries. Membrane technology can reduce the energy, carbon, and water intensity required in these separations significantly. However, these molecular separations often involve highly similar molecules, which reduces the efficacy of the membrane-based process. Some polymers used in organic solvent reverse osmosis (OSRO) have been known to suffer from plasticization, making them inefficient for these molecular separations. One method to avoid such effects is crosslinking the polymer. Here, we develop a method to uniformly crosslink a selective layer for a thin film composite (TFC) membrane to improve stability in organic solvents and tune separation performance. Traditional crosslinking methods are hindered by the diffusion of crosslinking agents into the polymer matrix, and this effect becomes more pronounced as crosslinking agents become bulkier. We have developed a method to introduce larger crosslinking agents that impart additional free volume to the membrane while overcoming diffusional limitations during the crosslinking infusion step. We utilize the commercial polymer Matrimid® 5218 and three crosslinking agents: para-xylylene diamine, 4,4'-biphenyldicarbontrile, and triptycene. We introduce a Boc protection method to the amines before dissolving into the membrane casting solution. After casting, the amines within the membrane are thermally deprotected to induce a solid-state crosslinking reaction. The resulting membranes were characterized by FT-IR, gel fraction, helium pycnometry, and TGA. The separations performance was tested in a crossflow OSRO permeation system for a toluene/triisopropyl benzene system and a multicomponent organic solvent mixture. The crosslinked TFC membranes showed excellent stability in organic solvents and tunable separation characteristics based on crosslinking agent and concentration of the crosslinking agent.

644d Mass Transport Modeling in Membrane-Based Chemical Separations

Akshay Deshmukh, Mathew M. Swisher, Nicolas G. Hadjiconstantinou, John H. Lienhard

Increasing the supply of renewable chemicals and fuels is critical to the deep decarbonization of a multitude of industrial products from polymers to pharmaceuticals. Biomass-derived feedstocks can simultaneously meet the growing demand for sustainable chemicals while reducing our reliance on fossil fuels and minimizing industrial greenhouse gas (GHG) emissions. However, the current production of chemicals can be energy intensive. Separation and purification accounts for over 40% of the energy consumed during the production of key chemicals ranging from ethanol and phenol to acetic acid and styrene.[1,2] Globally, separations processes account for a substantial portion of industrial energy consumption and GHG emissions. Thermal distillation forms an integral part of many chemical separations, separating mixtures by the volatility of their constituent components. Although widely used, distillation is inherently energy intensive, requiring large amounts of heat to drive vaporization, and accounting for 25% of US industrial energy consumption.[1–5]

Membrane-based processes, such as organic solvent reverse osmosis (OSRO), separate mixtures based on the size and charge of their constituent components, rather than their volatility. OSRO utilizes hydraulic pressure to create a chemical potential gradient to drive permeation through a selective membrane, splitting the feed into high and low permeability fractions. By avoiding the energy intensive vapor-liquid phase change, OSRO can drastically reduce the energy footprint of chemical separations.[6,7] Developing reliable thermodynamic models combined with a thorough understanding of transmembrane mass transport phenomena can accelerate future membrane design. However, quantifying chemical potential and mass transfer rates in industrially relevant mixtures is challenging due to the large number of components and functional groups.

In this investigation, we study the separation of benzene and its derivatives using nanoporous nanosheet membranes. We begin by building a computational platform to calculate the chemical potential driving force for the transmembrane permeation of non-ideal liquid-phase mixtures. An equation of state is combined with an excess Gibbs free energy model using the Wong-Sandler mixing rules, to quantify liquid-phase fugacity across a large pressure range. Interaction parameters are estimated through the nonlinear regression of fluid phase equilibrium and activity coefficient data. We then develop a computational model for multicomponent transmembrane transport, using the Maxwell-Stefan framework to capture the thermodynamic and diffusive coupling between molecular species in the diffusive boundary layer. Finally, we integrate transmembrane mass transport expressions to calculate how feed flow rate and composition vary along a membrane module. By elucidating the impact of membrane permeability and selectivities on energy consumption and separation efficacy in OSRO, we endeavor to guide future membrane materials and process development.

References:

1. S. Brueske, C. Kramer and A. Fisher, *Bandwidth Study on Energy Use and Potential Energy Saving Opportunities in U.S. Chemical Manufacturing*, 2015.
2. S. Brueske, R. Sabouni, C. Zach and H. Andres, *U.S. Manufacturing Energy Use and Greenhouse Gas Emissions Analysis: Petroleum Refining Sector*, 2012.
3. D. S. Sholl and R. P. Lively, *Nature*, 2016, **532**, 435–437.
4. R. P. Lively and D. S. Sholl, *Nature Materials*, 2017, **16**, 276–279.
5. R. P. Lively, *AIChE Journal*, 2021, **67**, 1–11.

6. D.-Y. Koh, B. A. McCool, H. W. Deckman and R. P. Lively, *Science*, 2016, **353**, 804–807.
7. K. A. Thompson, R. Mathias, D. Kim, J. Kim, N. Rangnekar, J. R. Johnson, S. J. Hoy, I. Bechis, A. Tarzia, K. E. Jelfs, B. A. McCool, A. G. Livingston, R. P. Lively and M. G. Finn, *Science*, 2020, **369**, 310–315.

644e Tunable Bottlebrush Membranes for Organic Solvent Nanofiltration

Pranav Ramesh, Mirco Sorci, Bratin Sengupta, Miao Yu, James (Chip) Kilduff, Georges Belfort

Abstract

Polymer membranes offer several advantages such as ease of synthesis and low cost. However, commercial application of these membranes is limited to separations that feature an order of magnitude difference in size between constituents. This can be attributed to inherent limitations of polymer materials such as pore flexibility and a log normal pore size distribution.

Here, we attempt to overcome these limitations and expand the use of polymeric membranes for organic solvent nanofiltration applications. A series of monomers are graft polymerized using Single Electron Transfer-Living Radical Polymerization onto a crosslinked commercial polyimide membrane (support). These modified bottlebrush structures form a new class of membranes whose structure and hence performance can be tailored for a given separation. Membrane properties can be controlled by synthesis conditions such as polymerization time, choice of monomer and crosslinker type and concentrations. To characterize their filtration performance, permeabilities and selectivities were obtained for feeds containing a series of pure alcohols, a series of five organic dyes in ethanol, and mixtures consisting of toluene-triisopropyl benzene (TIPB), isobutyraldehyde- isobutanol and methanol-toluene. Due to stability, tunability and ease of scale up, bottlebrush membranes show promise for applications requiring the separation of organic molecules of similar size.

644f Highly Interconnected Pores in Hybrid Nanofilms for Ultrafast Solvent Transport with Precise Molecular Separation

Bratin Sengupta, Qiaobei Dong, Dinesh Behera, Fanglei Zhou, Ji Jiang, Georges Belfort, Miao Yu

Materials with rigid, precisely controlled nanopores that are within 1 nm and stable in harsh conditions have potential to be fabricated into membranes for organic solvent nanofiltration (OSN). To achieve high permeance through these membranes, the traditional route is to reduce the selective layer thickness and thus lower the transport resistance. Although membranes as thin as one atom have been prepared, this thinness often challenges the membrane's potential for defect free scale-up. Another route, often overlooked, is to increase the pore interconnectivity of the membrane material, described by the ratio of porosity and tortuosity (ε/τ), which greatly influences the transport through a membrane. In this work, we fabricated membranes using a hybrid nanofilm with highly interconnected pores, with ε/τ of ~ 0.2, almost doubling that of the current OSN membranes, and obtained methanol permeance as high as 267 L m^{-2} h^{-1} bar^{-1}, > 2.5 times higher than the reported membranes with similar molecular weight cut-off (MWCO). Precise tuning of MWCO in the range of 200-1,000 Da with a step change as small as approximately 100 Da was realized by adjusting the membrane synthesis and calcination conditions. Molecular dynamics (MD) simulation shows the formation of nanometer-sized pores in hybrid material, while pore size and interconnectivity were realized as a function of the calcination conditions. These membranes have rigid pores that are stable in various organic solvents and at elevated temperatures, ensuring stable molecular separation under challenging industrial conditions.

644g Structure Design of Nanoporous Multilayer Graphene Membrane for Organic Solvent Nanofiltration

Dae Woo Kim

Graphene-based membranes are promising candidates for efficient organic solvent nanofiltration (OSN) processes because of their unique structural characteristics, such as mechanical/chemical stability and precise molecular sieving enabled by nanopore and interlayer spacing. Recently, to improve organic solvent permeance and selectivity, nanopores have been fabricated on graphene planes *via* chemical and physical methods. The nanopores serve as an additional channel for facilitating ultrafast solvent permeation while filtering organic molecules by size exclusion.

For single-layer graphene prepared by chemical vapor deposition, various top-down methods, such as focused ion beam, focused electron beam, ultraviolet-induced oxidative etching, ion bombardment followed by chemical etching, and oxygen plasma etching, have been used. Because single-layer or few-layer graphene is used as the selective layer, the membrane thickness can be significantly decreased to the atomic scale. However, as such processes are performed under high-vacuum conditions and post-treated in a chamber and graphene is required to be transferred on a porous support. Therefore, large-scale fabrication is still challenging with precise pore-size control.

As an alternative method, the nanopores can be generated in graphene oxide (GO) on a large scale by post-treatment of GO powder and GO film. Intrinsic defects and holes are present in the original GO, formed during the synthesis of GO, but the size and density of the pores are negligible and not controlled. Because post-treatments generate nanopores by the decomposition of oxygen-functional groups or defective sp2 carbon on the GO plane, dense sp3 carbons are formed near the pores, and oxygen-functional groups are also partially present. The defective structure is critical to tuning the interlayer spacing to enable the permeation of large organic solvents through the nanochannel. Moreover, because nanoporous graphene is soluble in water and organic solvents, it is beneficial to fabricate laminated membranes by solution processes such as vacuum filtration, spin coating, bar coating, spray coating, and slot-die coating.

In this talk, the post-thermal treatment method will be discussed to generate nanopores in graphene, and the membrane fabrication method will be handled as well. Particularly, nanopores can be generated on the basal plane of graphene by rapid thermal annealing of GO. The size of the nanopores is also tuned from micropores to nanopores by adjusting the activation temperatures. And the density of sp^2 domains can be controllable by additional microwave-assisted reduction. The optimized nanoporous graphene membranes show ultrafast alcohol and organic solvent permeance and a sharp molecular weight cut-off for molecules in nanoscale dimension. The membrane is feasible to separate mixed organic molecules dissolved in various organic solvents.

[REFERENCES]

Kim, D.W. et al., Carbon, 191, 563 (2022)

Kim, D.W. et al., J. Membr. Sci. 637, 119620 (2021)

Kim, D.W. et al., J. Membr. Sci. 618, 118638 (2021)

Kim, D.W. et al., J. Membr. Sci. 612, 118454 (2020)

Kim, D.W. et al., J. Mater. Chem. A 8, 8292 (2020)

643a Novel Electrospun Membrane Structures for Reverse Osmosis, CO_2 Stripping and Membrane Distillation

Sandra Kentish, Daniel Heath, Seungju Kim

Novel electrospun membranes have been developed that show great promise across a range of different applications. Electrospun membranes are generally composed of randomly orientated nanofibers that form a highly porous structure. In the present case we form these fibres into an ordered crosshatched structure that both increases mechanical strength and reduces tortuosity, leading to increased flux. We also use electrospinning to customise the surface of the membrane, forming resilient, superhydrophobic surfaces that are long lasting.

In our initial studies we formed ultrathin fibres of ~50nm diameter from polysulfone within a unique crosshatched structure. The use of polysulfone, coupled with a polydopamine addition to increase hydrophilicity, facilitated the addition of a standard polyamide active layer using interfacial polymerisation. The resulting membrane was able to withstand transmembrane pressures of up to 55 bar during reverse osmosis, while the random electrospun structure failed at 14 bar. The flux through the crosshatched structure was similarly 150 to 180% higher than that of a random nanofibre membrane and more than 300% higher than a standard asymmetric membrane. The salt rejection through all three systems was comparable, as this is dictated by the polyamide active layer.

We have further tested electrospun structures for carbon dioxide stripping from a chemical solvent. This is a challenging application, requiring a superhydrophobic, highly porous membrane that is resistant to temperatures of 100°C and above. We spin 200 nm fibres from poly(vinylidene fluoride-co-hexafluoro propylene), a polymer that is hydrophobic, but has greater alkali resistance than poly(vinylidene fluoride), which is needed for this application. Further, we form 'bead-on-string' and microparticle structures on the surface of the membrane to further increase hydrophobicity. All structures provide enhanced CO_2 flux relative to a standard asymmetric membrane, but the bead-on-string structure provides greater stability, with strong performance for up to ten days at 100 °C. In a further extension to this work, we spin cross-hatched and random nanofibre structures from hydroxyl polyimide and once formed, use thermal treatment to form TR nanofibers that also show strong performance in this application.

Most recently we have extended this approach to form cross-hatched nanofibre membranes formed from polytetrafluoroethylene (PTFE). The use of the cross-hatched structure and a microparticulate surface layer leads to very high water fluxes of close to 100 kg m^{-2} hr^{-1} during membrane distillation, readily exceeding that for both a commercial PTFE membrane and other electrospun structures in the literature, while maintain strong salt rejection.

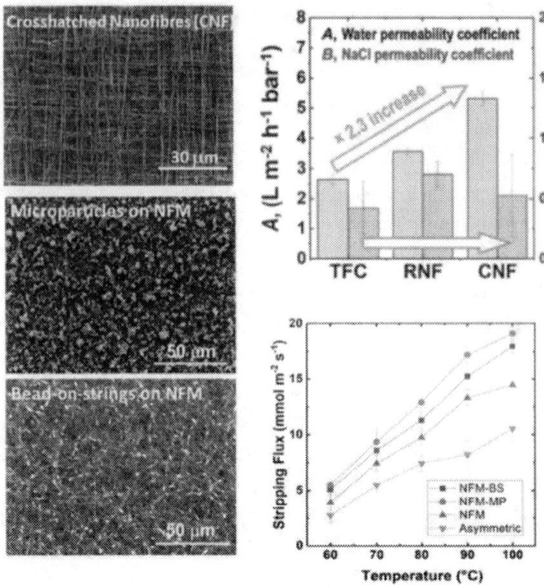
Crosshatched Nanofibres (CNF)
30 µm
Microparticles on NFM
50 µm
Bead-on-strings on NFM
50 µm
A, Water permeability coefficient
B, NaCl permeability coefficient
x 2.3 increase
A, (L m-2 h-1 bar-1)
B, (L m-2 h-1)
TFC
RNF
CNF
Stripping Flux (mmol m-2 s-1)
Temperature (°C)
NFM-BS
NFM-MP
NFM
Asymmetric

643b Optimizing Fabrication of Polysulfone Ultrafiltration Membranes Using a Green Solvent, Cyrene

Cannon Hackett, David Hale, Ranil Wickramasinghe, Xianghong Qian, Audie Thompson

Even though membranes can lead to more environmentally sustainable separation processes, membrane casting involves the use of toxic organic solvents. The demand for filtration membranes made with nontoxic green solvents has increased over the past several years. Cyrene (dihydrolevoglucosenone) is a solvent that can be used for fabricating membranes by phase inversion, and it is less toxic than traditional solvents such as n-methylpyrrolidone (NMP). Nearly all previous studies that used Cyrene for fabricating membranes used water as the nonsolvent in the coagulation bath. However, other nonsolvents, such as ethanol, have been investigated in conjunction with traditional solvents and sometimes led to enhancements in membrane performance. In this study, polysulfone ultrafiltration membranes were fabricated by nonsolvent-induced phase separation (NIPS), using Cyrene as the solvent and an ethanol/water mixture as the nonsolvent coagulation bath. Membranes were optimized by adjusting the ethanol content of the coagulation bath. Performance was tested by dead-end filtration. Membranes cast from Cyrene solution into pure water had low pure water flux (40.7 $L/m^2/h$) and bovine serum albumin (BSA) rejection (18.1%). When 70 v% ethanol was used in the coagulation bath, pure water flux increased to 106.4 $L/m^2/h$ and BSA rejection increased to 84.7%. Membranes were characterized to better understand the performance increase. Fourier-transform infrared spectroscopy (FTIR) analysis of membrane samples showed that membranes cast into pure water retained significant residual Cyrene even after 24 hours of storage in water. The inability of pure water to adequately remove Cyrene led to diminished membrane performance. However, precipitation baths with 25 v% or more ethanol fully removed Cyrene from the membranes. The closer hydrogen bonding solubility parameters of ethanol and Cyrene (vs. water and Cyrene) contributed to the improved Cyrene removal, as ethanol and Cyrene have better miscibility. Scanning electron microscopy (SEM) showed that membranes cast from Cyrene solution into 70 v% ethanol solution had a flatter and more porous surface compared to membranes cast into pure water, which had a crumpled surface morphology. All membranes cast from Cyrene solution had sponge-like pores, indicative of delayed demixing. Dissipative particle dynamics simulations are being used to further elucidate how differences in solvent and coagulation bath composition affect pore formation. Understanding the relationship between coagulation bath composition and performance parameters can lead to the determination of an optimal membrane fabrication protocol. The result of this study is a pathway for fabricating polysulfone ultrafiltration membranes with adequate performance, reduced toxicity, and guidance for further optimization.

643c 3D Printing of Thin Film Composite Membranes for Nanofiltration and Beyond

Jeffrey McCutcheon, Xin Qian, Mayur Ostwal, Edward Wazer

Making membranes through additive manufacturing (i.e. "3D printing") has emerged as a popular field of study in membrane science over the past half-decade. Additive manufacturing confers a number of benefits when making membranes including customizability, thickness control, morphology control, and hierarchical design. Alternative manufacturing techniques also enable the processing of emergent materials into high performance membranes. This is important because while the number of new materials in membranes has increased dramatically, the options for processing those new materials has not.

Our work has identified a new approach to making membranes through additive manufacturing using electrospray deposition. This method enables the processing of classic and emerging polymer materials into thin film composite (TFC) membranes. Unlike current TFC membranes, we do not need to form these selective layers *in-situ.*

Our work demonstrates the ability to print membranes for a variety of separations using a number of emergent polymers, such as zwitterionic copolymer (nanofiltration), polyether epoxy (chlorine tolerant nanofiltration), and PIM-1 (gas separations). Traditionally, these polymers would be processed using a phase inversion process, leading to an integrated asymmetric membrane with relatively low permeance. Our work demonstrates the versatility of the method to make TFC membranes with different polymers and yield selective yet substantially higher permeance due to our ability to make thinner, defect free membranes.

We contextualize this work in the broader perspective of membrane manufacturing which has seen little innovation in the past 4 decades. Innovations in membrane manufacturing will enable the use of the increasing number of new membrane materials and could help realize the benefits promised by these materials.

643d Innovative Manufacturing of Porous Oxide Hollow Fiber Membranes

Chen Zhang

Inorganic membranes (e.g., zeolite, ultramicroporous silica, ion-transport perovskite) can outperform polymer membranes by providing high productivity, high selectivity, and excellent thermal and chemical stability. Inorganic membranes often rely on porous oxide hollow fibers as substrates, which are traditionally manufactured by high-temperature methods. For example, the phase inversion-sintering method is based on sintering of polymer hollow fibers consisting of preformed nonporous alumina particles at 1500 °C. These high-temperature manufacturing processes are expensive and have limited inorganic membranes to a few niche applications.

In this talk, we will present innovative manufacturing of porous oxide hollow fiber membranes via petrification of polymer hollow fiber templates.[1] This novel petrification method represents a breakthrough in inorganic membrane manufacturing and consists of three simple steps of (1) silane soaking, (2) moisture exposure, and (3) air heating. The method allows formation of porous oxide hollow fiber membranes, which we name "petrified hollow fiber membranes", at temperatures far below the traditional phase inversion-sintering method. The lower temperature reduces the manufacturing cost of inorganic membranes, thereby potentially broadening the scope of their applications.

These petrified hollow fiber membranes have a unique hierarchical micro-/meso-/macroporous structure, giving both high surface area (650 m^2/g) and large pore volume (2.5 cm^3/g). Interestingly, like petrified wood reproduces the geometry and cell structure of wood, petrified hollow fiber membranes reproduce the hollow fiber geometry and asymmetric macropore structure of polymer hollow fiber templates. Their asymmetric macropore structure allows the petrified hollow fiber membranes to efficiently reject macromolecular solutes (e.g., protein) in cross-flow filtration tests. They are also highly gas permeable (helium permeance ~30,000 GPU), and therefore are potentially suitable as substrates of inorganic membranes for gas and vapor separations.

References:

[1]Liu, L.; Ku, C.-E.; Zhang, C., Petrified Hollow Fiber Membranes with Hierarchical Pores. *ACS Materials Letters* **2022**, 938-943.

643f Prospect of Using Machine Learning for Evaluating Gas Separation Membranes' Transport Properties and Assisted Fabrication

Arash Tayyebi, Ali Alshami

Polymeric membranes are widely used to separate various mixed species due to their high removal ability and economic cost; however, membrane fabrication requires significant research to determine appropriate polymers and solvents and test characterization. Machine Learning methods are computational approaches that can be applied to membrane fabrication and facilitate this process. ML techniques used in membrane fabrication are categorized into two types. The first type breaks the polymer structures and defines each monomer using numerical measurements. Each polymer has a fingerprint, and this feature can be input into the ML algorithm as legible data by using molecular descriptors or molecular fingerprints. Molecular descriptors are experimentally measured or theoretically derived properties of a molecule compared to the molecular fingerprint, which is usually in the form of bits or uses vector elements. Molecular fingerprints specify the existence or frequencies of specific fragments, while molecular descriptors are used to illustrate a molecule's physical, chemical, or topological characteristics. These methods represent molecules as a sequence of bits that computers can recognize: longer bit strings are more reliable through the similarity search since each significant bond in a molecule is defined separately as a bit and has more information about the molecular structure.

The second type is mapping a way to link the fingerprinted input and the target values. Classification, regression, dimensionality reduction, and clustering are the most common types of algorithms for chemical separation that can be applied to process data and predict the relationships between the input and output parameters using ML techniques. The ML technique is chosen and used based on the prediction model's target demand, the current dataset's size, and various ML method features. Artificial neural networks (ANNs), deep learning (DL), support vector machine (SVM), Gaussian process regression (GPR), and random forest (RF) are the ML methods that can be used to analyze non-linear relationships and can be employed in prediction models when the interactions between components are unclear, such as with membrane fabrication.

This research aims to explore the application of an ML algorithm to polymeric membrane creation, review recent research that used ML to fabricate polymeric membranes for gas component separation and address the critical needs. Training Heteropolymers instead of Homopolymers, producing novel polymers by an inverse design approach, and using reliable datasets that are created under the same conditions, are the most crucial necessities that should be investigated. We have summarized dataset acquisition and training algorithms, ML methods, and examined methods to verify the results. We also report on future development prospects for the ML-driven polymer-based membrane design methods.

643g Intrinsically Porous, Chemically Tunable, and Mixed Matrix 3D Printed Membranes for Targeted Purification

Ryan Mulvenna, Michael Scalzo, Abdollah Khosravanian, Huacheng Zhang, Benny D. Freeman, Timothy F. Scott, Matthew R. Hill

In the coming decades, the increasing and continued demand for low energy, cost effective separation of elemental lithium and Liquid Organic Hydrogen Carriers (LOHC) for a net-zero emissions economy remain a challenge. One critical aspect of this challenge is ensuring that the burgeoning demand for lithium in energy storage requires facile, scalable, and low cycle time processes of lithium extraction. Another critical aspect of this challenge is efficient separation of hydrogen rich and depleted LOHC fuels for on-demand catalytic generation of hydrogen for combustion.

Here, we will discuss our developed separation strategies to address these opportunities. In the process of lithium extraction, we will discuss our work of UV-curing thiolene polymers featuring intrinsically porous macromolecules of pillararenes. By facile re-functionalization of these moieties prior to UV-cured thiolene network formation, these materials enable selective transport of ions without the requirement of high-crystallinity, thus further enhancing their capability for scale-up and roll-to-roll processing. Secondly, in the process of LOHC fuel purification, we will discuss our work into the creation of 3-d printable Mixed Matrix Membranes (MMM). These composites utilize the selective transport of Metal Organic Frameworks (MOF) consisting of Zn_4O Cubane-1,4-dicarboxylic acid (1,4-H_2cdc) CUB-5 and its counterparts for selective transport of methylcyclohexane and toluene LOHC fuels. These selective MOFs may be utilized in a 3D-printable, semi-crystalline allylic disulfide ring opening polymer system as the filler for a chemically robust separation module.

519a Effect of Co Ions in Cross-Linked Poly(ethylene oxide) on O_2 Transport Characteristics

Taliehsadat Alebrahim, Narjes Esmaeili, Thien Tran, Haiqing Lin

Cobalt (Co)-based complexes are considered as attractive O_2-binding elements, which can reversibly and selectively interact with O_2 and affect O_2 gas transport properties. Herein, we investigate the effect of Co ions dissociated in amorphous cross-linked poly(ethylene oxide) (XLPEO) on physical properties and gas transport characteristics. Polymeric films containing Co salts were prepared by photo-polymerization of solutions containing poly(ethylene glycol) diacrylate (PEGDA), poly(ethylene glycol) methyl ether acrylate (PEGMEA), and different salts such as $Co(ClO_4)_2$, and $Co(BF_4)_2$. The salts can be fully dissociated, and the Co^{2+} ions form conjugations with the ethylene oxide in XLPEO, as validated by FTIR and XRD. Increasing the Co content increases the density (following the additive model) and glass transition temperature of XLPEO. For instance, adding 17 wt.% $Co(ClO_4)_2$ to XLPEO increases the T_g and density from -65 °C and 1.149 g/cm^3 to -26 °C and 1.304 g/cm^3, respectively. The O_2 and N_2 permeability are significantly impacted by adding high concentration of Co salts. Specifically, O_2 and N_2 permeability decreases from 26 and 10 to 0.24 and 0.14 Barrer, respectively, by adding 17 wt.% of $Co(BF_4)_2$. Interestingly, the CO_2/O_2 selectivity increases from 20 to 60 by addition of 17 wt.% of $Co(BF_4)_2$ presumably due to the retarded O_2 diffusion. Additionally, we found that O_2 solubility increases with increasing $Co(BF_4)_2$ content due to high affinity between Co ions and oxygen. For example, O_2 solubility increases from 0.2 to 6.4 cm^3(STP)/cm^3 atm at 35°C by addition of 6.4 wt.% $Co(BF_4)_2$ to XLPEO. Understanding the effect of Co ions-O_2 interactions on gas transport properties may provide guidance in designing advanced polymeric membranes with high CO_2/O_2 selectivity.

519b Polyformamidine As Fixed-Site Carrier for CO_2/N_2 Separation Membranes

Jingying Hu, Yang Han, Winston Ho

Amidine species was perceived as better alternatives for amine species in the facilitated transport membranes for having faster CO_2 reaction kinetics. The synthesis of poly(ethylene formamidine) (PFA) was carried out via the polycondensation of ethylenediamine and triethyl orthoformate catalyzed by acetic acid. The synthesized PFA possesses the highest amidine functional group density owing to the case using the shortest-chain diamine, ethylenediamine. Moreover, in comparison with mobile carriers, it can have a much higher boiling point, which practically eliminates the possibility of carrier volatility and leaching out. In order to obtain the preliminary gas permeation results, PFA (10 kDa MW) was blended with polyvinylalcohol (PVA, 1.1 MDa MW) and then coated onto nanoporous polyethersulfone (PES) substrates to form 170-nm thin membranes. Increasing the content of PFA from 36 wt.% to 66 wt.% in these membranes led to a substantial increase in membrane performance, especially for CO_2 permeance. Further increasing the PFA content to 83 wt.% still enhanced the CO_2 permeance but showed a moderate drop in CO_2/N_2 selectivity. Under an extremely high PFA content of 90 wt.%, the membrane showed a low CO_2 permeance of 1019 GPU but a high CO_2/N_2 selectivity of 236, indicating severe pore penetration. The membrane containing 83 wt.% PFA exhibited the best results with 2807 GPU CO_2 permeance and 72 CO_2/N_2 selectivity at 77°C. In view of the superior transport performance, it is necessary to increase the MW of PFA to make it capable of forming the membrane matrix alone to enable the incorporation of CO_2-reactive mobile carriers, thereby improving the overall CO_2 separation performance. By adopting the design of experiment technique, the synthesis conditions of PFA were improved in terms of the temperature profile, catalyst amount, and monomer concentrations, through which the MW was successfully increased to 3.3 MDa.

519c Transforming Barrier Polymers to High-Performance Gas Separation Membranes

Gaurav Iyer, Chen Zhang

Polymer membranes have excellent scalability and good separation performance for many gas pairs. However, owing to their permeability-selectivity tradeoff, their large-scale gas separation application has yet to evolve beyond air separation and natural gas purification. Carbon molecular sieve (CMS) membranes are derived by controlled pyrolysis of polymer membranes to provide a good balance of permeability, selectivity, and scalability. Formation of CMS membranes often relies on pyrolysis of advanced and expensive polymer precursors, such as polymers of intrinsic microporosity and fluorinated polyimides. Barrier polymers are much less expensive and used for beverage packaging at large scale. Barrier polymers, however, have been rarely considered for membrane-based gas separations due to their ultra-low gas permeabilities.

In this talk, we will present the transformation of impermeable barrier polymers to high-performance CMS membranes. An aromatic barrier polymer precursor was synthesized by condensation polymerization, which showed ultra-low CO_2 permeability (<0.002 Barrer) at room temperature. Pyrolysis of the barrier polymer precursor provided a novel CMS membrane that had over 100,000 times higher CO_2 permeability than the barrier polymer precursor. The barrier polymer-derived CMS membrane showed attractive selectivities, which put them above the polymer upper bounds for several commercially important gas pairs.

519d A Versatile Microporous Poly(Arylene Ether) Platform for Membrane-Based Gas Separation

Jing Ying Yeo, Sheng Guo, Francesco Maria Benedetti, Duha Syar, Timothy Swager, Zachary Smith

Membrane-based gas separations exhibit several advantages over conventional separation methods such as cryogenic distillation and adsorption technologies due to the absence of energy-intensive phase changes as well as the lack of toxic chemicals. However, it remains a challenge to synthesize materials suitable for membrane separations which demands high material processability and high gas separation performance. Microporous organic polymers (MOPs) is a class of materials that combine the gas sieving capability of microporous materials with the solution-processability of organic polymers. For this project, we present the rational design and synthesis of linear microporous poly(arylene ether)s (PAEs) via Pd-catalyzed C-O polymerization reactions. The scaffold of these new microporous polymers consists of rigid three-dimensional triptycene and highly stereocontorted spirobifluorene, generating a large internal free volume within the polymer matrix. This robust methodology for the PAEs synthesis allows for the facile incorporation of functionalities and branched linkers to enhance the membrane's permeation and mechanical properties, while remaining solution-processable for membrane fabrication. Such properties are rare in classic polymers of intrinsic microporosity (PIMs). Better CO_2 separation is also achieved by incorporating CO_2-philic groups, such as nitrile and tertiary amine groups, into this microporous polymeric scaffold. Furthermore, a branched polymer prepared using this synthetic strategy showed good gas separation performance and enhanced mechanical properties, which allowed for the formation of a submicron defect-free film with permeance-selectivity property sets that are comparable to high-performance ultrathin polymer membranes that have been optimized at industrial scale. The easily accessible PAE branching motif also endows these materials with plasticization resistance which enhances the longevity of the membrane, addressing a challenge facing many commercially available polymeric membranes. Results for the outstanding mechanical properties and gas transport performances will be presented for several gas separations of interest. The high physical stability, structural flexibility and and ease-of-processing suggest that this new platform of microporous polymers provide generalizable design strategies to address outstanding separation challenges for gas separation membranes. The methodology, implications, and future directions of applying this new class of PAEs to membrane-based gas separations will be outlined in this presentation.

519e Novel Iptycene-Based Polybenzimidazole Membranes for H_2/CO_2 Separation

Mengdi Liu, Ruilan Guo

H_2/CO_2 separation relatable to pre-combustion CO_2 capture from syngas after water-gas shift reactions demands robust polymer membrane materials with high thermal stability and strong size-sieving capability at high temperatures. Polybenzimidazole (PBI), such as *m*-PBI (Celazole®), has been the focus of increasing studies due to its good H_2/CO_2 separation properties and high thermal stability. However, *m*-PBI suffers from extremely low permeability. Although many previous studies were attempted to improve *m*-PBI's permeability by incorporating bulky moieties or large substituent groups, these strategies typically resulted in compromised size-sieving capability, particularly at high temperatures. On the other hand, modified PBI-based membranes, such as crosslinked PBIs, that were able to achieve ultrahigh selectivity often exhibited a significant loss of H_2 permeability. To address this issue, we developed a novel macromolecular design to incorporate shape-persistent iptycene (i.e., triptycene and pentiptycene) units into PBI structure to instill configurational free volume elements. The non-collapsible and well-defined microcavities originating from intrinsic molecular configuration of iptycene units are able to significantly enhance gas permeability as well as size sieving property. Moreover, by systematically varying types of iptycene units and pendent groups, structure-property relationships of these novel iptycene-based PBIs were investigated. In this work, a triptycene-based PBI (TPBI) and a pentiptycene-based PBI (PPBI) have been produced, and gas transport properties have been characterized. We demonstrated that by incorporating triptycene units, H_2 permeability was effectively improved by 300% relative to *m*-PBI. The bulkier pentiptycene units could disrupt chain packing more efficiently, exhibiting 30 times higher H_2 permeability than that of *m*-PBI. Post treatments including thermal treatment and crosslink via acid doping were performed to further tailor microporosity, leading to versatile and significantly improved separation performance. We found that thermal treatment at 350 ^oC favored rearrangement of TPBI chains to unblock more microcavities, resulting in further enhanced H_2 permeability by 200% relative to the untreated sample, which indicates the good temperature tolerance of configurational free volume. After crosslinked by doping with phosphoric acid, the packing of polymer chains was tightened due to proton transfer and hydrogen bonding interactions. As a result, the acid-doped TPBI exhibited largely improved H_2/CO_2 selectivity by 100%, outperforming the 2008 upper bound for H_2/CO_2 separation. This talk will discuss the preparation and characterization of novel iptycene-based PBIs membranes as well as the influence of configurational free volume, thermal treatment, and acid doping on their gas transport properties.

519f Thermal Cross-Linking of Ultra-Robust Membranes for Plasticization Resistance and Enhanced Permeability

Hyun Jung Yu, Jong Suk Lee

Polymers are the most widely used membrane materials for gas separations due to high processability and relatively low material cost. Most polymeric membranes, however, are susceptible to a significant selectivity loss upon exposure to condensable gases at high pressure due to undesirable plasticization. In this talk, I will present a novel method for enhancing both the permeability and plasticization tolerance of polymeric membranes via a combination of amidation and decarboxylation between a carboxylated polyimide (PI) and an amine-containing ladder-structured polysilsesquioxane (LAPSQ) [1]. Gas transport properties of a series of BTDA-Durene:DABA PI/LAPSQ membranes with different Durene:DABA molar ratios (3:2, 2:1, and 4:1) were examined with mechanical properties before and after thermal cross-linking. Also, molecular dynamics simulations were conducted to rationalize the enhanced CO_2 permeability as well as high plasticization resistance in the thermally cross-linked PI/LAPSQ (80/20) membranes. With that, a new approach of manufacturing hollow fiber membranes will be demonstrated for the scale-up of our highly plasticization-resistive membranes.

[1] H. J. Yu, C-H Chan, S. Y. Nam, S-J Kim, J. S. Yoo, J. S. Lee, Thermally cross-linked ultra-robust membranes for plasticization resistance and permeation enhancement – A combined theoretical and experimental study, J. Membr. Sci. 646 (2022), 120250.

519g Elucidating the Effect of Chlorination Versus Fluorination on the Transport Properties of Polymer Membranes for Gas, Vapor and Organic Solvent Separation

Jing Deng, William Box, Lucas Condes, Yoshiyuki Okamoto, Michele Galizia

Due to their peculiar transport properties, as well as their larger plasticization and physical aging resistance compared to conventional glassy polymers, perfluoropolymers garnered increasing attention in membrane science [1]. While their swelling resistance has been ascribed to the lower sorption capacity of traditional swelling agents, such as hydrocarbons, relative to hydrocarbon-based polymers, their superior physical aging resistance has been attributed to the presence of pendant rings directly attached to the polymer backbone, which hampers polymer chains to collapse [2,3]. Recent studies of gas transport in glassy polyimides exhibiting systematically varied degree of fluorination explored the effect of penetrant-perfluorinated moieties interactions, including CO_2, CH_4, O_2 and N_2, on gas solubility [4]. The multitude of results published since the early 1990s thus far indicates that while the role of fluorination on small molecule transport in polymer membranes has been the subject of a vast body of research, the role of other halogens, such as chlorine, on the transport properties of glassy polymers did not receive enough attention and, for this reason, it is still poorly understood. Some studies from Lin et al. indicate that while fluorine acts mostly on solubility, selectivity, physical aging rate and swelling propensity, chlorine acts essentially on the size-sieving ability, that is, on gas diffusivity [3]. The molecular origin of this phenomenon, however, is still unknown.

To fill these knowledge gaps, we propose a fundamental investigation of small molecule transport in a recently synthesized copolymer of perfluoro(2-methylene-4-methyl-1,3-dioxolane, PFMMD) and chlorotrifluoro ethylene (CTFE), namely poly (PFMMD-co-CTFE) [5]. The solubility and diffusivity of three model vapors, water, methanol (polar organic penetrant) and n-pentane (non-polar organic penetrant) were measured at multiple temperatures and activities, and compared to previously reported sorption and transport data in perfluoropolymers, such as Teflon® AF and Hyflon® AD. Interestingly, due to the larger polarity of the C-Cl bond compared to the C-F bond, poly (PFMMD-co-CTFE) exhibits a much larger hydrophilicity relative to chlorine-free perfluoropolymers, as well as much better swelling resistance in the presence of methanol, which makes poly (PFMMD-co-CTFE) an ideal candidate for the separation of organic solutions via organic solvent nanofiltration and reverse osmosis. The analysis of transport data, as well as heats of sorption and diffusion, provided relevant fundamental insights regarding the role of polymer-penetrant interactions, clustering and swelling, as well as the mechanism of small molecule transport in this class of materials, which helps individuate possible applications in membrane separations.

To put our result in a broader perspective, water, methanol and hydrocarbon transport in poly (PFMMD-co-CTFE) was compared to the transport of the same molecules in chlorinated organic liquids.

References

[1] Y. Okamoto et al., Perfluorodioxolane Polymers for Gas Separation Membrane Applications, Membranes 2020, 10, 394

[2] Y. Li, et al., Volumetric Properties and Sorption Behavior of Perfluoropolymers with Dioxolane Pendant Rings, Ind. Eng. Chem. Res. 2020, 59, 5276

[3] M. Yavari et al., The role of halogens in polychlorotrifluoroethylene (PCTFE) in membrane gas separations, J. Membr. Sci. 2018, 548, 380

[4] A. Wu et al., Non-equilibrium Lattice Fluid Modeling of Gas Sorption for Fluorinated Polyimides, Macromolecules 2020, 50, 6620

[5] M. Fang et al., High-performance perfluorodioxolane copolymer membranes for gas separation with tailored selectivity enhancement, J. Mater. Chem. A 2018, 6, 652

675a Combinatorial Screening of Ligand-Functionalized Microbeads for Unmethylated and Methylated DNA Binding Application

Subhadeep Dutta, Tyree Ratcliff, Curt Breneman, Kaushal Rege

Nucleic acids (such as DNA, RNA) are foundation for many downstream bioprocessing applications of gene therapies and vaccinations. Also DNA methylation patterns play a key role in evaluating gene expression in various diseases like cancer. Hence enrichment of nucleic acid become a challenging area of research where researchers are using different approaches to improve the process development and improve the recovery rate. Competing technologies to isolate methylated plasmids or respective gene of interest involve antibodies or designing DNA-binding proteins which are mainly geared towards research-scale applications; are time consuming and also expensive. Consequently, new approaches which leads to enrichment of methylated or unmethylated DNA from biological samples and bioprocessing media, in a rapid and cost-effective manner, can lead to transformative advances in these fields. We have developed aminoglycoside-derived microbeads, approximately 10μm in diameter, for binding DNA based on Coulomb interactions between DNA molecules and microbeads. These parental microbeads were further modified with different amino acids, chemotherapeutic drug ligands in order to obtain a diverse library of ligand-modified microbeads. The library of ligand-modified microbeads was screened for binding unmethylated and methylated DNA in order to identify lead candidates that showed preferential binding to methylated DNA over the unmethylated analog or vice versa. Adsorption isotherms were determined for both sets of leads and the data were fit to a Langmuir isotherm model. Higher degree (~70%) of pDNA elution was obtained after optimizing mixed buffer conditions with salt additives. Computational study with a novel DNA-Pixel (DIXEL) approach was used to demonstrate amino acid ligand interaction with different regions of methylated or unmethylated pDNA. The DIXEL approach allows for the evaluation of DNA sequences using a 2D representation and study its region-specific ligand interactions based on physical properties like electrostatic potential (EP) and lipophilicity. Our results conclude molecular engineering of a chemically diverse ligand-modified microbead platform; that can lead to identification of novel ligands for enriching methylated or unmethylated DNA for applications in bioseparation and rapid detection of diseases at low-resource settings.

675b Ultrafiltration of Monoclonal Antibodies: Combined Optimization of Formulation Parameters and Process Operating Conditions Under Uncertainty

Francesco Rossi, Fernanda Cunha, Eduardo Ximenes, Brian D. Bowes, Zhao Yu, Dennis T. Yang, Vincent Corvari, Michael Ladisch, Gintaras Reklaitis

Monoclonal antibodies (mAbs) are an important type of therapeutic agent that are effective treatments in several therapeutic areas, such as immunology, oncology and infectious disease. These immunoglobins owe their success to their higher efficacy, specificity and safety compared to conventional drugs, which is a direct consequence of their mechanism of action. Unfortunately, all these benefits come at a price, namely, mAbs are expensive to produce (Rudge and Ladisch, 2020; Maruthamuthu et al, 2020).

One of the reasons why monoclonal antibodies are extremely expensive is their complex and low-yield production process, which includes several challenging manufacturing steps, namely, an initial fed-batch fermentation followed by centrifugation, depth filtration, several chromatographic purifications, virus filtration and, finally, concentration by ultrafiltration. Therefore, one of the most direct ways to reduce the average cost of monoclonal antibodies is to optimize their production process, in terms of both processing sequence and individual manufacturing step yields.

Several authors have proposed optimal design configurations for mAb production processes (Liu et al., 2016) and computed the optimal operating conditions at which the fermentation and chromatographic separations should be carried out (Alhuthali et al., 2019; Papathanasiou et al., 2017). Some of these contributions have also analyzed the impact of model uncertainty on the results of their design/dynamic optimization calculations (Liu and Gunawan, 2017). Surprisingly, the ultrafiltration step has received very little attention so far, even though the typical yield of this operation may be as low as 85 %, in the absence of a recovery backflush (as the monoclonal antibody is being concentrated, a significant amount of it forms a hydrogel that sticks onto the ultrafiltration membrane). One of the very few studies, which deals with the systematic optimization of ultrafiltration systems for the concentration of monoclonal antibody solutions, has been conducted by Rossi et al. (2021). This contribution expands the scope of the aforementioned study by investigating the extent to which the combined optimization of formulation parameters (e.g., the buffer pH) and standard operational variables (e.g., the feed flowrates to the retentate and permeate channels of the ultrafiltration membrane cassette and the pressure levels at the retentate and permeate outlets) can help further enhance the overall performance of mAb ultrafiltration, measured in terms of final mAb concentration/recovery and operation time.

More specifically, we first update the hybrid mechanistic-statistical model of the ultrafiltration device, developed by Rossi et al. (2021), which consists of a first-principles-inspired model of the physical system, augmented with a Bayesian neural network, by re-training both model components with a more comprehensive set of experimental data (this training dataset includes physical property measurements for the mAb of interest, collected over an appropriate range of pHs, combined with ultrafiltration experiments, conducted at several different pHs). We then validate the updated hybrid model with an appropriate set of validation experiments, using both Bayesian and frequentist uncertainty propagation techniques. Finally, we utilize the validated hybrid model and multi-objective stochastic optimization techniques to identify the most appropriate mAb formulation, which we should use in the ultrafiltration step, and compute the optimal process conditions, at which we should operate, with the aim of maximizing the final mAb concentration/recovery and minimizing the operation time (under uncertainty).

The results of our calculations are very promising, in that we can further improve the final mAb concentration/recovery and reduce the ultrafiltration time, compared to Rossi et al. (2021), without violating any operational constraints, even in the presence of significant model uncertainty. This additional performance boost can be attributed to the very idea of optimizing formulation parameters and operational variables simultaneously.

References

Alhuthali, S., Fadda, S., andamp; Kontoravdi, C. (2019). Constrained optimisation of cell culture feeding strategy and temperature shift duration to enhance monoclonal antibody titre and purity. 2019 AIChE Annual Meeting, Orlando (FL).

Liu, Y., andamp; Gunawan, R. (2017). Bioprocess optimization under uncertainty using ensemble modeling. Journal of biotechnology, 244, 34-44.

Liu, S., Farid, S. S., andamp; Papageorgiou, L. G. (2016). Integrated optimization of upstream and downstream processing in biopharmaceutical manufacturing under uncertainty: a chance constrained programming approach. Industrial andamp; Engineering Chemistry Research, 55(16), 4599-4612.

Maruthamuthu, M. K., Rudge S. R., Ardekani, A. M., Ladisch, M. R., Verma, M. S. (2020). Process Analytical Technologies and Data Analytics for the manufacture of Monoclonal Antibodies, Trends in Biotechnology, 38(10), 1169-1186.

Papathanasiou, M. M., Quiroga-Campano, A. L., Steinebach, F., Elviro, M., Mantalaris, A., andamp; Pistikopoulos, E. N. (2017). Advanced model-based control strategies for the intensification of upstream and downstream processing in mAb production. Biotechnology progress, 33(4), 966-988.

Rossi, F., Zuponcic, J., Ximenes, E., Geng, S., Tao, Y., Corvari, V., Ladisch, M., Reklaitis, G. (2021). Dynamic optimization of an ultrafiltration system for the concentration of monoclonal antibody solutions under uncertainty. 2021 AIChE Annual Meeting, Boston (MA).

Rudge, S. R., and Ladisch, M. R. (2020). Industrial Challenges of Recombinant Proteins, Current Applications of Pharmaceutical Biotechnology, A. C. Silva, J. N. Moreira, J. M. S. Lobo, H. Almaida, editors, in Adv Biochem Eng Biotechnol, 171, 1 to 22.

675c Mechanistic Modeling of Tangential Flow Filtration to Guide Process Development

Michael Hartmann, Ariella Schwartz, John Welsh, Jennifer Pollard

Ultrafiltration (UF) and Diafiltration (DF) are key unit operations for bioprocess development that use tangential flow filtration (TFF) to provide a fast, efficient, and economical means of concentrating and buffer exchanging proteins. Each protein that undergoes TFF process development has buffer-dependent biophysical properties that requires unique operational conditions for efficient TFF performance. In addition, the application of TFF can be viscosity limited by active pharmaceutical ingredients (APIs) that require high concentration formulations (> 150 g/L), leading to additional process development. Here we present a modeling workflow, performed in early-stage development of UF and DF unit operations, that allows us to connect the physical components and solution properties of the drug substance (e.g. buffer composition and solution viscosity) to the operational set points of the experiment. Using the workflow, we can run a minimal number of experiments and use these results to simulate how the filtration will operate under a range of set points. The results from the simulations are then fed back into the experimental design to guide further process development. Connecting protein biophysics with TFF operation not only allows for quickly identifying molecule specific operational set points, but also minimizes material usage, ensures product yield and process robustness, and can inform other development decisions such as membrane selection.

675g A New Generation of Materials for Sustainable and Wearable Hemodialysis: Tests and Molecular Screening.

Thomas Fabiani, Maria Grazia De Angelis, Eleonora Ricci, Cristiana Boi, Simone Dimartino

Hemodialysis (HD) is a life-supporting treatment for 2 million patients with chronic kidney failure, affecting 697.5 million people worldwide. HD replaces kidney functionality, based on the exchange of uremic toxins (UT) present in blood through a semipermeable membrane with a buffered solution, called dialysate. The treatment is time-consuming and discontinuous causing health issues and a low quality of life. Furthermore, the water consumption is assessed around 400 L per session. The last 40 years have witnessed many attempts to design a wearable artificial kidney (WAK), which is a miniaturized HD unit where the dialysate is purified and recirculated back to the dialysis unit. Such a device, not yet on the market, would boost sustainability, accessibility, delivery of hemodialysis with huge improvement of the patients' lifestyle.

Previous attempts to build a wearable kidney involved a chemical conversion of urea, that is the most difficult toxin to remove from dialysate. However, such process releases by-products that can be harmful for the patient, while a technology involving only urea capture with adsorbent materials is intrinsically safe. Previous works have shown that adsorbents like activated carbon and zeolites could be efficiently embedded into porous polymeric supports made of cellulose acetate with optimal fluid dynamics and low energy consumption, building the so-called Mixed Matrix Membrane Adsorbers (MMMAs) [1-3]. However, the systems fabricated so far only employed conventional adsorbers like activated carbon and zeolites, with limited urea removal capacity.

In this work we propose to accelerate the deployment of adsorbent materials for uremic toxins removal with an *in silico* screening on existing and new crystal adsorbent structures. Such a methodology will allow to score promising materials for hemodialysis applications and contaminant removal, reducing the experimental efforts required and simultaneously providing better understanding of UTs adsorption on nanoporous materials.

The pipeline developed and presented mines crystal structures from databases applying constrains on selection of crystals according to rules of acceptability. Subsequently, a set of general or local descriptors of topological and physiochemical features is calculated, while binding capacity and strength toward UTs, together with selectivity, are investigated by means of molecular mechanics simulations. This protocol allows to score and rank the structures according to their suitability as adsorbers toward UTs. Furthermore, a quantitative structure-property relationship is devised for describing UTs adsorption by Covalent Organic Frameworks COFs, for the first time.

The in-house developed pipeline is versatile enough to describe other compounds and totally based on open-source tools and resources, such as RASPA as environment for molecular simulations. The optimal adsorbents selected in the screening will be embedded into porous polymeric supports to develop MMMAs for wearable hemodialysis.

[1] De Pascale, M. G. De Angelis and C. Boi, *Membranes*, *12*(2), 203 (2022) https://www.mdpi.com/2077-0375/12/2/203

[2] De Pascale, M. G. De Angelis and C. Boi, *Chemical Engineering Transactions*, 74, 781–786, (2019)

[3] De Pascale, 'Novel membranes for hemodialysis application,' PhD Thesis, University of Bologna, 202

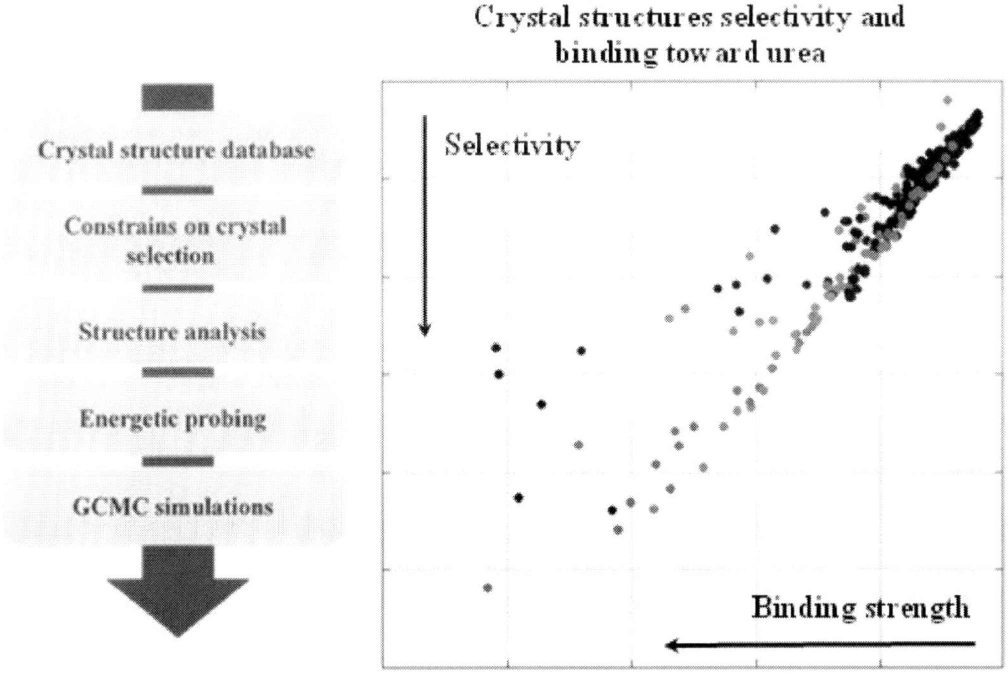

Figure 1 Schematic description of the scoring pipeline (on the left) and ranking of crystal structures according to selectivity and binding properties toward urea (on the right).

675e Partitioning and Mass Transport for Continuous Two-Phase Extraction of Viral Vaccines

Seth Kriz, Rehab Alhajjar, Ethan Burghardt, Christopher Kirschke, Sneha Singh, Pratik Joshi, Caryn Heldt

Biomanufacturing is trying to push past batch processing to continuous processes, an extremely beneficial technological leap that reduces costs, labor, and the footprint of the manufacturing plant. Aqueous two-phase extraction (ATPE) is a promising solution to enable this advance. The gentle environment afforded by ATPE is especially conducive to the purification of fragile whole-particle viral vaccines that are easily inactivated and too large to fit into the pores of conventional chromatography resins. Previously, we achieved over 80% recovery of two model viral products in the polymer phase of a PEG/citrate ATPE with high host cell protein and DNA removal in both batch and continuous modes. However, two major challenges to implementation exist: 1) optimization of efficient ATPEs for new viral vaccines is time and resource intensive, and 2) scale-up is hampered by a lack of understanding of the mass transport of viral particles in ATPE. To solve these problems, we designed two unique microfluidic channels. The first contains internal static mixer elements to increase interfacial area between the phases and hasten partitioning, allowing us to screen for optimum extraction conditions by changing the flow rates of our components. This greatly decreases development time and uses only microliters of solutions. A second microdevice allows direct observation of the transport of fluorescently tagged protein and virus particles across the interface of our ATPE. Using this device and a two-film transport model, we determined diffusion coefficients in each phase and the mass flux across the interface of porcine parvovirus. Combined, these tools represent a pathway to rapidly identify ATPEs for purification of new viral vaccine candidates and collect transport parameters required for scale up of continuous systems, all in a fraction of the time required for earlier manual optimizations. These findings will enable the design and scale-up of continuous downstream processes that provide the highest recovery possible using a small footprint and cost-effective equipment.

675f Advancing Real-Time Live Cell Imaging Technologies to Support the Development of Downstream Purification Processes for Therapeutic Viruses

Claire Velikonja, Landon Steenbakkers, Ian Gough, Elsa Sorman Paulsson, Christoffer Edlund, Rickard Sjogren, Kalpana Barnes, Brandon Corbett, Prashant Mhaskar, David Latulippe

Oncolytic virus therapies are a promising cancer treatment that uses viruses to selectively kill cancer cells, with the first FDA approved treatment (Imlygic for melanoma) released in 2015. However, widespread adoption of these treatments is limited due to the complexity of purifying viruses. It is generally well known that the development and optimization of novel purification processes has been greatly hindered due to the practical limitations of the current standard analytical methods (e.g., Hexon staining) for assessing virus infectivity titer. These analytical methods often require a minimum of 72 to 84 hours and significant manpower due to timed antibody washes, creating a major bottleneck within virus manufacturing research activities. Therefore, there needs to be a synergistic effort to develop analytical methods for determining virus infectivity titer that can process and analyze samples at the increased throughput of cutting-edge virus purification methods. Here, we demonstrate a wash-free adenovirus infectivity assay developed based on the production and concentration requirements of an adenovirus purification process with anion-exchange membrane absorbers. The assay was developed in collaboration with Sartorius and McMaster and leverages the Incucyte® Live-Cell Analysis System to reduce required manpower and increase assay throughput.

Adenovirus samples with a GFP-transgene were initially used, with the Incucyte® Live-Cell Analysis System and the Incucyte® AI Confluence software implemented to optimise cell culture regimes and cell confluence over time to ensure assay viability. Segmentation algorithms provided by the Incucyte® Base Analysis software were used to estimate virus titer values by the appearance of fluorescently labelled cells, which were compared to estimates obtained through traditional Hexon staining. Subsequently, adenovirus samples without a GFP transgene were used to extend the Incucyte's ability to provide label-free titer estimates using machine learning methodologies, with the results similarly validated using Hexon staining and compared. The comparative costs and labour were then tabulated, demonstrating the relative benefits of the Incucyte® based assay as the number of samples increases.

Overall, investigation of optimized virus purification techniques is essential as specialized biotherapeutics gain clinical prominence. However, development of analytical assays that can match increased purification throughput has lagged, making sample analysis the bottleneck in virus therapeutic production. The assay presented in this work has been specifically developed to handle the sampling requirements and virus titer range associated with virus membrane chromatography, demonstrating the potential of Incucyte® integrated live-cell analysis techniques in the analytical workflow to enable new separation processes.

675d Multi-Well Device for High-Throughput Screening of Any Flat Sheet Membrane Adsorber

Ana Arezina, Karina Kawka, Alexandra Jucan, Sarah Arnold, Boyang Zhang, David Latulippe

The development of membrane chromatography processes for biomanufacturing requires extensive optimization of the process conditions (e.g. solution pH, solution ionic strength). While many previous studies have demonstrated the advantages of performing optimization studies using 96-well filter plates (e.g. AcroPrep Advance from Pall[1]) either on manual systems or on robotic platforms, this approach is limited to the few membrane types that are sold in a filter plate format. There are many types of emerging membrane chromatography materials (e.g. 3D printed membranes[2], self-assembled block copolymer membranes[3]) that would greatly benefit from the ability to efficiently conduct a high-throughput screening (HTS) study of their separation performance.

Inspired by the recent trend of using rocker platforms for continuous perfusion process in organ-on-a-chip applications (e.g. IFlowPlate[4]), a HTS method for membrane chromatography involving laterally-fed, gravitational flow driven by a perfusion rocker was developed. The process is based on a custom multi-well plate with exterior dimensions and well-to-well spacing based on standard 96-well plate sizing, allowing for its integration with manual and automated liquid dispensers (e.g. multichannel pipettes). Each multi-well plate can be used to conduct up to 32 experiments in parallel, with each experiment requiring just a single small disc of membrane; additionally multiple devices can be stacked on the perfusion rocker to conduct more than 32 experiments at once. The continuous flow of solution back and forth across the membrane during each 'stage' of the chromatography process (i.e., bind, wash, elute) occurs due to passive liquid levelling in response to the interval rocking action. As a proof of principle, we first analyzed the binding capacity of commercial anion-exchange membranes (i.e. Mustang Q, Sartobind Q, Natrix Q) for typical biomolecules (e.g. BSA, DNA). Interestingly, we found a considerable disparity in the reported binding capacity values for these biomolecules in the current literature (e.g. BSA binding capacity for Sartobind Q has been reported as 18 and 29 mg/mL[5,6]) which provided additional motivation for an effective HTS tool for membrane chromatography processes.

Overall, the multi-well device developed in this study will allow for an improved understanding of biomolecule binding onto existing and emerging chromatographic membrane materials, which in turn will result in the development of separation processes that yield superior performance.

References

1. Pall. *AcroPrep^{TM} Advance 96-well Filter Plates*. Accessed April 7, 2022. https://shop.pall.com/us/en/products/filter-plates/96-well-filter-plates
2. Dong Z, Cui H, Zhang H, et al. 3D printing of inherently nanoporous polymers via polymerization-induced phase separation. *Nat. Commun.* 2021;12(247):1-12. doi: 10.1038/s41467-020-20498-1
3. Hahn J, Clodt JI, Filiz V, et al. Protein separation performance of self-assembled block copolymer membranes . Biotec*hnol Bioeng.* 2014;4(20):10252–10260. doi: 10.1039/c3ra47306f
4. Rajasekar S, Lin DSY, Abdul L, et al. IFlowPlate - A customized 384-well plate for the culture of perusable vascularized colon organoids. *Advanced Materials,* 2020. doi: 10.1002/adma.202002974
5. Sartorius. *Operating Instructions: Sartobind 96-well plates.* 2018. Accessed April 4, 2022.

https://www.sartorius.com/download/984502/manual-sartobind-96-well-plates-sl-6180-e-pdf-10450--data.pdf

6. MilliporeSigma. *Natrix Q Chromatography Membrane: Comparison of process-related impurity clearance for membrane adsorbers and resin-based chromatography technologies.* 2020. Accessed April 4, 2022. https://www.sigmaaldrich.com/CA/en/technical-documents/technical-article/pharmaceutical-and-biopharmaceutical-manufacturing/monoclonal-antibody-manufacturing/membrane-chromatography-vs-resin-chromatography

612a The iSKID Platform: At the Crossroads of Technical Development and Manufacturing Readiness

Jeff Salm

Integrated and continuous bioprocessing (ICB) has been a trendy topic in the biotechnology community for over a decade. Numerous presentations and publications have disclosed conceptual process designs, introduced new/enabling technologies, and advocated the benefits of ICB over currently accepted batch strategies. Some of these efforts have progressed to full, lab scale demonstrations producing data showing appropriate product quality and increased overall productivity further justifying the excitement and focus on ICB. Despite these numerous proof-of-concept ICB examples, relatively few have reported on advancing to implementation. Reasons for this disconnect might include capital limitations, implementation strategy, disruption to current paradigms, or perceived risk.

This talk will describe Pfizer's journey with the iSKID, an ICB platform that is currently being implemented for GMP manufacturing. Process design choices that helped enable implementation will be presented along with challenges encountered during the first at scale run. A high-level project plan for the first clinical molecule will be shown along with project mitigation strategies for implementation of this new technology. Product quality and process performance results from the at scale engineering run will be shown.

This talk will also look beyond the initial implementation of the iSKID to late-stage process development. Potential process control strategies that leverage multivariant models of the time dependent data series will be proposed. Finally, the value of the iSKID and ICB processes in general will be considered in the context of Pfizer's diverse portfolio and broader manufacturing network.

612b Continuous Purification of a Viral Vaccine Using Aqueous Two-Phase Extraction and Tangential Flow Filtration

Natalie Nold, Sheridan Waldack, Grace James, Trisha Colling, Seth Kriz, Caryn Heldt

Quick and reliable production of vaccines can have life-saving implications for patients. Continuous manufacturing, which is beginning to gain favor in the pharmaceutical industry, propels this quick and reliable production by enabling high product throughput for a smaller manufacturing footprint, reducing contamination risk during the production process, and increasing consistency of the product. However, traditional purification steps such as bind-and-elute affinity chromatography are difficult to adapt for continuous processing.

We have constructed a continuously-processing purification system for virus-based vaccines which leverages tangential flow filtration (TFF) for concentration and formulation as well as multiple aqueous two-phase extractions (ATPE) for purification. ATPE provides a more cost-effective and environmentally-friendly purification alternative to liquid chromatography, while being highly compatible with continuous manufacturing. ATPE is here composed of semi-miscible polyethylene glycol (PEG) polymer and citrate salt solutions. By leveraging hydrophobic and electrostatic interactions between the viral product and the ATPE components, the partitioning behavior of the viral product can be adjusted. The process begins by concentrating clarified cell culture material with a single-pass TFF. Next, during the first stage of ATPE, salting-out by high citrate concentrations and the amphiphilicity of the PEG-rich phase drive the virus into the PEG-rich phase, while the cell culture impurities remain in the citrate phase. However, the PEG solution is highly viscous so the product must be back-extracted into citrate solution to allow for further processing. Therefore, the virus-laden PEG is harvested for a second stage of ATPE, where low component concentrations coax the virus into a fresh citrate solution. From there, the purified product is formulated using a second single-pass TFF.

Using porcine parvovirus as a vaccine model, continuous recoveries so far correspond with batch experiments. Recoveries may further be improved by using an osmolyte additive to increase virus recovery during forward extraction. At this scale, from start to finish, the purification process is estimated to process 30 mL of concentrated PPV feedstock per hour. In future development, the concentration factor, fouling behavior, and a non-disruptive cleaning method will be determined for the TFFs. An interface sensing mechanism will also be developed to provide feedback for system control. This continuous purification framework presents a strategy for more efficient vaccine manufacturing at a lower cost compared to traditional methods, which could reduce cost and increase access to patients in need of therapies.

612c A Tubular Membrane-in-a-Shell Microextractor for Continuous Separation of Antibiotic Precursors

Michal Pribyl, Lukas Sauer, Zdenek Slouka

Continuous separation and purification of fine chemicals represent a challenging task of chemical and pharmaceutical engineering. Current slug-flow and parallel-flow microfluidic extractors suffer from several disadvantages such as the instability of the liquid-liquid interface or difficult realization of the counter-current arrangement. Hence, we designed and constructed a robust microfluidic extractor with a cylindrical membrane allowing for the integration with other unit operations to provide a continuous steady-state production. The microextractor was tested for the separation of 7-aminodeacetoxycephalosporanic acid (7-ADCA), an important precursor of β-lactam antibiotics. Aqueous two-phase system (ATPS) and a system of identical aqueous phases were used in a series of extraction experiments. We found that the counter-current arrangement is able to provide almost 100 % yield of 7-ADCA. In addition, we observed that the presented tube-in-a-shell microextractor does not require phase separation at the output and guarantees regular and reproducible flow along both sides of the cylindrical membrane. An enhanced enrichment of 7-ADCA expected in ATPS due to favorable partitioning was only reached at longer residence times. Lower mass transport resistances occurring in low-viscous aqueous phases turned out to have a primary effect on the extraction characteristics. A detailed analysis of transport resistances in the microextractor revealed that the contribution of the cylindrical membrane to the overall mass transport resistance is low. This important finding indicates that microcontactors with thin integrated membranes represent a promising option for micro- or milli-plants when fine chemicals would be continuously produced.

References

- Sauer L., Kralik D., Izak P., Slouka Z., Pribyl M.: Effects of aqueous systems and stabilization membranes on the separation of an antibiotic precursor in a microextractor, *Separation and Purification Technology* **2022**, *in press*

Acknowledgement: The authors thank for the support by the grant of the Czech Science Foundation (grant no. 20-09980S).

612d Effect of Membrane Module Geometry on the Critical Flux for IgG Precipitates

Mirko Minervini, Andrew Zydney

Recent increases in monoclonal antibody titer from Chinese Hamster Ovary cell culture has led to renewed interest in precipitation for the initial capture / purification of these high-value proteins. In addition, it is possible to implement precipitation (in a tubular flow reactor) with tangential flow filtration to develop a fully continuous low-cost capture step that could be directly integrated into a continuous biomanufacturing process. The membrane units used for dewatering and washing the precipitated protein must be able to operate for extended times without fouling. In this work, we examined the effect of the membrane module geometry on the critical flux and fouling mechanisms using human serum Immunoglobulin G (IgG) as a model protein. IgG was precipitated by addition of 10 mM zinc chloride and 7 wt % polyethylene glycol in a tubular precipitation reactor. The critical flux was then evaluated using flux stepping experiments with both flat sheet membranes in open channel cassettes and hollow fiber membrane modules with different inner diameter. The hollow fiber modules had relatively low critical flux due to clogging of the fibers at the module inlet, with this behavior confirmed using both SEM imaging and hydraulic permeability data for the fouled modules. In contrast, the open channel cassettes showed much less fouling, with no evidence of channel clogging. Instead, the fouling occurred on the membrane surface leading to a reduction in the hydraulic permeability. The open channel cassettes could be operated continuously for at least 24 hours by using a filtrate flux that was about 20% below the experimentally determined critical flux, without any evidence of significant fouling. These results provide important insights into the origin of the fouling behavior during tangential flow filtration of precipitated proteins and the appropriate design of membrane processes for continuous capture of monoclonal antibody products.

612e UV-Vis Spectroscopy-Based PLS Inferential Sensor for Detection of Host Cell Proteins in Concentrated Antibody Solution

Ian Gough, Claire Velikonja, Brandon Corbett, Thomas Kruse, David Latulippe, Prashant Mhaskar

Continuous in-line process analytical technology (PAT) provides the information necessary for model based controllers such as model predictive control (MPC) to enable optimizing process operation and overcome process disturbances utilizing feedback from process measurements. Where direct measurements are unavailable, inferential sensors, also known as soft sensors, such as those based on partial least squares (PLS) models using ultraviolet-visible (UV-Vis) light spectroscopy can determine the concentration of individual protein species within a mixture as the spectral fingerprint of each protein is unique to its amino acid composition. While previous studies have demonstrated the advantages of utilizing inferential sensors at the monoclonal antibody (mAb) affinity chromatography step, their designed experiments were limited to simple mixtures of a few model proteins rather than complex mixtures of real host cell proteins (HCPs) and they did not explore concentration ranges relevant to the subsequent ion-exchange polishing step.

Here we report the performance of PLS models created to estimate the concentration of HCP impurities down to 1 mg/L in the presence of concentrated immunoglobulin G (IgG, up to 25 g/L) using UV-Vis spectroscopy measurements and a D-optimal experimental design. The development and implementation of the PLS model resulted in the concentration of HCP being estimated with a root mean squared error (RMSE) of 6.2 mg/L and the concentration of IgG estimated with an RMSE under 1 g/L. A wavelength-selection algorithm using PLS models with up to five components showed the most informative wavelengths for estimating the HCP concentration were in the range of 600 – 700 nm, 800 – 900 nm and 230 – 300 nm. Notably, by tracking the RMSE throughout, the PLS model accuracy towards HCP estimations was observed to improve, where an initial subset of approximately 40 wavelengths is initially added as model predictors, and then adding wavelengths beyond this subset was found to increase the RMSE. The greatest improvement in RMSE was seen upon adding a third wavelength where the RMSE decreased from approximately 12 mg/L to 8 mg/L. We show that while UV-Vis-based inferential sensors do not meet the performance of enzyme-linked immunosorbent assays (ELISA) or mass spectroscopy (MS) for estimating trace levels of HCP impurities, there is potential for inferential sensors being useful prior to the final polishing step to detect process disturbance and enable feedforward MPC systems.

612f Continuous Downstream Process of Antibody Developed Based on the Process Analysis and Understanding

Fuminori Konoike, Noriko Yoshimoto, Shuichi Yamamoto

Downstream process (DSP) of monoclonal antibody (mAb) includes several batch chromatography steps. As the efficiency of DSP is low, various strategies have been developed to improve the efficiency such as continuous operation. In this paper, we will present a continuous DSP process of mAb developed by the MAB consortium*[1].

A typical platform DSP includes three chromatography steps. For the first "capture" step, protein A chromatography is normally used. A multi-column operation called "periodic counter-current chromatography (PCCC)" can be employed as a continuous operation for the capture step. The efficiency of PCCC is strongly dependent on the operating conditions, which should be determined properly [2, 3]. We decided to use a PCCC system with two-protein A columns.

The recovered pool of the capture step should be kept at low pH for a specified time for virus inactivation (VI). Although this step can be carried out continuously by using a flow reactor, our analysis showed that the scale-down is difficult. So we decided to use a batch reactor, which can work also as a surge tank.

The polishing step usually includes two chromatography columns of different separation modes. We decided to use a pseudo continuous operation method known as flow-through chromatography (FTC) [4]. In order to eliminate the buffer exchange between two FTC operations, the same buffer solution was employed so that the two columns (mixed mode and cation exchange) can be connected in series. Virus filter (VF) was also connected at the exit of the second column.

After 24-h operation of PCCC, the recovered pool was transferred to the batch VI process, which performed one hour incubation at an assigned low pH. The solution pH and the conductivity were then adjusted for the polishing FTC step. Then, the FTC step started. PCCC and FTC were operated in parallel. After each operation for 24 hours, the lines of PCCC setup were cleaned, and the columns and/or the VF were replaced for the polishing step.

The operating conditions including the mobile phase properties for PCCC and 2-column connected FTC processes were determined based on the experimental data. The continuous operations for five days were carried out with different mAb concentrations (1–3.2 g/L), feed flow rates (0.5–10 L/d) and column volumes (1–100 mL). High yield and purity were accomplished for all runs ; yield >80%, purity determined by SEC > 95% [1].

*Manufacturing Technology Association of Biologics http://cho-mab.or.jp/english/

References

[1] Konoike, F., Taniguchi, M., Yamamoto, S. Continuous downstream process of antibody - Process development, optimization and verification (in Japanese). 52th SCEJ Autumn Meeting, Paper No. LF120, 2021.

[2] Chen, C-S., Yoshimoto, N., Yamamoto, S. Prediction of the performance of capture chromatography processes of proteins and its application to the repeated cyclic operation optimization. J. Chem. Eng. Jpn., 53, 689-697 (2020)

[3] Chen，C-S., Konoike, F., Yoshimoto, N., Yamamoto, S. A regressive approach to the design of continuous capture process with multi-column chromatography for monoclonal antibodies J. Chromatogr. A 1658, 46260 (2021)

[4] Hasegawa. S., Chen，C-S., Yoshimoto, N., Yamamoto, S. Optimization of flow-through chromatography of proteins. J.Chem.Eng. Jpn., 53, 214-221(2020)

522b Evaluation of Preferred Binding Orientations of Protein Therapeutics in Multimodal Chromatographic Systems Using Protein Fingerprinting and Dewetting Simulations

Steven Cramer

The identification of preferred binding regions and specific residues on proteins that interact with ligands and ligand coated surfaces is important for understanding the molecular basis for the purification of biological products from product related impurities. In this presentation we will discuss a variety of biophysical and simulation tools that can provide insights into these interactions in multimodal chromatographic systems. NMR with labelled proteins and ligand coated nanoparticles is employed to identify preferred binding domains for the Fc domain of antibodies and to examine the impact of multimodal ligand chemistry and ligand densities on this interaction. Umbrella sampling molecular simulations are then used to provide information on the contributions of ligand clustering to these interactions. High throughput protein surface footprinting using covalent labeling followed by LC/MS analysis is then employed to identify preferred binding domains for a relatively large number of model proteins as well as several new classes of protein biotherapeutics in multimodal systems as a function of pH. Finally, this data set is examined using a variety of simulation tools including dewetting calculations of protein and chromatographic surfaces to shed light on the nature of selectivity in these complex bioseparation systems.

522c Electrochemical Separations for Bioprocessing Applications.

Sandra Kentish, Qiuyue Wang, Sahar Talebi, George Chen

Electro-driven membrane processes have gained an important place in the field of food engineering. In these systems, there are many charged organic molecules, such as proteins, polypeptides, chitosan oligomers and organic acids that need to be isolated.

In this presentation, an overview will be provided of a number of such electrochemical separation processes. This will include the use of electrodialysis for the removal of lactic acid from dairy whey. Capacitive deionisation will be explored as a contrast to this approach. Electrodialysis can also be combined with bipolar membranes to purify organic acids from fermentation broths and the use of monovalent selective membranes in this application will be described. Further, electrodialysis can be combined with filtration membranes to separate larger molecules such as immunoglobulins and lactoferrin.

The presentation will also give an overview of the mathematical models being developed to better understand the efficiency of these processes. In our own work, we have applied the Manning Model to explain the transfer behaviour of organic acid ions in anion-exchange membranes with some success. While this model was able to adequately fit experimental data for the lactate anion, it was unable to properly represent the acetate anion.

522d Developing an Electrospun Zwitterionic Membrane to Treat Produced Water Via Membrane Distillation

Jorge Almodovar

Membrane distillation is an emerging technology for treating oily wastewaters especially from oil and gas manufacturing operations. Here we focus on the development of fouling resistant membranes for treating hydraulic fracturing flow back and produced water. These wastewaters are highly fouling as they contain not only dissolved salts but also dissolved organic compounds and surfactants. We have developed a bilayer electrospun membrane that consists of an omniphobic bottom layer to prevent low-surface tension solution wicking and an under-water oleophobic antifouling surface to suppress foulant deposition. Silica nanoparticles were grown on the bottom layer through electrochemical interaction followed by coating with a fluorine monomer on the fiber surface. The zwitterionic co-polymer poly (glycidyl methacrylate-sulfobetaine methacrylate) (GS) was grafted using self-assembly between hydroxy group on the surface generated by alkaline treatment and anchor segment epoxy group on GS. The novel electrospun membrane have been challenged with industrial real hydraulic fracturing produce water. Preliminary results shows that fouling in our bilayered membrane is significantly reduced. This bilayer electrospun membrane may enable the commercialization of membrane distillation for treating highly impaired hydraulic fracture and flow back wastewaters.

522e Virus Clearance during Downstream Purification of Protein Therapeutics

Xianghong Qian, Solomon Isu, Ranil Wickramasinghe

Virus clearance is an essential component of the downstream purification of biopharmaceutical products. Virus filtration flux and throughput are largely product, solution condition and filter specific due to (1) the intrinsic properties of the product including its charge, hydrophobicity and tendency for aggregation at specific solution conditions; (2) product-related foulants such as fragments and variants; (3) impurities such as host cell proteins (HCPs), DNA and spiked virus particles. Our virus filtration results indicate that insolvable aggregates are not likely the major cause for the fouling of virus filters at a product titer of ~10 g/L. Instead, product variants with higher molecular weight (MW) and stronger hydrophobicity are likely the dominant foulants during virus filtration. These variants commonly exist in protein feed streams during cell culture production. Analysis of the feed, filtrate and elution fractions indicate that these higher MW variants tends to form soluble aggregates which cause significant decay of the filtration flux.

522f High Performance Countercurrent Membrane Systems for Bioprocessing

Andrew Zydney

Although membrane filtration is used throughout the downstream purification process for production of high value biopharmaceuticals, there are almost no applications of countercurrent membrane systems based on diffusive transport. This talk will examine two exciting applications of countercurrent membrane systems in bioprocessing: countercurrent dialysis for buffer exchange and High Performance Countercurrent Membrane Purification (HPCMP) for protein purification. Both processes use low-cost commercially-available hollow fiber membranes originally developed for hemodialysis. Countercurrent dialysis experiments demonstrated 99.9% impurity removal with less buffer required than for traditional batch diafiltration processes. HPCMP can effectively remove small host cell proteins, with the diffusive transport providing higher selectivity than pressure-driven membrane filtration while the countercurrent contacting reduces buffer requirements below those for traditional chromatographic processes. In addition, both systems can provide truly continuous operation, making them highly attractive for implementation as part of integrated systems for continuous biomanufacturing. These results clearly demonstrate the potential of developing countercurrent membrane systems for enhanced bioprocessing.

522g Multifunctional to Bioinspired Nanostructured Membranes

Dibakar Bhattacharyya, Rollie Mills, Francisco Leniz, Lindell Ormsbee

The synthesis and applications of responsive, multifunctional materials and membranes for environmental and bio-related use requires a high level of control of both the characteristics of the base polymeric or inorganic support layer, as well as, its corresponding surface properties. The special features for membrane processes that make them attractive for industrial applications are their compactness, ease of fabrication, operation, and modular design. Although membrane processes such as, Reverse Osmosis, Nanofiltration, Ultrafiltration, and Microfiltration have provided many successful applications ranging from high quality water production to material recovery, but incorporation of nanostructured aspects with tunable properties (learning from life sciences, such as enzymes, and porins)) in membranes has added immense value in the area of separations, reactions, water applications, and green synthesis. The use of macromolecules, such as, poly-acrylic acid (PAA), poly-glutamic acid (PLGA) provide pH responsive behavior through helix-coil transitions, whereas poly-N-isopropylacrylamide (pNIPAAm) provides temperature responsive behavior. Further, one can synthesize catalytic metal nanoparticles in the functionalized membrane for water remediation applications. The presentation will include bio-inspired functionalized membranes with tunable and responsive properties, layer-by layer assembly for enzyme catalysis, synthesis of metal nanoparticles and metal films in membranes, and industry related applications. Special acknowledgement goes to his recent students (A. Aher, H. Wan, A. Saad, S. Islam, M. Detisch). In addition to various industrial collaborators, NIH-NIEHS-SRP and NSF-EAGER, and NSF-RAPID program provided high level financial support.

676a Development and Application of a Quick Image Analysis Calibration Procedure for in-Situ Crystal Size Measurement

Wei-Lee Wu, Madeline Mills, Erik Larmore, Ulrich Schacht, Vaso Vlachos, Zoltan Nagy

Crystallization is a purity and particle control unit operation commonly used in industries such as pharmaceuticals, agrochemicals, and energetics. Often, the crystal mean size, morphology, and distribution can impact the critical quality attributes of the final product. Traditionally, offline measurements via image analysis or laser diffraction methods are used to analyze the crystal product. It is often beneficial to extract crystal samples intermittently during the crystallization process to better understand the complex dynamics in batch and continuous crystallization. However, frequent invasive product sampling can lead to undesired disturbances in the process dynamics. Non-invasive solutions to address data scarcity in particle size is to obtain estimated crystal sizes with in-situ measurement tools such as chord length measurements or in-situ imaging (Acevedo et al., 2021 and Larsen et al., 2006). These particle images of various solid density, opaqueness, size, and morphology are often analyzed with generic algorithms and/or complex mathematical models which can have discrepancies with offline measurement data or lead to long computational time.

To obtain a fast and accurate in-situ image analysis measurement of particle size for a high aspect ratio nature crystallization system, a systematic off-line image analysis calibration methodology was performed to model offline Malvern Morphologi image analysis results. Grey-scale image analysis algorithms via ImageJ were calibrated to extract size distribution data from in-situ images of varying size and solid density. Different sets of algorithms with varying parameters in the image analysis methods were tested and optimized by minimizing the size distribution error between the model and offline image analysis measurement. Trends observed in the calibrated parameters were then fitted to continuous functions varying in solid density to be able to adapt to changes in solid loading. The algorithms were then validated with a different particle data set with known solid loading for algorithm predictability. Lastly, the usability of the algorithms was tested on a dynamic data set to gauge the processing time and predict the size distribution with varying solid loading throughout the crystallization process due to dissolution, nucleation, and growth.

References:

(1) Acevedo, D.; Wu, W. L.; Yang, X.; Pavurala, N.; Mohammad, A.; O'Connor, T. F. Evaluation of Focused Beam Reflectance Measurement (FBRM) for Monitoring and Predicting the Crystal Size of Carbamazepine in Crystallization Processes. *CrystEngComm* **2021**, *23* (4), 972–985. https://doi.org/10.1039/d0ce01388a.

(2) Larsen, P. A.; Rawlings, J. B.; Ferrier, N. J. An Algorithm for Analyzing Noisy, in Situ Images of High-Aspect-Ratio Crystals to Monitor Particle Size Distribution. *Chem. Eng. Sci.* **2006**, *61* (16), 5236–5248. https://doi.org/10.1016/j.ces.2006.03.035.

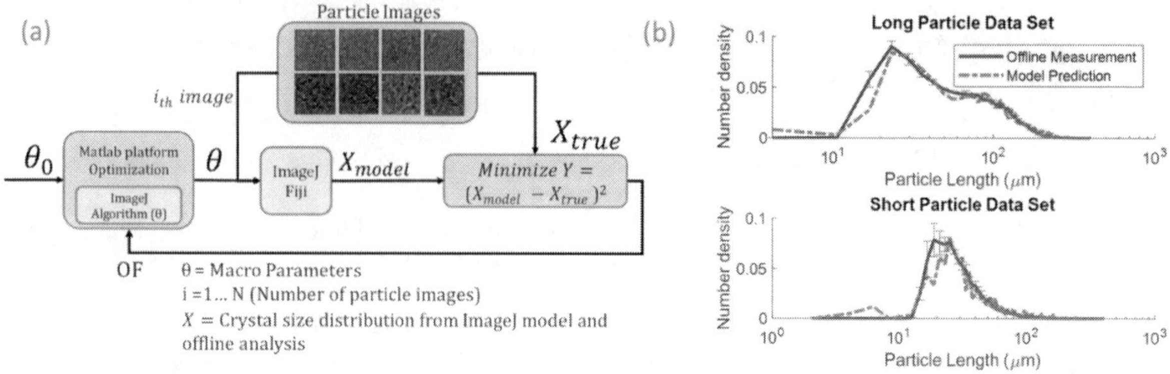

Figure 1. (a) ImageJ algorithm optimization illustrated as training loop to minimize error between model predictions and offline measurement, **(b)** Nominal distribution (red dashed) fits between two different data sets at 0.1% solid density (blue solid) compared to offline measured distribution (blue solid).

676b Characterization and Modelling Reactive Protein Crystallization: Defining the End of a Process

Isabella Jul-Jørgensen, Ryan Oliver, Anna Carnerup, Krist V. Gernaey, Christian A. Hundahl

This is a case study on how PAT's can be used for enhanced process understanding to improve a crystallization process and thus increase capacity in pharmaceutical manufacturing. The emergence and improvement of new PAT's enables insights into crystallization processes, that were not previously possible. Therefore, these technologies can be used to characterize and optimize crystallization processes that were introduced into manufacturing decades ago.

The studied crystallization process is a reactive crystallization used in the formulation of a drug product, where x% of the API is crystallized and (100-x)% of the API is dissolved. Upon formulation some of the API precipitates into amorphous particles from which the crystals grow. It had been assumed that the concentration of both API and preservatives in the dissolved phase stayed constant throughout the process. After crystallization the product is filled.

This study consists of three parts:
The first part is calibration and validation of a single in-line PAT probe to predict critical quality attributes (CQA) such as amount of API crystallinity, API and preservative concentration in solution and particle size distribution. The probe uses three different measurement techniques: Raman spectroscopy, high-dynamic range turbidity and advanced chord-length-distributions. The calibrations are performed with chemometrics with reference values from off-line HPLC for API and preservative concentration, on-line small-angle x-ray scattering for degree of crystallinity and automated microscopy for particle size distribution. Segregation of the crystal size distribution from the particle size distribution is difficult, as amorphous particles and small crystals can be difficult to distinguish due to similar aspect ratios and size. However, modelling using the crystallinity amount from the orthogonal experimental techniques will improve this segregation.

The second part explore the crystallization dynamics during the current formulation process. The data showed that the concentration of preservatives in the dissolved phase increased until the formation of the first crystals. Furthermore, the API concentration in the dissolved phase also increased during crystallization then reached a plateau after the crystallization start, but before the end of the conversion of amorphous to crystals. The particle size distribution shows the different phases in the crystallization process, but also that the crystals keep growing beyond the end of conversion of amorphous to crystals. This is most likely due to Ostwald Ripening. Furthermore, cryogenic transmission electron microscopy images were taken at critical timepoints for qualitative assessment of the particles at very high resolution. The data indicates that different subprocesses within the process have different endpoints and therefore raises the question of whether the process could be defined as finished earlier without affecting the CQA's. In extension of this another interesting question arises: "what happens if the product is filled if some processes, but not all, are finished?".

Thus, the third part investigates the effect of the filling process on the CQA's, as the product requires constant recirculation, which exerts a large amount of shear stress on the product. Shear stress can affect the crystallization process, and future experiments will investigate if the CQA's are affected when the drug product is filled even though one or several sub-processes have not ended yet. Preliminary results indicate that the amount of API crystallinity is increased while decreasing API concentration in the dissolved phase when adding shear stress. Furthermore, the aspect ratio of the particles increases as well. However, the particle size distribution does not appear to be affected if the shear stress is only applied before the onset of crystallization growth.

Furthermore, hybrid predictive modelling of the processes will be used to understand differences in nucleation, growth, agglomeration and breakage phenomena when applying shear stress during the different phases of crystallization. Also, the model can potentially be used for in silico development of crystallization process control strategies during manufacturing.

676c Digital Design Framework for the Continuous Crystallization of Diphenhydramine

Yash Barhate, Hemalatha Kilari, Wei-Lee Wu, Jaron Mackey, Haotang Li, Zoltan Nagy

In the pharmaceutical industry, the quality of the drug produced is a top priority because of the direct impact on human lives. Regulatory guidance for manufacturing processes has been provided by the U.S. Food and Drug Administration (FDA) to make sure that medications consistently meet quality standards. A key step in the purification of active pharmaceutical ingredients (APIs) is crystallization; a process that is stochastic and difficult to control. Through the use of process analytical technology (PAT) tools and modeling approaches, we aim to develop strategies for designing robust crystallization processes.

In this work, we demonstrate the application of combined model-free quality-by-control (mfQbC) and a model-based digital design approach for the design and optimization of crystallization of Diphenhydramine (DPH). DPH is an anti-histamine commonly found in Benadryl. While the continuous synthesis of DPH has been studied by multiple research groups in the past ten years, research literature lacks information on continuous crystallization development with modeling for the compound.[1–4] The PAT-based mfQbC framework implementation reduces both the number of experiments and time needed to design crystallization processes, but often results in sub-optimal crystallization design. However, we can augment the mfQbC results to guide model-based digital design, by identifying key mechanisms to be included in the model and providing high quality data for parameter estimation, which would reduce the overall crystallization development time and lead to an optimal design.[5]

Under the mfQBC approach, initial solvent screening experiments were conducted to guide solvent selection and gain information about the meta-stable zone width (MSZW). Preliminary batch experiments provided valuable insights about the initial parameter estimation and crystallization mechanisms for model selection and discrimination. Continuous crystallization process alternatives were screened and evaluated according to the suspension ability and degree of fouling of each configuration, in addition to product size and variability. The crystallizer configurations evaluated in this study include mixed suspension-mixed product removal (MSMPR) and continuous oscillatory baffled crystallizer (COBC). Initial parameter estimation estimates were validated in the selected configuration, and improved estimates were used for model development. Under mbQbC framework, a population balance-based model is developed for selected configuration and is used to run the simulation studies. With the selection of best crystallization system for the compound and model development in place, robust-optimization studies have been performed to identify optimal trajectories and design parameters in the design space of the system. The optimal conditions were experimentally validated to complete the cycle of designing a robust continuous crystallization system for Diphenhydramine.

Funding for this paper was made possible by the United States Food and Drug Administration.

References:

1. Snead, D. R. andamp; Jamison, T. F. End-to-end continuous flow synthesis and purification of diphenhydramine hydrochloride featuring atom economy, in-line separation, and flow of molten ammonium salts. *Chemical Science* **4**, 2822–2827 (2013).

2. Loren, B. P. *et al.* Mass spectrometric directed system for the continuous-flow synthesis and purification of diphenhydramine. *Chemical Science* **8**, 4363–4370 (2017).

3. Adamo, A. *et al.* On-demand continuous-flow production of pharmaceuticals in a compact, reconfigurable system. *Science* **352**, 61–67 (2016).

4. Collins, N. *et al.* Fully Automated Chemical Synthesis: Toward the Universal Synthesizer. *Organic Process Research and Development* **24**, 2064–2077 (2020).

5. Szilagyi, B., Eren, A., Quon, J. L., Papageorgiou, C. D. andamp; Nagy, Z. K. Application of Model-Free and Model-Based Quality-by-Control (QbC) for the Efficient Design of Pharmaceutical Crystallization Processes. *Crystal Growth and Design* **20**, 3979–3996 (2020).

676d Process Analytical Technology Controlled Crystallization of Energetic Materials

Mouhcine Doukkali, Eric Gauthier

The high explosive crystal quality, size and morphology can significantly affect their initiation sensitivity. Properties such as crystal shape, size, voids and inclusions can be directed by controlling the crystallization conditions. Improved control of process parameters is required to achieve the desired crystal characteristics since the control of these properties is difficult at the production scale of high explosives and results in large particle size distribution, requiring milling to achieve different class sizes and other potential material rework. When optimizing the size and shape of the crystals during batch crystallization, it is often preferred to operate within the metastable zone to avoid undesired nucleation. In this work explosive X seeds are generated using ultrasound at different saturation temperature within the MSZ. Those seeds are controlled to follow a supersaturation profile within the MSZ to promote the growth rate and keep the rate of nucleation low. Process Analytical Technology (PAT) was used to control the crystallization process through real-time measurements using ATR-FTIR, FBRM and PVM tools.

676e Improved Particle Characterisation from in-Line PAT: Comparison of Deep Learning and White-Box Methods

Christopher Boyle, Cameron Brown, Jan Sefcik, Javier Cardona

Introduction

In-line Process Analytical Technologies (PAT) are useful for measurement of particle characteristics (e.g. particle size distribution, PSD) in a non-destructive manner and with high time resolution inaccessible with off-line techniques. These measurements can be useful for process monitoring and accurate population balance modelling.

This work is concerned with assessing the performance of using in-line image and chord length distribution (CLD) analysis for PSD measurement. Imaging is obtained through in-line microscopy (e.g. BlazeMetrics' Blaze 400, or Mettler Toledo's Particle Vision and Measurement, PVM), while CLD can be obtained via light-scattering methods (e.g. Mettler Toledo's Focused Beam Reflectance Measurement, FBRM). Sensor data is analysed through different methods to yield PSDs. Traditional white-box (*i.e.* not machine learning) analyses can involve length "tuning" steps wherein the analysis is adapted to work with the system (Cardona *et al.* 2018), or are derived mathematically and are restricted to a small domain (Agimelen *et al.* 2015). Despite these drawbacks, these techniques are demonstrably effective, once tuned or when applied to the appropriate domain. Deep learning (black box) models have potentially increased performance and flexibility over white box techniques. This is in part due to their large number of parameters and due to modern breakthroughs in model design (He *et al.* 2017). Deep learning models are widely used in image analysis, with further applicability to other domains such as CLD transformation.

Different PAT sensors perform differently under different conditions. For image-based measurement: small particles may not be measurable with great accuracy due to resolution issues, concentrated systems will impact measurement of the particles due to particle overlaps, and large particles are more likely to interact with the edge of the image field of view and therefore are less likely to be sized resulting in left-skewed PSDs. For FBRM-based measurements: translucent particles cause a phenomenon called "chord splitting" wherein a single chord is mis-registered as multiple smaller chords resulting in modes appearing at lower bins than expected on the PSD. In addition, low concentration systems (small number of particles) can result in few counts in the CLD and present as noise and thus impacting the transformation from CLD to PSD.

Methods

In-line measurements were taken with the Mettler Toledo FBRM G400 (CLD) and PVM V819 (Imaging) probes of Polystyrene Standard Spheres (PolyS), a mixture of Polystyrene Spheres and Ellipsoids (PolyE), and of Lactose particles (Figure 3, (b)-(e)). In-line-derived PSDs from a range of crystal sizes and concentrations were compared with a ground truth either from off-line microscopy (PolyE, Lactose) or from manufacturer specifications (PolyS).

Two image analysis methods were employed. The first is a traditional approach (ImagingApp (Cardona *et al.*2018)) wherein a grayscale image is converted (in a process called segmentation) to a black and white mask (black pixels representing background and white being part of a particle) by applying edge detection and binarising filters. The second image analysis is a machine learning approach. We used the Detectron 2 (Wu *et al.* 2019) framework to train a Mask-RCNN (He *et al.* 2017) image segmentation model. The model was used pre-trained on the Common Objects in COntext (COCO) dataset and fine-tuned on a dataset composed of examples of PVM images from each system (PolyS, PolyE, Lactose) annotated to identify particles. Images are segmented (Figure 1) to find particles, which are sized by fitting a rotated rectangle. Sizes are collected, histogrammed, and normalised to yield the number density particle size distribution.

Mirroring image analysis, two CLD analyses are employed: a white-box approach and a black-box approach. The white-box approach (Agimelen *et al.* 2015) relies on a probabilistic model. The probabilistic model yields a transformation matrix for the forward problem of converting a PSD to its CLD. The inverse problem is complicated as the matrix is singular. An optimisation scheme is employed to search for the PSD most likely to correspond to the given CLD. A machine learning approach is also employed, using a model based on the popular ResNet (He *et al.* 2015) model used widely in image analysis. This model is trained on a subset of the measured CLD (PolyS, PolyE, Lactose) linked with offline-microscopy-derived PSDs (Morphologi G3). An example of applying these approaches is shown on Figure 2.

This gives four tools for obtaining PSDs: two analyses for each of two sensors, a sensor and analysis combination constitutes a "tool". In order to choose between tools, they need to be evaluated.

Tool evaluation is performed by comparing PSDs obtained using each tool with ground truth PSDs obtained from offline microscopy (for PolyE, Lactose) or manufacturer specification (for PolyS). Comparison is made through use of two metrics (distance functions): Root Mean Square Error (RMSE) and Integral Absolute Error of Cumulative Density Function (IAE of CDF). The RMSE metric (square root of the mean of the square difference in two PSDs) is well known and commonly used, however it saturates for large discrepancies in PSDs making it only suitable where the two inputs have some crossover. The second metric in use is the CDF of IAE, related to the Wasserstein Distance, and defined as the integral of the absolute error between the cumulative sum of the input PSDs. IAE of CDF gives an indication of how different two PSDs are but, importantly, the measure does not saturate.

Results and Discussion

Statistical (white-box) CLD analysis is found to be sensitive to non-uniform particle shape distributions, and to the artefacts introduced by the sensor itself. This is due to the physical basis of the transformation model and complicated by the non-trivial optimisation problem of finding the likely solution. In general, this method performs best with smaller particle sizes, but does not have an apparent trend with respect to concentration. The size dependency is understandable as our test materials for this method (polystyrene particles) are translucent and are therefore prone to chord splitting. The shape of the distribution should be preserved even at lower concentrations and the shape is important in the transformation (more so than raw count values). Low counts due to low concentration could yield noisier CLD, but this was not seen in our data (possibly due to large acquisition time making up for the low number of particles).

ML (black-box) CLD analysis yields improved results: output is very close to the ground truth (manufacturer specification, offline microscopy, or offline laser diffraction) but is limited by the training data: a much more varied training set is required for broader applicability.

Both white- and black-box approaches for image analysis show a similar trend in error with size: the largest (>30% of frame) and smallest (<10% of frame) particle sizes have higher error. This effect is shown for the ML approach on Figure 3 (a). This is due to resolution impacting measurement of small particles, while larger particles are less likely to be sampled at all. However, the white-box approach shows a trend in concentration: higher concentrations give higher error. This is not seen in the machine learning approach as the Mask-RCNN model is able to segment particles even in overlap (at least the particle in front, if not the background particle).

Conclusions

In this work we demonstrate a process for evaluation of in-line PAT through two in-line sensors, PVM and FBRM, each analysed using two techniques (a deterministic/white-box approach and a machine-learning/black box approach). These evaluations show trends for error with size and concentration. Image analysis has higher error for the largest particle sizes (> 30% of the frame) and smallest particle sizes (< 10% of the frame), for the PVM V819 sensor employed here. White-box CLD to PSD transformation has a strong reliance on size (directly proportional), but for our data shows no concentration dependence. This could inform the choice of using image analysis for middling sizes, and CLD transformation for small sizes. Larger particles may require the use of a larger imaging sensor, lesser magnification, or another in-line measurement technique not considered here.

Acknowledgements

This work was funded jointly by AstraZeneca, Bayer, Chiesi, Eli Lilly, GlaxoSmithKline, Novartis, Pfizer, Roche, Takeda, and UCB and was carried out within the CMAC Future Manufacturing Research Hub (Grant ref: EP/P006965/1). We are grateful to Francesca Perciballi (AstraZeneca), Neda Nazemifard (Takeda), Vaclav Svoboda (Pfizer), Chris Burcham (Eli Lilly), and Jan-Sebastiaan Uyttersprot (UCB) for project guidance.

References

Agimelen, O. S., Hamilton, P., Haley, I., Nordon, A., Vasile, M., Sefcik, J., and Mulholland, A. (2015) Estimate of particle size distribution and aspect ratio of non-spherical particles from chord length distribution. Chemical Engineering Science, 123: 629-640.

Cardona, J., Ferreira, C., McGinty, J., Hamilton, A., Agimelen, O. S., Cleary, A., Atkinson, R., Michie, C., Marshall, S., Chen, Y.-C., Sefcik, J., Andonovic, I., and Tachtatzis, C. (2018) Image analysis framework with focus evaluation for in-situ characterisation of particle size and shape attributes. Chemical Engineering Science, 191: 208-231.

He, K., Xiangyu Z., Shaoqing R., Jian S. (2015) Deep Residual Learning for Image Recognition. https://doi.org/10.48550/arXiv.1512.03385

He, K., Gkioxari, G., Dollár, P., Girshick, R. (2017) Mask R-CNN. IEEE International Conference on Computer Vision (ICCV). 2980-2988

Wu, Y., Kirillov, A., Massa, F., Lo, W.-Y., and Girshick, R. (2019) Detectron 2. https://github.com/facebookresearch/detectron2

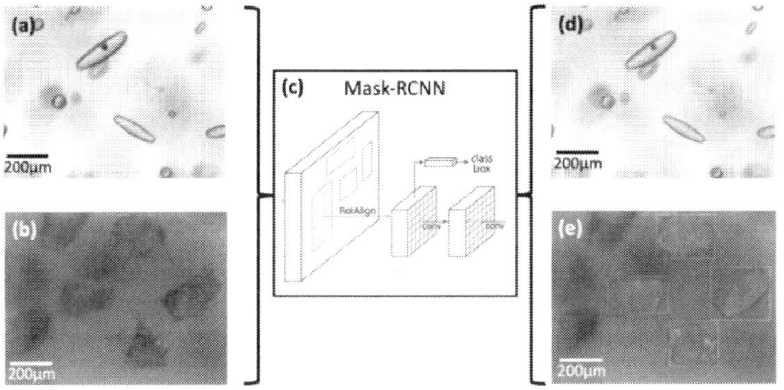

Figure 1. Example segmentation performed by Mask-RCNN. Example PVM images for (a) polystyrene spheres and ellipsoids 0-90µm at 10wt% and (b): lactose particles 212-250µm at 10wt%. (c) schematic of Mask-RCNN (He et al 2017). (d) and (e) results of applying Mask-RCNN to images (a) and (b) respectively. Mask-RCNN segmentation is overlaid on the original PVM.

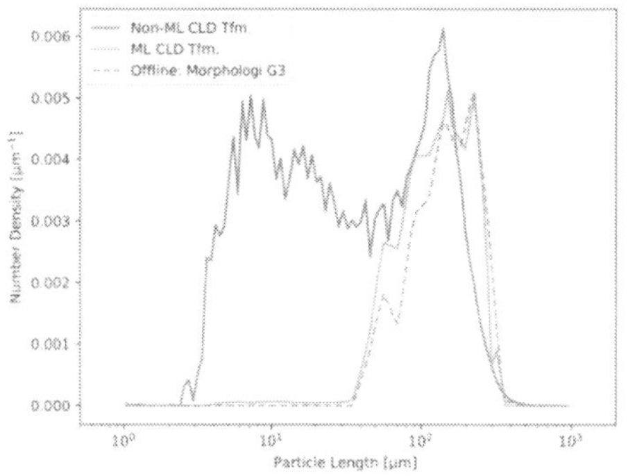

Figure 2. Example of results of CLD to PSD transformation for system of polystyrene spheres and ellipsoids sieved to 250-355µm at 10wt% using Agimelen's white box analysis (blue line), using our machine learning analysis (orange line), and compared to offline microscopy results (green dashed line).

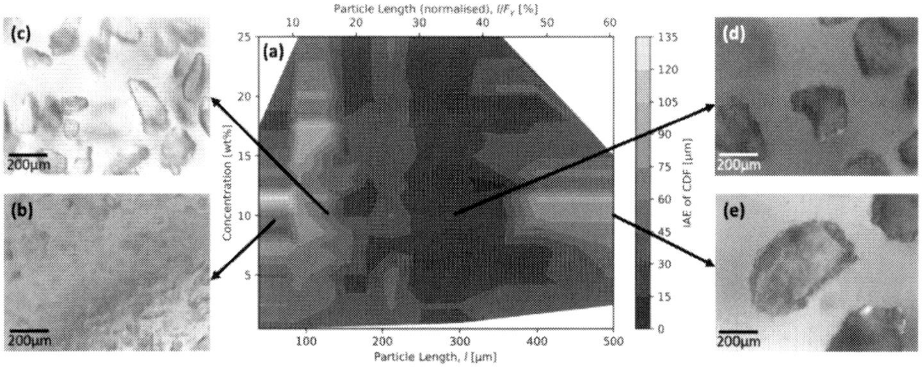

Figure 3. (a) Contour plot showing error (IAE of CDF) in PSDs obtained via Mask-RCNN image analysis of PVM images of lactose particles, plotted against particle length and concentration. Along the top, particle length is expressed as a percentage of the field of view height (short edge). Error value as in colour bar. White areas on plot have no data. (b)-(e) examples of PVM images of lactose particles with their position on (a) indicated by arrows. (b) 38-75µm, 0.5wt%; (c) 125-180µm, 10wt%; (d) 250-300µm, 10wt%; (e) 400-500µm, 10wt%.

674a Safer Solvents for Active Pharmaceutical Ingredients Purification Using Column Chromatography

Christian Ayafor, Toren Burton, Gregory Morose, Hsi-Wu Wong

Dichloromethane (DCM) is a widely used solvent in the polymer, chemical, and pharmaceutical sectors for the separation of chemicals using column chromatography. The annual production of DCM in the United States in 2012 was 260 million pounds. The United States remains highly dependent on this toxic chemical since it is extensively used as an extractant for the decaffeination of coffee in the food industry, as a coating remover, and other applications. In the pharmaceutical sector, DCM has often been utilized in chromatographic applications, particularly liquid chromatography, for the synthesis of active pharmaceutical ingredients (API). For instance, approximately 554,000 pounds of DCM were used by Johnson Matthey in 2019 for API purification. DCM has been linked to a variety of health problems, including cancer and the damage of the central nervous system. Furthermore, DCM is designated as a Higher Hazard Substance under the Massachusetts' Toxics Use Reduction Act, as a GreenScreen® Benchmark score of BM-1 and has a GSK rating of 4. Replacing this toxic solvent with a safer alternative(s) will provide significant environmental, health, and safety benefits. The major challenge is the performance and cost of the alternative solvents which cannot be compromised so that the businesses can stay competitive.

Replacing DCM by a safer solvent (SS) and/or a safer solvent blend (SSB) alternative has remained a hurdle to researchers over the past several decades. Jessop et al. (2012) reported DCM as one of the problematic solvents which safer substitute has not yet been found. Taygerly et al. (2012) developed a solvent guide to help determine alternatives to DCM in chromatographic applications based on safety, environmental impact, toxicity, and other factors. Chen et al. (2021) developed a methodology, that utilized Hansen Solubility Parameter (HSP), dissolution power, safety, and price, to down-select solvents that were then applied to thin-layer chromatography (TLC) to identify SS and SSB that could serve as an alternative to DCM in the separation of widely used APIs (acetaminophen, aspirin, and ibuprofen) from caffeine. The goal of this study is to use the SS and SSB identified by Chen et al. to determine, through column chromatography, which of them could be safer alternatives to DCM for the separation of widely used APIs from caffeine.

During this research, a Synthware™ Chromatography Column of 13.4 mm in diameter and 203 mm in length, with Standard Taper Joint and 1000 ml reservoir was used for the laboratory-scale experiments. Silica Gel 60, 0.060-0.2mm (70-230 mesh) was used as the polar stationary phase, while the baseline solvent (DCM) and each of the SS (ethyl acetate, methyl acetate, 1, 3 dioxolane, and dimethyl adipate) as well as the SSB (dimethyl adipate and ethyl acetate, dimethyl adipate and 1, 3 dioxolane, methyl acetate and ethyl acetate) are used as the mobile phases.

The stationary phase is wet packed while the APIs including caffeine are injected into the column when dissolved into the mobile phase being used to run the column. The compounds (API + caffeine) to silica gel ratio for every experiment was 1:40. The volumes collected for the baseline solvent (DCM) was 80 ml, with 50 mg calibration standards, while those collected for the SS and SSB were 3 ml, with 1 mg calibration standards.

Initial results depict a recovery factor (RF) of 0.66 and 0.73, respectively, for the separation of aspirin and ibuprofen from caffeine with DCM. Both APIs were 100% pure while no caffeine was recovered. When separating acetaminophen from caffeine, no analyte was recovered. All experiments using DCM required large amounts of solvent.

Safer solvents depicted RFs ranging from 0.91 – 0.99 for the separation of ibuprofen from caffeine, 0.75 – 0.93 for the separation of aspirin from caffeine, and 0.2 – 0.29 for the separation of acetaminophen from caffeine. The purity of each API was 99% and higher, except for the cases when 1, 3 dioxolane was used as mobile phase where the purity was 98% and higher. Far smaller amounts of safer solvents were used for the separation of APIs from caffeine compared to the amounts used for DCM.

The RFs for safer solvent blends ranged from 0.68 – 0.89 for the separation of Ibuprofen from Caffeine, 0.63 – 0.97 for the separation of aspirin from caffeine, and 0.23 – 0.28 for the separation of acetaminophen from caffeine. The purity of each API was 99% and higher, and far smaller amounts of SSBs were used for the separation of APIs from caffeine compared to the amounts used for DCM.

The results suggest that the dependence on DCM by pharmaceutical industry can be overcome by redesigning chromatographic applications to SS and/or SSB used in this study as the mobile phase, as more APIs shall be recovered with far less solvent quantities utilized for their recovery. The importance of using these SS/SSB to replace DCM cannot be overemphasized as it will greatly improve the health and safety of employees of this sector, and considerably limit the proliferation of toxic chemicals in the environment.

674b Use of SMB Technology to Extract Oxygenates from Dilute Fermentation Broth

Deepak Sharma

The principles of process intensification if used on existing processes can lead to development and redesign of existing processes to become more sustainable. This presentation will discuss a process and an adsorbent for the separation of ethanol associated oxygenates from a dilute mixture of ethanol and associated oxygenates in water in the presence of organic compounds derived from a biofermentation process. After pretreatment, the separation is carried out in a simulated moving bed adsorption system employing a stationary phase adsorbent comprising fluorinated carbon or modified C18 silica gel selective for the adsorption of ethanol and associated oxygenates, such as 2,3-butanediol, with a mobile phase desorbent selected from the group consisting of methanol, ethanol, propanol, and methyl tertiary butyl ether.

An improvement to the above SMB process is also discussed in which increasing the number of raffinate streams alone or in combination with a hot regeneration zone within the SMB cycle can significantly reduce the capital and operating costs associated with the incorporation of the SMB process to produce ethanol and butanediol from biofermentation effluent. This process is an improvement over conventional distillation which is very energy intensive and not practical for dilute fermentation broths.

The process is useful for removing water from dilute aqueous mixtures of organic compounds comprising ethanol in dilute concentration in water and produced by fermentation, biomass extraction, biocatalytic, and enzymatic processes which are not economically recoverable by conventional distillation methods.

674c Gas-Phase Simulated Moving Bed for Methane/Nitrogen Separation Using a Commercial Activated Carbon

Rafael Dias, Alexandre Ferreira, Alírio E. Rodrigues, Ana M. Ribeiro

With the rising demand of natural gas, the exploration of unconventional methane-rich sources has been gradually increasing over the years due to the depletion of the conventional wells. These sources are often contaminated with several impurities, among which nitrogen is the most difficult to be removed, owing to the similar properties to those of methane. When its content is above 3%, its removal becomes a necessity to ensure that the calorific value of the stream is not affected by the presence of this inert gas. The main technology used for this purpose is cryogenic distillation, which has very high energy costs associated and is unsuitable for medium and small-scale purification of methane streams. When compared to this technology, adsorption-based processes are very appealing thanks to their easy scalability and lower energy requirements. However, the performance of the solid materials for CH_4/N_2 separation becomes an issue due to their low selectivity. The use of a gas-phase simulated moving bed (SMB) process overcomes this challenge since it can separate mixtures with selectivities close to unity, by introducing a desorbent species to displace the adsorbed phase, essentially transforming a difficult separation into two easier ones.

The design and development of a gas-SMB process for the separation of methane and nitrogen mixtures is the main focus of this work, using a commercial activated carbon (BPL) as the adsorbent material and two potential desorbent gases – argon and carbon dioxide. The performance of the material was evaluated by measuring the adsorption equilibrium data for the four adsorbates and the dynamic behavior of single and multicomponent adsorption through fixed-bed experiments.

The pure component isotherms of N_2, CH_4, Ar, and CO_2, presented in Figure 1, were measured at temperatures of 303, 323, and 343 K in a pressure range of 0 - 2.5 bar using a volumetric apparatus. The data was successfully regressed with the Dual-Site Langmuir (DSL) model. The results show that CO_2 has the highest affinity to the stationary phase and Ar the lowest, over the entire range of temperature and pressure studied.

Single, binary, and ternary breakthrough experiments were carried out at 303 K and 1.5 bar. A mathematical model that encompasses mass, energy, and momentum balances was implemented in the gPROMS® software and utilized to predict the observed adsorption dynamics. Analysis of the ternary fixed bed experiments (Figure 2) show that the desorbent is easily displaced by the mixture and is equally able to displace the adsorbed phase, which is an essential step for SMB operation. The simulation results are in good agreement with the experimental data.

Finally, two simulated moving bed (SMB) cycles were employed to separate an equimolar CH_4/N_2 mixture using each of the desorbent gases to evaluate the impact of the desorbent strength in the process. The experiments were carried out in open loop, with no desorbent recycle. A 2-3-2-1 four-zone system configuration was considered, with a switching time of 50 s. The SMB unit was operated at 303 K, with the backpressure, located at the end of zone IV, set at 1.5 bar. A feed composition of 50% CH_4 and 50% N_2 was used, with a feed stream of 0.190 SLPM for the argon experiment and 0.060 SLPM for the CO_2 experiment. The internal profiles for each case are presented in Figure 3. In zone II, the more adsorbed species, methane, is retained onto the material, enriching the extract stream, while the less adsorbed species, nitrogen, is carried by the eluent. In section III, the nitrogen content increases as it moves with the fluid phase to the raffinate port. The flow rate in section I is high enough to guarantee that the adsorbent is cleaned, ensuring that section IV is not contaminated after the switch. In section IV, the nitrogen content decreases, with pure desorbent being obtained at the outlet, which can be recycled back into section I when operating in closed loop. Both cycles were capable of producing a high purity methane stream (96.2% and 97.4% for the Argon and CO_2 experiment, respectively) with a high recovery (> 92%). When argon is used as the desorbent gas, the extract product stream is obtained with a productivity of 13.9 $kg \cdot m^{-3}{}_{ads} \cdot h^{-1}$.

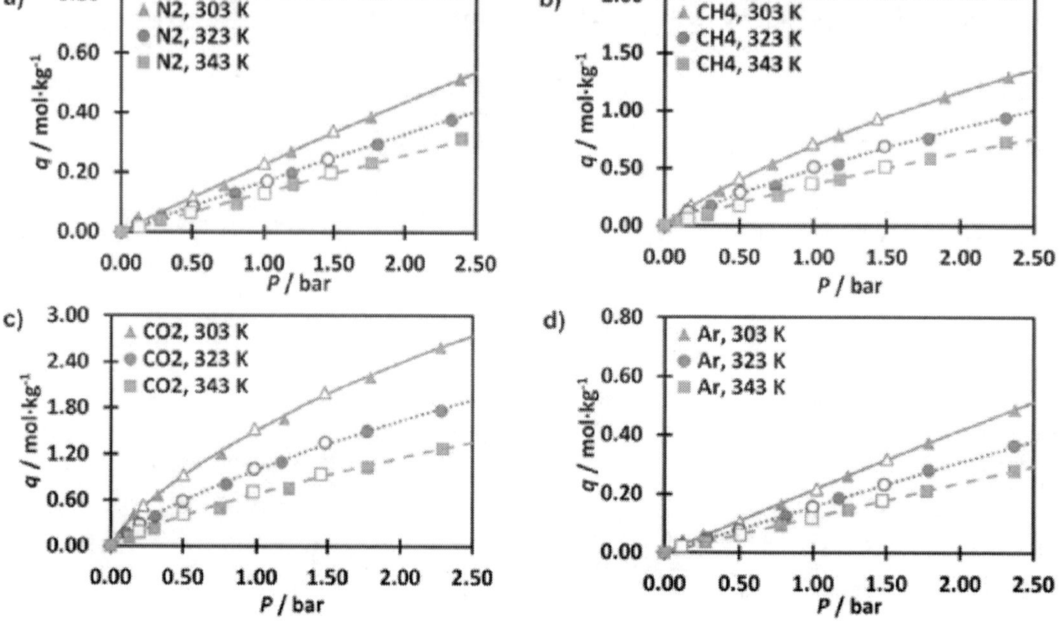

Figure 1 - Amount adsorbed of a) N_2, b) CH_4, c) CO_2 and d) Ar on BPL carbon at 303 (▲), 323 (●) and 343 K (■). Closed symbols correspond to adsorption and open symbols to desorption. Lines represent the DSL fitting.

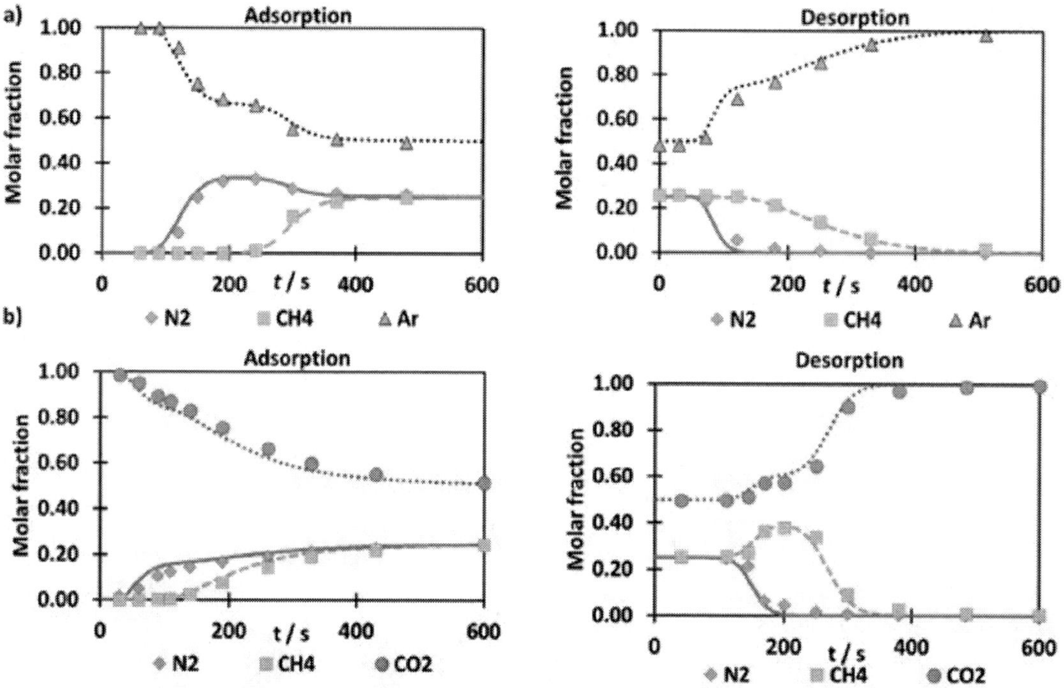

Figure 2 – Ternary fixed-bed experiments: a) $N_2/CH_4/Ar$ mixture (25%/25%/50%) over a BPL fixed bed initially saturated with Ar; b) $N_2/CH_4/CO_2$ mixture (25%/25%/50%) over a BPL fixed bed initially saturated with CO_2. Symbols represent experimental points and lines the simulation results.

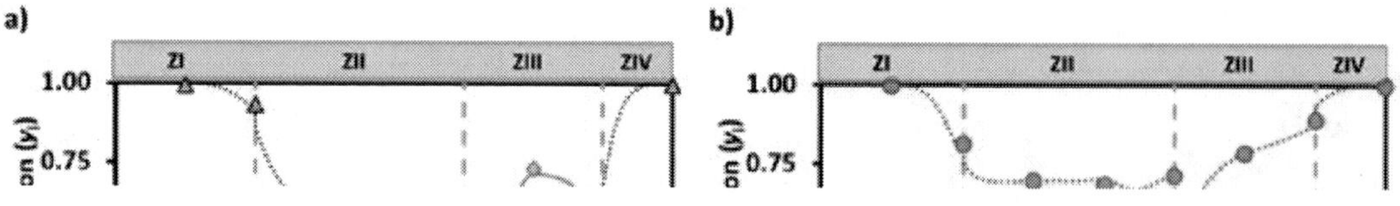

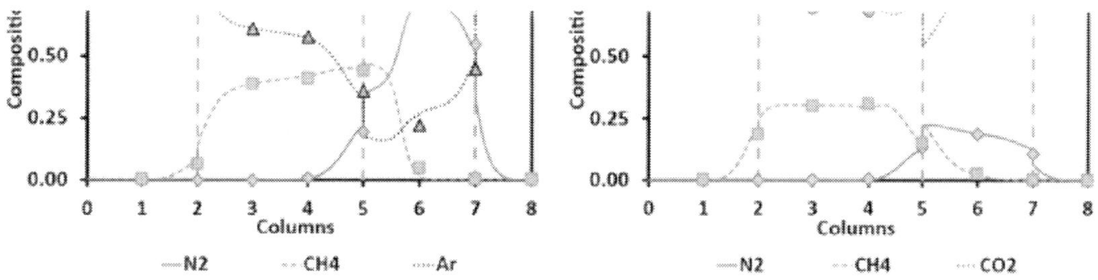

Figure 3 - Internal profile obtained at half switching time for: a) Argon as desorbent and b) CO_2 as desorbent, both at cyclic steady state. Points correspond to experimental data and lines represent the simulation results.

674d Parameter Estimation for Reactive Chromatography Model By Bayesian Inference and Parallel Sequential Monte Carlo

Hikari Sugiyama, Yota Yamamoto, Kensuke Suzuki, Tomoyuki Yajima, Yoshiaki Kawajiri

Reactive chromatography simultaneously performs reaction synthesis and separation of the product. This principle has been shown to improve the reaction conversion rate, enable the downsizing of equipment, and reduce operating costs. For industrial-scale applications, a simulated moving bed reactor (SMBR) can be a promising technique in which multiple columns are connected to continuously feed reactants to carry out reactions while performing separations simultaneously. This process has been investigated for various applications, including biodiesel production[i] and sugar inversion [ii]. The principle of simulated moving bed has been commercialized for many large-scale separations in the petrochemical industry and sugar production[iii].

Reliable model development is essential for designing complex processes such as SMBRs, which require operating conditions to be determined, so that product requirements are met. In the model-based development of such processes, experimental data are needed to estimate parameters in the model. However, analyzing the uncertainty in the estimated model parameters and quantifying the influence on the predictive performance remains a challenge.

This study aims to quantify the uncertainty in model parameters and develop reliable models to support the development of reactive chromatography. To this end, Bayesian estimation was used to estimate model parameters in this study. In the proposed approach, parameter uncertainty is shown as probability distributions. Since it is challenging to obtain parameters analytically, the sequential Monte Carlo method is applied, a sampling method for approximation. This method allows random sampling in parallel using multiple cores, which reduces the computation time compared to the conventional methods such as Markov chain Monte Carlo[iv]. Furthermore, several thousand points were sampled from the posterior probability distributions of the parameters, which were used in numerical simulations to evaluate predictions.

This study considers the synthesis of methyl acetate and water from acetic acid and methanol as a case study. Experimental data under various conditions were utilized, and parameters were estimated using multiple data sets. Experimental conditions that can successfully reduce parameter uncertainties were also investigated.

References:

[i] Fukumura T, Chiba R, Ono Y, Kubo M. Continuous synthesis of ethyl esters from triglycerides using simulated moving bed chromatographic reactor packed with solid acid catalysts. Kagaku Kogaku Ronbunshu. 2018;44(3):177-184. doi:10.1252/kakoronbunshu.44.177

[ii] Hashimoto K, Adachi S, Noujima H, Ueda Y. A new process combining adsorption and enzyme reaction for producing higher-fructose syrup. Biotechnol Bioeng. 1983;25(10):2371-2393. doi:10.1002/bit.260251008

[iii] Rodrigues AE, Pereira CSM, Santos JC. Chromatographic Reactors. Chem Eng Technol. 2012;35(7):1171-1183. doi:10.1002/ceat.201100696

[iv] Yamamoto Y, Yajima T, Kawajiri Y. Uncertainty Quantification for Chromatography Model Parameters by Bayesian Inference using Sequential Monte Carlo Method. *Chem Eng Res Des.*

2021;175:223-237. doi:10.1016/j.cherd.2021.09.003

674e Can a Computer 'Learn' Non-Linear Chromatography?: Physics-Based Deep Neural Networks for Chromatographic Separations

Sai Gokul Subraveti, Zukui Li, Vinay Prasad, Arvind Rajendran

The dynamics of chromatographic pulses are represented by hyperbolic partial differential equations along with non-linear equations describing the adsorption equilibrium. In this work we describe a physics-based artificial neural network framework for adsorption and chromatography emulation (PANACHE) that uses deep neural networks constrained by the underlying physics of the process [1,2]. The addition of physics-based constrained allows us to substantially reduce training efforts and also provides excellent description of the pulse propagation in the entire spatio-temporal domain although only initial and boundary data is used for the training itself. The methodology is demonstrated through simulated examples using the four combinations of the generalized Langmuir model, Langmuir (L)-Langmuir, Anti-Langmuir (AL)- Anti-Langmuir, L-AL and AL-L. We further demonstrate that it is possible to use only experimentally measured chromatographic elution curves to train PANACHE, i.e., without requiring explicit measurements of the adsorption isotherm. This method can have applications for systems whose single/competitive isotherms that are complex to be described.

1. Subraveti, S.G., et al., 'Can a computer "learn" nonlinear chromatography?: Physics-based deep neural networks for simulation and optimization of chromatographic processes', J. Chromatogr A, in Press

2. Subraveti, S.G., et al., 'Physics-based neural networks for simulation and synthesis of cyclic adsorption processes', Ind. Engg. Chem. Res., in Press

674f Mechanistic Modeling and Control of Column-Free Continuous Chromatography Process

Frederik Doerr, Anish Dighe, Amos E. Lu, Eben Crawford, Oleg Shinkazh, Richard D. Braatz

Chromatography is a critical process used to purify biological molecules. The dominant configuration employs resin particles in columns typically operated in batch mode. The resin either captures the biological molecule of interest or allows the biological molecule to flow through while removing impurities (aka "polishing"). As bioreactors have become more productive, the operating costs have increasingly been shifted to downstream purification, with the cost of goods reported as high as 75% for the chromatography columns for some biotherapeutics such as monoclonal antibodies. Continuous configurations that oscillate between multiple columns have become commercially available, reducing the operating costs of downstream purification, which are referred to as "continuous chromatography."[1]

This work involves an alternative continuous chromatography design that does not employ columns and does not have an oscillatory operation. This Continuous Countercurrent Tangential Chromatography (CCTC) technology more efficiently uses resin (by far the most expensive part of the process for some biotherapeutics, such as monoclonal antibodies) while operating in quasi-steady conditions so that the material leaving the process has greater consistency.[2] This technology is also able to capture and release biological molecules in less than ten minutes, which is essential for those molecules that have poor stability when adsorbed on or within the resin particles.

This work derives the first distributed parameter model for the CCTC process. The model is validated experimentally for the most expensive step in the purification of monoclonal antibodies (mAbs), which is the capture of mAbs from clarified bioreactor material. The separation is achieved due to the adsorption of mAb from the feed into the surface and pores of resin particles suspended in an aqueous buffer solution.[3-4] The mechanistic model developed for this process relies on the kinetic parameters, including the binding rate constant, pore diffusion coefficient, and film transfer coefficient. A full Bayesian inference is performed using the Markov-chain Monte-Carlo (MCMC) method that incorporates prior knowledge and quantifies parameter uncertainties. The model accurately predicts the concentration of purified mAb as a function of residence time. Furthermore, we derive a model-based control system for the automated operation of the system. The concentration profile of mAb shows stable output and quickly responds to variations in the feed composition. The robustness, scalability, and high yield for large biomolecules with very short residence times enable the capture and elution of poorly stable biotherapeutic molecules.

References:

1. Behere, K.; Yoon, S., Chromatography bioseparation technologies and in-silico modelings for continuous production of biotherapeutics. *J Chromatogr A* **2020,** *1627*, 461376.
2. Shinkazh, O.; Kanani, D.; Barth, M.; Long, M.; Hussain, D.; Zydney, A. L., Countercurrent tangential chromatography for large-scale protein purification. *Biotechnol Bioeng* **2011,** *108* (3), 582-91.
3. Dutta, A. K.; Tran, T.; Napadensky, B.; Teella, A.; Brookhart, G.; Ropp, P. A.; Zhang, A. W.; Tustian, A. D.; Zydney, A. L.; Shinkazh, O., Purification of monoclonal antibodies from clarified cell culture fluid using Protein A capture continuous countercurrent tangential chromatography. *J Biotechnol* **2015,** *213*, 54-64.

4. Dutta, A. K.; Fedorenko, D.; Tan, J.; Costanzo, J. A.; Kahn, D. S.; Zydney, A. L.; Shinkazh, O., Continuous countercurrent tangential chromatography for mixed mode post-capture operations in monoclonal antibody purification. *J Chromatogr A* **2017,** *1511*, 37-44.

674g Process Modeling and Techno-Economic Optimization of a Moving Bed Contactor for CO_2 Capture Using a Diamine-Appended Metal–Organic Framework

Ryan Hughes, Goutham Kotamreddy, Debangsu Bhattacharyya, Stephanie Didas, Surya T. Parker, Jeffrey R. Long, Benjamin P. Omell, Michael S. Matuszewski

Aqueous solvents are the current leading commercial technologies for post-combustion CO_2 capture, but they suffer from several drawbacks that include degradation, large energy penalties, and low working capacities. Solid sorbents are viable alternatives to these aqueous solvents. Most notably, a novel class of functionalized metal-organic frameworks (MOF's) have been shown to exhibit promising characteristics for post-combustion capture from the flue gas in the fossil-fired power plants[1–3]. The MOF studied in this work, Mg_2(dobpdc) (dobpdc^{4-} = 4,4′-dioxidobiphenyl-3,3′-dicarboxylate), contains one-dimensional hexagonal channels with unsaturated Mg^{2+} sites and is functionalized by binding the diamine 2,2-dimethyl-1,3-diaminopropane (dmpn) to these sites which then chemisorb CO_2 in cooperative mechanism to form carbamate and carbamic acid species. This MOF exhibits a step-shaped isotherm under low partial pressures of CO_2 which leads to higher working capacities under similar temperature and pressure intervals when compared to traditional sorbents. Along with high working capacities, this functionalized MOF also shows excellent long-term stability and maintains performance under humid conditions, both desirable attributes for flue gas CO_2 capture.[1]

To realize the economic potential of this new class of MOFs, both the contactor technology and their operating conditions are crucial. While there has been much work on optimizing the design of the towers, their internals, and operating conditions for the solvent-based capture processes, there are very few publications in the open literature on optimal design of the contactor technology for solid-sorbent systems. Previous work has focused on economic analysis of a fixed bed process[4], but a moving bed (MB) contactor can offer several advantages over other solid-gas contactors, such as a fixed bed or fluidized bed. The counter-current flow pattern exhibited in a MB results in larger driving forces when compared to a fixed bed contactor while the moderate hydrodynamics also will reduce particle attrition when compared to a fluidized bed. Since MOF particles typically do not process high attrition resistance characteristics, moving bed systems offer an excellent option for these materials.

In this work, a dynamic, one-dimensional, non-isothermal MB reactor model is developed and used to model a full scale temperature swing adsorption process using the functionalized MOF. The model simultaneously solves a set of partial differential equations for the mass and energy balances of the system, and accounts for internal and external mass transfer resistances, reaction kinetics, and heat transfer between the gas and solid phase. A hydrodynamic model is also developed. The model also accounts for heat transfer between the system and an embedded heat exchanger which can provide/remove a large amount of heat and facilitate operation close to the optimal temperature profile as the capture target change or the flue gas flowrate/composition changes. A cost model is also developed incorporating both capital and operating costs. A mathematical programming problem is then solved for simultaneous optimization of the design and operating variables of the MB process. The study shows that the MB system can considerably improve the economics of the MOF technology compared to the fixed bed processes.

<u>References:</u>

(1) Milner, P. J.; Siegelman, R. L.; Forse, A. C.; Gonzalez, M. I.; Runčevski, T.; Martell, J. D.; Reimer, J. A.; Long, J. R. A Diaminopropane-Appended Metal–Organic Framework Enabling Efficient CO_2 Capture from Coal Flue Gas via a Mixed Adsorption Mechanism. *J. Am. Chem. Soc.* **2017**, *139* (38), 13541–13553. https://doi.org/10.1021/jacs.7b07612.

(2) McDonald, T. M.; Mason, J. A.; Kong, X.; Bloch, E. D.; Gygi, D.; Dani, A.; Crocellà, V.; Giordanino, F.; Odoh, S. O.; Drisdell, W. S.; Vlaisavljevich, B.; Dzubak, A. L.; Poloni, R.; Schnell, S. K.; Planas, N.; Lee, K.; Pascal, T.; Wan, L. F.; Prendergast, D.; Neaton, J. B.; Smit, B.; Kortright, J. B.; Gagliardi, L.; Bordiga, S.; Reimer, J. A.; Long, J. R. Cooperative Insertion of CO_2 in Diamine-Appended Metal-Organic Frameworks. *Nature* **2015**, *519* (7543), 303–308. https://doi.org/10.1038/nature14327.

(3) Forse, A. C.; Milner, P. J.; Lee, J.-H.; Redfearn, H. N.; Oktawiec, J.; Siegelman, R. L.; Martell, J. D.; Dinakar, B.; Porter-Zasada, L. B.; Gonzalez, M. I.; Neaton, J. B.; Long, J. R.; Reimer, J. A. Elucidating CO_2 Chemisorption in Diamine-Appended Metal–Organic Frameworks. *J. Am. Chem. Soc.* **2018**, *140* (51), 18016–18031. https://doi.org/10.1021/jacs.8b10203.

(4) Hughes, R.; Kotamreddy, G.; Ostace, A.; Bhattacharyya, D.; Siegelman, R. L.; Parker, S. T.; Didas, S. A.; Long, J. R.; Omell, B.; Matuszewski, M. Isotherm, Kinetic, Process Modeling, and Techno-Economic Analysis of a Diamine-Appended Metal–Organic Framework for CO_2 Capture Using Fixed Bed Contactors. *Energy Fuels* **2021**, *35* (7), 6040–6055. https://doi.org/10.1021/acs.energyfuels.0c04359.

520a Advances in Surface Chemistry Assessment of Adsorbents By Combining Adsorption, Liquid Intrusion and NMR Spectroscopy

Carlos Cuadrado Collados, Dorothea Wisser, Jakob Söllner, Markus Terlinden, Martin Hartmann, Matthias Thommes

In this work, we focus on the development of a comprehensive strategy/methodology for assessing the effective surface chemistry of nanoporous materials by combining advanced adsorption studies, novel liquid intrusion techniques and solid-state NMR spectroscopy.

For our study we have chosen the highly ordered mesoporous silica molecular sieve SBA-15 and SBA-15 silica functionalized with methyl groups (methylation with hexamethyldisiazane) as model materials allowing for a surface characteristics with different degree of surface functionalization. A full textural characterization of the materials could be obtained by high-resolution argon and nitrogen adsorption experiments at 87 K and 77 K, respectively, coupled with the application of dedicated methods based on non-local-density functional theory.

The applied strategy for surface chemistry characterization is based on the use of adsorptives exhibiting different wettability such as argon adsorption at 87 K, which leads to a complete wetting adsorbate (effective contact angel is zero) independent of the effective surface chemistry, and water adsorption at various temperatures (273 K - 323 K), which is sensitive to the underlying surface chemistry. Depending on the effective hydrophilicity/hydrophobicity of the adsorbent surface, water may adsorb at pressures P smaller than the bulk saturation pressure P_0 (i.e., at P/P_0 £ 1) in case the water adsorbate is completely or at least partially wetting (i.e. effective contact angle < 90°). In case of a fully methylated surface, argon at 87 K still forms complete wetting films, while water adsorption leads to a non-wetting adsorbate (effective contact angle > 90°) indicative of pore filling occurring at pressures larger than the bulk saturation pressure (i.e., at $P/P_0 > 1$). In the latter case we investigate the pore filling of water at pressures > P_0 by the application of novel water intrusion/extrusion experiments.

Coupled with detailed knowledge about the pore structure (textural characterization) and isoteric heats of adsorption data (obtained from argon and water adsorption measurements at various temperatures), the obtained results allowed us to develop a methodology for quantifying the effective hydrophilicity/hydrophobicity (e.g., hydrophilicity index) and the corresponding effective contact angles.

By coupling these results with advanced, dedicated ^{1}H, ^{13}C and ^{29}Si MAS NMR studies, we are able to assess the chemical nature and surface density of different functional groups, as well as to determine the location of the functional groups (outer surface vs. mesopores or micropores).

Our results and the developed strategy/methodology are essential for the surface characterization of siliceous nanoporous materials for applications in heterogeneous catalysis and separation such as chromatography.

520c Molecular Simulation of Adsorption and Diffusion in Rigid Nanoporous Amorphous Materials

Raghuram Thyagarajan, David Sholl

The development of publicly available materials databases for materials including zeolites (IZA-SC), metal organic frameworks (CoRE-MOF), and inorganic materials (The Materials Project) has been a key enabler of high-throughput computational screening and data-driven discovery of materials for potential use in new technologies. We recently introduced a similar database for structures of porous rigid amorphous materials[1], an important class of materials for which no such resource was available. The database includes atomically detailed structures of disordered materials like amorphous carbons, kerogens, hyper-cross-linked polymers etc. generated using a wide range of simulation techniques by multiple research groups. We present extensive computational analyses for material characterization by calculating a series of scalar (e.g., accessible surface area) and vector (e.g., pore size distribution) descriptors. A variety of gas adsorption isotherms for both single component and binary mixtures are predicted for each structure.We also discuss the agreement between binary adsorption data and predictions from the Ideal Adsorbed Solution Theory.

In addition to adsorption isotherms, we have computed the diffusion coefficients of CH_4 and CO_2 in several of these structures. We present a diverse collection of molecular diffusivities in amorphous materials and examine their concentration dependence by comparing data from adsorption and diffusion simulations. We also discuss the correlations of computed self-diffusivities with scalar descriptors and Knudsen diffusivities.

References:

1. Thyagarajan, R.; Sholl, D. S. A Database of Porous Rigid Amorphous Materials. *Chem. Mater.* 2020, *32*, 8020-8033.

520d Philippine Natural Zeolites and Its Potential for Air Quality Management

Bryan Alamani, Jose Aldrin Vea, Eleanor Olegario, Ramuel John Tamargo

Air pollution is one of the factors that affect the climate, public and individual health in recent years. One of the most common air pollutants, the tiny particles in the air called particulate matter, contain not only chemical contaminants but also biological pollutants like bacteria and viruses. These can cause respiratory and cardiovascular diseases, reproductive and central nervous system dysfunctions, and cancer. Furthermore, the climate change resulting from the chemical contaminants in the air affects the geographical distribution of infectious diseases. In this global pandemic, air is one of the ways in the transmission of the virus and clean air is imperative to achieving a safe and healthy environment. Adsorption is one of the many processes that can be used to contain these pollutants and reduce their levels to an extent that does not pose environmental concerns. Here, we shall present our work on transforming natural zeolite (Philippine Natural Zeolites) into particles exhibiting sufficient surface area and functionality to reduce pollutant content of ambient air. Furthermore, we shall discuss the properties of these recrystallized and functionalized natural zeolites. Insights on how the transformation of natural zeolites can modulate its properties shall also be present. It is envisioned that this preliminary work can highlight various interventions that enhances the engineering value of natural zeolites for air quality management.

520e Advanced Characterization of Disordered Mesoporous Solids

Henry R. N. B. Enninful

Mesoporous solids have found wide uses in catalysis, adsorption, molecular separations, gas and energy storage, among others. Indeed, precise structure-function relations of guest molecules in pores are rightly obtained when accurate knowledge of the confining space is available. At present, most approaches for pore characterization consider disordered porous materials as individual collection of pores where fluid phase transition occurring in one pore has no effect on adjacent pores. However, pore interconnectivity renders fluid phase behavior in one pore dependent on the phase state in its neighboring pore.

We show systematically how a model of statistical chain of pores can be applied for solid-liquid phase transitions of porous solids. By creating a kernel-based approach incorporating a variable non-frozen layer thickness between solid core and pore wall and eliminating any a priori assumption of phase transition occurring by metastable or equilibrium transition, we refine the thermoporometry characterization technique.

For verification, we show how this approach works well with ordered materials like MCM-41 porous silica and reveals disorder in SBA-15 materials. This approach can be extended to other phase transition phenomena such as gas adsorption.

References

[1] Schneider D.; Kondrashova D.; Valiullin R., 2017, "Phase transitions in disordered mesoporous solids", Scientific Reports, 7, 7216.
[2] Schneider, D. and Valiullin R., 2019, "Capillary condensation and evaporation in irregular channels: Sorption isotherm for serially connected pore model", Journal of Physical Chemistry C, 123, 16239.
[3] Enninful H.R.N.B., Schneider D., Hoppe A., König S., Fröba M., Enke D. and Valiullin R., 2019, "Comparative gas sorption and cryoporometry study of mesoporous glass structure: Application of the serially connected pore model", Frontiers in Chemistry, doi: 10.3389/fchem.2019.00230.
[4] Enninful H.R.N.B., Schneider D., Kohns R., Enke D. and Valiullin R., 2020, "A novel approach for advanced thermoporometry characterization of mesoporous solids: Transition kernels and the serially connected pore model", Microporous and Mesoporous Materials 309, 110534.
[5] Enninful H.R.N.B., Schneider D., Enke D. and Valiullin R., 2021, "Impact of Geometrical Disorder on Phase Equilibria of Fluids and Solids Confined in Mesoporous Materials", Langmuir, 37, 3521-3537.

520f Characterization of LTA Zeolites Based on Pore Typification

Sebastiao M. P. Lucena, Jose Carlos A. Oliveira, Daniel V. Gonçalves, Diana C. S. Azevedo, Moisés Bastos-Neto

Despite the great industrial importance of zeolites, the characterization of these materials based on adsorption is still very limited. Adsorption has traditionally been used to determine the pore size of zeolites. However, the available models are semi-empirical, based on an average surface potential to calculate the gas-solid interactions. Unlike carbonaceous materials, which intensively apply molecular simulation and density functional theory methods in their characterization, zeolites did not have an equivalent development. In addition, several zeolites are unsuitable for characterization with N_2 or Ar due to diffusional limitations of these gases. In a recent study, Dantas and Neymark (2020)* have introduced a methodology – the pore type distribution (PTD) – for pore structure characterization based on matching a theoretical isotherm generated by the combination of isotherms of adsorption in individual pore types to an experimental isotherm. Here we propose to extend this methodology to zeolites. We selected Na-LTA (4A) zeolite as an application example. Based on the 8 local isotherms, one from each supercage of the LTA unit cell, we developed a kernel that estimates the most likely energy distribution levels among supercages from the best fits of experimental isotherms. The method was applied in the characterization of LTA commercial and laboratory synthesized samples. The pore type distribution of the samples, showed that at least 4 supercages with strong interaction potential are a probable signature of the LTA zeolites studied. The new simulated isotherm based on this pore type distribution considerably improved the agreement between simulated and experimental isotherms. The heat of adsorption variation in supercages is related to the relative positioning of cations between the S3 and S2 sites. The theoretical average volume obtained was used to assess the level of defects of samples obtained from unconventional precursors raw material or those that underwent hydrothermal degradation. The new methodology enables the rational use of adsorptive methods in the characterization of small pore zeolites, which can be extended to other classes of zeolites, allowing new information to be obtained and expanding the use of adsorption as a characterization tool.

(*) Dantas, S.; Neimark, A. V. Coupling Structural and Adsorption Properties of Metal-Organic Frameworks: From Pore Size Distribution to Pore Type Distribution. ACS Appl. Mater. Interfaces 2020, 12, 15595.

Individual supercages

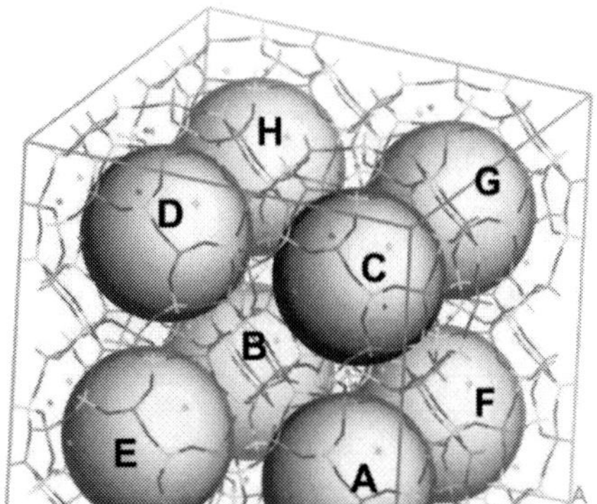

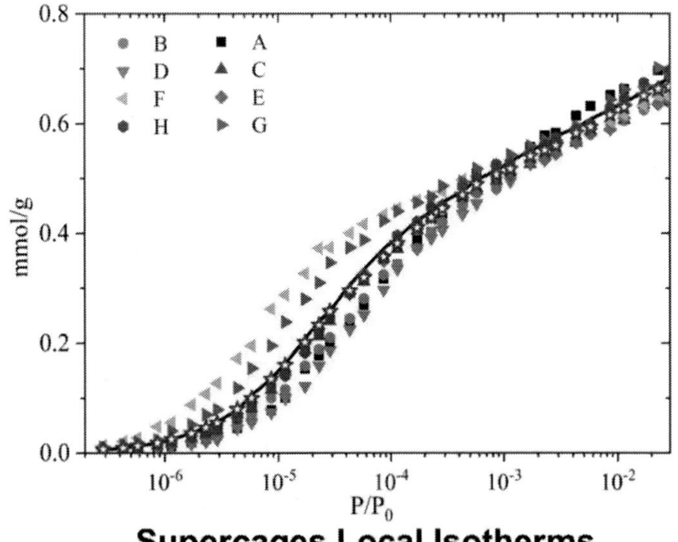

Supercages Local Isotherms

520g Post Spinning Modification of Porous Organic Cage Fiber Sorbents for CO_2 Capture

Isaiah Borne, Jennel McGriff, Christopher W. Jones, Ryan P. Lively

It is clear that CO_2 capture from point sources and direct capture from ambient air is necessary to achieve global climate goals and avoid excessive global temperature increases. One method to develop effective adsorbents for CO_2 is to impregnate mesoporous or microporous materials with polyamines like branched polyethylenimine (PEI) or tetraethylenepentamine (TEPA). Here a new class microporous materials, porous organic cage (POCs) are explored as a base material for PEI impregnation. CC3 is the POC used in this work and it has been previously spun into polymeric fiber sorbents for adsorptive separation applications. CC3 is composed of chemically weak imine bonds which should be vulnerable to degradation from primary amines present in PEI. Unexpectedly, CC3 maintains its microporosity and chemical structure after extended PEI exposure. The CC3-PEI fibers are characterized via TGA, SEM, FTIR, N_2 physisorption at 77K, and CO_2 isotherms. CC3 fibers impregnated with PEI consistently showed higher CO_2 uptakes at low pressures showing that PEI was contributing to CO_2 capture along with CC3. Breakthrough studies were conducted with these CC3-PEI fibers and compared to mesoporous silica-PEI fibers to compare the kinetics of CO_2 uptake. This work shows that imine based POCs can be combined with PEI to make them effective adsorbents for CO_2 capture.

520h The Use of Zeolites and Carbons for the Separation of Refrigerant R-410A: Thermodynamic Modeling of Pure Gas and Binary Sorption

Andrew Yancey, David R. Corbin, Mark Shiflett

Worldwide legislation is currently phasing out the use and production of hydrofluorocarbons (HFCs) due to large global warming potentials (GWPs). The most recent is the U.S. American Innovation and Manufacturing (AIM) act in December 2020, which calls for a 15% reduction of HFC production and use by 2035. The United Nations recently published a report in Summer 2021 that cited HFCs as a contributor toward anthropogenic global climate change. The report further states that if atmospheric HFCs, along with other potent greenhouse gases (GHGs), are not reduced during the 21st century, this will result in a 1.5-2.0 K global temperature increase, further perpetuating extreme weather patterns. As HFCs are phased out of use, an estimated 2,800 ktons of global refrigerant supply will need to be effectively and responsibly disposed of. Instead of venting or incinerating the HFCs, which would perpetuate global climate change, the unused HFCs should be reclaimed, recycled, and repurposed.

Recycling HFC refrigerants is a challenging task since many are azeotropic or near-azeotropic HFC refrigerant blends. In order to effectively recycle these refrigerants, the HFC blends must first be separated so that the constituent species can be dealt with individually. Traditional distillation techniques cannot separate azeotropes. Adsorption-based separation is an alternative to traditional distillation that has proven to effectively separate zeotropic, azeotropic, and isomeric fluorocarbon mixtures. Furthermore, adsorption-based separations have the potential to require less energy and produce a higher purity of products than traditional distillation. The following presentation will focus on the use of basic zeolites, acidic zeolites, and activated carbons for the separation of refrigerant R-410A (50/50 wt% HFC-32/HFC-125).

In order to effectively design any separation process, reliable thermodynamic models are required. Thermodynamic measurements have been made using a Hiden Isochema XEMIS gravimetric microbalance for pure gas sorption of various basic and acidic zeolites, as well as activated carbons. Resulting isotherms were fit to pure gas isotherm models, which were used in Ideal Adsorbed Solution Theory (IAST) to predict binary sorption behavior and selectivity trends. Binary sorption measurements were made using a separate Hiden Isochema gravimetric microbalance that uses the integral mass balance (IMB) method to calculate sorption of HFC-125 and HFC-32 on each sorbent. Using experimental data, sorbed phase activity coefficients were calculated, fit to vapor-adsorption equilibrium (VAE) activity coefficient models, and integrated into Real Adsorbed Solution Theory (RAST) sorption models to improve IAST predictions. Thermodynamic modeling development, as well as molecular level interpretations of results will be discussed. Conclusions will be drawn regarding the potential of each adsorbent to separate refrigerant R-410A.

677a Predicting and Mitigating Retro-Condensation in TSA Dryers Using Aspen Adsorption

Robert Broekhuis, Pranit S. Metkar

Dryers based on the temperature swing adsorption (TSA) concept, using molecular sieves such as zeolite 4A, are commonly used in the process industries to remove water from process streams beyond what can be achieved by condensation and knockout. While the technology is mature and generally well understood, choices of design parameters and details of cyclic operation are usually presented in general terms, using heuristic recommendations, because the detailed modeling of these cyclic systems is complex and time consuming, often requiring specialized tools. Yet the heuristic guidelines do not apply universally to all TSA configurations and operations, so there is an incentive to model specific implementations. Aspen Adsorption® (AA), a tool available to many process engineers, provides a convenient means of designing and understanding cyclic adsorption processes. However, one phenomenon that does not lend itself to ready modeling in AA is retro-condensation occurring during the hot regeneration step of a TSA cycle, in which water removed from a hot part of the bed travels to a cooler part of the bed and condenses to form a liquid phase. Depending on the extent of condensation and the temperature at which it occurs, this phenomenon can cause progressive damage to the sorbent, limiting its useful service life. Since AA does not account for phase transitions within the sorbent bed, the tool must be used creatively to represent retro-condensation. In a case study of air drying, we model water condensation using a custom extension of the adsorption isotherm, which allows estimation of the liquid holdup (and associated temperature) as it varies in time and space during regeneration. Starting with a base case featuring significant retro-condensation, we show how changing regeneration parameters such as regeneration gas flowrate, pressure, and the progression of its temperature can reduce the severity of (or ultimately eliminate) the phenomenon.

677b Feasibility Study of Pressure Swing Adsorption (PSA) Processes for CO_2 Capture and H_2 Purification in Blue Hydrogen Processes and Hydrogen Deblending in the Gas Network

Yan Chen, Hyungwoong Ahn

A methane reforming-based hydrogen has a distinctive advantage over green hydrogen in that it is dispatchable when needed under any climate conditions. Accordingly, the conventional hydrogen is expected to play a key role in the future hydrogen supply chain, even after green hydrogen processes are widely deployed. However, the fossil fuel-based hydrogen production process must be decarbonised so that it can also contribute to achieving the net zero emission target.

An adsorption process has been actively studied for CO_2 capture mainly from two locations of the H_2 production process: One is the synthesis gas generated by reforming of methane or light hydrocarbon or gasification of biomass, coals or heavy hydrocarbons, and the other is the H_2 PSA tail gas of a hydrogen production process following. It is still uncertain which feed stream would be more advantageous for CO_2 capture. In this study, a feasibility study was carried out to answer this question [1].

The second part of this study is to redesign the H_2 purification PSA for producing a high-purity H_2 from the CO_2-depleted synthesis gas downstream of a CO_2 capture unit applied to the synthesis gas and to estimate the H_2 PSA performance [2].

It is imperative to find a way to supply both hydrogen and methane at the same time to meet the demands of household and industry for both fuel gases during the transition period. To this end, it is likely that hydrogen is blended with methane and the gas mixture is transported through the existing gas network. The gas mixture must be separated into each of two gases near the final stage of the gas distribution network. A PSA process designed for separation of hydrogen from methane was evaluated at various mixing ratios of the two gases.

1. Chen and Ahn, Feasiblity Study of Vacuum Pressure Swing Adsorption for CO_2 Capture from an SMR Hydrogen Plant: Comparison between Synthesis Gas Capture and Tail Gas Capture, *Frontiers in Chemical Engineering* **3**, 742963, 2021.

2. Luberti et al., Design of a H_2 PSA for Cogeneration of Ultrapure Hydrogen and Power at an Advanced Integrated Gasification Combined Cycle with Pre-combustion Capture, *Adsorption* **20(2-3)**, 511-524, 2014.

677c Variability of Adsorption Properties on HiSiV3000 Individual Pellets

Stylianos Kalaitzopoulos, Enzo Mangano, Stefano Brandani

The use of kinetic and equilibrium information extracted from mg-scale measurements to scale up adsorption processes is often considered questionable, due to the lack of homogeneity on commercial adsorbents from the same batch. This inhomogeneity may be caused mainly by large equipment and/or inappropriate treatment of materials during the manufacturing process[1]. In this work we present a methodology for the quantification of variability and reproducibility through the measurement of the effective diffusivity of individual commercial pellets.

In this analysis, Zero Length Column (ZLC) measurements[2] are carried out on individual commercial HiSiv3000 extrudates from UOP, a Honeywell company. ZLC responses with ppm-levels of n-Pentane as a probe molecule are used for the characterization of the effective diffusivity following the recommended experimental protocol[3]. This allows the unequivocal determination of the effective macropore diffusivity and the Henry law constant through a campaign of experiments at different flowrates, using partial and full equilibration of the sample over a temperature range of 150-200°C.

To determine intra-batch variability, a number of randomly chosen pellets were examined under identical conditions with the ZLC. The moving average of the kinetic time and Henry law constants for each additional sample was monitored until stable values were obtained for both parameters. This completes the iterative process and the average values can be then extracted.

To demonstrate that these are representative of the properties of the larger batch, the average values were used to fully predict a breakthrough experiment using a sample 100 times larger carried out under the same conditions.

The results show an excellent match between the experimental curve and the model allowing also to determine the maximum deviation from the average values. Our approach confirms that for adsorbent materials of good quality single pellet characterization using the ZLC can be used to predict satisfactorily the behavior of a scaled-up process through a proper model implementation[4] even in the worst possible scenario, i.e. for the individual pellet properties furthest from the average. The approach can be applied as a rapid tool to monitor the batch homogeneity during the quality control of manufactured adsorbents.

[1] Wilkins N. S., Rajendran A. andamp; Farooq S. *Dynamic column breakthrough experiments for measurement of adsorption equilibrium and kinetics*. Adsorption **27**, 397–422 (2021).

[2] Eic M. and Ruthven D.M. *A new experimental technique for measurement of intracrystalline diffusivity*. Zeolites **8,** 40–45 (1988).

[3] Brandani S. andamp; Mangano E. *The Zero Length Column Technique to Measure Adsorption Equilibrium and Kinetics: Lessons Learnt from 30 Years of Experience*. Adsorption **27**, 319–351 (2021).

[4] Brandani F., Rouse A., Brandani S. andamp; Ruthven, D. M. *Adsorption Kinetics and Dynamic Behavior of a Carbon Monolith*. Adsorption **10**, 99–109 (2004).

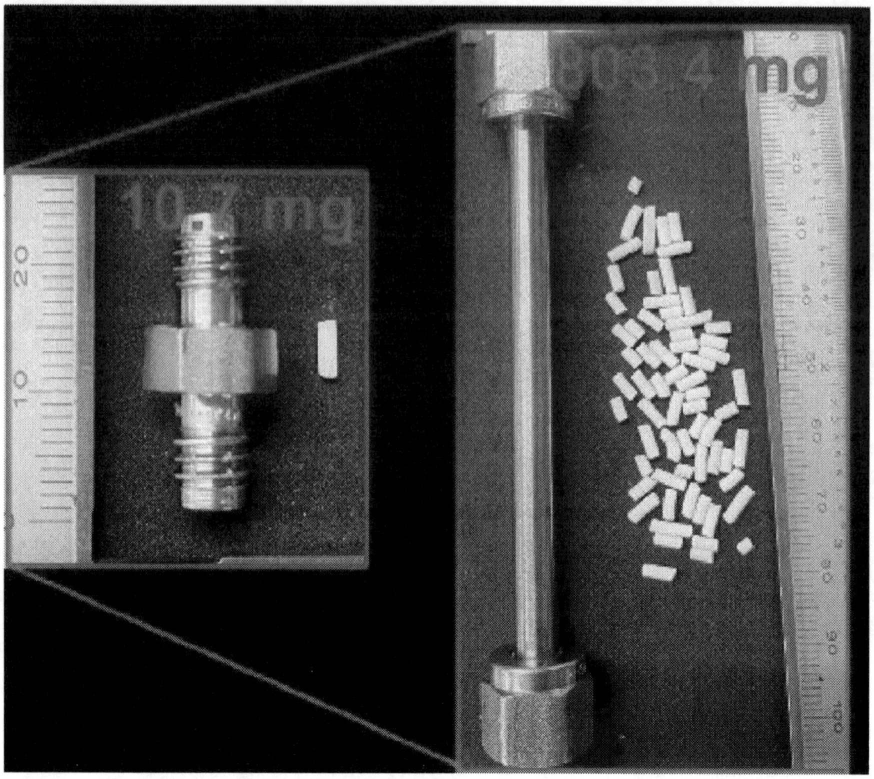
10.7 mg
808.4 mg

677d Controlled Morphology, Post-Synthetic Modification, and Scale-up of MOFs for Direct Air Capture of Carbon Dioxide

Xakin Ramirez Isunza, Brittany Bonnett, Connor Farrell, Amanda J. Morris, Stephen Martin

Net-zero emissions technologies have been identified by the IPCC as an important tool to achieve the goal of a less than 2 °C global average temperature rise over pre-industrial levels. Several companies are currently working to commercialize direct-air capture (DAC) technologies to remove CO_2 directly from the atmosphere, but significant strides need to be made in the development of sorbent materials to make this an economically viable approach. Sorbent materials need to exhibit high CO_2 capacity at low CO_2 concentrations (i.e. at atmospheric concentrations of ~400 ppm), rapid sorption and desorption kinetics to facilitate capture and regeneration cycles, low regeneration energy to minimize operating cost and carbon footprint, and low material capital costs. Sorbent materials also need to exhibit high stability over numerous sorption/regeneration cycles to increase the lifetime of the materials.

We have studied a series of metal organic framework (MOF) materials with the aim of optimizing and scaling them up as solid sorbents for direct air capture of CO_2. Specifically, we are interested in tuning the particle size of the MOFs and the porosity of both the nano- and meso- porous structures present in a packed-bed geometry. Previously, a test series of controlled UiO-66 materials with well-characterized defect-levels and particles sizes was examined, and additional MOFs were selected based on high capacity (mmol/g) of CO_2 while also considering the isosteric heat of adsorption (Q_{st}). Herein we will present on DAC using Mg-MOF-74 and MIL-101-Cr based sorbents.

Modulated syntheses of Mg-MOF-74 using different modulators (benzoic and salicylic acid) as well as a variety of reaction scales (100 mL, 250 mL, and 500 mL) resulted in variable and controllable particle morphologies, indicating that Mg-MOF-74 is highly favorable for morphological and size control studies relating to synthesis and gas sorption performance. MIL-101-Cr was synthesized in 25 mL and 265 mL reaction vessels using acetic acid as a modulator, resulting in negligible crystallographic or morphological difference between the samples. Additionally, a fraction of each synthesized MOF sample underwent post-synthetic ethylenediamine grafting to increase chemisorption site density. The resulting MOF samples were studied using isothermal gravimetric and volumetric sorption techniques. Based on these results, a subset of the MOFs was studied under multiple sorption/regeneration cycles in a packed bed sorption apparatus under DAC operating conditions.

677e Trace Metal Removal By Adsorption for Ultra-High Purity Amines

Yujun Liu, Towhid Hasan, Xue (Ida) Chen, Jerrod Ruddick, James York, Amarnath Singh

For many applications such as electronic and pharmaceutical production process, ultra-high purity organic solvents, such as amines, are always desired with single digit ppb level metal ion impurities, and the demands are ever-increasing. However, metal removal by adsorption processes from organic solutions are rarely reported and available literature is normally for removing metal ions from aqueous solution. Recently, we evaluated several chelation ion-exchange resin adsorbents with different functional groups such as Aminomethylphosphonic, Sulfonic acid/phosphonic acid, and Iminodiacetic for purifying N-Methylethanolamine containing a number of metal ions at tens to hundred ppb levels. The evaluation started with adsorbent screening by measuring the adsorption capacities, all conducted in a Clean Laboratory. Twenty-four trace metals were analyzed with ICP-MS. The best adsorbent was identified among those screened according to the major metal impurity removing efficiency, namely for Fe and Zn. The adsorption isotherm and the breakthrough curve were then measured for the identified adsorbent and a preliminary absorber design was conducted.

677f Selective Adsorption of Chlorofluorocarbons(CFC) and Hydrochlorofluorocarbons(HCFC) in Hydrofluorocarbon's (HFC) Using Adsorbent Supported Metal Oxide.

Anup Kumar Doraiswamy

Hydrofluorocarbons (HFC) are used in a variety of applications such as blowing agents, refrigerants and plasma etching gases for the semiconductor industry and are often commercially manufactured from fluorination of chlorofluorocarbons (CFC) or hydrochlorofluorocarbons (HCFC). As a result, these products may contain undesired impurities which may be toxic, have high global warming or ozone depletion potential, as well as impact the end market process use of the gas. Some of these impurities form azeotropic combinations with the matrix gas leading to difficulties in separation using conventional distillation. A novel purification method to selectively remove these undesired impurities (CFC's and HCFC's) from hydrofluorocarbons using metal oxide based adsorption bed along with specific pre-treatment processes will be discussed. Conventionally, these adsorbent beds have been employed to completely remove or scrub halocarbons from various gas streams for environmental purposes. In this application, we are taking advantage of the relatively higher affinity of chlorine containing compounds towards metals and metal oxides compared to the fluorocarbons for selective adsorption thereby producing ultrahigh purity electronic gases. Examples will be presented including 245fa (1,1,1,3,3-Pentafluoropropane) purification utilizing a metal oxide adsorption bed where some key CFC/HCFC impurities including 1,1,2 trichlorotrifluoroethane and 2,2 dichloro-1,1,1trifluoroethane were observed to be reduced from 284 ppm and 16 ppm to << 1ppm respectively when measured using GC/MS/TCD.

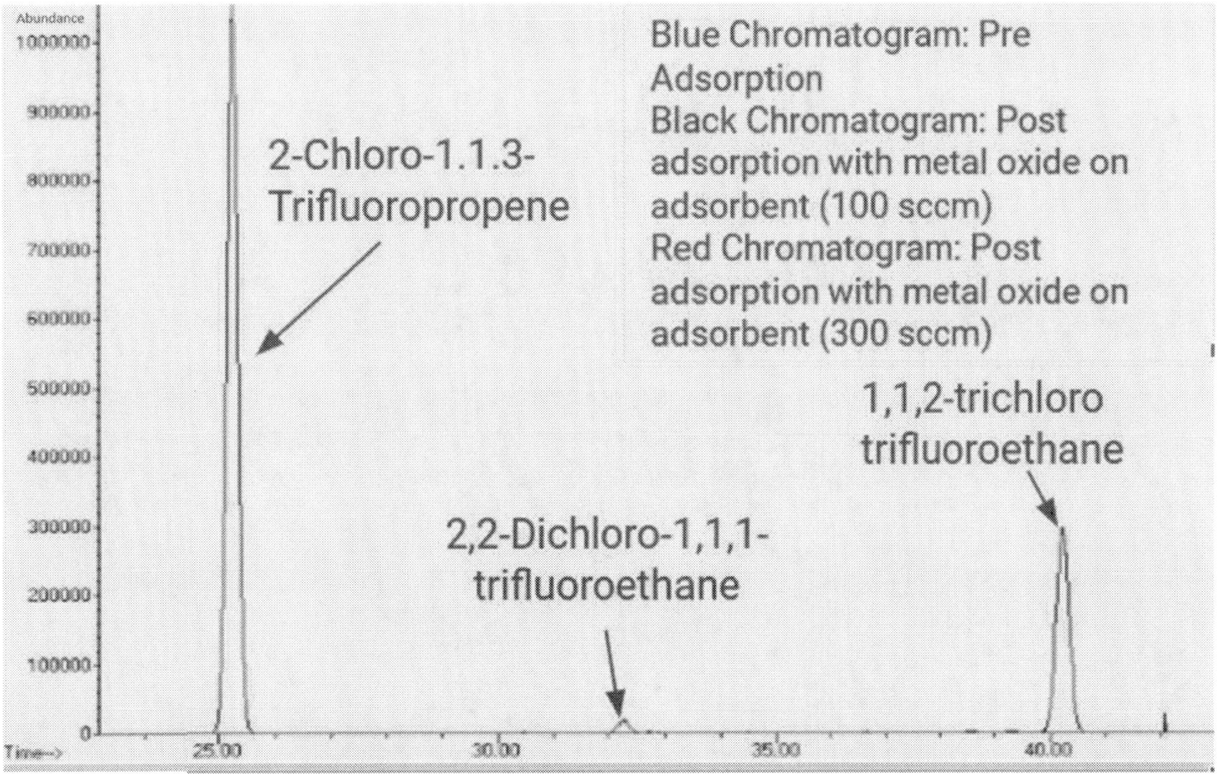

568a Simultaneous Impregnation and Encapsulation of Silica Gel with Salt Hydrates for Thermal Energy Storage Applications

Suboohi Shervani, Curtis Strong, F Handan Tezel

Several materials have been tested for thermal energy storage so far. The energy storage density for salt hydrates in host matrix is high. However, the salt hydrates, especially LiCl and $CaCl_2$ are not stable in host matrix due to low deliquescence relative humidity. The current paper focuses on the simultaneous impregnation and encapsulation of silica gel matrix with salt hydrates for thermal energy storage applications. Methyl cellulose was used as the encapsulating agent to encapsulate salt hydrate and silica gel. The thermal energy storage properties were measured using a lab scale thermal energy storage setup. An energy storage density of 241 kWh/m^3 (0.87 GJ/m^3) and specific energy of 630 Wh/kg (2268 kJ/kg) were achieved. The cyclic properties of the prepared materials were also tested and they were found to be very stable. These composites were also characterized with Scanning Electron Microscopy, which showed the stable morphology and porous structure.

568b Non-Vapor Compression Adsorptive Heating and Cooling Technology

James A. Ritter, K. James Hay, Armin Ebner

Global warming is perhaps the largest threat facing mankind in the 21st century. There are only a limited number of ways this global warming threat can be mitigated. One such way is to eliminate the use of HFC-134 as a refrigerant in air conditioning systems in various military, commercial and residential platforms. It is purported that water vapor can be used instead of HFC-134 as a refrigerant, but not as a replacement refrigerant in existing air condition systems. An entirely new non-vapor compression air conditioning system is proposed based on proven concepts borrowed from pressure swing adsorption (PSA) technology. This new technology is referred to simply as a PSA heat pump (HP).

PSA is commonly used in industry for the separation and purification of gas streams, like in the production of oxygen from air and in gas dehydration. Commonly, in PSA, an adsorbate fluid is moved into a substrate adsorbent. Gas species adsorb more to an adsorbent at higher pressures and desorb reversibly from the adsorbent at lower pressures. If the adsorbent is selective for one gas species over another, a separation or purification can be achieved. The same principle of adsorption to and reversible desorption from an adsorbent can be applied to the pressure swing cycling of a pure gas or vapor for energy storage or transfer. Hence, a PSA HP system utilizes water vapor under vacuum for cooling and heating applications with high efficiency.

It must be emphasized that this PSA HP system is not based on the traditional concept of an adsorption heat pump, where no mechanical pump is used. In that kind of system, the transport of the coolant (working fluid) is carried out instead by heating/desorbing at high pressure in one bed and by cooling/adsorbing at low pressure in another bed, thereby eliminating the need for a compressor or pump. The major problem with this traditional approach is the excessively small coefficient of performance (COP) (< 1.0) mainly due to electrical energy being required for bed regeneration during the heating/desorbing cycle step. In this new PSA HP technology, the adsorption/desorption phenomena do not replace the role of the pump; on the contrary, it requires and thus takes advantage of a pump being present. Hence, electric power is needed only to operate the pump, not for electrically heating the bed. What is identified as isothermal condensation in a traditional vapor compression HP system is carried out by adsorption at high temperature and pressure in this new PSA HP system, and what is identified as isothermal evaporation in a traditional vapor compression HP system is carried out by desorption at low temperature and pressure in this new PSA HP system.

This new non-vapor compression PSA HP cooling system cycles an adsorbate fluid such as water vapor between two adsorbent beds such as BPL activated carbon in a closed loop under vacuum. A vacuum pump is utilized to swing pressure between a high-pressure water vapor adsorbed state and a low-pressure water vapor non-adsorbed state. During the pressure swing process, when water transitions from a fluid state to an adsorbed state, heat is released (exothermically) during the adsorption process which can be ejected from the system to heat an object or space or be ejected to a thermal dump such as the atmosphere. During the pressure swing process, when water transitions from an adsorbed state to a fluid state, heat is absorbed (endothermically) during the desorption process which can be extracted from an object or space to cool the object or space. This presentation will describe this highly efficient means to transfer thermal energy to and from spaces for cooling and heating by cycling water vapor back and forth from a series of heat exchanger tubes filled with just BPL activated carbon via simulation with the UofSC dynamic adsorption process simulator (DAPS).

568c Use of Oat Hulls for Adsorption Based Thermal Energy Storage Applications

Suboohi Shervani, F Handan Tezel

Thermal energy storage applications by adsorption have attracted a lot of attention recently from researchers since it is a clean technology that will promote the use of renewable sources of energy. Different classes of adsorbent materials have been tested for thermal energy storage applications in the literature. However, among those, biomass is catching more attention due to its sustainable and renewable characteristics. Oat hulls have been used in generating electricity and were found to be very promising and cheap. On the other hand, salt hydrates based composites have been attractive options for space heating and domestic hot water applications, due to their high energy density values, their optimal operating temperature ranges and lack of toxic chemicals. This paper focuses on the thermal energy storage properties of oat hulls based composite adsorbents with salt hydrates. The composites were prepared and tested for structural characterization by Scanning Electron Microscopy. Thermal energy storage properties of these composites were measured by a lab scale setup for moisture adsorption from air at different relative humidities. The pure oat hulls have been observed to be stable for multiple cycles with 46 kWh/m^3 of energy storage density at 50% Relative Humidity at the inlet after a regeneration temperature of 80°C.

568e Bioenergy Production with Carbon Capture: An Optimization of Materials

Jackson Faylor, Edgar Ramirez Contreras, Kathy Tong, David Wagner

Bioenergy with carbon capture and storage (BECCS) is a carbon negative process for energy production. BECCS has the potential to become a major process in the fight against climate change and has been increasingly recognized as such since its inception in 1998. One thing that is particularly intriguing about BECCS is that, in addition to producing energy while removing carbon from the atmosphere, it can produce solid carbon which can be easily stored and used for industrial applications. BECCS is a promising process with a low-energy demand.

We have used a design of experiments to quantify the energy production of biomass devolatilization coupled with the adsorption capacity of the resulting biochar. Tested factors include temperature, heating rate, particle size, fuel/biomass type, and gas atmosphere. There are many properties of biochar that can be affected by the temperature at which thermal decomposition occurs. As the temperature increases, we see a reduction in volatile matter due to dehydration and devolatilization. Most of the organic material is volatile, therefore increasing the temperature will evolve more potential energy, however, there is a trade-off between energy and carbon capture potential. Aside from the temperature, heating rate is also an important parameter because that can determine biochar yield. The higher the heating rate, the lower the biochar yield, but the better the CO2 adsorption capacity. Determining the optimal heating rate for the biomass is a goal of this study.

A design of experiments approach was utilized to determine specific adsorption capacity of CO_2. Specific capacities were in excess of 1.5 mmol CO_2 per gram of biochar for temperature cycles between 30°C and 80°C. Additionally, activation energies and exothermic peaks are determined via thermogravimetry and calorimetry. Scoping studies show a consistent activation energy curve as conversion increases to approximately 60%, where a peak indicates the devolatilization of most of the organic matter. We suspect that this presents the limit to energy production and shows the approximate peak of carbon capture potential for BECCS applications.

144a Molecular Simulation and Experimental Study for the Separation of Azeotropic Hydrofluorocarbon Refrigerant Mixtures Using Silicalite.

Eliseo Marin Rimoldi, Andrew Yancey, Edward Maginn, Mark B. Shiflett

Hydrofluorocarbons (HFCs) are a type of fluorinated refrigerant used extensively in residential and industrial refrigeration systems. These substances were developed to replace chlorofluorocarbons due to their high ozone depletion potential (ODP). While HFCs have a zero ODP, they have a high global warming potential (GWP). Various government regulations targeting global warming effects have increased the interest to find new technologies to reduce fluorinated greenhouse gas emissions from refrigeration systems. Among HFCs, R-410A is a widely used refrigerant that is composed of 50.0% mass HFC-32 and 50.0 % mass HFC-125. There is a pressing need to find a sustainable separation process for this mixture that allows for its future recycling. This study examines the zeolite silicalite as a separation agent of R-410A using grand canonical Monte Carlo molecular simulations and experiments using a Xemis gravimetric microbalance. Thermodynamic properties such as isosteric heats, selectivities and adsorption isotherms are presented for HFC-32 and HFC-125 and compared with experimental data. Additionally, simulations provide molecular insights on the adsorption mechanism. Mixed-gas adsorption isotherms are presented and compared to ideal adsorbed solution theory.

144b Separation and Purification of High Value Chemicals Extracted from Food Waste

Yagya Gupta, Laura Elizabeth Beckett, Sunitha Sadula, Vibin Vargheese, LaShanda Korley, Dionisios Vlachos

The profound challenge of food waste (FW) is evidenced by the fact that approximately a third of global food production (*i.e.,* 1.3 billion tons) ends up as waste. Traditional FW treatment methods, such as disposal into landfills and composting, are environmentally unfriendly. Intensive exploration of other methods that selectively convert the diverse components of FW to high commercial value chemicals is a key to alleviating this challenge. Extraction of antioxidant and antimicrobial activity providing chemicals (flavonoids, phenolic acids, etc.) from FW feedstocks is one such economically lucrative option to valorize FW. However, separation and purification of these high commercial value chemicals is challenging owing to their similar chemical nature, high boiling points and low concentrations in FW. To the best of our knowledge, currently there is no method to separate and purify phenolic compounds post extraction from FW or non-food biomass. Our study directly addresses this issue by proposing a scalable and extensible methodology for separation of the target molecules using molecular imprinted polymers. Synthesis of polymers with imprinted functionality is key in optimizing the selectivity and performance of the material. Computational methods were used to screen through 28 monomers to expose their interaction with the target molecule. Experimental investigations revealed that itaconic acid with chlorogenic acid as the template provides the highest separation factor (higher than the literature standard, acrylamide). The performance of the polymer at separating structurally similar was tested in various extraction solvents. Complementing IR studies were done to understand the underlying interactions. The application of the polymer for separation of polyphenols extracted from real food waste and standard mixtures with known composition is demonstrated.

144c Microwave Swing Adsorption for Rapid Cycle CO_2 Capture: Effect of Material and Process Parameters

Yamid Alí Gómez Rueda, Eduardo Pérez-Botella, Brieuc Verougstraete, Mohsen Gholami, Joeri Denayer

In Thermal Swing Adsorption, typically a preheated gas is used to regenerate the adsorbent. Alternative, electrified, heating methods such as Joule heating (ESA), Induction heating and Microwave heating allow for a direct and contactless heating of the adsorbent, allowing independent control of the heating process and purge gas flow. Such electrified methods therefore allow to obtain the product at higher purity (less or no dilution by a purge gas) and, if available, can make use of renewable electricity, reducing the carbon footprint of the adsorption process. Disadvantages of the aforementioned methods include the need of specific materials that are compatible with the heating method and problems associated to scale up.

In the present work, we have studied the regeneration of adsorbents using microwave irradiation. Carbon based adsorbents were formulated under the form of extrudates or as structured adsorbent. A novel experimental setup, using a 915MHz Microwave generator, was built for cyclic adsorption-desorption experiments. The effect of adsorbent shape and dimensions on the efficiency of the heating process was evaluated. Temperature distribution and evolution was monitored as a function of microwave power, irradiation time and material properties. The efficiency of the conversion of microwave energy into heat generated within the adsorbent was determined. Adsorbent bed regeneration using Microwave Swing Adsorption was regeneration was evaluated for 2 important industrial applications.

144e Investigation of Hydrothermal Carbonization and Chemical Activation Process Conditions on Hydrogen Storage in Loblolly Pine-Derived Superactivated Hydrochars

Al Ibtida Sultana, Toufiq Reza

While the challenge of storing hydrogen in inexpensive and renewable adsorbents is relentlessly pursued by researchers all over the world, application of hydrochar derived from biomass is also gaining attention as it can be subsequently chemically activated using activating agents like KOH in order to tailor the development of favorable porosity. However, the synergistic effect of hydrothermal carbonization (HTC) process conditions as well as KOH activating conditions on the development of surface morphology is required to be assessed with the application of such porous superactivated hydrochars in hydrogen storage application. In this study, highly porous superactivated hydrochars were fabricated from inexpensive and abundant loblolly pine. Loblolly pine was hydrothermally carbonized at 180 °C, 220 °C and 260 °C and the hydrochars were then activated at different experimental conditions of 700 °C, 800°C and 900 °C using solid KOH to loblolly pine hydrochar ratio of 2:1, 3:1 and 4:1 to produce superactivated hydrochars. Superactivated hydrochars as well as loblolly pine and its corresponding hydrochars underwent physicochemical analysis as well as surface morphology analysis by SEM and nitrogen adsorption isotherms at 77 K in order to investigate the effect on BET, pore volume, and pore size distribution due to various process conditions. The superactivated hydrochars were then analyzed to quantify total hydrogen storage capacity of these materials at 77 K and up to pressure of 55 bar. Porosity of superactivated hydrochars were as high as 3666 m^2/g of BET specific surface area (SSA), total pore volume of 1.56 cm^3/g and micropore volume of 1.32 cm^3/g with the hydrogen storage capacity of 10.2 wt.% at 77 K and 55 bar. It was conclusive from principal component analysis that higher HTC temperature with moderate activation condition demonstrated favorability in developing porous superactivated hydrochars for hydrogen storage applications.

144f Effect of Moisture on Sub-Ambient DAC with MIL-101(Cr)-Amine Sorbents

Guanhe Rim, Mingyu Song, Pranjali Priyadarshini, Fanhe Kong, Ryan P. Lively, Christopher W. Jones

The level of humidity in air is a crucial factor that impacts the performance of solid sorbents in direct air capture (DAC) of CO_2, impacting sorption kinetics, thermodynamics and energy consumption in sorbent regeneration. There may be advantages to operating DAC technologies in regions where the absolute humidity is low. As the absolute humidity dramatically reduces with decreasing temperature, it may be advantageous to operate DAC plants in cold or temperate climates. Unfortunately, very few DAC sorbents have been studies under "sub-ambient" conditions (< 20 °C). In this study, amine-impregnated (TEPA, tetraethylenepentamine or PEI, poly(ethyleneimine)) MIL-101(Cr) metal-organic frameworks (MOFs) are prepared, characterized, and used as model sorbents for sub-ambient DAC. Their 400 ppm CO_2 adsorption behavior is investigated at sub-ambient and humid conditions with CO_2/H_2O temperature programmed desorption (TPD) and *in situ* FTIR spectroscopy, supplemented by cyclic sorption/desorption cycling to probe the potential for low temperature DAC operation.

144g Impact of Functional Groups on LiCl Impregnation and Performance of LiCl@ MIL-101 Analogs for Atmospheric Water Harvesting

Mengjiao Wu, Krista Walton

Supporting hygroscopic salt with porous materials is a promising strategy for atmospheric water harvesting. The impregnated hygroscopic salt enables high water uptake at low relative humidity, while the porous structure allows for water uptake at higher relative humidity, also preventing salt aggregation and leaching after sorption cycles. Previous studies have focused on LiCl loading into large pore water-stable MOFs. Studies on functionalized MOF materials are, however, have been missing. In this work, LiCl was impregnated into MIL-101, and into its amino or hydroxyl functionalized analogs at various salt solution concentrations. The amine groups were introduced by using a modified linker, whereas hydroxyl groups were introduced by immobilizing ethanolamine at the open Cr metal sites of the parent MIL-101. The concentrations of aqueous LiCl solutions were optimized for each MOF material. It has been found that the introduction of the amine functional group promotes salt impregnation, while the hydroxyl group leads to a decreased amount of salt impregnated. The final water uptake depends on the balance between the salt loading and the remaining pore volume for water adsorption. The best performing sample was cycled, and the performance was retained, indicating the stability of these composites. This composite sorbent was then immobilized onto porous Ni foam for future water-harvesting device integration. By using concentrated LiCl and MIL-101 solutions, mechanically stable LiCl@MIL-101/Ni foams were obtained. The initial water uptakes were higher than for powders due to external LiCl excess. The leaching of extra framework salt eventually leads to water uptake comparable to that of the tested powders. Performance was retained after several cycles.

144h Application of ZIF-8 and Zeolite Y in a Pressure-Temperature Swing Adsorption Process for Simultaneous H_2 Purification and CO_2 Capture

Hassan Azzan, David Danaci, Camille Petit, Ronny Pini

Abstract:

By enabling carbon-neutral hydrogen (H_2), the coupling of fossil fuel-based H_2 production with carbon capture and storage (CCS) is one of the greatest opportunities to fast-track the decarbonisation of Europe's energy system[1]. At the industrial scale, H_2 recovery and purification is carried out using a Pressure Swing Adsorption (PSA) process, whereby additional parallel adsorbers can be added to capture carbon dioxide (CO_2) from the tail gas[2]. Recent studies show that co-purification of CO_2 and H_2 at high purities and recoveries can be achieved in a single process unit deploying a single vacuum PSA (VPSA) cycle[3,4]. However, these novel schemes use only one adsorbent and are limited to cycles driven solely by pressure and vacuum. We contend that opportunities for process improvement exist, when one considers: (i) pressure-temperature swing adsorption cycles as well as (ii) novel adsorbents or adsorbent combinations that can handle the impurities present in the feed gas. In this endeavor, a major limitation is the scarcity of adsorption equilibrium data for CO_2, H_2 and the relevant impurities (e.g., CH_4, N_2) over a sufficiently wide range of temperature (25-120° C) and pressure (< 30 bar), and for a range of adsorbent materials. These data are not only essential input parameters for process simulators, but can also be used for rapid material screening purposes within equilibrium-based process modelling and optimization workflows[5].

In this study, we systematically compare the ability of adsorbents to co-purify of H_2 and CO_2 within a single PTSA process. To this end, we have measured equilibrium adsorption isotherms of CO_2, H_2, CH_4 and N_2 on a commercial zeolitic imidazolate framework, ZIF-8, and Y-type zeolites (both H^+ and cation exchanged forms) in a Rubotherm magnetic suspension balance in the pressure range 0 – 30 bar and at temperatures between 25 – 120° C. We show that these materials present favorable characteristics for a PTSA process, because they feature a 'gentle' uptake at low pressure, while providing sufficient working capacity at higher pressures.

We have also assessed the potential of these adsorbents, and of those for which literature data are available, for H_2 purification with CO_2 capture using an adsorption-equilibrium model combined with multi-objective optimization. To this end, we have considered a range of feed gas conditions corresponding to different fossil fuel based H_2 production processes. A simple process configuration is chosen that includes any combination of the following steps: adsorption, blowdown, heating, evacuation and pressurization. As such, process optimization has been carried out by considering both VPSA and VPTSA cycle schemes by using the evacuation and blowdown pressure as well as the desorption temperature as adjustable variables. We have evaluated material performance in terms of purity and recovery of CO_2 and H_2, working capacity and energy consumption. We show that for adsorbents, such as zeolite 13X and activated carbon, the addition of a temperature swing step can reduce the energy consumption up to 30%, while maintaining the required working capacity. The improvement is more evident for ZIF-8 because it possesses a higher adsorption capacity at the feed pressure, while featuring a moderately steep isotherm at low pressure, thereby reducing the vacuum work. Dynamic adsorption studies will be conducted on these promising adsorbents to provide experimental demonstration of their performance as well as the parameterization needed for detailed process modelling and scale up.

References

1. IEA. The Future of Hydrogen. *Futur. Hydrog.* (2019) doi:10.1787/1e0514c4-en.
2. Sircar, S. andamp; Golden, T. C. Purification of hydrogen by pressure swing adsorption. *Sep. Sci. Technol.* **35**, 667–687 (2000).
3. Streb, A., Hefti, M., Gazzani, M. andamp; Mazzotti, M. Novel Adsorption Process for Co-Production of Hydrogen and CO2 from a Multicomponent Stream. *Ind. Eng. Chem. Res.* **58**, 17489–17506 (2019).
4. Streb, A. andamp; Mazzotti, M. Novel Adsorption Process for Co-Production of Hydrogen and CO_2 from a Multicomponent Stream-Part 2: Application to Steam Methane Reforming and Autothermal Reforming Gases. *Ind. Eng. Chem. Res.* **59**, 10093–10109 (2020).
5. Danaci, D., Bui, M., Mac Dowell, N. andamp; Petit, C. Exploring the limits of adsorption-based CO2 capture using MOFs with PVSA-from molecular design to process economics. *Mol. Syst. Des. Eng.* **5**, 212–231 (2020).

144i Adsorption of Pfas in Heteroatom-Doped Mesoporous Carbons: Experiment and Simulation

Dipendu Saha, Zachary Romisher, Scott Van Bramer, Sandip Khan, Swasti Medha, Joanna Weyrich

Per- and polyfluoroalkyl substances (PFAS) are class of fluorinated alkyl compounds that includes about 5000 individual compounds. They are ubiquitous and found in a diverse spectrum of industrial and household objects. PFAS have contaminated the water streams of most developed and developing countries and are considered an emerging pollutant. These anthropogenic compounds do not have any natural sources and they are extremely difficult to degrade. There are no known natural occurrences or microbial activity that break down this class of materials. As a result, they are often referred to as 'forever chemicals'. PFAS bioaccumulates by direct or passive exposure and causes serious illnesses in humans. Adsorption is one of the most promising strategies for removal of PFOS from drinking water and wastewater. In this research fluorine and nitrogen-doped mesoporous carbons were synthesized with different proportions of heteroatoms. These carbons were characterized and the adsorptive removal of perfluorooctane sulfonic acid (PFOS, one type of prominent PFAS) in different water matrices was studied. In addition, molecular dynamics (MD) simulations were performed to calculate thermodynamics parameters of PFOS adsorption. The results confirmed that this carbon demonstrated an excellent removal capacity of PFOS from different types of water matrices compared to the commercially available granular activated carbon (GAC).
Per- and polyfluoroalkyl substances (PFAS) are class of fluorinated alkyl compounds that includes about 5000 individual compounds. They are ubiquitous and found in a diverse spectrum of industrial and household objects. PFAS have contaminated the water streams of most developed and developing countries and are considered an emerging pollutant. These anthropogenic compounds do not have any natural sources and they are extremely difficult to degrade. There are no known natural occurrences or microbial activity that break down this class of materials. As a result, they are often referred to as 'forever chemicals'. PFAS bioaccumulates by direct or passive exposure and causes serious illnesses in humans. Adsorption is one of the most promising strategies for removal of PFOS from drinking water and wastewater. In this research fluorine and nitrogen-doped mesoporous carbons were synthesized with different proportions of heteroatoms. These carbons were characterized and the adsorptive removal of perfluorooctane sulfonic acid (PFOS, one type of prominent PFAS) in different water matrices was studied. In addition, molecular dynamics (MD) simulations were performed to calculate thermodynamics parameters of PFOS adsorption. The results confirmed that this carbon demonstrated an excellent removal capacity of PFOS from different types of water matrices compared to the commercially available granular activated carbon (GAC).

277a Bridging the Interfacial Gap in Mixed-Matrix Membranes for Precise Molecular Sieving in Organic Solvents

Gyorgy Szekely

Organic solvent separations based on membrane technology offer a lower energy alternative compared to traditional thermal separations. However, it requires materials that operate reliably in chemically aggressive environments. To obtain solvent-resistant membranes, crosslinking is commonly required, usually by immersing membranes in a solution of reactive crosslinking agents. In quest to achieve highly selective membranes, we investigated the mixed matrix membranes (MMMs) containing metal–organic framework (MOF) fillers, owing to the regularity and ordered pore arrangement in the MOF structure [1]. However, the poor interfacial adhesion between the polymer matrix and the MOF filler has emerged as a bottleneck, creating defects and random voids on the membrane's surface. In this study, we bridged the gap between their material constituents. We prepared a series of novel membranes comprising MOF nanoparticles with similar chemical and morphological properties but having various pore sizes (UiO-66–68-NH_2). The MOFs' surfaces were covalently grafted with poly(*N*-isopropylacrylamide) (PNIPAM) chains, which could then become entangled with the membranes' polymer matrix. This strategy was inspired by the natural phenomenon of strong interfacial adhesion between the calcified cementum of the tooth and the surrounding tissues provided by intermingled collagen fibers. Morphological investigations and organic solvent nanofiltration (OSN) tests revealed that membranes with PNIPAM-grafted fillers do not suffer from the formation of pinholes at the filler–matrix interface that are detrimental to the filtration performance. The experimental results showed an excellent match with a predictive model of nanofiltration built around the premise of liquid transport through the highly ordered pores of the MOF filler. The OSN test revealed the increase in the flux and molecular weight cut-off values with increasing MOF aperture size which confirm the key role that internal MOF pores play in the filtration.

[1] L. Cseri *et al.*, J. Mater. Chem. A, 2021, 9, 23793–23801.

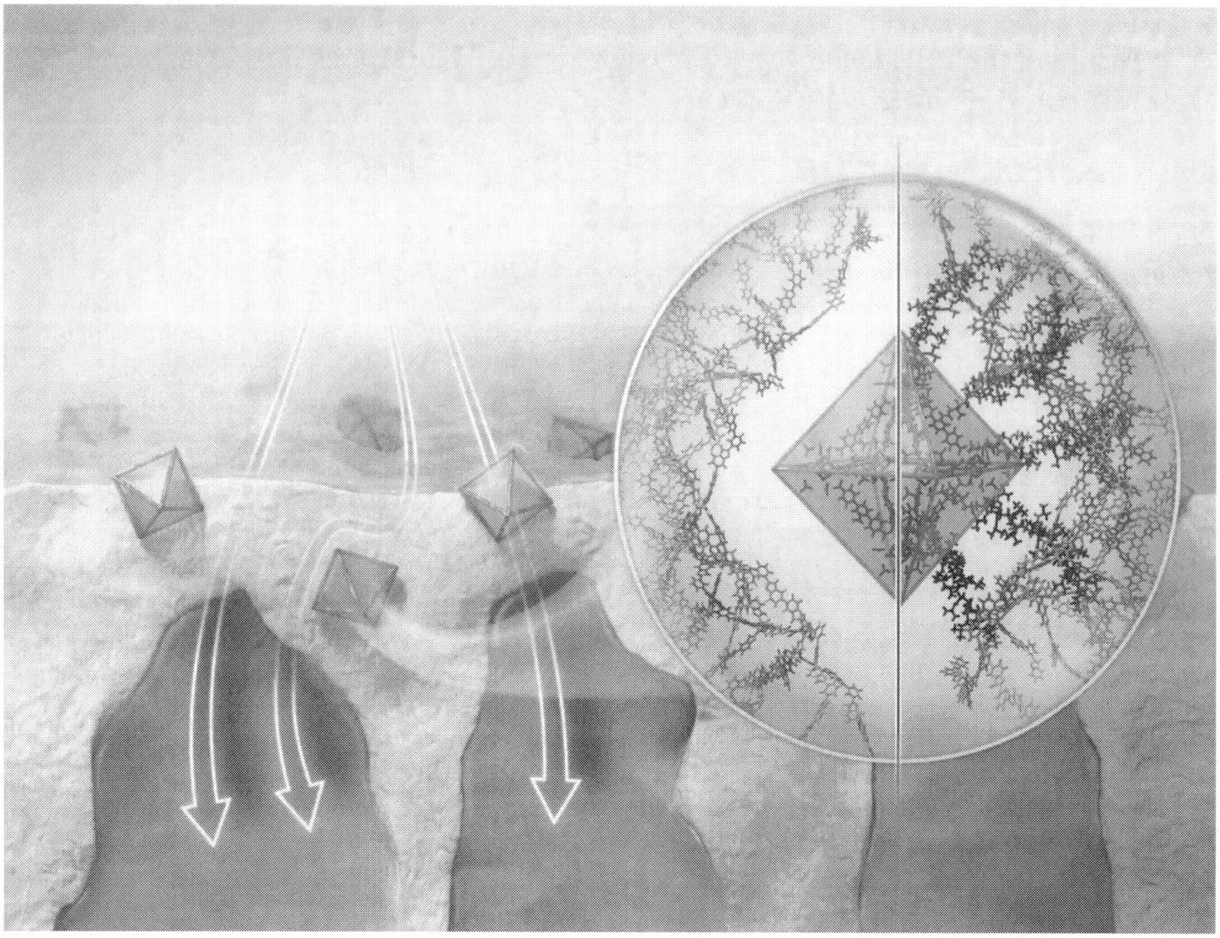

277h Isoporous Membranes as Support Membranes for Composite Membranes

J. G. Wijmans

As composite membranes are made with ever thinner top layers, the influence of the support membrane on composite membrane performance increases. This influence goes beyond the transport resistance of the support membrane becoming a rate limiting factor. We will show that the surface structure of the support membranes just by itself can reduce membrane permeance. The analysis shows that isoporous surfaces are uniquely suited to serve as support membranes. Experimental data will be presented plus the results of analytical methods that provide insights into the formation of isoporous surfaces.

277c Development of High-Performance Polymeric Membrane for Natural Gas Dehydration

Eyad Qasem, Suzana Nunes

Abstract

As the world is shifting toward renewable energy, natural gas consumption has increased since it emits less carbon compared to coal and oil. To meet this increasing demand, natural gas industries are largely investing in upgrading gas processing infrastructure. Gas processing is required to eliminate impurities, primarily by using absorption, adsorption, and cryogenic processes. Conventionally, the natural gas is dehydrated by the glycol absorption unit, which has complex process and maintenance issues; hence, alternative technology needs to be developed.

Membrane technology may be a viable option for sour gas treatment, hydrocarbon dewpointing, and dehydration in natural gas treatment. The membrane process has a significant potential for gas dehydration as it is small, durable, energy-saving, and commercially marketable. In the case of natural gas dehydration, water vapor is separated from a mixture of methane, carbon dioxide, and heavy hydrocarbons. Although most polymeric membranes have a very high selectivity of water vapor over methane, recent research has proved that this specific membrane application is limited by methane loss due to mass transfer limitations. Therefore, intensive study for lab-scale membranes under a harsh processing environment is required to develop membrane materials for this application.

This work aims to develop a high-performance polymeric

membrane for natural gas dehydration by enhancing gas separation properties of mechanically and thermally stable polymers blended with highly hydrophilic polymers. Dense films were fabricated with polymer blends and characterized by several techniques to understand the blending behaviors and the intrinsic gas separation properties. To further improve the performance for natural gas dehydration, thin-film composite membranes were fabricated and characterized with mixed gases, and their stability was evaluated under harsh stream conditions.

277d Cross-Linked Block Copolymer Membranes

Radoslaw Gorecki, Sandra Aristizabal Guerrero, Saibal Bhaumik, Konstantinos Ntetsikas, Nikos Hadjichristidis, Suzana Nunes

Porous and dense polymer-based films are easily manufactured by conventional large-scale methods and are widely used as membranes. Polymer membranes find their applications in various separation processes in the industry, medicine, or biotechnology. They are mainly produced based on commercially available homopolymers, such as polysulfone, polyacrylonitrile, and polydimethylsiloxane. However, particularly convenient properties can be achieved using block copolymers (BCPs), including commercial ones like polyether-*b*-polyamide (Pebax®), and sulfonated pentablock copolymers (Nexar™) [1,2].

Depending on the composition of the BCP counterparts, such as their length, solvophilicity, and solvent used, different morphologies are formed via self-assembly of these materials in solution. These phenomena can be directly used to control the pore size distribution at the nanoscale when producing the membrane by solution casting and phase inversion [3]. When BCPs are used as dense membranes or coatings, a plethora of possibilities is opened by combining materials with different properties into a single membrane, allowing to finely tune the resulting membrane characteristics. The BCPs can be terminated with different active groups to allow cross-linking either of the self-assembled structure or cross-linking with other matrixes, which can yield a confined membrane structure providing enhanced thermal, mechanical, and chemical resistance.

Here, we present a new strategy demonstrated with polystyrene-*block*-poly(ethylene oxide) (PS-*b*-PEO) BCP with a block ratio of ~70 % PS [4]. We study the PS-*b*-PEO in two versions - double-terminated by functional groups (α, ω-difunctionalized), and monoterminated by functional groups (α-functionalized). The double termination allows cross-linking of the BCP. We explore how the self-assembly changes upon the induced cross-linking in different organic solvents and how it affects the resulting membrane morphology and stability. We utilize this knowledge to produce stable BCP cross-linked membranes to target applications such as gas separation using dense cross-linked BCP films, but also to form porous materials to be applied for liquid processing or as scaffolds for biotechnology and medicine. The scanning electron microscopy micrographs of prepared porous membranes and their respective thermal stability curve obtained by thermogravimetric analysis are presented in Figure 1.

References:

[1] S.P. Nunes, A. Car, From charge-mosaic to micelle self-assembly: Block copolymer membranes in the last 40 years, Industrial and Engineering Chemistry Research. 52 (2013) 993–1003. https://doi.org/10.1021/ie202870y.

[2] L. Upadhyaya, A.Y. Gebreyohannes, F.H. Akhtar, G. Falca, V. Musteata, D.K. Mahalingam, R. Almansoury, K.C. Ng, S.P. Nunes, NEXARTM-coated hollow fibers for air dehumidification, Journal of Membrane Science. 614 (2020). https://doi.org/10.1016/j.memsci.2020.118450.

[3] S.P. Nunes, Block Copolymer Membranes for Aqueous Solution Applications, Macromolecules. 49 (2016) 2905–2916. https://doi.org/10.1021/acs.macromol.5b02579.

[4] G. Polymeropoulos, G. Zapsas, K. Ntetsikas, P. Bilalis, Y. Gnanou, N. Hadjichristidis, 50th Anniversary Perspective: Polymers with Complex Architectures, Macromolecules. 50 (2017) 1253–1290. https://doi.org/10.1021/acs.macromol.6b02569.

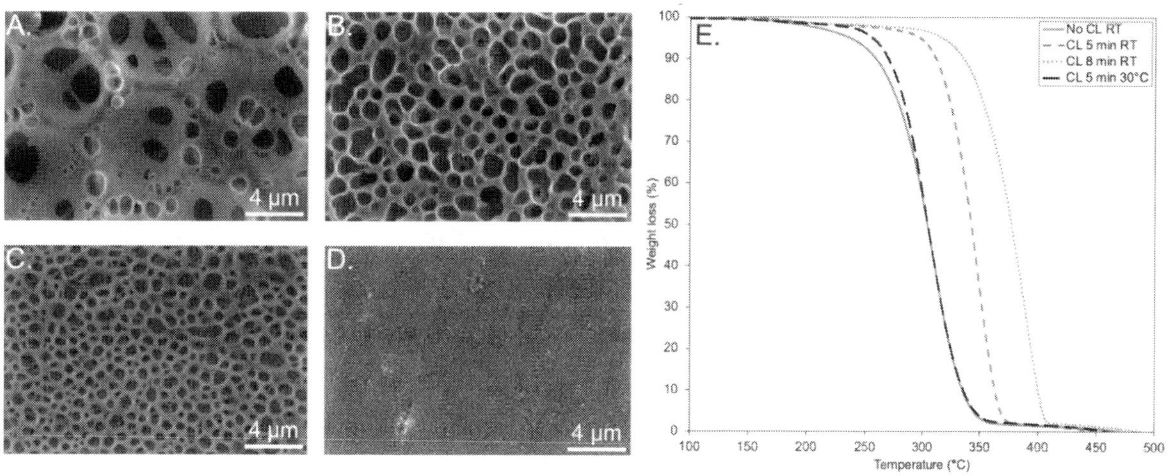

Figure 1. Scanning electron microscopy micrographs of PS-b-PEO porous membranes: A. Non-cross-linked, cast at room temperature (No CL RT); B. Cross-linked for 5 minutes at room temperature (CL 5 min RT); C. Cross-linked for 8 minutes at room temperature (CL 8 min RT); D. Cross-linked for 5 minutes at 30°C (CL 5 min 30°C). E. Thermogravimetric analysis of the membranes.

277e Polymeric Membranes for H_2 Separation

Tai-Shung Chung

Clean water, clean energy, global warming and affordable healthcare are four major concerns globally resulting from clean water shortages, high fluctuations of oil prices, climate changes and high costs of healthcare. Clean water and public health are highly related, while clean energy is essential for sustainable prosperity. Among many potential solutions, advances in membrane materials and technology are one of the most direct, effective and feasible approaches to solve these sophisticated issues. Membrane technology is a fully integrated science and engineering which consists of materials science and engineering, chemistry and chemical engineering, separation and purification phenomena, environmental science and sustainability, molecular simulation, process and product design. In this presentation, we will introduce and summarize our efforts on polymeric nanomaterials for membrane development in the fields of H_2 purification and separation. Various material and fabrication strategies to enhance membrane performance will be discussed.

277g Realisation of the Potential of Practical Graphene Oxide Membranes for Molecular Separations

Mainak Majumder

Membrane-based separation technologies have many advantages over traditional separation methods such as adsorption, and distillation in terms of energy- and cost-efficiency and modular deployment of technology. In practice, these advantages can be meaningfully harnessed only if advanced membranes with properties such as high permeance, tailorable selectivity [1-2], chlorine, pH and solvent resistance [3-4], low-fouling characteristics [1], long-term stability under operational conditions [5,6] alongside green and sustainable manufacturability [6] are demonstrated. In the last 5-6 years, our research group has taken rapid strides to realise these properties in membranes made from graphene-oxide. In this talk, I will summarize this journey reflecting on our work including some aspects of translation to industry and future directions.

[1] *Nat Commun* **7,** 10891 (2016), [2]*Phil. Trans. R. Soc. A.*374**:** 20150028, [3] *ACS Appl. Mater. Interfaces* 2018, 10, 2, 2067–2074, [4] *ACS Appl. Bio Mater.* 2020, 3, 1, 584–592, [5]*J. Mater. Chem. A*, 2020,**8**, 24800-24811, [6] *ACS Sustainable Chem. Eng.* 2021, 9, 32, 10846–10856

277f Effect of Packing Nonuniformity on the Performance of Hollow Fiber Membrane Gas Separation Modules Fabricated from Fiber Tows

Lili Sun, Grigorios Panagakos, Glenn Lipscomb

Membrane processes are the preferred option for many gas separations. Hollow fiber membranes commonly are used in these processes. Commercial hollow fiber membrane modules often are fabricated from small groups of fibers (i.e., a tow) that are arranged into a larger cylindrical bundle.

The use of fiber tows can lead to packing non-uniformity in two ways: 1) the spacing between tows and the surrounding case and 2) the spacing between tows within the bundle. This nonuniformity can result in nonideal fluid distribution in the shell and be detrimental to module performance.

Full three-dimensional models of flow and mass transfer within fiber bundles formed from tows have been developed. These models account for the detailed fiber arrangement within modules and are solved to evaluate gas separation module performance in terms of stage cut and required membrane area as a function product gas purity for carbon capture and nitrogen production applications.

Results are presented illustrating the effects of varying fiber packing within a tow and the packing of tows. The results suggest smaller tows are preferred over larger tows, at fixed overall fiber packing, and the inter-tow packing should be comparable to the intra-tow fiber packing.

To reduce the computational cost of simulating large fiber bundles, the use of an equivalent planar bundle is proposed. A simple procedure for determining the geometry of the equivalent planar bundle is described and validated through comparisons of calculated performance metrics with results obtained from full three-dimensional simulations.

342a Design and Synthesis of Polymer Membranes Based on Theoretical Principles

Mirco Sorci, Surya Karla, Chenyu Guan, Corey Woodcock, Dustin J. Andersen, Ali Reza Behzad, Suzana Nunes, Joseph Hersey, Lauren Tice, Sal Giglia, Joel Plawsky, Georges Belfort

To optimize the filtration performance of synthetic polymer membranes, their internal microstructure needs to be designed and synthesized based on theoretical principles and not empirically, as is done at present, using phase inversion and interfacial polymerization. Here, we present a rational approach in which we develop an *in-silico* tool to (a) simulate separation of particles with different properties (i.e., chemistry, size, shape, flexibility, etc.), and (b) propose, test, and optimize novel microstructures. To validate these predictions, experiments are required. The *in-silico* tool comprises 2D computational fluid and particle drag mechanics combined with particle and membrane force measurements in ionic solutions, modeled by the DLVO and the extended DLVO (xDLVO) theories, to study particle intrusion and capture in and escape through microporous polymer membranes. We have previously demonstrated[1] that (a) the *in-silico* tool predictions for a commercial membrane show substantial flow channeling and qualitatively agree with experimental filtration measurements using scanning electron microscopy with particle tracking via energy dispersive X-ray spectroscopy, and (b) tear-drop microstructures of different orientation weakly separate particles that differ in size by a factor of 2.

In this work, we have conducted extensive intermolecular force measurements between streptavidin, a stable protein coated on a polystyrene sphere, and two commercial microfiltration poly(ether sulfone) and poly(vinylidene difluoride) membranes and obtained requisite parameters such as the Hamaker constants, surface potentials, hydrogen-bonding free energies of cohesion between the water molecules and characteristic decay lengths of water (l=1.0 nm at 20°C) from theoretical fits of DLVO and xDLVO theories[2] for 100s of experimental runs at various solution conditions (pH and ionic strength). We have also used the *in-silico* tool to obtain filtration performance predictions with membranes comprising a series of spherical microstructures of one size and in mixtures of different sizes to separate suspensions of particles that differ in size by a factor of 10 (i.e., the holy grail of membrane filtration). These microstructures allow adjustment of the tortuosity of the flow path and variation of the reactivity of the surface. Experiments are also underway testing these spherical particle microstructures. To our knowledge, this is the first attempt combining particle drag mechanics with intermolecular force measurements to design and synthesize microfiltration membranes based on theoretical principles. The *in-silico* tool can be used to characterize membranes for separation performance and guide improved design, synthesis, and testing of new microporous membranes.

References

1. Sorci, M.; Wookcock, C. C.; Andersen, D. J.; Behzad, A. R.; Nunes, S.; Plawsky, J.; Belfort, G., Linking microstructure of membranes and performance. *J. Membr. Sci.* **2020,** *594*, 1-9.
2. van Oss, C., Long-range and short-range mechanisms of hydrophobic attraction and hydrophilic repulsion in specific and aspecific interactions. *Journal of Molecular Recognition* **2003,** *16*, 177-190.

342b Novel Strategy to Control the Pore Structure and Surface Chemistry of Cellulose-Based Carbon Molecular Sieve Membranes

Tiago Araújo, Francisco Barbosa, Gabriel Bernardo, Adélio Mendes

Introduction

The hastening of climate change and global warming caused by the greenhouse gas emissions to the atmosphere is a pressing issue for our society. The anthropogenic nature of this problem requires the development of new technological solutions to achieve the carbon neutrality goals implemented by the European Union for 2050 [1]. Hydrogen or biomethane are considered the vectors for this technological transition in the energy, transport and industrial systems [2]. When compared with the conventional separation technologies for CO_2 capture, air enrichment or H_2 recovery, such as cryogenic distillation, chemical absorption or pressure-swing adsorption (PSA), membrane separation technologies are the most energy efficient, versatile, low capital cost and environmentally friendly [3-5]. Carbon molecular sieve (CMS) membranes appear the most promising alternative not only for their better trade-off between permeability and selectivity but also for their chemical robustness: they work very well even in extremely harsh chemical, environmental, high-pressure and temperature applications [5-7]. CMS membranes possess a rigid pore structure with a bimodal pore size distribution: the micropores (< 2 nm) provide the sorption sites and the ultramicropores (< 0.7 nm) enable the molecular sieving mechanism [8]. This pore structure, when tuned properly for the desired separation, is responsible for the membrane's high permeability and selectivity [5]. CMS membranes are fabricated by the controlled carbonization of a polymeric precursor. Rodrigues *et al.* [9] reported for the first time a tailor-made precursor from cellulose dissolved in a mixture of ionic liquid/dimethyl sulfoxide (DMSO). These CMS membranes withstand permeate steams with a relative humidity close to 100 % without suffering pore blockage, even at room temperature. In addition to the proper selection of polymeric precursor, selection of adequate carbonization conditions and post-carbonization treatments, the optimization of the precursor preparation conditions and pre-carbonization treatments can also be used to improve the performance of the final CMS membranes. For addressing the challenge of preparing CMS membranes with the desired pore size distribution and with very high porosity, this work proposes the development of different strategies to control the structural and surface chemistry of the cellulose-based CMS membranes to the desired different separations.

Experimental

CMS membranes were fabricated by controlled carbonization conditions at 550 °C from an optimized ionic liquid cellulose-based precursor [10]. Small molecules with chemical affinity to cellulose were initially added to the cellulose/IL/DMSO polymeric precursor solution. Furthermore, the cellulose polymers were immersed in baths containing plasticizers. By immersing CMS membrane precursors (which are saturated in water) in a solution with a plasticizer, an exchange of solvents takes place. These solvents should be released during drying and carbonization, altering the micropore structure. The separation performance of such membranes was evaluated with pure gases at 25 °C, following a procedure described elsewhere [10].

Results and discussion

The results of the single-component gas permeation of the prepared CMS membranes with the additives added to the precursor solution are shown in **Figure 1A**. The addition of any additive increased the permeability of the CMS membranes to hydrogen, but the same behavior was not observed for the other gases. The addition of 1 wt.% urea produced CMS membranes with higher permeabilities than the CMS membrane without additive. Compared with the addition of propylene glycol and glycerol, it appears that the molecule with the highest molecular mass and boiling point (glycerol) produced CMS membranes with higher permeabilities than propylene glycol. This result leads us to conclude that these additives can function as functionalizing agents for CMS membranes as well as porogenic agents, as their incorporation can change the membrane gas permeabilities. Regarding the permselectivities of the CMS membranes produced (**Figure 1B**), it is verified that the addition of propylene glycol increased 19-fold the H_2/CH_4

selectivity relative to the CMS membrane without additive and increased 4-fold the CO_2/CH_4 permselectivity. Therefore, the concentration of propylene glycol in the polymeric precursor solution was increased and the CMS membranes produced presented permeabilities higher than the membrane with 1 wt.% of propylene glycol. However, the selectivities of the CMS membrane produced with 2 wt.% propylene glycol decreased. These results indicate that increasing the concentration of the additive acts as a porogenic agent and allows for the adjustment of the size of the ultramicropores for the desired separations.

Additionally, the prepared cellulose films (without any additive on the precursor solution) were immersed in different soften baths containing ethylene glycol, triethylene glycol, urea and xylitol. In **Figure 1C** are presented the results of single gas permeation of CMS membranes prepared from cellulose films with a dip-coating of 5 wt.% additives for 30 minutes. Membranes prepared with ethylene glycol presents higher permeabilities. CMS membranes prepared with the other additives have lower permeabilities to carbon dioxide and oxygen than membranes without the additive. Therefore, it turns out that the addition of different additives produces CMS membranes with different performances. The permselectivities and the respective Robeson Index of the CMS membranes prepared are listed in **Table** 1. This table highlights several promising results: the membranes prepared with urea presents an O_2/N_2 permselectivity of 198; while the membrane prepared with ethylene glycol has very exciting CO_2/CH_4 and CO_2/N_2 permselectivities. CMS membranes prepared with xylitol shows interesting results for carbon dioxide removal from flue gas. These plasticizers seam to act as protector of the membrane pores from collapsing during drying and carbonization. As described by other authors, glycerol acts as a membrane pore radius-maintaining agent by filling the micropores [11].

Conclusions

The two strategies proposed demonstrate to be extremely effective in controlling the structure and surface chemistry of membranes for various separations. Both techniques are simple and inexpensive to perform. Applying the different protocols described above, it is possible to tune the pores in order to make the prepared materials more permeable and/or selective for the desired separation. In the presentation of this work, an in-depth study of the characterization of the materials prepared will be carried out with the aim of understanding and optimizing the action of different additives in the formulation of the polymeric precursor of carbon membranes. Furthermore, the permeation results at high temperatures and different pressures of the prepared CMS membranes will be present.

Acknowledgments

Araújo is grateful to the Portuguese Foundation for Science and Technology (FCT) for the doctoral grant (reference SFRH/BD/143598/2019) supported by POPH/FSE. G.Bernardo thanks the Portuguese Foundation for Science and Technology (FCT) for the financial support of his work contract through the Scientific Employment Stimulus Individual Call – (CEEC_IND/02039/2018). This work was financially supported by LA/P/0045/2020 (ALiCE), UIDB/00511/2020 and UIDP/00511/2020 (LEPABE), funded by national funds through FCT/MCTES (PIDDAC).

References

[1] A. Sikora, European Green Deal – legal and financial challenges of the climate change, ERA Forum, 21 (2020) 681-697.

[2] D.S. Sholl, R.P. Lively, Seven chemical separations to change the world, Nature, 532 (2016) 435-437.

[3] R.W. Baker, Future Directions of Membrane Gas Separation Technology, Industrial andamp; Engineering Chemistry Research, 41 (2002) 1393-1411.

[4] G. Bernardo, T. Araújo, T. da Silva Lopes, J. Sousa, A. Mendes, Recent advances in membrane technologies for hydrogen purification, International Journal of Hydrogen Energy, 45 (2020) 7313-7338.

[5] L. Lei, L. Bai, A. Lindbråthen, F. Pan, X. Zhang, X. He, Carbon membranes for CO2 removal: Status and perspectives from materials to processes, Chemical Engineering Journal, 401 (2020) 126084.

[6] T. Araújo, G. Bernardo, A. Mendes, Cellulose-Based Carbon Molecular Sieve Membranes for Gas Separation: A Review, Molecules, 25 (2020).

[7] L. Lei, F. Pan, A. Lindbråthen, X. Zhang, M. Hillestad, Y. Nie, L. Bai, X. He, M.D. Guiver, Carbon hollow fiber membranes for a molecular sieve with precise-cutoff ultramicropores for superior hydrogen separation, Nature Communications, 12 (2021) 268.

[8] M. Rungta, G.B. Wenz, C. Zhang, L. Xu, W. Qiu, J.S. Adams, W.J. Koros, Carbon molecular sieve structure development and membrane performance relationships, Carbon, 115 (2017) 237-248.

[9] S.C. Rodrigues, M. Andrade, J. Moffat, F.D. Magalhães, A. Mendes, Preparation of carbon molecular sieve membranes from an optimized ionic liquid-regenerated cellulose precursor, Journal of Membrane Science, 572 (2019) 390-400.

[10] T. Araújo, M. Andrade, G. Bernardo, A. Mendes, Stable cellulose-based carbon molecular sieve membranes with very high selectivities, Journal of Membrane Science, 641 (2022) 119852.

[11] L. Lei, A. Lindbråthen, X. Zhang, E.P. Favvas, M. Sandru, M. Hillestad, X. He, Preparation of carbon molecular sieve membranes with remarkable CO2/CH4 selectivity for high-pressure natural gas sweetening, Journal of Membrane Science, 614 (2020) 118529.

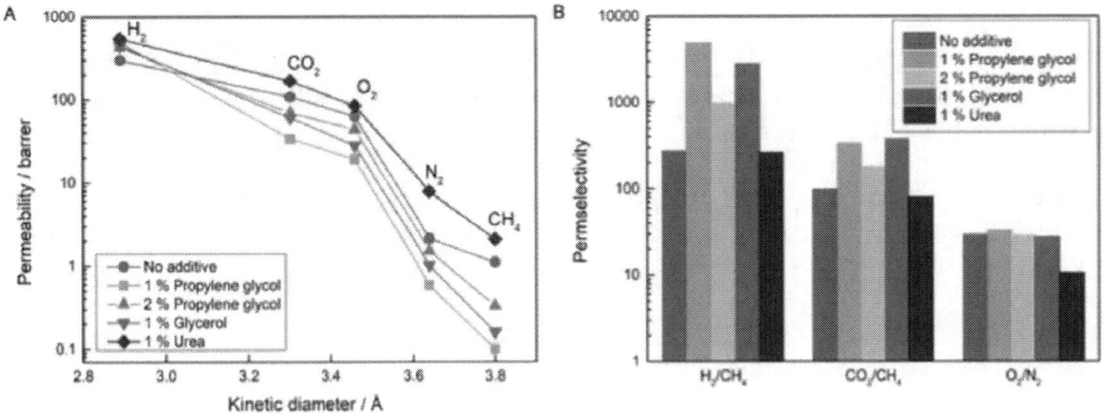

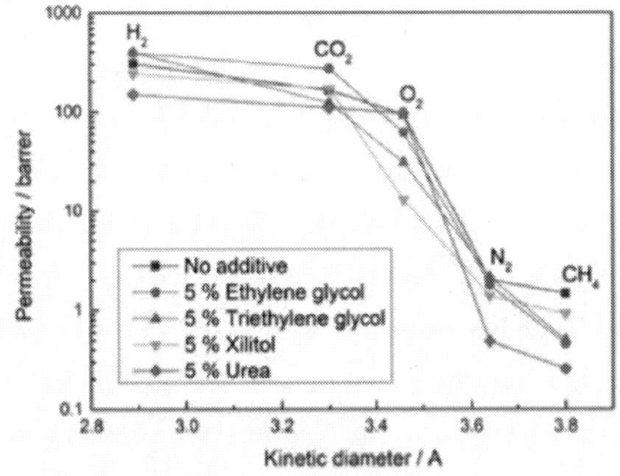

Table 1. CMS membranes prepared with 5 wt.% of dip-coated additives permsectivity (α) and respective Robeson Index (θ) for O_2/N_2, CO_2/CH_4, CO_2/N_2 and H_2/CH_4.

	O_2/N_2		CO_2/CH_4		CO_2/N_2		H_2/CH_4	
	α	θ	α	θ	α	θ	α	θ
No additive	47.0	8.60	112	2.20	82.0	1.20	208	3.60
Ethylene glycol	34.8	5.90	590	13.8	152.5	2.70	853	18.5
Triethylene glycol	14.3	2.20	245	4.20	57.0	0.80	795	17.5
Xylitol	9.50	1.20	186	3.70	125	2.80	263	3.70
Urea	198	36.6	428	7.10	225	2.90	583	5.30

Figure 1. Influence of the additive addition on the CMS polymer precursor preparation on **(A)** single gas permeation and **(B)** ideal selectivity; **(C)** CMS membranes gas permeability to different gas species of CMS membranes prepared with 5 wt.% of dip-coated additives.

342c Cellulose-Cellulose Composite Membranes for Ultrafiltration

Mathias Ulbricht, Duc Hoa Tran

Membranes made from cellulose were in history and are also currently important for the development of membrane technologies. Alternative fabrication methods for completely biopolymer-based separation membranes are of high interest with view on improved sustainability [1].

This project explores a new kind of composite membrane, with a thin selective cellulose layer on filter paper as porous support. Suited ionic liquids (IL) in combination with dimethylsulfoxide (DMSO) can be used to dissolve cellulose under mild conditions and subsequently to cast regenerated cellulose materials via shaping and phase separation [2]. The novel fabrication method of completely renewable and biodegradable cellulose-cellulose composite membranes for ultrafiltration (UF) is therefore based on film casting of a solution of cellulose in an IL/DMSO mixture on filter paper as support and subsequent phase separation in water. Typical composite membranes have a cellulose film thickness of a few μm on filter paper with a pore size in the range of several μm. The separation performance of the obtained membranes can be specifically adjusted by fabrication parameters such as cellulose concentration, structure of IL and DMSO co-solvent content in the casting solution as well as coagulation bath conditions. When using the dry support, it is partially penetrated by the cellulose solution during the film casting, leading to too low permeance. The separation performance can be improved by pre-wetting the support with various liquids to prevent pore penetration, but that can also weaken the lamination between cellulose barrier layer and paper support, as shown by adhesion tests. It was found that prewetting of the filter paper with DMSO can solve the problem, leading to composite membranes with a strongly attached cellulose film on the paper support and superior, tunable UF properties for aqueous feeds. Such membranes have also been evaluated in terms of their UF separation performance in a wide range of organic solvents.

[1] S. P. Nunes, P. Z. Culfaz-Emecen, G. Z. Ramon, T. Visser, G. H. Koops, W. Jin, M. Ulbricht, Thinking the future of membranes: Perspectives for advanced and new membrane materials and manufacturing processes, Journal of Membrane Science 598 (2020) 117761.

[2] A. S. M. Wittmar, D. Koch, O. Prymak, M. Ulbricht, Factors affecting the non-solvent induced phase separation of cellulose from ionic liquid-based solutions, ACS Omega 5 (2020) 27314-27322.

342d Self-Assembly of Zwitterionic and Charged Random Copolymers for Membranes with Exceptional Performance

Ayse Asatekin

Membranes offer a highly energy-efficient, simple to operate, scalable and portable separation method for applications that range from water treatment to oil and gas processing to pharmaceutical manufacturing. Yet, their broader use is typically limited by their limited selectivity and/or fouling in the presence of complex feeds. Polymer self-assembly is a powerful tool for developing new membranes with controlled nanostructure and pore chemistry. Many efforts in this field have focused on block copolymers, resulting in membranes with impressive features and performance. This presentation will instead focus on the self-assembly of random copolymers that combine hydrophobic repeat units with zwitterionic or charged monomers. Due to the very high incompatibility between these units, self-assembly occurs despite the very short segment lengths. The resultant materials feature a bicontinuous network of hydrophilic domains through which water and solutes permeate, held together by the hydrophobic regions. The random architecture, coupled with post-treatment steps such as cross-linking, enable membranes with extremely small effective pore sizes, from ~1.5 nm pores that separate dyes and salts, down to 0.9 nm pores the perform ion separations. Using cross-linkable zwitterionic amphiphilic copolymers (ZACs), we have developed membranes that can separate not only based on ion size and charge, but also based on the favorability of ion-zwitterion interactions, leading to unprecedented selectivity between like-sized anions. More recently, we have also shown that amphiphilic polyampholytes, defined as terpolymers that combine hydrophobic, anionic, and cationic monomers, can behave similar performance to ZACs, exhibiting tunable, controlled pore size, excellent fouling resistance, and high permeance. Importantly, this new material family offers us broader freedom to design the chemical environment in the nanopores, addressing new precision separations.

342e 3D Printed Nanofiltration Composite Membranes with Reduced Concentration Polarisation

Davide Mattia

Concentration polarisation (CP) occurs when non-interacting solutes build up near the membrane-feed interface, resulting in a higher solute concentration than that in the bulk solution. This build up can induce water flux decline, reduce quality of produced water, increase power consumption and operating costs in membrane-based water treatment, including reverse osmosis (RO) and nanofiltration (NF) processes. There is an extensive literature on methods to reduce the impact of CP through modification of the membranes' surface chemistry, structure, or a combination of the two. However, these measures have shown limited success since they also negatively impact membrane flux and, moreover, the uncertainty around their long-term stability and scalability has resulted in their low uptake by industry. Therefore, there is intense interest in methods that would effectively reduce CP without negatively affecting permeance and ensuring stable performance over time.

Here, we present an innovative approach to minimising concentration polarisation using 3D printed nanofiltration (NF) composite membranes consisting of a NF polydopamine-coated polyvinylidene fluoride (PVDF/PDA) selective layer on a 3D printed asymmetric wavy (patterned) support. The result is a wavy composite membrane with pure water permeance of 14 ± 2 LMH bar^{-1} and molecular weight cut-off of ~550 Da, measured using a crossflow NF setup at a transmembrane pressure of 2 bar for Reynolds number (Re) of 700, using a range of dyes (rejection > 95% and mass balance >97% for all tests). The CP behaviour of the composite membranes was assessed by filtration of Congo red (CR) dye solution (0.01 g L^{-1}), showing that the wavy pattern significantly reduced the impact of CP compared to flat membranes, with a nearly tripling of the mass transfer coefficient and a 57% decline of the CP factor. Computational fluid dynamics showed that these significant performance improvements were due to improved hydrodynamics, with the maximum surface shear stress induced by the wavy structure (1.35 Pa) an order of magnitude higher than that of the flat membranes (0.18 Pa) at Re = 700. These results demonstrate that 3D printing is a viable technology route to reducing concentration polarisation in membrane nanofiltration applications.

References:

- Mazinani, S. *et al., 3D printed nanofiltration composite membranes with reduced concentration polarisation.* **J Membrane Sci – 2022**, https://doi.org/https://doi.org/10.1016/j.memsci.2021.120137.

- Al-Shimmery, *et al., 3D printed composite membranes with enhanced anti-fouling behaviour.* **J Membrane Sci – 2019**, https://doi.org/https://doi.org/10.1016/j.memsci.2018.12.058,

- Mazinani, S., *et al., 3D Printed Fouling-Resistant Composite Membranes.* **ACS Appl Mater Interf – 2019**,. https://doi.org/https://doi.org/10.1021/acsami.9b07764.

- Low, Z.-X., *et al., Perspective on 3D printing of separation membranes and comparison to related unconventional fabrication techniques.* **J Membrane Sci – 2017**. https://doi.org/http://dx.doi.org/10.1016/j.memsci.2016.10.006.

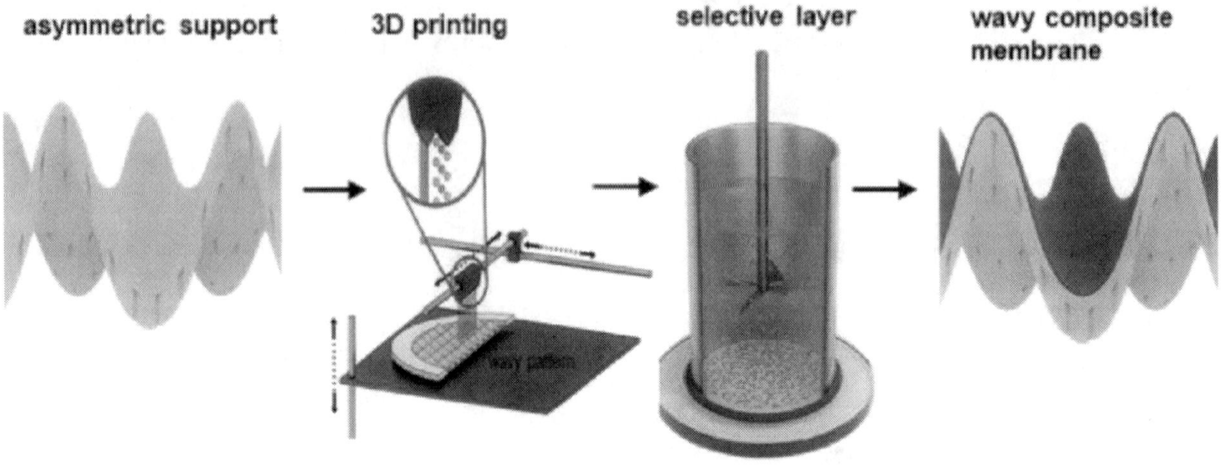
asymmetric support
3D printing
selective layer
wavy composite membrane
wavy pattern

342f Challenges to Tackle with Membrane Science for a More Sustainable Future

Suzana Nunes

We are facing environmental challenges of concerning dimensions. Furthermore, in times of political uncertainty, the reasons for an independency from fossil fuels becomes evident. More energy-efficient processes are needed. The history tells us that times of crisis have seeded disrupting technological advances. Separation processes consume a large portion of energy worldwide, since they are mostly thermally driven. Membrane technology has a unique opportunity to catalyze the transition to more sustainable separations and to fully sustainable energy, enabling also holistic approaches of rarifying material resources recovery. Important is to invest in processes that would be realistically profitable in terms of the overall energy impact. We are in an exciting situation compared to decades ago concerning the versatility of materials available for membrane preparation. The solutions must be scalable and the materials for the membrane fabrication must be easily processable. Critical analysis of the results and strategic feasibility is required. Considering sustainability goals, my group has dedicated in the past more than a decade to hydrogen technology, mainly to the development of fuel cell polymeric membranes and to the separation of CO_2 and other gases in coal plants. Fuel cell and electrolyzers (also operating with ion-exchange polymeric membranes, like fuel cells) shall have an important role in our sustainable future, by enabling energy conversion with green hydrogen. However, CO_2 separation should be a transitory solution, hopefully to a time in which coal and oil will not be the main energy resource. CO_2 capture technologies should be carefully evaluated, since in many cases their operation might require a substantial energy consumption to succeed, compromising part of its benefit. At least as relevant for our future is the opportunity of applying membranes to enable process intensification and integration with lower carbon footprint and the opportunity to enable more efficient purifications, fractionation and material resources in the chemical, petrochemical, and pharmaceutical industry. By making a separation efficient, the number of steps for achieving a satisfactory level of purity can be reduced with consequent energy reduction. By considering all these aspects, the main research focuses of my group are currently the development of the following topics (relevant for different sectors):

- High-performance polymeric membranes for nanofiltration and pervaporation application in organic solvent medium and at high temperatures (100-500°C) (chemical, petrochemical separations)
- Highly selective nanofiltration membranes for precise ions (mining) and for chiral separations (pharmaceutical separations)
- Membranes with high permeance for dehumidification (energy-efficient air condition) and natural gas dehydration (petrochemical separations)
- Sustainable membrane fabrication (green solvents, natural polymers, recycling, and circular chemistry)

The choice of materials is essential to fulfil the requirements to provide the needed progress in the different sectors. A deep understanding of the polymeric material chemistry, the mechanism of membrane formation, considerations on processability and scalability are relevant. A variety of materials have been considered in the last decade for the membrane fabrication as flat-sheet and hollow fibers, including high-performance functionalized polymers, like polytriazole and polyoxadiazole, submitted to different chemical and thermal crosslinking strategies, self-assembled block copolymers, stimuli responsive covalent-organic networks, macrocycles, organic cages and different 2D fillers. Each of them is more suitable for a different manufacturing method, such as phase inversion, dip-coating or interfacial polymerization. By the incorporation of building blocks with intrinsic selectivity, it is essential to avoid diluting their contribution to the membrane properties. Common issues are lack of filler-matrix adhesion, lack of percolation, undesired aggregation, and polymer filler permeability mismatch, which have affected the success of approaches like mixed-matrix membranes which are being proposed for decades without a real breakthrough. More effective approaches to secure the benefit of advanced porous selective building blocks are therefore important and are topic of research in our group keeping in mind the possibility of preparation in continuous machines and integration in modules with the necessary mechanical stability.

While we see challenges, we see also daring initiatives in previously unexpected parts of the world. An example is the announcement of a series of strategies to diversify the economy of Gulf countries from oil and gas to renewables. There is now the awareness that oil should be rather considered to produce more valuable chemicals in a sustainable way, there is a consideration that desalination should not only provide water but should be seen as an opportunity for potentially harvesting ions for holistic industrial processes. The completely renewable city NEOM shall integrate technologies that only now are becoming reality. The market for pharmaceutical and biotech industries is fast growing in this part of the world. The chances to extend the use of membrane technology might be larger than in long established industrial economies. It would be more beneficial to build a sustainable industry with newly available technology and easier to do it than to substitute long operating analogous systems in other countries. New great chances for membrane technology development and industrial implementation are there to be explore. It is the responsibility of membrane scientist community to invest in the right challenges and propose the most competitive solutions.

AUTHOR INDEX